LINEAR ALGEBRA

GEORGI E. SHILOV

Professor of Mathematics
Moscow University

Revised English Edition
Translated and Edited by

Richard A. Silverman

DOVER PUBLICATIONS, INC., NEW YORK

Published in Canada by General Publishing Company, Ltd., 30 Lesmill Road, Don Mills, Toronto, Ontario.

This Dover edition, first published in 1977, is an unabridged and corrected republication of the English translation originally published by Prentice-Hall, Inc., in 1971.

International Standard Book Number: 0-486-63518-X
Library of Congress Catalog Card Number: 77-075267

Manufactured in the United States of America
Dover Publications, Inc.
180 Varick Street
New York, N.Y. 10014

PREFACE

This book is intended as a text for undergraduate students majoring in mathematics and physics. It presents the material ordinarily covered in a course on linear algebra and subsequently drawn upon in various branches of mathematical analysis. However, it should be noted that the term "linear algebra" has for some time ceased to describe the actual content of the course, representing as it does a synthesis of various ideas from algebra, geometry and analysis. And although analysis in the strict sense of the term (i.e., the branch of mathematics concerned with limits, differentiation, integration, etc.) plays only a background role in this book, it is in fact the actual organizing principle of the course, since the problems of "linear algebra" can be regarded both as "finite-dimensional projections" and as the "support" for the basic problems of analysis.

The text stems in part from my previous book *An Introduction to the Theory of Linear Spaces* (Prentice-Hall, 1961), henceforth denoted by *LS*. Briefly, the differences between *LS* and the present book are the following: *LS* is entirely concerned with real spaces, while this book considers spaces over an arbitrary number field, with the real and complex spaces being considered as closely related special cases of the general theory. A chapter has been introduced on the Jordan canonical form of the matrix of a linear operator in a real or complex space. Moreover, we also study the canonical form of the matrix of a normal operator in a complex space equipped with a scalar product, deducing as special cases the canonical forms of the matrices of Hermitian, anti-Hermitian and unitary operators and their real analogues.

The final lengthy chapter in *LS* on the geometry of infinite-dimensional Hilbert space has been omitted, since a more systematic treatment of this topic (in a functional analysis context) is available in a number of other books. Instead, further new material bearing directly on the basic content of the course has been added, namely Chapter 11 on the structure of matrix algebras (written at my request by A. Y. Khelemski) and an appendix on the structure of matrix categories, based on my article with I. M. Gelfand (Vestnik MGU, Ser. Mat. Mekh., No. 4 (1963), pp. 27–48). Chapter 11 and the appendix, although completely elementary in method, are nevertheless somewhat higher in level than the rest of the book (as indicated by the asterisks) and represent advanced developments in the theory of linear algebra.

Each chapter is equipped with a set of problems, and hints and answers to these problems appear at the end of the book. To a certain extent, the problems help to develop necessary technical skill, but they are primarily intended to illustrate and amplify the material in the text. Certain groups of problems can serve as the basis for seminar discussions. The same is true of Chapter 11 and the appendix, as well as of the starred sections (the latter contain ancillary material that can be omitted on first reading).

It is my pleasant duty to acknowledge the painstaking efforts of M. S. Agranovich, the editor of the book, and to thank him for a number of valuable suggestions. I also wish to thank I. Y. Dorfman for checking the solutions to all the problems.

G. E. S.

CONTENTS

*chapter 11

FINITE-DIMENSIONAL ALGEBRAS AND THEIR REPRESENTATIONS 312

*Appendix

CATEGORIES OF FINITE-DIMENSIONAL SPACES 335

HINTS AND ANSWERS 361

BIBLIOGRAPHY 379

INDEX 381

chapter 1

DETERMINANTS

1.1. Number Fields

1.11. Like most of mathematics, linear algebra makes use of number systems (number fields). By a *number field* we mean any set K of objects, called "numbers," which, when subjected to the four arithmetic operations again give elements of K. More exactly, these operations have the following properties (field axioms):

a. To every pair of numbers α and β in K there corresponds a (unique) number $\alpha + \beta$ in K, called the *sum* of α and β, where

1) $\alpha + \beta = \beta + \alpha$ for every α and β in K (*addition is commutative*);
2) $(\alpha + \beta) + \gamma = \alpha + (\beta + \gamma)$ for every α, β, γ in K (*addition is associative*);
3) There exists a number 0 (*zero*) in K such that $0 + \alpha = \alpha$ for every α in K;
4) For every α in K there exists a number (*negative element*) γ in K such that $\alpha + \gamma = 0$.

The solvability of the equation $\alpha + \gamma = 0$ for every α allows us to carry out the operation of subtraction, by defining the *difference* $\beta - \alpha$ as the sum of the number β and the solution γ of the equation $\alpha + \gamma = 0$.

b. To every pair of numbers α and β in K there corresponds a (unique) number $\alpha \cdot \beta$ (or $\alpha\beta$) in K, called the *product* of α and β, where

5) $\alpha\beta = \beta\alpha$ for every α and β in K (*multiplication is commutative*);
6) $(\alpha\beta)\gamma = \alpha(\beta\gamma)$ for every α, β, γ in K (*multiplication is associative*);

7) There exists a number 1 ($\neq 0$) in K such that $1 \cdot \alpha = \alpha$ for every α in K;

8) For every $\alpha \neq 0$ in K there exists a number (*reciprocal element*) γ in K such that $\alpha\gamma = 1$.

c. *Multiplication is distributive over addition*, i.e.,

9) $\alpha(\beta + \gamma) = \alpha\beta + \alpha\gamma$ for every α, β, γ in K.†

The solvability of the equation $\alpha\gamma = 1$ for every $\alpha \neq 0$ allows us to carry out the operation of division, by defining the *quotient* β/α as the product of the number β and the solution γ of the equation $\alpha\gamma = 1$.

The numbers 1, $1 + 1 = 2$, $2 + 1 = 3$, etc. are said to be *natural*; it is assumed that none of these numbers is zero.‡ By the *integers* in a field K we mean the set of all natural numbers together with their negatives and the number zero. By the *rational numbers* in a field K we mean the set of all quotients p/q, where p and q are integers and $q \neq 0$.

Two fields K and K' are said to be *isomorphic* if we can set up a one-to-one correspondence between K and K' such that the number associated with every sum (or product) of numbers in K is the sum (or product) of the corresponding numbers in K'. The number associated with every difference (or quotient) of numbers in K will then be the difference (or quotient) of the corresponding numbers in K'.

1.12. The most commonly encountered concrete examples of number fields are the following:

a. The *field of rational numbers*, i.e., of quotients p/q where p and $q \neq 0$ are the ordinary integers subject to the ordinary operations of arithmetic. (It should be noted that the integers by themselves *do not form a field*, since they do not satisfy axiom 8).) It follows from the foregoing that every field K has a subset (subfield) isomorphic to the field of rational numbers.

b. The *field of real numbers*, having the set of all points of the real line as its geometric counterpart. An axiomatic treatment of the field of real numbers is achieved by supplementing axioms 1)–9) with the axioms of order and the least upper bound axiom.§

---

† Note that axioms 5) and 9) also imply $(\alpha + \beta)\gamma = \alpha\gamma + \beta\gamma$.

‡ Given two elements N and E, say, we can construct a field by the rules $N + N = N$, $N + E = E$, $E + E = N$, $N \cdot N = N$, $N \cdot E = N$, $E \cdot E = E$. Then, in keeping with our notation, we should write $N = 0$, $E = 1$ and hence $2 = 1 + 1 = 0$. To exclude such number systems, we require that all natural field elements be nonzero.

§ For a detailed treatment of real numbers, see, for example, G. H. Hardy, *Pure Mathematics*, ninth edition, The Macmillan Co., New York (1945), Chap. 1.

c. The *field of complex numbers* of the form $a + ib$, where a and b are real numbers (i is not a real number), equipped with the following operations of addition and multiplication (Hardy, *op. cit.*, Chap. 3):

$$(a_1 + ib_1) + (a_2 + ib_2) = (a_1 + a_2) + i(b_1 + b_2),$$

$$(a_1 + ib_1)(a_2 + ib_2) = (a_1a_2 - b_1b_2) + i(a_1b_2 + a_2b_1).$$

For numbers of the form $a + i0$, these operations reduce to the corresponding operations for real numbers; briefly we write $a + i0 = a$ and call complex numbers of this form *real*. Thus it can be said that the field of complex numbers has a subset (subfield) isomorphic to the field of real numbers. Complex numbers of the form $0 + ib$ are said to be (*purely*) *imaginary* and are designated briefly by ib. It follows from the multiplication rule that

$$i^2 = i \cdot i = (0 + i1)(0 + i1) = -1.$$

1.13. Henceforth we will designate the field of real numbers by R and the field of complex numbers by C. According to the "fundamental theorem of algebra" (Hardy, *op. cit.*, Appendix II, p. 492), we can not only carry out the four arithmetic operations in C but also solve any algebraic equation

$$z^n + a_1z^{n-1} + \cdots + a_n = 0.$$

The field R of real numbers does not have this property. For example, the equation $z^2 + 1 = 0$ has no solutions in the field R.

Many of the subsequent considerations are valid for any number field. In what follows, we will use the letter K to denote an *arbitrary* number field. If some property is true for the field K, then it is automatically true for the field R and the field C, which are special cases of the general field K.

1.2. Problems of the Theory of Systems of Linear Equations

In this and the next two chapters, we shall study systems of linear equations. In the most general case, such a system has the form

$$\begin{aligned} a_{11}x_1 + a_{12}x_2 + \cdots + a_{1n}x_n &= b_1, \\ a_{21}x_1 + a_{22}x_2 + \cdots + a_{2n}x_n &= b_2, \\ \cdots\cdots\cdots\cdots\cdots\cdots\cdots\cdots \\ a_{k1}x_1 + a_{k2}x_2 + \cdots + a_{kn}x_n &= b_k. \end{aligned} \tag{1}$$

Here $x_1, x_2, \ldots, x_n$ denote the unknowns (elements of the field K) which are to be determined. (Note that we do not necessarily assume that the number of unknowns equals the number of equations.) The quantities $a_{11}, a_{12}, \ldots, a_{kn}$, taken from the field K, are called the *coefficients* of the system. The first index of a coefficient indicates the number of the equation in which the coefficient appears, while the second index indicates the number of the unknown with which the coefficient is associated.† The quantities $b_1, b_2, \ldots, b_k$ appearing in the right-hand side of (1), taken from the same field K, are called the *constant terms* of the system; like the coefficients, they are assumed to be known. By a *solution* of the system (1) we mean any set of numbers $c_1, c_2, \ldots, c_n$ from the same field K which, when substituted for the unknowns $x_1, x_2, \ldots, x_n$ turns all the equations of the system into identities.‡

Not every system of linear equations of the form (1) has a solution. For example, the system

$$\begin{aligned} 2x_1 + 3x_2 &= 5, \\ 2x_1 + 3x_2 &= 6 \end{aligned} \tag{2}$$

obviously has no solution at all. Indeed, whatever numbers c_1, c_2 we substitute in place of the unknowns x_1, x_2, the left-hand sides of the equations of the system (2) are the same, while the right-hand sides are different. Therefore no such substitution can simultaneously convert both equations of the system into identities.

A system of equations of the form (1) which has at least one solution is called *compatible*; a system which does not have solutions is called *incompatible*. A compatible system can have one solution or several solutions. In the latter case, we distinguish the solutions by indicating the number of the solution by a superscript in parentheses; for example, the first solution will be denoted by $c_1^{(1)}, c_2^{(1)}, \ldots, c_n^{(1)}$, the second solution by $c_1^{(2)}, c_2^{(2)}, \ldots, c_n^{(2)}$, and so on. The solutions $c_1^{(1)}, c_2^{(1)}, \ldots, c_n^{(1)}$ and $c_1^{(2)}, c_2^{(2)}, \ldots, c_n^{(2)}$ are regarded as *distinct* if at least one of the numbers $c_i^{(1)}$ does not coincide with the corresponding numbers $c_i^{(2)}$ $(i = 1, 2, \ldots, n)$. For example, the system

$$\begin{aligned} 2x_1 + 3x_2 &= 0, \\ 4x_1 + 6x_2 &= 0 \end{aligned} \tag{3}$$

has the distinct solutions

$$c_1^{(1)} = c_2^{(1)} = 0 \quad \text{and} \quad c_1^{(2)} = 3,\ c_2^{(2)} = -2$$

(and also infinitely many other solutions). If a compatible system has a unique solution, the system is called *determinate*; if a compatible system has at least two different solutions, it is called *indeterminate*.

---

† Thus, for example, the symbol a_{34} should be read as "*a* three four" and not as "*a* thirty-four."

‡ We emphasize that the set of numbers $c_1, c_2, \ldots, c_n$ represents *one* solution of the system and not n solutions.

We can now formulate the basic problems which arise in studying the system (1):

a) *To ascertain whether the system* (1) *is compatible or incompatible*;

b) *If the system* (1) *is compatible, to ascertain whether it is determinate*;

c) *If the system* (1) *is compatible and determinate, to find its unique solution*;

d) *If the system* (1) *is compatible and indeterminate, to describe the set of all its solutions.*

The basic mathematical tool for studying linear systems is *the theory of determinants*, which we consider next.

1.3. Determinants of Order n

1.31. Suppose we are given a *square matrix*, i.e., an array of n^2 numbers a_{ij} $(i, j = 1, 2, \ldots, n)$, all elements of a field K:

$$\left\| \begin{matrix} a_{11} & a_{12} & \cdots & a_{1n} \\ a_{21} & a_{22} & \cdots & a_{2n} \\ \cdot & \cdot & \cdots & \cdot \\ a_{n1} & a_{n2} & \cdots & a_{nn} \end{matrix} \right\|. \tag{4}$$

The number of rows and columns of the matrix (4) is called its *order*. The numbers a_{ij} are called the *elements* of the matrix. The first index indicates the row and the second index the column in which a_{ij} appears. The elements $a_{11}, a_{22}, \ldots, a_{nn}$ form the *principal diagonal* of the matrix.

Consider any product of n elements which appear in different rows and different columns of the matrix (4), i.e., a product containing *just one element from each row and each column.* Such a product can be written in the form

$$a_{\alpha_1 1} a_{\alpha_2 2} \cdots a_{\alpha_n n}. \tag{5}$$

Actually, for the first factor we can always choose the element appearing in the first column of the matrix (4); then, if we denote by α_1 the number of the row in which the element appears, the indices of the element will be α_1, 1. Similarly, for the second factor we can choose the element appearing in the second column; then its indices will be α_2, 2, where α_2 is the number of the row in which the element appears, and so on. Thus, the indices $\alpha_1, \alpha_2, \ldots, \alpha_n$ are the numbers of the rows in which the factors of the product (5) appear, when we agree to write the column indices in increasing order. Since, by hypothesis, the elements $\alpha_{\alpha_1 1}, a_{\alpha_2 2}, \ldots, a_{\alpha_n n}$ appear in *different* rows of the matrix (4), one from each row, then the numbers $\alpha_1, \alpha_2, \ldots, \alpha_n$ are all different and represent some permutation of the numbers $1, 2, \ldots, n$.

By an *inversion* in the sequence $\alpha_1, \alpha_2, \ldots, \alpha_n$, we mean an arrangement of two indices such that the larger index comes before the smaller index. The total number of inversions will be denoted by $N(\alpha_1, \alpha_2, \ldots, \alpha_n)$. For example, in the permutation 2, 1, 4, 3, there are two inversions (2 before 1, 4 before 3), so that

$$N(2, 1, 4, 3) = 2.$$

In the permutation 4, 3, 1, 2, there are five inversions (4 before 3, 4 before 1, 4 before 2, 3 before 1, 3 before 2), so that

$$N(4, 3, 1, 2) = 5.$$

If the number of inversions in the sequence $\alpha_1, \alpha_2, \ldots, \alpha_n$ is even, we put a plus sign before the product (5); if the number is odd, we put a minus sign before the product. In other words, we agree to write in front of each product of the form (5) the sign determined by the expression

$$(-1)^{N(\alpha_1, \alpha_2, \ldots, \alpha_n)}.$$

The total number of products of the form (5) which can be formed from the elements of a given matrix of order n is equal to the total number of permutations of the numbers $1, 2, \ldots, n$. As is well known, this number is equal to $n!$.

We now introduce the following definition:

By the ***determinant*** *D of the matrix* (4) *is meant the algebraic sum of the n! products of the form* (5), *each preceded by the sign determined by the rule just given, i.e.,*

$$D = \sum (-1)^{N(\alpha_1, \alpha_2, \ldots, \alpha_n)} a_{\alpha_1 1} a_{\alpha_2 2} \cdots a_{\alpha_n n}. \tag{6}$$

Henceforth, the products of the form (5) will be called the *terms* of the determinant D. The elements a_{ij} of the matrix (4) will be called the *elements* of D, and the order of (4) will be called the *order* of D. We denote the determinant D corresponding to the matrix (4) by one of the following symbols:

$$D = \begin{vmatrix} a_{11} & a_{12} & \cdots & a_{1n} \\ a_{21} & a_{22} & \cdots & a_{2n} \\ \cdot & \cdot & \cdots & \cdot \\ a_{n1} & a_{n2} & \cdots & a_{nn} \end{vmatrix} = \det \|a_{ij}\|. \tag{7}$$

For example, we obtain the following expressions for the determinants of orders two and three:

$$\begin{vmatrix} a_{11} & a_{12} \\ a_{21} & a_{22} \end{vmatrix} = a_{11}a_{22} - a_{21}a_{12},$$

$$\begin{vmatrix} a_{11} & a_{12} & a_{13} \\ a_{21} & a_{22} & a_{23} \\ a_{31} & a_{32} & a_{33} \end{vmatrix} = a_{11}a_{22}a_{33} + a_{21}a_{32}a_{13} + a_{31}a_{12}a_{23} - a_{31}a_{22}a_{13} - a_{21}a_{12}a_{33} - a_{11}a_{32}a_{23}.$$

We now indicate the role of determinants in solving systems of linear equations, by considering the example of a system of two equations in two unknowns:

$$a_{11}x_1 + a_{12}x_2 = b_1,$$
$$a_{21}x_1 + a_{22}x_2 = b_2.$$

Eliminating one of the unknowns in the usual way, we can easily obtain the formulas

$$x_1 = \frac{b_1a_{22} - b_2a_{12}}{a_{11}a_{22} - a_{21}a_{12}}, \qquad x_2 = \frac{a_{11}b_2 - a_{21}b_1}{a_{11}a_{22} - a_{21}a_{12}},$$

assuming that these ratios have nonvanishing denominators. The numerators and denominators of the ratios can be represented by the second-order determinants

$$a_{11}a_{22} - a_{21}a_{12} = \begin{vmatrix} a_{11} & a_{12} \\ a_{21} & a_{22} \end{vmatrix},$$

$$b_1a_{22} - b_2a_{12} = \begin{vmatrix} b_1 & a_{12} \\ b_2 & a_{22} \end{vmatrix},$$

$$a_{11}b_2 - a_{21}b_1 = \begin{vmatrix} a_{11} & b_1 \\ a_{21} & b_2 \end{vmatrix}.$$

It turns out that similar formulas hold for the solutions of systems with an arbitrary number of unknowns (see Sec. 1.7).

1.32. The rule for determining the sign of a given term of a determinant can be formulated somewhat differently, in geometric terms. Corresponding to the enumeration of elements in the matrix (4), we can distinguish two natural positive directions: from left to right along the rows, and from top to bottom along the columns. Moreover, the slanting lines joining any two elements of the matrix can be furnished with a direction: we shall say that the line segment joining the element a_{ij} with the element a_{kl} has *positive slope* if its right endpoint lies lower than its left endpoint, and that it has *negative slope* if its right endpoint lies higher than its left endpoint.† Now imagine that in the matrix (4) we draw all the segments with *negative* slope joining pairs of elements $a_{\alpha_1 1}, a_{\alpha_2 2}, \ldots, a_{\alpha_n n}$ of the product (5). Then we put a plus sign before the product (5) if the number of all such segments is even, and a minus sign if the number is odd.

---

† This definition of "slope" is not to be confused with the geometric notion with the same name. In fact, the sign convention adopted here is the opposite of that used in geometry.

For example, in the case of a fourth-order matrix, a plus sign must be put before the product $a_{21}a_{12}a_{43}a_{34}$, since there are two segments of negative slope joining the elements of this product:

$$\begin{Vmatrix} a_{11} & a_{12} & a_{13} & a_{14} \\ a_{21} & a_{22} & a_{23} & a_{24} \\ a_{31} & a_{32} & a_{33} & a_{34} \\ a_{41} & a_{42} & a_{43} & a_{44} \end{Vmatrix}.$$

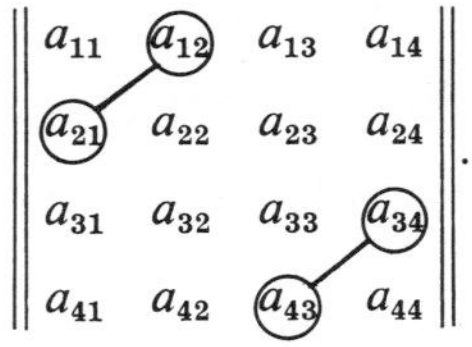

However, a minus sign must be put before the product $a_{41}a_{32}a_{13}a_{24}$, since in the matrix there are five segments of negative slope joining these elements:

$$\begin{Vmatrix} a_{11} & a_{12} & a_{13} & a_{14} \\ a_{21} & a_{22} & a_{23} & a_{24} \\ a_{31} & a_{32} & a_{33} & a_{34} \\ a_{41} & a_{42} & a_{43} & a_{44} \end{Vmatrix}.$$

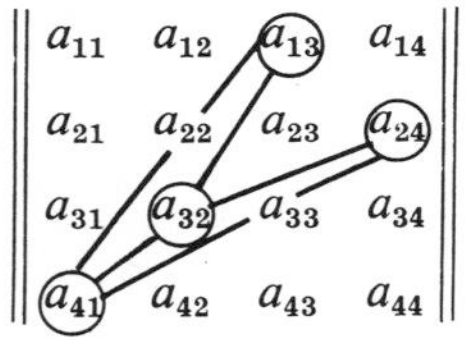

In these examples, the number of segments of negative slope joining the elements of a given term equals the number of *inversions* in the order of the first indices of the elements appearing in the term. In the first example, the sequence 2, 1, 4, 3 of first indices has two inversions; in the second example, the sequence 4, 3, 1, 2 of first indices has five inversions.

We now show that *the second definition of the sign of a term in a determinant is equivalent to the first.* To show this, it suffices to prove that the number of inversions in the sequence of first indices of a given term (with the second indices in natural order) is always equal to the number of segments of negative slope joining the elements of the given term in the matrix. But this is almost obvious, since the presence of a segment of negative slope joining the elements $a_{\alpha_i i}$ and $a_{\alpha_j j}$ means that $\alpha_i > \alpha_j$ for $i < j$, i.e., there is an inversion in the order of the first indices.

1.4. Properties of Determinants

1.41. The transposition operation. The determinant

$$\begin{vmatrix} a_{11} & a_{21} & \cdots & a_{n1} \\ a_{12} & a_{22} & \cdots & a_{n2} \\ \cdot & \cdot & \cdots & \cdot \\ a_{1n} & a_{2n} & \cdots & a_{nn} \end{vmatrix} \tag{8}$$

obtained from the determinant (7) by interchanging rows and columns with the same indices is said to be the *transpose* of the determinant (7). We now show that *the transpose of a determinant has the same value as the original determinant.* In fact, the determinants (7) and (8) obviously consist of the same terms; therefore it is enough for us to show that identical terms in the determinants (7) and (8) have identical signs. Transposition of the matrix of a determinant is clearly the result of rotating it (in space) through 180° about the principal diagonal $a_{11}, a_{22}, \ldots, a_{nn}$. As a result of this rotation, every segment with negative slope (e.g., making an angle $\alpha < 90°$ with the rows of the matrix) again becomes a segment with negative slope (i.e., making the angle $90° - \alpha$ with the rows of the matrix). Therefore the number of segments with negative slope joining the elements of a given term does not change after transposition. Consequently the sign of the term does not change either. Thus the signs of all the terms are preserved, which means that the value of the determinant remains unchanged.

The property just proved establishes the equivalence of the rows and columns of a determinant. Therefore further properties of determinants will be stated and proved only for columns.

1.42. The antisymmetry property. By the property of being *antisymmetric with respect to columns*, we mean the fact that a determinant changes sign when two of its columns are interchanged. We consider first the case where two adjacent columns are interchanged, for example columns j and $j + 1$. The determinant which is obtained after these columns are interchanged obviously still consists of the same terms as the original determinant. Consider any of the terms of the original determinant. Such a term contains an element of the jth column and an element of the $(j + 1)$th column. If the segment joining these two elements originally had negative slope, then after the interchange of columns, its slope becomes positive, and conversely. As for the other segments joining pairs of elements of the term in question, each of these segments does not change the character of its slope after the column interchange. Consequently the number of segments with negative slope joining the elements of the given term changes by one when the two columns are interchanged; therefore each term of the determinant, and hence the determinant itself, changes sign when the columns are interchanged.

Suppose now that two nonadjacent columns are interchanged, e.g., column j and column k $(j < k)$, where there are m other columns between. This interchange can be accomplished by successive interchanges of adjacent columns as follows: First column j is interchanged with column $j + 1$, then with columns $j + 2, j + 3, \ldots, k$. Then the column $k - 1$ so obtained (which was formerly column k) is interchanged with columns $k - 2, k - 3, \ldots, j$. In all, $m + 1 + m = 2m + 1$ interchanges of adjacent columns are required, each of which, according to what has just been proved, changes the sign of the determinant. Therefore, at the end of the process, the determinant will have a sign opposite to its original sign (since for any integer m, the number $2m + 1$ is odd).

1.43. COROLLARY. *A determinant with two identical columns vanishes.*

Proof. Interchanging the columns does not change the determinant D. On the other hand, as just proved, the determinant must change its sign. Thus $D = -D$, which implies that $D = 0$. ▮†

1.44. The linear property of determinants. This property can be formulated as follows:

a. THEOREM. *If all the elements of the jth column of a determinant D are "linear combinations" of two columns of numbers, i.e., if*

$$a_{ij} = \lambda b_i + \mu c_i \qquad (i = 1, 2, \ldots, n)$$

where λ *and* μ *are fixed numbers, then D is equal to a linear combination of two determinants:*

$$D = \lambda D_1 + \mu D_2. \tag{9}$$

Here both determinants D_1 *and* D_2 *have the same columns as the determinant D except for the jth column; the jth column of* D_1 *consists of the numbers* b_i, *while the jth column of* D_2 *consists of the numbers* c_i.

Proof. Every term of the determinant D can be represented in the form

$$\begin{aligned} a_{\alpha_1 1}a_{\alpha_2 2}\cdots a_{\alpha_j j}\cdots a_{\alpha_n n} &= a_{\alpha_1 1}a_{\alpha_2 2}\cdots(\lambda b_{\alpha_1} + \mu c_{\alpha_j})\cdots a_{\alpha_n n} \\ &= \lambda a_{\alpha_1 1}a_{\alpha_2 2}\cdots b_{\alpha_j}\cdots a_{\alpha_n n} + \mu a_{\alpha_1 1}a_{\alpha_2 2}\cdots c_{\alpha_j}\cdots a_{\alpha_n n}. \end{aligned}$$

Adding up all the first terms (with the signs which the corresponding terms have in the original determinant), we clearly obtain the determinant D_1, multiplied by the number λ. Similarly, adding up all the second terms, we obtain the determinant D_2, multiplied by the number μ. ▮

It is convenient to write this formula in a somewhat different form. Let D be an arbitrary fixed determinant. Denote by $D_j(p_i)$ the determinant which is obtained by replacing the elements of the jth column of D by the numbers p_i $(i = 1, 2, \ldots, n)$. Then (9) takes the form

$$D_j(\lambda b_i + \mu c_i) = \lambda D_j(b_i) + \mu D_j(c_i).$$

b. The linear property of determinants can easily be extended to the case where every element of the jth column is a linear combination not of two terms but of any other number of terms, i.e.

$$a_{ij} = \lambda b_i + \mu c_i + \cdots + \tau f_i.$$

† The symbol ▮ means Q.E.D. and indicates the end of a proof.

In this case,

$$D_j(a_{ij}) = D_j(\lambda b_i + \mu c_i + \cdots + \tau f_i)$$
$$= \lambda D_j(b_i) + \mu D_j(c_i) + \cdots + \tau D_j(f_i). \qquad (10)$$

1.45. Corollary. *Any common factor of a column of a determinant can be factored out of the determinant.*

Proof. If $a_{ij} = \lambda b_i$, then by (10) we have

$$D_j(a_{ij}) = D_j(\lambda b_i) = \lambda D_j(b_i). \quad \blacksquare$$

1.46. Corollary. *If a column of a determinant consists entirely of zeros, then the determinant vanishes.*

Proof. Since 0 is a common factor of the elements of one of the columns, we can factor it out of the determinant, obtaining

$$D_j(0) = D_j(0 \cdot 1) = 0 \cdot D_j(1) = 0. \quad \blacksquare$$

1.47. Addition of an arbitrary multiple of one column to another column.

a. Theorem. *The value of a determinant is not changed by adding the elements of one column multiplied by an arbitrary number to the corresponding elements of another column.*

Proof. Suppose we add the kth column multiplied by the number λ to the jth column ($k \neq j$). The jth column of the resulting determinant consists of elements of the form $a_{ij} + \lambda a_{ik}$ ($i = 1, 2, \ldots, n$). By (9) we have

$$D_j(a_{ij} + \lambda a_{ik}) = D_j(a_{ij}) + \lambda D_j(a_{ik}).$$

The jth column of the second determinant consists of the elements a_{ik}, and hence is identical with the kth column. It follows from Corollary 1.43† that $D_j(a_{ik}) = 0$, so that

$$D_j(a_{ij} + \lambda a_{ik}) = D_j(a_{ij}). \quad \blacksquare$$

b. Naturally, Theorem 1.47a can be formulated in the following more general form: *The value of a determinant is not changed by adding to the elements of its jth column first the corresponding elements of the kth column multiplied by λ, next the elements of the lth column multiplied by μ, etc., and finally the elements of the pth column multiplied by τ ($k \neq j, l \neq j, \ldots, p \neq j$).*

1.48. Because of the invariance of determinants under transposition (Sec. 1.41), all the properties of determinants proved in this section for columns remain valid for rows as well.

---

† Corollary 1.43 refers to the (unique) corollary in Sec. 1.43, Theorem 1.47a to the theorem in Sec. 1.47a, etc.

1.5. Cofactors and Minors

1.51. Consider any column, the jth say, of the determinant D. Let a_{ij} be any element of this column. Add up all the terms containing the element a_{ij} appearing in the right-hand side of equation (6)

$$D = \sum(-1)^{N(\alpha_1, \alpha_2, \ldots, \alpha_n)} a_{\alpha_1 1} a_{\alpha_2 2} \cdots a_{\alpha_n n},$$

and then factor out the element a_{ij}. The quantity which remains, denoted by A_{ij}, is called the *cofactor of the element* a_{ij} of the determinant D.

Since every term of the determinant D contains an element from the jth column, (6) can be written in the form

$$D = a_{1j}A_{1j} + a_{2j}A_{2j} + \cdots + a_{nj}A_{nj}, \tag{11}$$

called the *expansion of the determinant D with respect to the* (*elements of the*) *jth column*. Naturally, we can write a similar formula for any *row* of the determinant D. For example, for the ith row we have the formula

$$D = a_{i1}A_{i1} + a_{i2}A_{i2} + \cdots + a_{in}A_{in}. \tag{12}$$

This gives the following

THEOREM. *The sum of all the products of the elements of any column* (*or row*) *of the determinant D with the corresponding cofactors is equal to the determinant D itself.*

Equations (11) and (12) can be used to calculate determinants, but first we must know how to calculate cofactors. We will show how this is done in Sec. 1.53.

1.52. Next we note a consequence of (11) and (12) which will be useful later. Equation (11) is an identity in the quantities $a_{1j}, a_{2j}, \ldots, a_{nj}$. Therefore it remains valid if we replace a_{ij} $(i = 1, 2, \ldots, n)$ by any other quantities. The quantities $A_{1j}, A_{2j}, \ldots, A_{nj}$ remain unchanged when such a replacement is made, since they do not depend on the elements a_{ij}. Suppose that in the right and left-hand sides of the equality (11) we replace the elements $a_{1j}, a_{2j}, \ldots, a_{nj}$ by the corresponding elements of any other column, say the kth. Then the determinant in the left-hand side of (11) will have two identical columns and will therefore vanish, according to Corollary 1.43. Thus we obtain the relation

$$a_{1k}A_{1j} + a_{2k}A_{2j} + \cdots + a_{nk}A_{nj} = 0 \tag{13}$$

for $k \neq j$. Similarly, from (12) we obtain

$$a_{l1}A_{i1} + a_{l2}A_{i2} + \cdots + a_{ln}A_{in} = 0 \tag{14}$$

for $l \neq i$. Thus we have proved the following

THEOREM. *The sum of all the products of the elements of a column* (*or row*) *of the determinant D with the cofactors of the corresponding elements of another column* (*or row*) *is equal to zero.*

1.53. If we delete a row and a column from a matrix of order n, then, of course, the remaining elements form a matrix of order $n - 1$. The determinant of this matrix is called a *minor* of the original nth-order matrix (and also a minor of its determinant D). If we delete the ith row and the jth column of D, then the minor so obtained is denoted by M_{ij} or $M_{ij}(D)$.

We now show that the relation

$$A_{ij} = (-1)^{i+j} M_{ij} \tag{15}$$

holds, so that the calculation of cofactors reduces to the calculation of the corresponding minors. First we prove (15) for the case $i = 1$, $j = 1$. We add up all the terms in the right-hand side of (6) which contain the element a_{11}, and consider one of these terms. It is clear that the product of all the elements of this term except a_{11} gives a term c of the minor M_{11}. Since in the matrix of the determinant D, there are no segments of negative slope joining the element a_{11} with the other elements of the term selected, the sign ascribed to the term $a_{11}c$ of the determinant D is the same as the sign ascribed to the term c in the minor M_{11}. Moreover, by suitably choosing a term of the determinant D containing a_{11} and then deleting a_{11}, we can obtain any term of the minor M_{11}. Thus the algebraic sum of all the terms of the determinant D containing a_{11}, with a_{11} deleted, equals the product M_{11}. But according to Sec. 1.51, this sum is equal to the product A_{11}. Therefore, $A_{11} = M_{11}$ as required.

Now we prove (15) for arbitrary i and j, making essential use of the fact that the formula is valid for $i = j = 1$. Consider the element $a_{ij} = a$, appearing in the ith row and the jth column of the determinant D. By successively interchanging adjacent rows and columns, we can move the element a over to the upper left-hand corner of the matrix; to do this, we need

$$i - 1 + j - 1 = i + j - 2$$

interchanges. As a result, we obtain the determinant D_1 with the same terms as those of the original determinant D multiplied by

$$(-1)^{i+j-2} = (-1)^{i+j}.$$

The minor $M_{11}(D_1)$ of the determinant D_1 is clearly identical with the minor $M_{ij}(D)$ of the determinant D. By what has been proved already, the sum of the terms of the determinant D_1 which contain the element a, with a deleted, is equal to $M_{11}(D_1)$. Therefore the sum of the terms of the original determinant D which contain the element $a_{ij} = a$, with a deleted, is equal to

$$(-1)^{i+j} M_{11}(D_1) = (-1)^{i+j} M_{ij}(D).$$

According to Sec. 1.51, this sum is equal to A_{ij}. Consequently

$$A_{ij} = (-1)^{i+j} M_{ij},$$

which completes the proof of (15).

1.54. Formulas (11) and (12) can now be written in the following commonly used variants:

$$D = (-1)^{1+j} a_{1j} M_{1j} + (-1)^{2+j} a_{2j} M_{2j} + \cdots + (-1)^{n+j} a_{nj} M_{nj}, \quad (11')$$

$$D = (-1)^{i+1} a_{i1} M_{i1} + (-1)^{i+2} a_{i2} M_{i2} + \cdots + (-1)^{i+n} a_{in} M_{in}. \quad (12')$$

1.55. *Examples*

a. A third-order determinant has six distinct expansions, three with respect to rows and three with respect to columns. For example, the expansion with respect to the first row is

$$\begin{vmatrix} a_{11} & a_{12} & a_{13} \\ a_{21} & a_{22} & a_{23} \\ a_{31} & a_{32} & a_{33} \end{vmatrix} = a_{11} \begin{vmatrix} a_{22} & a_{23} \\ a_{32} & a_{33} \end{vmatrix} - a_{12} \begin{vmatrix} a_{21} & a_{23} \\ a_{31} & a_{33} \end{vmatrix} + a_{13} \begin{vmatrix} a_{21} & a_{22} \\ a_{31} & a_{32} \end{vmatrix}.$$

b. An nth-order determinant of the form

$$D_n = \begin{vmatrix} a_{11} & 0 & 0 & \cdots & 0 \\ a_{21} & a_{22} & 0 & \cdots & 0 \\ a_{31} & a_{32} & a_{33} & \cdots & 0 \\ \cdot & \cdot & \cdot & \cdots & \cdot \\ a_{n1} & a_{n2} & a_{n3} & \cdots & a_{nn} \end{vmatrix}$$

is called *triangular*. Expanding D_n with respect to the first row, we find that D_n equals the product of the element a_{11} with the triangular determinant

$$D_{n-1} = \begin{vmatrix} a_{22} & 0 & \cdots & 0 \\ a_{32} & a_{33} & \cdots & 0 \\ \cdot & \cdot & \cdots & \cdot \\ a_{n2} & a_{n3} & \cdots & a_{nn} \end{vmatrix}$$

of order $n - 1$. Again expanding D_{n-1} with respect to the first row, we find that

$$D_{n-1} = a_{22} D_{n-2},$$

where D_{n-2} is a triangular determinant of order $n - 2$. Continuing in this way, we finally obtain

$$D = a_{11}a_{22} \cdots a_{nn},$$

i.e., a triangular determinant equals the product of the elements appearing along its principal diagonal.

c. Calculate the *Vandermonde determinant*

$$W(x_1, \ldots, x_n) = \begin{vmatrix} 1 & 1 & \cdots & 1 \\ x_1 & x_2 & \cdots & x_n \\ x_1^2 & x_2^2 & \cdots & x_n^2 \\ \cdot & \cdot & \cdots & \cdot \\ x_1^{n-1} & x_2^{n-1} & \cdots & x_n^{n-1} \end{vmatrix}.$$

Solution. $W(x_1, \ldots, x_n)$ is a polynomial of degree $n - 1$ in x_n, with coefficients depending on $x_1, \ldots, x_{n-1}$. This polynomial vanishes if x_n takes any of the values $x_1, x_2, \ldots, x_{n-1}$, since then the determinant has two identical columns. Hence, by a familiar theorem of elementary algebra, the polynomial $W(x_1, \ldots, x_n)$ is divisible by the product $(x_n - x_1) \cdots (x_n - x_{n-1})$, so that

$$W(x_1, \ldots, x_n) = a(x_1, \ldots, x_{n-1}) \prod_{k=1}^{n-1} (x_n - x_k).$$

The quantity $a(x_1, \ldots, x_{n-1})$ is the leading coefficient of the polynomial $W(x_1, \ldots, x_n)$. Expanding the Vandermonde determinant with respect to the last column, we see that this coefficient is just $W(x_1, \ldots, x_{n-1})$. It follows that

$$W(x_1, \ldots, x_n) = W(x_1, \ldots, x_{n-1}) \prod_{k=1}^{n-1} (x_n - x_k).$$

Similarly,

$$W(x_1, \ldots, x_{n-1}) = W(x_1, \ldots, x_{n-2}) \prod_{j=1}^{n-2} (x_{n-1} - x_j),$$

$$\cdots\cdots\cdots\cdots\cdots\cdots\cdots\cdots$$

$$W(x_1, x_2) = W(x_1)(x_2 - x_1),$$

and obviously

$$W(x_1) = 1.$$

Multiplying all these equalities together, we get the desired result

$$W(x_1, \ldots, x_n) = \prod_{1 \leqslant i < m \leqslant n} (x_m - x_i).$$

In particular, *if the quantities* $x_1, \ldots, x_n$ *are all distinct, then*

$$W(x_1, \ldots, x_n) \neq 0.$$

1.6. Practical Evaluation of Determinants

1.61. Formula (12) takes a particularly simple form when all the elements of the ith row vanish except one element, say a_{ik}. In this case

$$D = a_{ik}A_{ik}, \tag{16}$$

and the calculation of the determinant D of order n reduces at once to the calculation of a determinant of order $n - 1$. If in addition to a_{ik}, there is another nonzero element a_{ij} in the ith row, then multiplying the kth column by $\lambda = a_{ij}/a_{ik}$ and subtracting it from the ith column, we obtain a determinant which is equal to the original one (cf. Sec. 1.47) but which now has a zero in the ith row and jth column. By a sequence of similar operations, we change any determinant with a nonzero element a_{ik} in the ith row into a determinant in which all the elements of the ith row equal zero except a_{ik}. This new determinant can then be evaluated by (16). Of course, similar operations can also be performed on the *columns* of a determinant.

1.62. *Example.* Calculate the following determinant of order five:

$$D = \begin{vmatrix} -2 & 5 & 0 & -1 & 3 \\ 1 & 0 & 3 & 7 & -2 \\ 3 & -1 & 0 & 5 & -5 \\ 2 & 6 & -4 & 1 & 2 \\ 0 & -3 & -1 & 2 & 3 \end{vmatrix}.$$

Solution. There are already two zeros in the third column of this determinant. In order to obtain two more zeros in this column, we multiply the fifth row by 3 and add it to the second row and then multiply the fifth row by 4 and subtract it from the fourth row. After performing these operations and expanding the determinant with respect to the third column, we obtain

$$D = \begin{vmatrix} -2 & 5 & 0 & -1 & 3 \\ 1 & -9 & 0 & 13 & 7 \\ 3 & -1 & 0 & 5 & -5 \\ 2 & 18 & 0 & -7 & -10 \\ 0 & -3 & -1 & 2 & 3 \end{vmatrix} = (-1)^{3+5}(-1) \begin{vmatrix} -2 & 5 & -1 & 3 \\ 1 & -9 & 13 & 3 \\ 3 & -1 & 5 & -5 \\ 2 & 18 & -7 & -10 \end{vmatrix}$$

$$= - \begin{vmatrix} -2 & 5 & -1 & 3 \\ 1 & -9 & 13 & 7 \\ 3 & -1 & 5 & -5 \\ 2 & 18 & -7 & -10 \end{vmatrix}.$$

The simplest thing to do now is to produce three zeros in the first column; to do this, we add twice the second row to the first row, subtract three times the second row from the third row and subtract twice the second row from the fourth row:

$$D = -\begin{vmatrix} -2 & 5 & -1 & 3 \\ 1 & -9 & 13 & 7 \\ 3 & -1 & 5 & -5 \\ 2 & 18 & -7 & -10 \end{vmatrix} = -\begin{vmatrix} 0 & -13 & 25 & 17 \\ 1 & -9 & 13 & 7 \\ 0 & 26 & -34 & -26 \\ 0 & 36 & -33 & -24 \end{vmatrix}$$

$$= -(-1)^{1+2}\begin{vmatrix} -13 & 25 & 17 \\ 26 & -34 & -26 \\ 36 & -33 & -24 \end{vmatrix}.$$

To simplify the calculation of the third-order determinant just obtained, we try to decrease the absolute values of its elements. To do this, we factor the common factor 2 out of the second row, add the second row to the first and subtract twice the second row from the third row:

$$D = 2\begin{vmatrix} -13 & 25 & 17 \\ 13 & -17 & -13 \\ 36 & -33 & -24 \end{vmatrix} = 2\begin{vmatrix} 0 & 8 & 4 \\ 13 & -17 & -13 \\ 10 & 1 & 2 \end{vmatrix}$$

$$= 2 \cdot 4\begin{vmatrix} 0 & 2 & 1 \\ 13 & -17 & -13 \\ 10 & 1 & 2 \end{vmatrix}.$$

There is already one zero in the first row. To obtain still another zero, we subtract twice the third column from the second column. After this, the evaluation of the determinant is easily completed.

$$D = 8\begin{vmatrix} 0 & 2 & 1 \\ 13 & -17 & -13 \\ 10 & 1 & 2 \end{vmatrix} = 8\begin{vmatrix} 0 & 0 & 1 \\ 13 & 9 & -13 \\ 0 & -3 & 2 \end{vmatrix} = 8(-1)^{1+3}\begin{vmatrix} 13 & 9 \\ 10 & -3 \end{vmatrix}$$

$$= 8 \cdot 3\begin{vmatrix} 13 & 3 \\ 10 & -1 \end{vmatrix} = 8 \cdot 3(-13 - 30) = -8 \cdot 3 \cdot 43 = -1032.$$

1.7. Cramer's Rule

1.71. We are now in a position to solve systems of linear equations. First we consider a system of the special form

$$\begin{aligned} a_{11}x_1 + a_{12}x_2 + \cdots + a_{1n}x_n &= b_1, \\ a_{21}x_1 + a_{22}x_2 + \cdots + a_{2n}x_n &= b_2, \\ \cdots\cdots\cdots\cdots\cdots\cdots\cdots\cdots \\ a_{n1}x_1 + a_{n2}x_2 + \cdots + a_{nn}x_n &= b_n, \end{aligned} \tag{17}$$

i.e., a system which has the same number of unknowns and equations. The coefficients a_{ij} $(i, j = 1, 2, \ldots, n)$ form the *coefficient matrix* of the system; we assume that the determinant of this matrix is different from zero. We now show that such a system is always *compatible* and *determinate*, and we obtain a formula which gives the unique solution of the system.

We begin by assuming that $c_1, c_2, \ldots, c_n$ is a solution of (17), so that

$$\begin{aligned} a_{11}c_1 + a_{12}c_2 + \cdots + a_{1n}c_n &= b_1, \\ a_{21}c_1 + a_{22}c_2 + \cdots + a_{2n}c_n &= b_2, \\ \cdots\cdots\cdots\cdots\cdots\cdots\cdots\cdots \\ a_{n1}c_1 + a_{n2}c_2 + \cdots + a_{nn}c_n &= b_n. \end{aligned} \tag{18}$$

We multiply the first of the equations (18) by the cofactor A_{11} of the element a_{11} in the coefficient matrix, then we multiply the second equation by A_{21}, the third by A_{31}, and so on, and finally the last equation by A_{n1}. Then we add all the equations so obtained. The result is

$$\begin{aligned} &(a_{11}A_{11} + a_{21}A_{21} + \cdots + a_{n1}A_{n1})c_1 \\ &+ (a_{12}A_{11} + a_{22}A_{21} + \cdots + a_{n2}A_{n1})c_2 + \cdots \\ &+ (a_{1n}A_{11} + a_{2n}A_{21} + \cdots + a_{nn}A_{n1})c_n = b_1A_{11} + b_2A_{21} + \cdots + b_nA_{n1}. \end{aligned} \tag{19}$$

By Theorem 1.51, the coefficient of c_1 in (19) equals the determinant D itself. By Theorem 1.52, the coefficients of all the other c_j $(j \neq 1)$ vanish. The expression in the right-hand side of (19) is the expansion of the determinant

$$D_1 = \begin{vmatrix} b_1 & a_{12} & \cdots & a_{1n} \\ b_2 & a_{22} & \cdots & a_{2n} \\ \cdot & \cdot & \cdots & \cdot \\ b_n & a_{n2} & \cdots & a_{nn} \end{vmatrix}$$

with respect to its first column. Therefore (19) can now be written in the form

$$Dc_1 = D_1,$$

so that

$$c_1 = \frac{D_1}{D}.$$

In a completely analogous way, we can obtain the expression

$$c_j = \frac{D_j}{D} \qquad (j = 1, 2, \ldots, n), \tag{20}$$

where

$$D_j = \begin{vmatrix} a_{11} & a_{12} & \cdots & a_{1,j-1} & b_1 & a_{1,j+1} & \cdots & a_{1n} \\ a_{21} & a_{22} & \cdots & a_{2,j-1} & b_2 & a_{2,j+1} & \cdots & a_{2n} \\ \cdot & \cdot & \cdots & \cdot & \cdot & \cdot & \cdots & \cdot \\ a_{n1} & a_{n2} & \cdots & a_{n,j-1} & b_n & a_{n,j+1} & \cdots & a_{nn} \end{vmatrix} = D_j(b_i)$$

is the determinant obtained from the determinant D by replacing its jth column by the numbers $b_1, b_2, \ldots, b_n$. Thus we obtain the following result:

If a solution of the system (17) *exists, then* (20) *expresses the solution in terms of the coefficients of the system and the numbers in the right-hand side of* (17). *In particular, we find that if a solution of the system* (17) *exists, it is unique.*

1.72. We must still show that a solution of the system (17) always exists. Consider the quantities

$$c_j = \frac{D_j}{D} \qquad (j = 1, 2, \ldots, n),$$

and substitute them into the system (17) in place of the unknowns $x_1, x_2, \ldots, x_n$. Then this reduces all the equations of the system (17) to identities. In fact, for the ith equation we obtain

$$\begin{aligned} a_{i1}c_1 + a_{i2}c_2 + \cdots + a_{in}c_n &= a_{i1}\frac{D_1}{D} + a_{i2}\frac{D_2}{D} + \cdots + a_{in}\frac{D_n}{D} \\ &= \frac{1}{D}[a_{i1}(b_1A_{11} + b_2A_{21} + \cdots + b_nA_{n1}) \\ &\quad + a_{i2}(b_1A_{12} + b_2A_{22} + \cdots + b_nA_{n2}) + \cdots \\ &\quad + a_{in}(b_1A_{1n} + b_2A_{2n} + \cdots + b_nA_{nn})] \\ &= \frac{1}{D}[b_1(a_{i1}A_{11} + a_{i2}A_{12} + \cdots + a_{in}A_{1n}) + \cdots \\ &\quad + b_2(a_{i1}A_{21} + a_{i2}A_{22} + \cdots + a_{in}A_{2n}) + \cdots \\ &\quad + b_n(a_{i1}A_{n1} + a_{i2}A_{n2} + \cdots + a_{in}A_{nn})]. \end{aligned}$$

By Theorems 1.51 and 1.52, only one of the coefficients of the quantities $b_1, b_2, \ldots, b_n$ is different from zero, namely the coefficient of b_1, which is equal to the determinant D itself. Consequently, the above expression reduces to

$$\frac{1}{D} b_i D = b_i,$$

i.e., is identical with the right-hand side of the ith equation of the system.

1.73. Thus the quantities c_j ($j = 1, \ldots, n$) actually constitute a solution of the system (17), and we have found the following prescription (*Cramer's rule*) for obtaining solutions of (17):

If the determinant of the system (17) *is different from zero, then* (17) *has a unique solution, namely, for the value of the unknown* x_j ($j = 1, \ldots, n$) *we take the fraction whose denominator is the determinant* D *of* (17) *and whose numerator is the determinant obtained by replacing the jth column of* D *by the column consisting of the constant terms of* (17), *i.e., the numbers in the right-hand sides of the system.*

Thus finding the solution of the system (17) reduces to calculating determinants. Ways of solving more general systems (with vanishing determinants, or with a number of equations different from the number of unknowns) will be given in the next two chapters.

1.74. ***Remark***. One sometimes encounters systems of linear equations whose constant terms are not numbers but vectors, e.g., in analytic geometry or in mechanics. Cramer's rule and its proof remain valid in this case as well; one must only bear in mind that the values of the unknowns $x_1, x_2, \ldots, x_n$ will then be vectors rather than numbers. For example, the system

$$x_1 + x_2 = \mathbf{i} - 3\mathbf{j},$$
$$x_1 - x_2 = \mathbf{i} + 5\mathbf{j}$$

has the unique solution

$$c_1 = \mathbf{i} + \mathbf{j}, \qquad c_2 = -4\mathbf{j}.$$

1.8. Minors of Arbitrary Order. Laplace's Theorem

1.81. Theorem 1.54 on the expansion of a determinant with respect to a row or a column is a special case of a more general theorem on the expansion of a determinant with respect to a whole set of rows or columns. Before formulating this general theorem (Laplace's theorem), we introduce some new notation.

Suppose that in a square matrix of order n we specify any $k \leqslant n$ different rows and the same number of different columns. The elements appearing at the intersections of these rows and columns form a square matrix of order k. The determinant of this matrix is called a *minor of order* k of the original matrix of order n (also a minor of order k of the determinant D); it is denoted by

$$M = M^{i_1, i_2, \ldots, i_k}_{j_1, j_2, \ldots, j_k},$$

where $i_1, i_2, \ldots, i_k$ are the numbers of the deleted rows, and $j_1, j_2, \ldots, j_k$ are the numbers of the deleted columns.

If in the original matrix we delete the rows and columns which make up the minor M, then the remaining elements again form a square matrix, this time of order $n - k$. The determinant of this matrix is called the *complementary minor* of the minor M, and is denoted by the symbol

$$\bar{M} = \bar{M}^{i_1, i_2 \ldots, i_k}_{j_1, j_2, \ldots, j_k}.$$

In particular, if the original minor is of order 1, i.e., is just some element a_{ij} of the determinant D, then the complementary minor is the same as the minor M_{ij} discussed in Sec. 1.53.

Consider now the minor

$$M_1 = M^{1,2 \ldots, k}_{1,2 \ldots, k}$$

formed from the first k rows and the first k columns of the determinant D; its complementary minor is

$$M_2 = \bar{M}_1 = \bar{M}^{1,2, \ldots, k}_{1,2, \ldots, k}.$$

In the right-hand side of equation (6), p. 6 group together all the terms of the determinant whose first k elements belong to the minor M_1 (and thus whose remaining $n - k$ elements belong to the minor M_2). Let one of these terms be denoted by c; we now wish to determine the sign which must be ascribed to c. The first k elements of c belong to a term c_1 of the minor M_1. If we denote by N_1 the number of segments of negative slope corresponding to these elements, then the sign which must be put in front of the term c_1 in the minor M_1 is $(-1)^{N_1}$. The remaining $n - k$ elements of c belong to a term c_2 of the minor M_2; the sign which must be put in front of this term in the minor M_2 is $(-1)^{N_2}$, where N_2 is the number of segments of negative slope corresponding to the $n - k$ elements of c_2. Since in the matrix of the determinant D there is not a single segment with negative slope joining an element of the minor M_1 with an element of the minor M_2, the total number of segments of negative slope joining elements of the term c equals the sum $N_1 + N_2$. Therefore the sign which must be put in front of the term c is given by the expression $(-1)^{N_1+N_2}$, and hence is equal to the product of the signs of the terms c_1 and c_2 in the minors M_1 and M_2. Moreover, we note that the product of any term of the minor M_1 and any term of the minor M_2 gives us one of the terms of the determinant D that have been grouped together. It follows that the sum of all the terms that we have grouped together from the expression for the determinant D given by (6) is equal to the *product of the minors M_1 and M_2*.

Next we solve the analogous problem for an arbitrary minor

$$M_1 = M^{i_1, i_2, \ldots, i_k}_{j_1, j_2, \ldots, j_k},$$

with complementary minor M_2. By successively interchanging adjacent rows and columns, we can move the minor M_1 over to the upper left-hand corner of the determinant D; to do so, we need a total of

$$(i_1 - 1) + (i_2 - 2) + \cdots + (i_k - k) \\ + (j_1 - 1) + (j_2 - 2) + \cdots + (j_k - k)$$

interchanges. As a result, we obtain a determinant D_1 with the same terms as in the original determinant but multiplied by $(-1)^{i+j}$, where

$$i = i_1 + i_2 + \cdots + i_k, \qquad j = j_1 + j_2 + \cdots + j_k.$$

By what has just been proved, the sum of all the terms in the determinant D_1 whose first k elements appear in the minor M_1 is equal to the product M_1M_2. It follows from this that the sum of the corresponding terms of the determinant D is equal to the product

$$(-1)^{i+j}M_1M_2 = M_1A_2,$$

where the quantity

$$A_2 = (-1)^{i+j}M_2$$

is called the *cofactor of the minor* M_1 in the determinant D. Sometimes one uses the notation

$$A_2 = \bar{A}^{i_1, i_2, \ldots, i_k}_{j_1, j_2, \ldots, j_k},$$

where the indices indicate the numbers of the deleted rows and columns.

Finally, let the rows of the determinant D with indices $i_1, i_2, \ldots, i_k$ be fixed; some elements from these rows appear in every term of D. We group together all the terms of D such that the elements from the fixed rows $i_1, i_2, \ldots, i_k$ belong to the columns with indices $j_1, j_2, \ldots, j_k$. Then, by what has just been proved, the sum of all these terms equals the product of the minor

$$M^{i_1, i_2, \ldots, i_k}_{j_1, j_2, \ldots, j_k}$$

with the corresponding cofactor. In this way, all the terms of D can be divided into groups, each of which is characterized by specifying k columns. The sum of the terms in each group is equal to the product of the corresponding minor and its cofactor. Therefore the entire determinant can be represented as the sum

$$D = \sum M^{i_1, i_2, \ldots, i_k}_{j_1, j_2, \ldots, j_k} \bar{A}^{i_1, i_2, \ldots, i_k}_{j_1, j_2, \ldots, j_k}, \tag{21}$$

where the indices $i_1, i_2, \ldots, i_k$ (the indices selected above) are fixed, and the sum is over all possible values of the column indices $j_1, j_2, \ldots, j_k$ $(1 \leqslant j_1 < j_2 < \cdots < j_k \leqslant n)$. The expansion of D given by (21) is called *Laplace's theorem.* Clearly, Laplace's theorem constitutes a generalization of the formula for expanding a determinant with respect to one of its rows (derived in Sec. 1.54). There is an analogous formula for expanding the determinant D with respect to a fixed set of columns.

1.82. *Example.* The determinant of the form

$$D = \begin{vmatrix} a_{11} & \cdots & a_{1k} & 0 & \cdots & 0 \\ a_{21} & \cdots & a_{2k} & 0 & \cdots & 0 \\ \cdot & \cdots & \cdot & \cdot & \cdots & \cdot \\ a_{kk} & \cdots & a_{kk} & 0 & \cdots & 0 \\ a_{k+1,1} & \cdots & a_{k+1,k} & a_{k+1,k+1} & \cdots & a_{k+1,n} \\ \cdot & \cdots & \cdot & \cdot & \cdots & \cdot \\ a_{n1} & \cdots & a_{nk} & a_{n,k+1} & \cdots & a_{nn} \end{vmatrix}$$

such that all the elements appearing in both the first k rows and the last $n - k$ columns vanish, is called *quasi-triangular.* To calculate the determinant, we expand it with respect to the first k rows by using Laplace's theorem. Only one term survives in the sum (21), and we obtain

$$D = \begin{vmatrix} a_{11} & \cdots & a_{1k} \\ \cdot & \cdots & \cdot \\ a_{k1} & \cdots & a_{kk} \end{vmatrix} \times \begin{vmatrix} a_{k+1,k+1} & \cdots & a_{k+1,n} \\ \cdot & \cdots & \cdot \\ a_{n,k+1} & \cdots & a_{nn} \end{vmatrix}.$$

1.9. Linear Dependence between Columns

1.91. Suppose we are given m columns of numbers with n numbers in each:

$$A_1 = \left\| \begin{matrix} a_{11} \\ a_{21} \\ \cdot \\ \cdot \\ \cdot \\ a_{n1} \end{matrix} \right\|, \quad A_2 = \left\| \begin{matrix} a_{12} \\ a_{22} \\ \cdot \\ \cdot \\ \cdot \\ a_{n2} \end{matrix} \right\|, \ldots, \quad A_m = \left\| \begin{matrix} a_{1m} \\ a_{2m} \\ \cdot \\ \cdot \\ \cdot \\ a_{nm} \end{matrix} \right\|.$$

We multiply every element of the first column by some number λ_1, every element of the second column by λ_2, etc., and finally every element of the last (mth) column by λ_m; we then add corresponding elements of the columns.

As a result, we get a new column of numbers, whose elements we denote by $c_1, c_2, \ldots, c_n$. We can represent all these operations schematically as follows:

$$\lambda_1 \begin{Vmatrix} a_{11} \\ a_{21} \\ \cdot \\ \cdot \\ \cdot \\ a_{n1} \end{Vmatrix} + \lambda_2 \begin{Vmatrix} a_{21} \\ a_{22} \\ \cdot \\ \cdot \\ \cdot \\ a_{n2} \end{Vmatrix} + \cdots + \lambda_m \begin{Vmatrix} a_{1m} \\ a_{2m} \\ \cdot \\ \cdot \\ \cdot \\ a_{nm} \end{Vmatrix} = \begin{Vmatrix} c_1 \\ c_2 \\ \cdot \\ \cdot \\ \cdot \\ c_n \end{Vmatrix},$$

or more briefly as

$$\lambda_1 A_1 + \lambda_2 A_2 + \cdots + \lambda_m A_m = C,$$

where C denotes the column whose elements are $c_1, c_2, \ldots, c_n$. The column C is called a *linear combination of the columns* $A_1, A_2, \ldots, A_m$, and the numbers $\lambda_1, \lambda_2, \ldots, \lambda_m$ are called the *coefficients* of the linear combination. As special cases of the linear combination C, we have the *sum* of the columns if $\lambda_1 = \lambda_2 = \cdots = \lambda_m = 1$ and the *product* of a column by a number if $m = 1$.

Suppose now that our columns are not chosen independently, but rather make up a determinant D of order n. Then we have the following

THEOREM. *If one of the columns of the determinant D is a linear combination of the other columns, then $D = 0$.*

Proof. Suppose, for example, that the qth column of the determinant D is a linear combination of the jth, kth, . . . , pth columns of D, with coefficients $\lambda_j, \lambda_k, \ldots, \lambda_p$, respectively. Then, according to Sec. 1.47, by subtracting from the qth column first the jth column multiplied by λ_j, then the kth column multiplied by λ_k, etc., and finally the pth column multiplied by λ_p, we do not change the value of the determinant D. However, as a result, the qth column consists of zeros only, from which it follows that $D = 0$. ∎

It is remarkable that the converse is also true, i.e., *if a given determinant D is equal to zero, then (at least) one of its columns is a linear combination of the other columns.* The proof of this theorem requires some preliminary considerations, to which we now turn.

1.92. Again suppose we have m columns of numbers with n elements in each. We can write them in the form of a matrix

$$A = \begin{Vmatrix} a_{11} & a_{12} & \cdots & a_{1m} \\ a_{21} & a_{22} & \cdots & a_{2m} \\ \cdot & \cdot & \cdots & \cdot \\ a_{n1} & a_{n2} & \cdots & a_{nm} \end{Vmatrix}$$

with n rows and m columns. If k columns and k rows of this matrix are held fixed, then the elements appearing at the intersections of these columns and rows form a square matrix of order k, whose determinant is a minor of order k of the original matrix A (see p. 21); this determinant may either be vanishing or nonvanishing. If, as we shall always assume, not all of the a_{ik} are zero, then we can always find an integer r which has the following two properties:

1) The matrix A has a minor of order r which does not vanish;

2) Every minor of the matrix A of order $r + 1$ and higher (if such actually exist) vanishes.

The number r which has these properties is called the *rank* of the matrix A. If all the a_{ik} vanish, then the rank of the matrix A is considered to be zero ($r = 0$). Henceforth we shall assume that $r > 0$. The minor of order r which is different from zero is called the *basis minor* of the matrix A. (Of course, A can have several basis minors, but they all have the same order r.) The columns which contain the basis minor are called the *basis columns.*

1.93. Concerning the basis columns, we have the following important

THEOREM (***Basis minor theorem***). *Any column of the matrix A is a linear combination of its basis columns.*

Proof. To be explicit, we assume that the basis minor of the matrix is located in the first r rows and first r columns of A. Let s be any integer from 1 to m, let k be any integer from 1 to n, and consider the determinant

$$D = \begin{vmatrix} a_{11} & a_{12} & \cdots & a_{1r} & a_{1s} \\ a_{21} & a_{22} & \cdots & a_{2r} & a_{2s} \\ \cdot & \cdot & \cdots & \cdot & \cdot \\ a_{r1} & a_{r2} & \cdots & a_{rr} & a_{rs} \\ a_{k1} & a_{k2} & \cdots & a_{kr} & a_{ks} \end{vmatrix}$$

of order $r + 1$. If $k \leqslant r$, the determinant D is obviously zero, since it then has two identical rows. Similarly, $D = 0$ for $s \leqslant r$. If $k > r$ and $s > r$, then the determinant D is also equal to zero, since it is then a minor of order $r + 1$ of a matrix of rank r. Consequently $D = 0$ for any values of k and s.

We now expand D with respect to its last row, obtaining the relation

$$a_{k1}A_{k1} + a_{k2}A_{k2} + \cdots + a_{kr}A_{kr} + a_{ks}A_{ks} = 0, \tag{22}$$

where the numbers $A_{k1}, A_{k2}, \ldots, A_{kr}, A_{ks}$ denote the cofactors of the elements $a_{k1}, a_{k2}, \ldots, a_{kr}, a_{ks}$ appearing in the last row of D. These cofactors do not depend on the number k, since they are formed by using elements a_{ij} with $i \leqslant r$. Therefore we can introduce the notation

$$A_{k1} = c_1, A_{k2} = c_2, \ldots, A_{kr} = c_r, A_{ks} = c_s.$$

Substituting the values $k = 1, 2, \ldots, n$ in turn into (22), we obtain the system of equations

$$\begin{aligned} c_1a_{11} + c_2a_{12} + \cdots + c_ra_{1r} + c_sa_{1s} &= 0, \\ c_1a_{21} + c_2a_{22} + \cdots + c_ra_{2r} + c_sa_{2s} &= 0, \\ \cdots\cdots\cdots\cdots\cdots\cdots\cdots\cdots \\ c_1a_{n1} + c_2a_{n2} + \cdots + c_ra_{nr} + c_sa_{ns} &= 0. \end{aligned} \tag{23}$$

The number $c_s = A_{ks}$ is different from zero, since A_{ks} is a basis minor of the matrix A. Dividing each of the equations (23) by c_s, transposing all the terms except the last to the right-hand side, and denoting $-c_j/c_s$ by λ_j ($j = 1, 2, \ldots, r$), we obtain

$$\begin{aligned} a_{1s} &= \lambda_1a_{11} + \lambda_2a_{12} + \cdots + \lambda_ra_{1r}, \\ a_{2s} &= \lambda_1a_{21} + \lambda_2a_{22} + \cdots + \lambda_ra_{2r}, \\ \cdots\cdots\cdots&\cdots\cdots\cdots\cdots\cdots \\ a_{ns} &= \lambda_1a_{n1} + \lambda_2a_{n2} + \cdots + \lambda_ra_{nr}. \end{aligned} \tag{24}$$

These equations show that the sth column of the matrix A is a linear combination of the first r columns of the matrix (with coefficients $\lambda_1, \lambda_2, \ldots, \lambda_r$). The proof of the theorem is now complete, since s can be any number from 1 to m. ▮

1.94. We are now in a position to prove the converse of Theorem 1.91 (already mentioned at the end of Sec. 1.91):

THEOREM. *If the determinant D vanishes, then it has at least one column which is a linear combination of the other columns.*

Proof. Consider the matrix of the determinant D. Since $D = 0$, the basis minor of this matrix is of order $r < n$. Therefore, after specifying the r basis columns, we can still find at least one column which is not one of the basis columns. By the basis minor theorem, this column is a linear combination of the basis columns. Thus we have found a column of the determinant D which is a linear combination of the other columns. ▮

Note that we can include all the remaining columns of the determinant D in this linear combination by assigning them zero coefficients (say).

1.95. The results just obtained can be formulated in a somewhat more symmetric way. If the coefficients $\lambda_1, \lambda_2, \ldots, \lambda_m$ of a linear combination of m columns $A_1, A_2, \ldots, A_m$ (see Sec. 1.91) are equal to zero, then obviously the linear combination is just the *zero column*, i.e., the column consisting entirely of zeros. But it may also be possible to obtain the zero column from the given columns by using coefficients $\lambda_1, \lambda_2, \ldots, \lambda_m$ which are not all equal to zero. In this case, the given columns $A_1, A_2, \ldots, A_m$ are called *linearly dependent*. For example, the columns

$$A_1 = \left\|\begin{matrix} 1 \\ 2 \\ 3 \\ 4 \end{matrix}\right\|, \qquad A_2 = \left\|\begin{matrix} 2 \\ 4 \\ 6 \\ 8 \end{matrix}\right\|, \qquad A_3 = \left\|\begin{matrix} 1 \\ 1 \\ 1 \\ 1 \end{matrix}\right\|$$

are linearly dependent, since the zero column can be obtained as the linear combination

$$2 \cdot A_1 - 1 \cdot A_2 + 0 \cdot A_3.$$

A more detailed statement of the definition of linear dependence is the following: The columns

$$A_1 = \left\|\begin{matrix} a_{11} \\ a_{21} \\ \cdot \\ \cdot \\ \cdot \\ a_{n1} \end{matrix}\right\|, \qquad A_2 = \left\|\begin{matrix} a_{12} \\ a_{22} \\ \cdot \\ \cdot \\ \cdot \\ a_{n2} \end{matrix}\right\|, \quad \ldots, \qquad A_m = \left\|\begin{matrix} a_{m1} \\ a_{m2} \\ \cdot \\ \cdot \\ \cdot \\ a_{nm} \end{matrix}\right\|$$

are called *linearly dependent* if there exist numbers $\lambda_1, \lambda_2, \ldots, \lambda_m$, not all equal to zero, such that the system of equation

$$\begin{aligned} \lambda_1 a_{11} + \lambda_2 a_{12} + \cdots + \lambda_m a_{1m} &= 0, \\ \lambda_1 a_{21} + \lambda_2 a_{22} + \cdots + \lambda_m a_{2m} &= 0, \\ \cdots\cdots\cdots\cdots\cdots\cdots\cdots\cdots \\ \lambda_1 a_{n1} + \lambda_2 a_{n2} + \cdots + \lambda_m a_{nm} &= 0 \end{aligned}$$

is satisfied, or equivalently such that

$$\lambda_1 A_1 + \lambda_2 A_2 + \cdots + \lambda_m A_m = 0,$$

where the symbol 0 on the right-hand side denotes the zero column. *If one of the columns $A_1, A_2, \ldots, A_m$,* (e.g., the last column) *is a linear combination of the others, i.e.,*

$$A_m = \lambda_1 A_1 + \lambda_2 A_2 + \cdots + \lambda_{m-1} A_{m-1}, \tag{25}$$

then the columns $A_1, A_2, \ldots, A_m$ are linearly dependent. In fact, (25) is equivalent to the relation

$$\lambda_1 A_1 + \lambda_2 A_2 + \cdots + \lambda_{m-1} A_{m-1} - A_m = 0.$$

Consequently, there exists a linear combination of the columns $A_1, A_2, \ldots, A_m$, whose coefficients are not equal to zero (e.g., with the last coefficient equal to -1) whose sum is the zero column; this just means that the columns $A_1, A_2, \ldots, A_m$ are linearly dependent.

Conversely, *if the columns $A_1, A_2, \ldots, A_m$ are linearly dependent, then (at least) one of the columns is a linear combination of the other columns.* In fact, suppose that in the relation

$$\lambda_1 A_1 + \lambda_2 A_2 + \cdots + \lambda_{m-1} A_{m-1} + \lambda_m A_m = 0 \tag{26}$$

expressing the linear dependence of the columns $A_1, A_2, \ldots, A_m$, the coefficient λ_m, say, is nonzero. Then (26) is equivalent to the relation

$$A_m = -\frac{\lambda_1}{\lambda_m} A_1 - \frac{\lambda_2}{\lambda_m} A_2 - \cdots - \frac{\lambda_{m-1}}{\lambda_m} A_{m-1},$$

which shows that the column A_m is a linear combination of the columns $A_1, A_2, \ldots, A_{m-1}$. Thus, finally, *the columns $A_1, A_2, \ldots, A_m$ are linearly dependent if and only if one of the columns is a linear combination of the other columns.*

1.96. Theorems 1.91 and 1.94 show that the determinant D vanishes if and only if one of its columns is a linear combination of the other columns. Using the results obtained in Sec. 1.95, we have the following

THEOREM. *The determinant D vanishes if and only if there is linear dependence between its columns.*

1.97. Since the value of a determinant does not change when it is transposed (see Sec. 1.41), and since transposition changes columns to rows, we can change columns to rows in all the statements made above. In particular, *the determinant D vanishes if and only if there is linear dependence between its rows.*

PROBLEMS

1. With what sign do the terms

a) $a_{23}a_{31}a_{42}a_{56}a_{14}a_{65}$,

b) $a_{32}a_{43}a_{14}a_{51}a_{66}a_{25}$

appear in the determinant of order 6?

2. Write down all the terms appearing in the determinant of order four which have a minus sign and contain the factor a_{23}.

3. With what sign does the term $a_{1n}a_{2,n-1}\cdots a_{n1}$ appear in the determinant of order n?

4. Show that of the $n!$ terms of a determinant of order n, exactly half ($n!/2$) have a plus sign according to the definition of Sec. 1.3, while the other half have a minus sign.

5. Use the linear property of determinants (Sec. 1.44) to calculate

$$\Delta = \begin{vmatrix} am + bp & an + bq \\ cm + dp & cn + dq \end{vmatrix}.$$

6. The numbers 20604, 53227, 25755, 20927 and 78421 are divisible by 17. Show that the determinant

$$\begin{vmatrix} 2 & 0 & 6 & 0 & 4 \\ 5 & 3 & 2 & 2 & 7 \\ 2 & 5 & 7 & 5 & 5 \\ 2 & 0 & 9 & 2 & 7 \\ 7 & 8 & 4 & 2 & 1 \end{vmatrix}$$

is also divisible by 17.

7. Calculate the determinants

$$\Delta_1 = \begin{vmatrix} 246 & 427 & 327 \\ 1014 & 543 & 443 \\ -342 & 721 & 621 \end{vmatrix}, \qquad \Delta_2 = \begin{vmatrix} 2 & 1 & 1 & 1 & 1 \\ 1 & 3 & 1 & 1 & 1 \\ 1 & 1 & 4 & 1 & 1 \\ 1 & 1 & 1 & 5 & 1 \\ 1 & 1 & 1 & 1 & 6 \end{vmatrix}.$$

8. Calculate the determinant

$$P(x) = \begin{vmatrix} 1 & 1 & 2 & 3 \\ 1 & 2 - x^2 & 2 & 3 \\ 2 & 3 & 1 & 5 \\ 2 & 3 & 1 & 9 - x^2 \end{vmatrix}.$$

9. Calculate the nth-order determinant

$$\Delta = \begin{vmatrix} x & a & a & \cdots & a \\ a & x & a & \cdots & a \\ a & a & x & \cdots & a \\ \cdot & \cdot & \cdot & \cdots & \cdot \\ a & a & a & \cdots & x \end{vmatrix}.$$

10. Prove that

$$\begin{vmatrix} 1 & \cdot & \cdots & \cdot \\ x_1 & x_2 & \cdots & x_n \\ x_1^2 & x_2^2 & \cdots & x_n^2 \\ \cdot & \cdot & \cdots & \cdot \\ x_1^{n-2} & x^{n-2} & \cdots & x_n^{n-2} \\ x_1^n & x_2^n & \cdots & x_n^n \end{vmatrix} = \begin{vmatrix} 1 & 1 & \cdots & 1 \\ x_1 & x_2 & \cdots & x_n \\ x_1^2 & x_2^2 & \cdots & x_n^2 \\ \cdot & \cdot & \cdots & \cdot \\ x_1^{n-2} & x_2^{n-2} & \cdots & x_n^{n-2} \\ x_1^{n-1} & x_2^{n-2} & \cdots & x_n^{n-1} \end{vmatrix} \times \sum_{k=1}^{n} x_k.$$

11. Solve the system of equations

$$\begin{aligned} x_1 + 2x_2 + 3x_3 + 4x_4 + 5x_5 &= 13, \\ 2x_1 + x_2 + 2x_3 + 3x_4 + 4x_5 &= 10, \\ 2x_1 + 2x_2 + x_3 + 2x_4 + 3x_5 &= 11, \\ 2x_1 + 2x_2 + 2x_3 + x_4 + 2x_5 &= 6, \\ 2x_1 + 2x_2 + 2x_3 + 2x_4 + x_5 &= 3. \end{aligned}$$

12. Formulate and prove the theorem which bears the same relation to Laplace's theorem as Theorem 1.52 bears to Theorem 1.51.

13. Construct four linearly independent columns of four numbers each.

14. Show that if the rows of a determinant of order n are linearly dependent, then its columns are also linearly dependent.

chapter 2

LINEAR SPACES

2.1. Definitions

2.11. In analytic geometry and mechanics one uses vectors (directed line segments) subject to certain suitably defined operations. The reader is undoubtedly already familiar with the meaning of the sum of two vectors and the product of a vector and a real number, operations obeying the usual laws of arithmetic.†

The concept of a linear space generalizes that of the set of all vectors. The generalization consists first in getting away from the concrete nature of the objects involved (directed line segments) without changing the properties of the operations on the objects, and secondly in getting away from the concrete nature of the admissible numerical factors (real numbers). This leads to the following definition: A set $\mathbf{K}$ is called a *linear* (or *affine*) *space over a field K* if

a) Given any two elements $x, y \in \mathbf{K}$, there is a rule (the addition rule) leading to a (unique) element $x + y \in \mathbf{K}$, called the *sum* of x and y;‡

---

† For the time being, we are not concerned with the other vector operations, namely scalar and vector products. In any event, these two products cannot play as basic a role as that played by the product of a vector and a real number. In fact, the scalar product of two vectors is no longer a vector, while the operation of forming a vector product, although leading to a vector, is noncommutative.

‡ Here and subsequently, we use some notation from set theory. By $a \in A$ we mean that the element a belongs to the set A; by $B \subset A$ we mean that the set B is a subset of the set A (B may coincide with A). The two relations $B \subset A$ and $A \subset B$ are equivalent to the assertion that the sets A and B coincide. The symbols $\in$ and $\subset$ are called *inclusion relations*. The fact that $a \in A$ (or $A \subset B$) is sometimes written $A \ni a$ (or $B \supset A$). By $a \notin A$ we mean that the element a does *not* belong to the set A.

b) Given any element $x \in \mathbf{K}$ and any number $\lambda \in K$, there is a rule (the rule for multiplication by a number) leading to a (unique) element $\lambda x \in \mathbf{K}$, called the *product* of the element x and the number λ;

c) These two rules obey the axioms listed below in Secs. 2.12 and 2.13.

The elements of a linear space will be called *vectors*, regardless of the fact that their concrete nature may be quite unlike the more familiar directed line segments. The geometric notions associated with the term "vector" will help us explain and often anticipate important results, as well as find a direct geometric interpretation (which would otherwise not be obvious) of various facts from algebra and analysis. In particular, in the next chapter we will obtain a simple geometric characterization of all the solutions of a homogeneous or nonhomogeneous system of linear equations.

2.12. The addition rule has the following properties:

1) $x + y = y + x$ for every $x, y \in \mathbf{K}$;

2) $(x + y) + z = x + (y + z)$ for every $x, y, z \in \mathbf{K}$;

3) There exists an element $0 \in \mathbf{K}$ (the *zero vector*) such that $x + 0 = x$ for every $x \in \mathbf{K}$;

4) For every $x \in \mathbf{K}$ there exists an element $y \in \mathbf{K}$ (the *negative element*) such that $x + y = 0$.

2.13. The rule for multiplication by a number has the following properties:

5) $1 \cdot x = x$ for every $x \in \mathbf{K}$;

6) $\alpha(\beta x) = (\alpha\beta)x$ for every $x \in \mathbf{K}$ and every $\alpha, \beta \in K$;

7) $(\alpha + \beta)x = \alpha x + \beta x$ for every $x \in \mathbf{K}$ and every $\alpha, \beta \in K$;

8) $\alpha(x + y) = \alpha x + \alpha y$ for every $x, y \in \mathbf{K}$ and every $\alpha \in K$.

2.14. Axioms 1)–8) have a number of simple implications:

a. THEOREM. *The zero vector in a linear space is unique.*

Proof. The existence of at least one zero vector is asserted in axiom 3). Suppose there are two zero vectors 0_1 and 0_2 in the space $\mathbf{K}$. Setting $x = 0_1$, $0 = 0_2$ in axiom 3), we obtain

$$0_1 + 0_2 = 0_1.$$

Setting $x = 0_2$, $0 = 0_1$ in the same axiom, we obtain

$$0_2 + 0_1 = 0_2.$$

Comparing the first of these relations with the second and using axiom 1), we find that $0_1 = 0_2$. ∎

b. THEOREM. *Every element in a linear space has a unique negative.*

Proof. The existence of at least one negative element is asserted in axiom 4). Suppose an element x has two negatives y_1 and y_2. Adding y_2 to both sides of the equation $x+y_1=0$ and using axioms 1)–3), we get

$$y_2 + (x + y_1) = (y_2 + x) + y_1 = 0 + y_1 = y_1,$$
$$y_2 + (x + y_1) = y_2 + 0 = y_2,$$

whence $y_1 = y_2$. ∎

c. THEOREM. *The relation*

$$0 \cdot x = 0$$

holds for every element x in a linear space.†

Proof. Consider the element $0 \cdot x + 1 \cdot x$. Using axioms 7) and 5), we get

$$0 \cdot x + 1 \cdot x = (0 + 1) \cdot x = 1 \cdot x = x,$$

$$0 \cdot x + 1 \cdot x = 0 \cdot x + x,$$

whence

$$x = 0 \cdot x + x.$$

Let y be the negative of x, and add y to both sides of the last equation. Then

$$0 = x + y = (0 \cdot x + x) + y = 0 \cdot x + (x + y) = 0 \cdot x + 0 = 0 \cdot x,$$

whence

$$0 = 0 \cdot x. \quad ∎$$

d. THEOREM. *Given any element x of a linear space, the element*

$$y = (-1) \cdot x$$

serves as the negative of x.

Proof. Form the sum $x + y$. Using the axioms and Theorem 2.14c, we find that

$$x + y = 1 \cdot x + (-1) \cdot x = (1 - 1) \cdot x = 0 \cdot x = 0. \quad ∎$$

---

† In the right-hand side of the equation, 0 denotes the zero vector, and in the left-hand side the number 0.

e. The negative of a given element x will now be denoted by $-x$, since Theorem 2.14d makes this a natural notation. The presence of a negative allows us to introduce the operation of subtraction, i.e., the *difference* $x-y$ is defined as the sum of x and $-y$. This definition agrees with the definition of subtraction in arithmetic.

2.15. A linear space over the field R of real numbers will be called *real* and denoted by the symbol $\mathbf{R}$. A linear space over the field C of complex numbers will be called *complex* and denoted by the symbol $\mathbf{C}$. If the nature of the elements $x, y, z, \ldots$ and the rules for operating on them are specified (where axioms 1)–8) must be satisfied), then we call the linear space *concrete*. As a rule, such spaces will be denoted by their own special symbols.

The following four kinds of concrete spaces will be of particular importance later:

a. *The space* V_3. The elements of this space are the free vectors studied in three-dimensional analytic geometry. Each vector is characterized by a length and a direction (with the exception of the zero vector, whose length is zero and whose direction is arbitrary). Addition of vectors is defined in the usual way by the parallelogram rule. Multiplication of a vector by a number λ is also defined in the usual way, i.e., the length of the vector is multiplied by $|\lambda|$, while its direction remains unchanged if $\lambda > 0$ and is reversed if $\lambda < 0$. It is easily verified that all the axioms 1)–8) are satisfied in this case. We denote the analogous sets of two-dimensional and one-dimensional vectors, which are also linear spaces, by V_2 and V_1, respectively; V_1, V_2 and V_3 are linear spaces over the field R of real numbers.

b. *The space* K_n. An element of this space is any ordered n-tuple

$$x = (\xi_1, \xi_2, \ldots, \xi_n)$$

of n numbers from the field K. The numbers $\xi_1, \xi_2, \ldots, \xi_n$ are called the *components* of the element x. The operations of addition and multiplication by a number $\lambda \in K$ are specified by the following rules:

$$(\xi_1, \xi_2, \ldots, \xi_n) + (\eta_1, \eta_2, \ldots, \eta_n) = (\xi_1 + \eta_1, \xi_2 + \eta_2, \ldots, \xi_n + \eta_n) \quad (1)$$

$$\lambda(\xi_1, \xi_2, \ldots, \xi_n) = (\lambda\xi_1, \lambda\xi_2, \ldots, \lambda\xi_n). \quad (2)$$

It is easily verified that axioms 1)–8) are satisfied. In particular, the element 0 is the n-tuple consisting of n zeros:

$$0 = (0, 0, \ldots, 0).$$

Actually, we dealt with elements of this space in Sec. 1.9, except that we wrote them there in the form of columns of numbers rather than rows of numbers. If K is the field R of real numbers, we write R_n instead of K_n, while if K is the field C of complex numbers, we write C_n instead of K_n.

c. *The space $R(a, b)$.* An element of this space is any continuous real function $x = x(t)$ defined on the interval $a \leqslant t \leqslant b$. The operations of addition of functions and multiplication of functions by real numbers are defined by the usual rules of analysis, and it is obvious that axioms 1)–8) are satisfied. In this case, the element 0 is the function which is identically zero. The space $R(a, b)$ is a linear space over the field R of real numbers.

d. Correspondingly, the space $C(a, b)$ is the space of all continuous complex-valued functions on the interval $a \leqslant t \leqslant b$. This space is a linear space over the field C of complex numbers.

2.16. We note that all the properties of elements of concrete spaces (e.g., the vectors of the space V_3) which are based only on axioms 1)–8) are also valid for the elements of an arbitrary linear space. For example, analyzing the proof of Cramer's rule for solving the system of linear equations

$$\begin{aligned} a_{11}x_1 + a_{12}x_2 + \cdots + a_{1n}x_n &= b_1, \\ a_{21}x_1 + a_{22}x_2 + \cdots + a_{2n}x_n &= b_2, \\ \cdots\cdots\cdots\cdots\cdots\cdots\cdots\cdots \\ a_{n1}x_1 + a_{n2}x_2 + \cdots + a_{nn}x_n &= b_n, \end{aligned}$$

we observe that insofar as the quantities $b_1, b_2, \ldots, b_n$ are concerned, the proof is based only on axioms 1)–8) and the fact that these quantities can be added and multiplied by numbers in K. As has already been pointed out in Sec. 1.74, this permits us to generalize Cramer's rule to systems in which the quantities $b_1, b_2, \ldots, b_n$ are vectors (elements of the space V_3). Furthermore, this permits us to assert that Cramer's rule is also valid for systems in which the elements $b_1, b_2, \ldots, b_n$ are elements of any linear space $\mathbf{K}$. We note only that then the values of the unknowns $x_1, x_2, \ldots, x_n$ are also elements of the space $\mathbf{K}$, and in fact can be expressed linearly in terms of the quantities $b_1, b_2, \ldots, b_n$.

2.17. ***Remark.*** In analytic geometry, it is sometimes convenient to consider vectors which are not free but have their initial points attached to the origin of coordinates. The convenience of this approach is that every vector is then associated with a point of space, namely its end point, and every point of space can be specified by giving the corresponding vector, called the *radius vector* of the point. With this picture in mind, we sometimes call the elements of a linear space *points* instead of vectors.† Of course, this change in terminology is not accompanied by any change whatsoever in the definitions, and merely appeals to our geometric intuition.

† We then talk of the "coordinates" of a point, rather than of the "components" of a vector.

2.2. Linear Dependence

2.21. Let $x_1, x_2, \ldots, x_k$ be vectors of the linear space **K** over a field K, and let $\alpha_1, \alpha_2, \ldots, \alpha_k$ be numbers from K. Then the vector

$$y = \alpha_1 x_1 + \alpha_2 x_2 + \cdots + \alpha_k x_k$$

is called a *linear combination* of the vectors $x_1, x_2, \ldots, x_k$, and the numbers $\alpha_1, \alpha_2, \ldots, \alpha_k$ are called the *coefficients* of the linear combination.

If $\alpha_1 = \alpha_2 = \cdots = \alpha_k = 0$, then $y = 0$ by Theorem 2.14c. However, there may exist a linear combination of the vectors $x_1, x_2, \ldots, x_k$ which equals the zero vector, even though its coefficients are not all zero. In this case, the vectors $x_1, x_2, \ldots, x_k$ are called *linearly dependent*. In other words, the vectors $x_1, x_2, \ldots, x_k$ are said to be linearly dependent if there exist numbers $\alpha_1, \alpha_2, \ldots, \alpha_k$, not all equal to zero, such that

$$\alpha_1 x_1 + \alpha_2 x_2 + \cdots + \alpha_k x_k = 0. \tag{3}$$

If (3) is possible only in the case where

$$\alpha_1 = \alpha_2 = \cdots = \alpha_k = 0,$$

the vectors $x_1, x_2, \ldots, x_k$ are said to be *linearly independent* (over K).

2.22. ***Examples***

a. In the linear space V_3, linear dependence of two vectors means that they are parallel to the same straight line. Linear dependence of three vectors means that they are parallel to the same plane. Any four vectors are linearly dependent.

b. We now explain what is meant by linear dependence of the vectors $x_1, x_2, \ldots, x_k$ of the linear space K_n. Let the vector x_i have components $\xi_1^{(i)}, \xi_2^{(i)}, \ldots, \xi_n^{(i)}$ $(i = 1, 2, \ldots, k)$. Then the linear dependence expressed by

$$\alpha_1 x_1 + \alpha_2 x_2 + \cdots + \alpha_k x_k = 0$$

means that the n equations

$$\begin{aligned} \alpha_1 \xi_1^{(1)} + \alpha_2 \xi_1^{(2)} + \cdots + \alpha_k \xi_1^{(k)} &= 0, \\ \alpha_1 \xi_2^{(1)} + \alpha_2 \xi_2^{(2)} + \cdots + \alpha_k \xi_2^{(k)} &= 0, \\ \cdots\cdots\cdots\cdots\cdots\cdots\cdots\cdots \\ \alpha_1 \xi_n^{(1)} + \alpha_2 \xi_n^{(2)} + \cdots + \alpha_k \xi_n^{(k)} &= 0 \end{aligned} \tag{4}$$

hold, where the constants $\alpha_1, \alpha_2, \ldots, \alpha_k$ are not all equal to zero. This is the same definition of linear dependence as that given in Sec. 1.95 for columns of numbers.

Thus the problem of whether or not the vectors $x_1, x_2, \ldots, x_k$ are linearly dependent reduces in the general case to the problem of whether or not there exists a nontrivial solution of the homogeneous system of equations (4),† with coefficients equal to the corresponding components of the given vectors. This problem will be solved completely in Sec. 3.21, where we will find a rule allowing us to decide whether or not given vectors in the space K_n are linearly dependent from an examination of their components.

c. In some cases, however, we can even now decide whether or not a given system of vectors is linearly dependent. For example, consider the n vectors

$$\begin{aligned} e_1 &= (1, 0, 0, \ldots, 0), \\ e_2 &= (0, 1, 0, \ldots, 0), \\ &\cdots\cdots\cdots\cdots \\ e_n &= (0, 0, 0, \ldots, 1) \end{aligned}$$

in the space K_n. For these vectors, the system (4) has the form

$$\begin{aligned} \alpha_1 \cdot 1 + \alpha_2 \cdot 0 + \alpha_3 \cdot 0 + \cdots + \alpha_n \cdot 0 &= 0, \\ \alpha_1 \cdot 0 + \alpha_2 \cdot 1 + \alpha_3 \cdot 0 + \cdots + \alpha_n \cdot 0 &= 0, \\ \cdots\cdots\cdots\cdots\cdots\cdots\cdots\cdots \\ \alpha_1 \cdot 0 + \alpha_2 \cdot 0 + \alpha_3 \cdot 0 + \cdots + \alpha_n \cdot 1 &= 0, \end{aligned}$$

and obviously has the unique solution

$$\alpha_1 = \alpha_2 = \cdots = \alpha_n = 0.$$

Thus the vectors $e_1, e_2, \ldots, e_n$ in the space K_n are linearly independent.

d. Linear dependence of the vectors

$$x_1 = x_1(t), x_2 = x_2(t), \ldots, x_k = x_k(t)$$

in the space $R(a, b)$ (or $C(a, b)$) means that the functions $x_1(t), x_2(t), \ldots, x_k(t)$ satisfy a relation of the form

$$\alpha_1 x_1(t) + \alpha_2 x_2(t) + \cdots + \alpha_k x_k(t) \equiv 0,$$

where the constants $\alpha_1, \alpha_2, \ldots, \alpha_k$ are not all equal to zero. For example, the functions

$$x_1(t) = \cos^2 t, \qquad x_2(t) = \sin^2 t, \qquad x_3(t) = 1$$

are linearly dependent, since the relation

$$x_1(t) + x_2(t) - x_3(t) \equiv 0$$

† Concerning the terms "homogeneous" and "nontrivial," see Sec. 2.42e.

holds. On the other hand, as we now show, the functions $1, t, t^2, \ldots, t^k$ are linearly independent. In fact, suppose there exists a relation

$$\alpha_0 \cdot 1 + \alpha_1 t + \cdots + \alpha_k t^k \equiv 0. \tag{5}$$

Then, by successively differentiating (5) k times, we obtain a system of $k + 1$ equations in the quantities $\alpha_0, \alpha_1, \ldots, \alpha_k$, with a determinant which is clearly different from zero (recall Sec. 1.55b). Solving this system by Cramer's rule (Sec. 1.75), we find that

$$\alpha_0 = \alpha_1 = \cdots = \alpha_k = 0.$$

Consequently, the functions $1, t, t^2, \ldots, t^k$ are linearly independent in the space $R(a, b)$, as asserted.

2.23. Next we note two simple properties of systems of vectors, both involving the notion of linear dependence.

a. LEMMA. *If some of the vectors $x_1, x_2, \ldots, x_k$ are linearly dependent, then the whole system $x_1, x_2, \ldots, x_k$ is also linearly dependent.*

Proof. Without loss of generality, we can assume that the vectors $x_1, x_2, \ldots, x_j$ $(j < k)$ are linearly dependent. Thus there is a relation

$$\alpha_1 x_1 + \alpha_2 x_2 + \cdots + \alpha_j x_j = 0,$$

where at least one of the constants $\alpha_1, \alpha_2, \ldots, \alpha_j$ is different from zero. By Theorem 2.14c and axiom 3) of Sec. 2.12, we have

$$\alpha_1 x_1 + \alpha_2 x_2 + \cdots + \alpha_j x_j + 0 \cdot x_{j+1} + \cdots + 0 \cdot x_k = 0.$$

But then the vectors $x_1, x_2, \ldots, x_k$ are also linearly dependent, since at least one of the constants $\alpha_1, \alpha_2, \ldots, \alpha_j, 0, \ldots, 0$ is different from zero. ∎

b. LEMMA. *The vectors $x_1, x_2, \ldots, x_k$ are linearly dependent if and only if one of the vectors can be expressed as a linear combination of the others.*

Proof. A similar statement has already been encountered; in fact, it was proved for columns of numbers in Sec. 1.95. Inspecting the proof given there, we see that it is based only on the possibility of performing on columns the operations of addition and multiplication by real numbers. Hence the proof can be carried through for the elements of any linear space, i.e., our lemma is valid for any linear space. ∎

2.3. Bases, Components, Dimension

2.31. By definition, a system of linearly independent vectors $e_1, e_2, \ldots, e_n$ in a linear space $\mathbf{K}$ over a field K is called a *basis* for $\mathbf{K}$ if, given any $x \in \mathbf{K}$, there exists an expansion

$$x = \xi_1 e_1 + \xi_2 e_2 + \cdots + \xi_n e_n \qquad (\xi_j \in K, j = 1, 2, \ldots, n). \tag{6}$$

It is easy to see that under these conditions *the coefficients in the expansion* (6) *are uniquely determined.* In fact, if we can write two expansions

$$x = \xi_1 e_1 + \xi_2 e_2 + \cdots + \xi_n e_n,$$

$$x = \eta_1 e_1 + \eta_2 e_2 + \cdots + \eta_n e_n$$

for a vector x, then, subtracting them term by term, we obtain the relation

$$0 = (\xi_1 - \eta_1)e_1 + (\xi_2 - \eta_2)e_2 + \cdots + (\xi_n - \eta_n)e_n,$$

from which, by the assumption that the vectors $e_1, e_2, \ldots, e_n$ are linearly independent, we find that

$$\xi_1 = \eta_1, \xi_2 = \eta_2, \ldots, \xi_n = \eta_n.$$

The uniquely defined numbers $\xi_1, \xi_2, \ldots, \xi_n$, are called the *components of the vector x with respect to the basis $e_1, e_2, \ldots, e_n$.*

2.32. *Examples*

a. A familiar basis in the space V_3 is formed by the three orthogonal unit vectors $\mathbf{i}, \mathbf{j}, \mathbf{k}$. The components ξ_1, ξ_2, ξ_3 of a vector $\mathbf{x}$ with respect to this basis are the projections of $\mathbf{x}$ along the coordinate axes.

b. An example of a basis in the space K_n is the system of vectors

$$\begin{aligned} e_1 &= (1, 0, \ldots, 0), \\ e_2 &= (0, 1, \ldots, 0), \\ &\cdots\cdots\cdots\cdots \\ e_n &= (0, 0, \ldots, 1), \end{aligned}$$

already considered in Sec. 2.22c. Indeed it is obvious that the relation

$$x = \xi_1(1, 0, \ldots, 0) + \xi_2(0, 1, \ldots, 0) + \cdots + \xi_n(0, 0, \ldots, 1)$$

holds for every vector

$$x = (\xi_1, \xi_2, \ldots, \xi_n) \in K_n.$$

This fact, together with the linear independence of the vectors $e_1, e_2, \ldots, e_n$ already proved, shows that these vectors form a basis in the space K_n. In particular, we see that the numbers $\xi_1, \xi_2, \ldots, \xi_n$ are just the components of the vector x with respect to the basis $e_1, e_2, \ldots, e_n$.

c. In the space $R(a, b)$ there does not exist a basis in the sense defined here. The proof of this statement will be given in Sec. 2.36c.

2.33. The fundamental significance of the concept of a basis for a linear space consists in the fact that when a basis is specified, the originally abstract linear operations in the space become ordinary linear operations with numbers, i.e., the components of the vectors with respect to the given basis. In fact, we have the following

THEOREM. *When two vectors of a linear space* **K** *are added, their components (with respect to any basis) are added. When a vector is multiplied by a number* λ, *all its components are multiplied by* λ.

Proof. Let

$$x = \xi_1 e_1 + \xi_2 e_2 + \cdots + \xi_n e_n,$$
$$y = \eta_1 e_1 + \eta_2 e_2 + \cdots + \eta_n e_n.$$

Then

$$x + y = (\xi_1 + \eta_1)e_1 + (\xi_2 + \eta_2)e_2 + \cdots + (\xi_n + \eta_n)e_n,$$
$$\lambda x = \lambda\xi_1 e_1 + \lambda\xi_2 e_2 + \cdots + \lambda\xi_n e_n,$$

by the axioms of Secs. 2.12 and 2.13. ▮

2.34. If in a linear space **K** we can find n linearly independent vectors while every $n + 1$ vectors of the space are linearly dependent, then the number n is called the *dimension* of the space **K** and the space **K** itself is called *n-dimensional.* A linear space in which we can find an arbitrarily large number of linearly independent vectors is called *infinite-dimensional.*

THEOREM. *In a space* **K** *of dimension n there exists a basis consisting of n vectors. Moreover, any set of n linearly independent vectors of the space* **K** *is a basis for the space.*

Proof. Let $e_1, e_2, \ldots, e_n$ be a system of n linearly independent vectors of the given n-dimensional space **K**. If x is any vector of the space, then the set of $n + 1$ vectors

$$x, e_1, e_2, \ldots, e_n$$

is linearly dependent, i.e., there exists a relation of the form

$$\alpha_0 x + \alpha_1 e_1 + \alpha_2 e_2 + \cdots + \alpha_n e_n = 0, \tag{7}$$

where at least one of the coefficients $\alpha_0, \alpha_1, \ldots, \alpha_n$ is different from zero. Clearly α_0 is different from zero, since otherwise the vectors $e_1, e_2, \ldots, e_n$ would be linearly dependent, contrary to hypothesis. Thus, in the usual way, i.e., by dividing (7) by α_0 and transposing all the other terms to the other side, we find that x can be expressed as a linear combination of the vectors $e_1, e_2, \ldots, e_n$. Since x is an arbitrary vector of the space **K**, we have shown that the vectors $e_1, e_2, \ldots, e_n$ form a basis for the space. ▮

2.35. The preceding theorem has the following converse:

THEOREM. *If there is a basis in the space* **K**, *then the dimension of* **K** *equals the number of basis vectors.*

Proof. Let the vectors $e_1, e_2, \ldots, e_n$ be a basis for **K**. By the definition of a basis, the vectors $e_1, e_2, \ldots, e_n$ are linearly independent; thus we already have n linearly independent vectors. We now show that any $n + 1$ vectors of the space **K** are linearly dependent.

Suppose we are given $n + 1$ vectors of the space **K**:

$$
\begin{aligned}
x_1 &= \xi_1^{(1)} e_1 + \xi_2^{(1)} e_2 + \cdots + \xi_n^{(1)} e_n, \\
x_2 &= \xi_1^{(2)} e_1 + \xi_2^{(2)} e_2 + \cdots + \xi_n^{(2)} e_n, \\
&\cdots\cdots\cdots\cdots\cdots\cdots\cdots\cdots\cdots \\
x_{n+1} &= \xi_1^{(n+1)} e_1 + \xi_2^{(n+1)} e_2 + \cdots + \xi_n^{(n+1)} e_n.
\end{aligned}
$$

Writing the components of each of these vectors as a column of numbers, we form the matrix

$$
A = \left\| \begin{matrix} \xi_1^{(1)} & \xi_1^{(2)} & \cdots & \xi_1^{(n+1)} \\ \xi_2^{(1)} & \xi_2^{(2)} & \cdots & \xi_2^{(n+1)} \\ \cdot & \cdot & \cdots & \cdot \\ \xi_n^{(1)} & \xi_n^{(2)} & \cdots & \xi_n^{(n+1)} \end{matrix} \right\|
$$

with n rows and $n + 1$ columns. The basis minor of the matrix A (see Sec. 1.92) is of order $r \leqslant n$. If $r = 0$, the linear dependence is obvious. Let $r > 0$. After specifying the r basis columns, we can still find at least one column which is not one of the basis columns. But then, according to the basis minor theorem, this column is a linear combination of the basis columns. Thus the corresponding vector of the space **K** is a linear combination of some other vectors among the given $x_1, x_2, \ldots, x_{n+1}$. But in this case, according to Lemma 2.23b, the vectors $x_1, x_2, \ldots, x_{n+1}$ are linearly dependent. ∎

a. The space V_3 is three-dimensional, since it has a basis consisting of the three vectors **i, j, k** (see Example 2.32a). Similarly, V_2 is two-dimensional and V_1 is one-dimensional.

b. The space K_n is n-dimensional, since it contains a basis consisting of the n vectors $e_1, e_2, \ldots, e_n$ (see Example 2.32b).

c. In each of the spaces $R(a, b)$ and $C(a, b)$, there is an arbitrarily large number of linearly independent vectors (see Example 2.22d), and hence these spaces are infinite-dimensional. Therefore neither space has a basis, for the presence of a basis would contradict Theorem 2.35.

d. Every complex linear space **C** is obviously a real space as well, since the domain of complex numbers contains the domain of real numbers. However, the dimension of **C** as a complex space does not coincide with that of **C** as a real space. In fact, if the vectors $e_1, \ldots, e_n$ are linearly independent in **C** regarded as a complex space, then the vectors $e_1, ie_1, \ldots, e_n, ie_n$ are linearly independent in **C** regarded as a real space. Hence the dimension of **C** regarded as a real space is twice as large as that of **C** regarded as a complex space (provided the dimension is finite).

2.4. Subspaces

2.41. Suppose that a set **L** of elements of a linear space **K** has the following properties:

a) If $x \in \mathbf{L}$, $y \in \mathbf{L}$, then $x + y \in \mathbf{L}$;
b) If $x \in \mathbf{L}$ and λ is an element of the field K, then $\lambda x \in \mathbf{L}$.

Thus **L** is a set of elements with linear operations defined on them. We now show that this set is also a *linear space*. To do so, we must verify that the set **L** with the operations a) and b) satisfies the axioms of Secs. 2.12 and 2.13, Axioms 1), 2) and 5)–8) are satisfied, since they hold quite generally for all elements of the space **K**. It remains to verify axioms 3) and 4). Let x be any element of **L**. Then, by hypothesis, $\lambda x \in \mathbf{L}$ for every $\lambda \in K$. First we choose $\lambda = 0$. Then, since $0 \cdot x = 0$ by Theorem 2.14c, the zero vector belongs to the set **L**, i.e., axiom 3) is satisfied. Next we choose $\lambda = -1$. Then, by Theorem 2.14d, $(-1)x$ is the negative of the element x. Thus, if an element x belongs to the set x, so does the negative of x. This means that axiom 4) is also satisfied, so that **L** is a linear space, as asserted. Consequently, every set $\mathbf{L} \subset \mathbf{K}$ with properties a) and b) is called a *linear subspace* (or simply a *subspace*) of the space **K**.

2.42. ***Examples***

a. The set whose only element is the zero vector of the space **K** is obviously the smallest possible subspace of **K**.

b. The whole space **K** is the largest possible subspace of **K**.

These two subspaces of **K**, the whole space and the set $\{0\}$ consisting of the zero vector alone, are sometimes called *trivial* subspaces. All the other subspaces of **K** are then said to be *nontrivial*.

c. Let $\mathbf{L}_1$ and $\mathbf{L}_2$ be two subspaces of the same linear space **K**. Then the set of all vectors $x \in \mathbf{K}$ belonging to both $\mathbf{L}_1$ and $\mathbf{L}_2$ forms a subspace called the *intersection* of the subspaces $\mathbf{L}_1$ and $\mathbf{L}_2$. The set of all vectors of the form $y + z$, where $y \in \mathbf{L}_1$, $z \in \mathbf{L}_2$ forms a subspace, denoted by $\mathbf{L}_1 + \mathbf{L}_2$ and called the *sum* of the subspaces $\mathbf{L}_1$ and $\mathbf{L}_2$.

d. All the vectors in the space V_3 parallel to a plane (or a line) form a subspace. If we talk about points rather than about vectors, as in Sec. 2.17, then the subspaces of V_3 are the sets of points lying on some plane (or line) passing through the origin of coordinates.

e. Consider the set L of all vectors $(\xi_1, \xi_2, \ldots, \xi_n)$ in the space K_n whose coordinates satisfy a system of linear equations of the form

$$\begin{aligned} a_{11}x_1 + a_{12}x_2 + \cdots + a_{1n}x_n &= 0, \\ a_{21}x_1 + a_{22}x_2 + \cdots + a_{2n}x_n &= 0, \\ \cdots\cdots\cdots\cdots\cdots\cdots \\ a_{k1}x_1 + a_{k2}x_2 + \cdots + a_{kn}x_n &= 0, \end{aligned} \tag{8}$$

with coefficients in the field K and constant terms equal to zero. Such a system is called a *homogeneous* linear system. A homogeneous linear system is always compatible, since it obviously has the "trivial" solution

$$x_1 = x_2 = \cdots = x_n = 0.$$

Let $c_1^{(1)}, c_2^{(1)}, \ldots, c_n^{(1)}$ and $c_1^{(2)}, c_2^{(2)}, \ldots, c_n^{(2)}$ be two solutions of this system, and form the numbers

$$c_1 = c_1^{(1)} + c_1^{(2)},\ c_2 = c_2^{(1)} + c_2^{(2)}, \ldots, c_n = c_n^{(1)} + c_n^{(2)}.$$

Then clearly $c_1, c_2, \ldots, c_n$ *is again a solution of the system* (8). In fact, substituting these numbers into the ith equation of the system, we obtain

$$\begin{aligned} a_{i1}c_1 + a_{i2}c_2 + \cdots + a_{in}c_n \\ &= a_{i1}(c_1^{(1)} + c_1^{(2)}) + a_{i2}(c_2^{(1)} + c_2^{(2)}) + \cdots + a_{in}(c_n^{(1)} + c_n^{(2)}) \\ &= (a_{i1}c_1^{(1)} + a_{i2}c_2^{(1)} + \cdots + a_{in}c_n^{(1)}) \\ &\quad + (a_{i1}c_1^{(2)} + a_{i2}c_2^{(2)} + \cdots + a_{in}c_n^{(2)}) = 0, \end{aligned}$$

as asserted; this solution will be called the *sum* of the solutions $c_1^{(1)}, c_2^{(1)}, \ldots, c_n^{(1)}$ and $c_1^{(2)}, c_2^{(2)}, \ldots, c_n^{(2)}$. Similarly, if $c_1, c_2, \ldots, c_n$ is an arbitrary solution of the system (8), then the numbers $\lambda c_1, \lambda c_2, \ldots, \lambda c_n$ also form a solution of (8) for every fixed $\lambda \in K$; this solution will be called the *product of the solution* $c_1, c_2, \ldots, c_n$ *and the number* λ. Thus *solutions of a homogeneous linear system* (8) *with coefficients and constant terms in a given field K can be added to one another and multiplied by numbers from the same field K, with the result still a solution of* (8). In other words, the set L is a subspace of the space K_n, and hence a linear space in its own right. We will call L the *solution space of the system* (8). In Sec. 3.41 we will calculate the dimension of this space and construct a basis for it.

2.43. We now consider some properties of subspaces which are related to the definitions of Secs. 2.2 and 2.3. First of all, we note that every linear relation which connects the vectors $x, y, \ldots, z$ in a subspace $\mathbf{L}$ is also valid in the whole space $\mathbf{K}$, and conversely. In particular, the fact that the vectors $x, y, \ldots, z \in \mathbf{L}$ are linearly dependent holds true simultaneously in the subspace $\mathbf{L}$ and in the space $\mathbf{K}$. For example, if every set of $n + 1$ vectors is linearly dependent in the space **K**, then this fact is true *a fortiori* in the subspace **L**. It follows that *the dimension of any subspace* **L** *of an n-dimensional space* **K** *does not exceed the number n.* According to Theorem 2.34, *in any subspace* $\mathbf{L} \subset \mathbf{K}$ *there exists a basis with the same number of vectors as the dimension of* **L**. Of course, if a basis $e_1, e_2, \ldots, e_n$ is chosen in **K**, then in the general case we cannot choose the basis vectors of the subspace **L** from the vectors $e_1, e_2, \ldots, e_n$, because none of these vectors may belong to **L**. However, it can be asserted that *if a basis* $f_1, f_2, \ldots, f_l$ *is chosen in the subspace* **L** (*which, to be explicit, is assumed to have dimension* $l < n$), *then additional vectors* $f_{l+1}, \ldots, f_n$ *can always be chosen in the whole space* **K** *such that the system* $f_1, f_2, \ldots, f_l, \ldots, f_n$ *is a basis for all of* **K**.

To prove this, we argue as follows: In the space **K** there are vectors which cannot be expressed as linear combinations of $f_1, f_2, \ldots, f_l$. Indeed, if there were no such vectors, then the vectors $f_1, f_2, \ldots, f_l$, which are linearly independent by hypothesis, would constitute a basis for the space **K**, and then by Theorem 2.35 the dimension of **K** would be l rather than n. Let f_{l+1} be any of the vectors that cannot be expressed as a linear combination of $f_1, f_2, \ldots, f_l$. Then the system $f_1, f_2, \ldots, f_l, f_{l+1}$ is linearly independent. In fact, suppose there were a relation of the form

$$\alpha_1 f_1 + \alpha_2 f_2 + \cdots + \alpha_l f_l + \alpha_{l+1} f_{l+1} = 0.$$

Then if $\alpha_{l+1} \neq 0$, the vector f_{l+1} could be expressed as a linear combination of $f_1, f_2, \ldots, f_l$, while if $\alpha_{l+1} = 0$, the vectors $f_1, f_2, \ldots, f_l$ would be linearly dependent. But both these results contradict the construction. If now every vector of the space **K** can be expressed as a linear combination of $f_1, f_2, \ldots, f_l, f_{l+1}$, then the system $f_1, f_2, \ldots, f_l, f_{l+1}$ forms a basis for **K** (and $l + 1 = n$), which concludes our construction. If $l + 1 < n$, then there is a vector f_{l+2} which cannot be expressed as a linear combination of $f_1, f_2, \ldots, f_l, f_{l+1}$, and hence we can continue the construction. Eventually, after $n - l$ steps, we obtain a basis for the space **K**.

2.44. We say that the vectors $g_1, \ldots, g_k$ are *linearly independent over the subspace* $\mathbf{L} \subset \mathbf{K}$ if the relation

$$\alpha_1 g_1 + \cdots + \alpha_k g_k \in \mathbf{L} \qquad (\alpha_1, \ldots, \alpha_k \in K)$$

implies

$$\alpha_1 = \cdots = \alpha_k = 0.$$

If **L** is the subspace consisting of the zero vector alone, then linear independence over **L** means ordinary linear independence. Linear dependence of the vectors $g_1, \ldots, g_k$ over the subspace **L** means that there exists a linear combination $\alpha_1 g_1 + \cdots + \alpha_k g_k$ belonging to **L**, where at least one of the coefficients $\alpha_1, \ldots, \alpha_k$ is nonzero.

The largest possible number of vectors of the space **K** which are linearly independent over the subspace $\mathbf{L} \subset \mathbf{K}$ is called the *dimension of* **K** *over* **L**.

If the vectors $g_1, \ldots, g_k$ are linearly independent over the space $\mathbf{L} \subset \mathbf{K}$ and if the vectors $f_1, \ldots, f_l$ are linearly independent in the subspace **L**, then the vectors $g_1, \ldots, g_k, f_1, \ldots, f_l$ are linearly independent in the whole space **K**. In fact, if there were a relation of the form

$$\alpha_1 f_1 + \cdots + \alpha_l f_l + \beta_1 g_1 + \cdots + \beta_k g_k = 0,$$

or equivalently

$$\beta_1 g_1 + \cdots + \beta_k g_k = -(\alpha_1 f_1 + \cdots + \alpha_l f_l) \in \mathbf{L},$$

then

$$\beta_1 = \cdots = \beta_k = 0,$$

by the assumed linear independence of the vectors $g_1, \ldots, g_k$ over **L**. It follows that $\alpha_1 = \cdots = \alpha_l = 0$, by the linear independence of the vectors $f_1, \ldots, f_l$.

The vectors $f_{l+1}, \ldots, f_n$ constructed in Sec. 2.43 are linearly independent over the subspace **L**. In fact, if there were a relation of the form

$$\alpha_{l+1} f_{l+1} + \cdots + \alpha_n f_n = \alpha_1 f_1 + \cdots + \alpha_l f_l$$

with at least one of the numbers $\alpha_{l+1}, \ldots, \alpha_n$ not equal to zero, then the vectors $f_1, \ldots, f_n$ would be linearly dependent, contrary to the construction. Hence the dimension of the space **K** over **L** is no less than $n - l$. On the other hand, this dimension cannot be greater than $n - l$, since if $n - l + 1$ vectors $h_1, \ldots, h_{n-l+1}$, say, were linearly independent over **L**, then the vectors $h_1, \ldots, h_{n-l+1}, f_1, \ldots, f_l$, of which there are more than n, would be linearly independent in **K**. Therefore the dimension of **K** over **L** is precisely $n - l$.

2.45. The direct sum. We say that a linear space **L** is the *direct sum* of given subspaces $\mathbf{L}_1, \ldots, \mathbf{L}_m \subset \mathbf{L}$ if

a) For every $x \in \mathbf{L}$ there exists an expansion

$$x = x_1 + \cdots + x_m,$$

where $x_1 \in \mathbf{L}_1, \ldots, x_m \in \mathbf{L}_m$;

b) This expansion is unique, i.e., if

$$x = x_1 + \cdots + x_m = y_1 + \cdots + y_m$$

where $x_j \in \mathbf{L}_j$, $y_j \in \mathbf{L}_j$ $(j = 1, \ldots, m)$, then

$$x_1 = y_1, \ldots, x_m = y_m.$$

However, the validity of condition b) is a consequence of the following simpler condition:

b′) If

$$0 = z_1 + \cdots + z_m$$

where $z_1 \in \mathbf{L}_1, \ldots, z_m \in \mathbf{L}_m$, then

$$z_1 = \cdots = z_m = 0.$$

In fact, given two expansions $x = x_1 + \cdots + x_m$, $x = y_1 + \cdots + y_m$, suppose b′) holds. Then subtracting the second expansion from the first, we get

$$0 = (x_1 - y_1) + \cdots + (x_m - y_m),$$

and hence $x_1 = y_1, \ldots, x_m = y_m$, because of b′). Conversely, b′) follows from b) if we set $x = 0$, $x_1 = \cdots = x_m = 0$.

It follows from condition b) that every pair of subspaces $\mathbf{L}_1, \ldots, \mathbf{L}_m$ has only the element 0 in common. In fact, if $z \in \mathbf{L}_j$ and $z \in \mathbf{L}_k$, then using b) and comparing the two expansions

$$z = z + 0, \qquad z \in \mathbf{L}_j, \qquad 0 \in \mathbf{L}_k,$$

$$z = 0 + z, \qquad 0 \in \mathbf{L}_j, \qquad z \in \mathbf{L}_k,$$

we find that $z = 0$.

Thus an n-dimensional space $\mathbf{K}_n$ is the direct sum of the n one-dimensional subspaces determined by any n linearly independent vectors. Moreover, the space $\mathbf{K}_n$ can be represented in various ways as a direct sum of subspaces not all of dimension 1.

2.46. Let $\mathbf{L}$ be a fixed subspace of an n-dimensional space $\mathbf{K}_n$. Then *there always exists a subspace $\mathbf{M} \subset \mathbf{K}_n$ such that the whole space $\mathbf{K}_n$ is the direct sum of* $\mathbf{L}$ *and* $\mathbf{M}$. To prove this, we use the vectors $f_{l+1}, \ldots, f_n$ constructed in Sec. 2.43, which are linearly independent over the subspace $\mathbf{L}$. Let $\mathbf{M}$ be the subspace consisting of all linear combinations of the vectors $f_{l+1}, \ldots, f_n$. Then $\mathbf{M}$ satisfies the stipulated requirement. In fact, since the vectors $f_1, \ldots, f_n$ form a basis in $\mathbf{K}_n$ (see Sec. 2.43), every vector $x \in \mathbf{L}$ has an expansion of the form

$$x = \alpha_1 f_1 + \cdots + \alpha_l f_l + \alpha_{l+1} f_{l+1} + \cdots + \alpha_n f_n = y + z,$$

where

$$y = \alpha_1 f_1 + \cdots + \alpha_l f_l \in \mathbf{L},$$

$$z = \alpha_{l+1} f_{l+1} + \cdots + \alpha_n f_n \in \mathbf{M}.$$

Moreover $x = 0$ implies $\alpha_1 = \cdots = \alpha_n = 0$, since the vectors $f_1, \ldots, f_n$ are linearly independent. Therefore conditions a)–b′) of Sec. 2.45 are satisfied, so that $\mathbf{K}_n$ is the direct sum of $\mathbf{L}$ and $\mathbf{M}$.

2.47. a. If the dimension of the space $\mathbf{L}_k$ equals r_k ($k = 1, \ldots, m$) and if r_k linearly independent vectors $f_{k1}, \ldots, f_{kr_k}$ are selected in each space $\mathbf{L}_k$, then every vector x of the sum $\mathbf{L} = \mathbf{L}_1 + \cdots + \mathbf{L}_k$ can be expressed as a linear combination of these vectors. Hence *the dimension of the sum of the spaces* $\mathbf{L}_1, \ldots, \mathbf{L}_k$ *does not exceed the sum of the dimensions of the separate spaces. If the sum* $\mathbf{L}_1 + \cdots + \mathbf{L}_k$ *is direct*, then the vectors $f_{11}, \ldots, f_{1r_1}, \ldots, f_{k1}, \ldots, f_{kr_k}, \ldots, f_{m1}, \ldots, f_{mr_m}$ are all linearly independent, so that in this case *the dimension of the sum is precisely the sum of the dimensions.*

b. In the general case, the dimension of the sum is related to the dimensions of the summands in a more complicated way. Here we consider only the problem of determining the dimension of the sum of *two* finite-dimensional subspaces **P** and **Q** of the space **K**, of dimensions p and q, respectively. Let **L** be the intersection of the subspaces **P** and **Q**, and let **L** have dimension l. First we choose a basis $e_1, e_2, \ldots, e_l$ in **L**. Then, using the argument of Sec. 2.43, we augment the basis $e_1, e_2, \ldots, e_l$ by the vectors $f_{l+1}, f_{l+2}, \ldots, f_p$ to make a basis for the whole subspace **P** and by the vectors $g_{l+1}, g_{l+2}, \ldots, g_q$ to make a basis for the whole subspace **Q**. By definition, every vector in the sum $\mathbf{P} + \mathbf{Q}$ is the sum of a vector from **P** and a vector from **Q**, and hence can be expressed as a linear combination of the vectors

$$e_1, \ldots, e_l, f_{l+1}, \ldots, f_p, g_{l+1}, \ldots, g_q. \tag{9}$$

We now show that these vectors *form a basis for the subspace* $\mathbf{P} + \mathbf{Q}$. To show this, it remains to verify their linear independence. Assume that there exists a linear relation of the form

$$\alpha_1 e_1 + \cdots + \alpha_l e_l + \beta_{l+1} f_{l+1} + \cdots + \beta_p f_p + \gamma_{l+1} g_{l+1} + \cdots + \gamma_q g_q = 0, \tag{10}$$

where at least one of the coefficients $\alpha_1, \ldots, \gamma_q$ is different from zero. We can then assert that at least one of the numbers $\gamma_{l+1}, \ldots, \gamma_q$ is different from zero, since otherwise the vectors

$$e_1, \ldots, e_l, f_{l+1}, \ldots, f_p$$

would be linearly dependent, which is impossible in view of the fact that they form a basis for the subspace **P**. Consequently the vector

$$x = \gamma_{l+1} g_{l+1} + \cdots + \gamma_q g_q \neq 0, \tag{11}$$

for otherwise the vectors $g_{l+1}, \ldots, g_q$ would be linearly dependent. But it follows from (10) that

$$-x = \alpha_1 e_1 + \cdots + \beta_p f_p \in \mathbf{P},$$

while (11) shows that $x \in \mathbf{Q}$. Thus x belongs to both **P** and **Q**, and hence belongs to the subspace **L**. But then

$$x = \gamma_{l+1} g_{l+1} + \cdots + \gamma_q g_q = \lambda_1 e_1 + \cdots + \lambda_l e_l,$$

and since the vectors

$$e_1, \ldots, e_l, g_{l+1}, \ldots, g_q$$

are linearly independent, we have

$$\gamma_{l+1} = \cdots = \gamma_q = 0.$$

This contradiction shows that the vectors (9) are actually linearly independent, and hence form a basis for the subspace $\mathbf{P} + \mathbf{Q}$. It follows from Theorem 2.35 that the dimension of $\mathbf{P} + \mathbf{Q}$ equals the number of basis vectors (9). But this number equals $p + q - l$. Thus, finally, *the dimension of the sum of two subspaces is equal to the sum of their dimensions minus the dimension of their intersection.*

c. COROLLARY. *Let $\mathbf{R}_p$ and $\mathbf{R}_q$ be two subspaces of dimensions p and q, respectively, of an n-dimensional space $\mathbf{R}_n$, and suppose $p + q > n$. Then the intersection of $\mathbf{R}_p$ and $\mathbf{R}_q$ is of dimension no less than $p + q - n$.*

2.48. Factor spaces

a. Given a subspace $\mathbf{L}$ of a linear space $\mathbf{K}$, an element $x \in \mathbf{K}$ is said to be *comparable* with an element $y \in \mathbf{K}$ (more exactly, *comparable relative to* $\mathbf{L}$) if $x - y \in \mathbf{L}$. Obviously, if x is comparable with y, then y is comparable with x, so that the relation of comparability is symmetric. Every element $x \in \mathbf{K}$ is comparable with itself. Moreover, if x is comparable with y and y is comparable with z, then x is comparable with z, since

$$x - z = (x - y) + (y - z) \in \mathbf{L}.$$

b. The set of all elements $y \in \mathbf{K}$ comparable with a given element $x \in \mathbf{K}$ is called a *class*, and is denoted by $\mathbf{X}$. As just shown, a class $\mathbf{X}$ contains the element x itself, and every pair of elements $y \in \mathbf{X}$, $z \in \mathbf{X}$ are comparable with each other. Moreover, if $u \notin \mathbf{X}$, then u is not comparable with any element of $\mathbf{X}$. Therefore two classes either have no elements in common or else coincide completely. The subspace $\mathbf{L}$ itself is a class. This class is denoted by 0, since it contains the zero element of the space $\mathbf{K}$.

c. The whole space $\mathbf{K}$ can be partitioned into a set of nonintersecting classes $\mathbf{X}, \mathbf{Y}, \ldots$. This set of classes will be denoted by $\mathbf{K}/\mathbf{L}$. We now introduce linear operations in $\mathbf{K}/\mathbf{L}$ as follows: Given two classes $\mathbf{X}$, $\mathbf{Y}$ and two elements α, β of the field K, we wish to define the class $\alpha\mathbf{X} + \beta\mathbf{Y}$. To do this, we choose arbitrary elements $x \in \mathbf{X}$, $y \in \mathbf{Y}$ and find the class $\mathbf{Z}$ containing the element $z = \alpha x + \beta y$. This class is then denoted by $\alpha\mathbf{X} + \beta\mathbf{Y}$. Clearly, $\alpha\mathbf{X} + \beta\mathbf{Y}$ is uniquely defined. In fact, suppose we choose another element x_1 of the class $\mathbf{X}$ and another element y_1 of the class $\mathbf{Y}$. Then

$$(\alpha x_1 + \beta y_1) - (\alpha x + \beta y) = \alpha(x_1 - x) + \beta(y_1 - y)$$

belongs to the space **L**, since $x_1 - x$ and $y_1 - y$ both belong to **L**. It follows that $\alpha x_1 + \beta y_1$ belongs to the same class as $\alpha x + \beta y$.

In particular, the above prescription defines addition of two classes **X** and **Y**, as well as multiplication of a class by a number $\alpha \in K$. We now show that these operations obey the axioms of a linear space, enumerated in Secs. 2.12 and 2.13. In fact, the validity of axioms 1) and 2) of Sec. 2.12 and axioms 5)–8) of Sec. 2.13 for classes follows at once from their validity for elements of the space **K**. Moreover, the zero element of the space **K**/**L** is the class 0 (consisting of all elements of the subspace **L**), while the inverse of the class **X** is the class consisting of all inverses of elements of the class **X**. Thus axioms 3) and 4) of Sec. 2.12 are also satisfied for the set of classes **K**/**L**. The resulting linear space **K**/**L** is called the *factor space* of the space **K** with respect to the subspace **L**.

2.49. THEOREM. *Let* $\mathbf{K} = \mathbf{K}_n$ *be an n-dimensional linear space over the field K, and let* $\mathbf{L} = \mathbf{L}_l \subset \mathbf{K}$ *be an l-dimensional subspace of* **K**. *Then the factor space* **K**/**L** *is of dimension* $n - l$.

Proof. Choose any basis $f_1, \ldots, f_l \in \mathbf{L}$, and augment it, as in Sec. 2.43, by vectors $f_{l+1}, \ldots, f_n$ to make a basis for the whole space **K**. Then the classes $\mathbf{X}_{l+1} \ni f_{l+1}, \ldots, \mathbf{X}_n \ni f_n$ form a basis in the space **K**/**L**. To see this, we note that given any $x \in \mathbf{K}$, there is a representation

$$x = \sum_{k=1}^{n} \alpha_k f_k,$$

and hence a representation

$$\mathbf{X} = \sum_{k=l+1}^{n} \alpha_k \mathbf{X}_k$$

for the class $\mathbf{X} \ni x$. Moreover, the classes $\mathbf{X}_{l+1}, \ldots, \mathbf{X}_n$ are linearly independent. In fact, if

$$\alpha_{l+1}\mathbf{X}_{l+1} + \cdots + \alpha_n \mathbf{X}_n = 0 \in \mathbf{K}/\mathbf{L}$$

for any $\alpha_{l+1}, \ldots, \alpha_n$ in K, then, in particular, there would be a relation

$$\alpha_{l+1} f_{l+1} + \cdots + \alpha_n f_n \in \mathbf{L}.$$

But $f_{l+1}, \ldots, f_n$ are linearly independent over **L** (see Sec. 2.44), and hence $\alpha_{l+1} = \cdots = \alpha_n = 0$, as required. Thus the $n - l$ classes $\mathbf{X}_{l+1}, \ldots, \mathbf{X}_n$ form a basis in **K**/**L**. It follows from Theorem 2.35 that **K**/**L** is of dimension $n - l$. ▌

2.5. Linear Manifolds

2.51. An important way of constructing subspaces is to form the linear manifold spanned by a given system of vectors. Let $x, y, z, \ldots$ be a system of vectors of a linear space **K**. Then by the *linear manifold spanned by* $x, y, z, \ldots$ is meant the set of all (finite) linear combinations

$$\alpha x + \beta y + \gamma z + \cdots \tag{12}$$

with coefficients $\alpha, \beta, \gamma, \ldots$ in the field K. It is easily verified that this set has properties a) and b) of Sec. 2.41. Therefore the linear manifold spanned by a system $x, y, z, \ldots$ is a subspace of the space **K**. Obviously, every subspace containing the vectors $x, y, z, \ldots$ also contains all their linear combinations (12). Consequently, *the linear manifold spanned by the vectors* $x, y, z, \ldots$ *is the smallest subspace containing these vectors.* The linear manifold spanned by the vectors $x, y, z, \ldots$ is denoted by $\mathbf{L}(x, y, z, \ldots)$.

2.52. *Examples*

a. The linear manifold spanned by the basis vectors $e_1, e_2, \ldots, e_n$ of a space **K** is obviously the whole space **K**.

b. The linear manifold spanned by two (noncollinear) vectors of the space V_3 consists of all the vectors parallel to the plane determined by the two vectors.

c. The linear manifold spanned by the system of functions $1, t, t^2, \ldots, t^k$ of the space $K(a, b)$ (K is R or C) consists of the set of all polynomials in t of degree no higher than k. The linear manifold spanned by the infinite system of functions $1, t, t^2, \ldots$ consists of all polynomials (of any degree) in the variable t with coefficients in the field K.

2.53. We now note two simple properties of linear manifolds.

a. LEMMA. *If the vectors* $x', y', \ldots$ *belong to the linear manifold spanned by the vectors* $x, y, \ldots$, *then the linear manifold* $\mathbf{L}(x, y, \ldots)$ *contains the whole linear manifold* $\mathbf{L}(x', y', \ldots)$.

Proof. Since the vectors $x', y', \ldots$ belong to the subspace $\mathbf{L}(x, y, \ldots)$ then all their linear combinations, whose totality constitutes the linear manifold $\mathbf{L}(x', y', \ldots)$, also belong to the subspace $\mathbf{L}(x, y, \ldots)$. ▌

b. LEMMA. *Every vector of the system* $x, y, \ldots$ *which is linearly dependent on the other vectors of the system can be eliminated without changing the linear manifold spanned by* $x, y, \ldots$.

Proof. If the vector x, say, is linearly dependent on the vectors $y, z, \ldots$, this means that $x \in \mathbf{L}(y, z, \ldots)$. It follows from Lemma 2.53a that

$$\mathbf{L}(x, y, z, \ldots) \subset \mathbf{L}(y, z, \ldots).$$

On the other hand, obviously

$$\mathbf{L}(y, z, \ldots) \subset \mathbf{L}(x, y, z, \ldots).$$

Together these two relations imply

$$\mathbf{L}(y, z, \ldots) = \mathbf{L}(x, y, z, \ldots). \quad \blacksquare$$

2.54. We now pose the problem of *constructing a basis for a linear manifold and determining the dimension of a linear manifold.* In solving this problem, we will assume that the number of vectors $x, y, \ldots$ spanning the linear manifold $\mathbf{L}(x, y, \ldots)$ is *finite*, although some of our conclusions do not actually require this assumption.

Suppose that among the vectors $x, y, \ldots$ spanning the linear manifold $\mathbf{L}(x, y, \ldots)$ we can find r linearly independent vectors $x_1, x_2, \ldots, x_r$, say, such that every vector of the system $x, y, \ldots$ is a linear combination of $x_1, x_2, \ldots, x_r$. Then *the vectors $x_1, x_2, \ldots, x_r$ form a basis for the space* $\mathbf{L}(x, y, \ldots)$. Indeed, by the very definition of a linear manifold, every vector $z \in \mathbf{L}(x, y, \ldots)$ can be expressed as a linear combination of a finite number of vectors of the system $x, y, \ldots$. But, by hypothesis, each of these vectors can be expressed as a linear combination of $x_1, x_2, \ldots, x_r$. Thus eventually the vector z can also be expressed as a linear combination of the vectors $x_1, x_2, \ldots, x_r$. This, together with the assumption that the vectors $x_1, x_2, \ldots, x_r$ are linearly independent, shows that $x_1, x_2, \ldots, x_r$ indeed form a basis, as asserted.

According to Theorem 2.35, the dimension of the space $\mathbf{L}(x, y, \ldots)$ is equal to the number r. Since there can be no more than r linearly independent vectors in an r-dimensional space, we can draw the following conclusions:

a. *If the number of vectors $x, y, \ldots$ spanning $\mathbf{L}(x, y, \ldots)$ is larger than the number r, then the vectors $x, y, \ldots$ are linearly dependent. If the number of these vectors equals r, then the vectors are linearly independent.*

b. *Every set of $r + 1$ vectors from the system $x, y, \ldots$ is linearly dependent.*

c. *The dimension of the space $\mathbf{L}(x, y, \ldots)$ can be defined as the maximum number of linearly independent vectors in the system $x, y, \ldots$.*

2.6. Hyperplanes

2.61. As already noted in Sec. 2.42d, if we adopt the "point" rather than the "vector" interpretation in the space V_3, then the geometric entity corresponding to the notion of a subspace is a plane (or a straight line) passing through the origin of coordinates. But it is also desirable to include in our scheme of things planes and straight lines which do not pass through the origin of coordinates. Noting that such planes and straight lines are obtained from planes and straight lines passing through the origin of coordinates by means of a parallel displacement in space, i.e., by a *shift*, we are led in a natural way to the following general construction:

Let **L** be a subspace of a linear space **K**, and let $x_0 \in \mathbf{K}$ be a fixed vector which in general does not belong to **L**. Consider the set **H** of all vectors of the form

$$x = x_0 + y$$

where the vector y ranges over the whole subspace **L**. Then **H** is called a *hyperplane*, more specifically, *the result of shifting the subspace* **L** *by the vector* x_0. We note that in general a hyperplane is itself not a linear space.

2.62. *Examples*

a. In the space V_3 the set of all vectors starting from the origin of coordinates and terminating on a plane γ forms a hyperplane. It is easily verified that this hyperplane is a subspace if and only if the plane γ passes through the origin of coordinates.

b. In the space K_n consider the set H consisting of the vectors $x = (\xi_1, \xi_2, \ldots, \xi_n)$ whose components satisfy the compatible nonhomogeneous system of linear equations

$$\begin{aligned} a_{11}x_1 + a_{12}x_2 + \cdots + a_{1n}x_n &= b_1, \\ a_{21}x_1 + a_{22}x_2 + \cdots + a_{2n}x_n &= b_2, \\ \cdots\cdots\cdots\cdots\cdots\cdots\cdots\cdots\cdots \\ a_{k1}x_1 + a_{k2}x_2 + \cdots + a_{kn}x_n &= b_k, \end{aligned} \tag{13}$$

and the set L consisting of the vectors $y = (\eta_1, \eta_2, \ldots, \eta_n)$ whose components satisfy the homogeneous system of linear equations with the same coefficients:

$$\begin{aligned} a_{11}y_1 + a_{12}y_2 + \cdots + a_{1n}y_n &= 0, \\ a_{21}y_1 + a_{22}y_2 + \cdots + a_{2n}y_n &= 0, \\ \cdots\cdots\cdots\cdots\cdots\cdots\cdots\cdots\cdots \\ a_{k1}y_1 + a_{k2}y_2 + \cdots + a_{kn}y_n &= 0. \end{aligned} \tag{13$'$}$$

As we already know from Example 2.42e, the set L is a subspace of the space K_n. Let $x_0 = (\xi_1^{(0)}, \xi_2^{(0)}, \ldots, \xi_n^{(0)})$ be a solution of the system (13). Then the set H is identical with the set of all sums $x_0 + y$ where y ranges over the whole subspace L. In fact, if $y = (\eta_1, \eta_2, \ldots, \eta_n)$ is a solution of the system (13′), then the vector

$$x = x_0 + y = (\xi_1^{(0)} + \eta_1, \xi_2^{(0)} + \eta_2, \ldots, \xi_n^{(0)} + \eta_n)$$

is obviously a solution of the system (13), i.e., belongs to the set H. Conversely, if x is any vector of the set H, then the difference $y = x - x_0$ certainly satisfies the system (13′), i.e., the vector y belongs to the subspace

L. In view of the definition given above, the set H is a hyperplane, namely the result of shifting the space L by the vector x_0.

2.63. We can assign a dimension to every hyperplane, even if it is not a subspace, i.e., we consider the *dimension* of the hyperplane $\mathbf{H}$ to be equal to the dimension of the subspace $\mathbf{L}$ from which $\mathbf{H}$ was obtained by shifting. For this definition to be suitable, we must show that the given hyperplane $\mathbf{H}$ can be obtained as a shift of *only one* subspace. To prove this, suppose $\mathbf{H}$ is both the result of shifting the subspace $\mathbf{L}$ by the vector x_0 and the result of shifting the subspace $\mathbf{L}'$ by the vector x_0'. Then for any $z \in \mathbf{H}$ we have both $z = x_0 + y$ where $y \in \mathbf{L}$ and $z = x_0' + y'$ where $y' \in \mathbf{L}'$. It follows that $\mathbf{L}'$ is the set of vectors of the form $y' = (x_0 - x_0') + y$ where y is an arbitrary vector in $\mathbf{L}$, i.e., the subspace $\mathbf{L}'$ is the result of shifting the subspace $\mathbf{L}$ by the vector $x_1 = x_0 - x_0'$. Clearly x_1 belongs to the subspace $\mathbf{L}$. In fact, the zero vector, just like any other element of the space $\mathbf{L}'$, can be represented in the form $x_1 + y_1$ where $y_1 \in \mathbf{L}$ (since $\mathbf{L}'$ is the subspace $\mathbf{L}$ shifted by the vector x_1). Therefore $x_1 = -y_1$, so that $x_1 \in \mathbf{L}$, as asserted. But then every vector $y' \in \mathbf{L}'$ also belongs to the subspace $\mathbf{L}$, since y' is the sum of a vector $x_1 \in \mathbf{L}$ and a vector $y \in \mathbf{L}$. It follows that $\mathbf{L}' \subset \mathbf{L}$. Because of the complete symmetry of the hypothesis, we can prove similarly that $\mathbf{L} \subset \mathbf{L}'$. Together with $\mathbf{L}' \subset \mathbf{L}$, this implies $\mathbf{L} = \mathbf{L}'$, as required.

In what follows, hyperplanes of dimension 1 will be called *straight lines*, and hyperplanes of dimension 2 will be called *planes*.

2.7. Morphisms of Linear Spaces

2.71. Let ω be a rule which assigns to every given vector x' of a linear space $\mathbf{K}'$ a vector x'' in a linear space $\mathbf{K}''$. Then ω is called a *morphism* (or *linear operator*)† if the following two conditions hold:

a) $\omega(x' + y') = \omega(x') + \omega(y')$ for every $x', y' \in \mathbf{K}'$;
b) $\omega(\alpha x') = \alpha\omega(x')$ for every $x' \in \mathbf{K}'$ and every $\alpha \in K$.

A morphism ω mapping the space $\mathbf{K}'$ onto the whole space $\mathbf{K}''$ is called an *epimorphism*. A morphism ω mapping $\mathbf{K}'$ onto part (or all) of $\mathbf{K}''$ in a one-to-one fashion (so that $x' \neq y'$ implies $\omega(x') \neq \omega(y')$) is called a *monomorphism*. A morphism ω mapping $\mathbf{K}'$ onto all of $\mathbf{K}''$ in a one-to-one fashion (i.e., a morphism which is both an epimorphism and a monomorphism) is called an *isomorphism*, and the spaces $\mathbf{K}'$ and $\mathbf{K}''$ themselves are said to be *isomorphic* (more exactly, *K-isomorphic*). The usual notation for a morphism is

$$\omega : \mathbf{K}' \to \mathbf{K}''.$$

† More exactly, a *morphism of* $\mathbf{K}'$ *into* $\mathbf{K}''$ (or a *linear operator mapping* $\mathbf{K}'$ *into* $\mathbf{K}''$).

2.72. *Examples*

a. Let **L** be a subspace of a space **K**. Then the mapping ω which assigns to every vector $x \in \mathbf{L}$ the same vector $x \in \mathbf{K}$ is a morphism of **L** into **K**, and in fact a monomorphism (but not an epimorphism if $\mathbf{L} \neq \mathbf{K}$). This morphism is said to *embed* **L** in **K**.

b. Let **L** be a subspace of a space **K**, and let $\mathbf{K}/\mathbf{L}$ be the factor space of **K** with respect to **L** (see Sec. 2.48). Then the mapping ω which assigns to every vector $x \in \mathbf{K}$ the class $\mathbf{X} \in \mathbf{K}/\mathbf{L}$ containing x is a morphism of ω into $\mathbf{K}/\mathbf{L}$, and in fact an epimorphism (but not a monomorphism if $\mathbf{L} \neq 0$). This morphism ω is called the *canonical mapping of* **K** *onto* $\mathbf{K}/\mathbf{L}$.

2.73. a. Let the space $\mathbf{K}'$ be n-dimensional with basis $e_1', \ldots, e_n'$, and choose n arbitrary vectors $e_1'', \ldots, e_n''$ in $\mathbf{K}''$. With every given vector

$$x' = \sum_{k=1}^{n} \xi_k e_k'$$

in $\mathbf{K}'$ we associate the vector

$$\omega(x') = x'' = \sum_{k=1}^{n} \xi_k e_k''$$

in $\mathbf{K}''$ with the same components ξ_k $(k = 1, \ldots, n)$. Then the mapping $\omega(x') = x''$ is a morphism of the space $\mathbf{K}'$ into the space $\mathbf{K}''$. In fact, given any two vectors

$$x' = \sum_{k=1}^{n} \xi_k e_k', \qquad y' = \sum_{k=1}^{n} \eta_k e_k'$$

in $\mathbf{K}'$, it follows from Theorem 2.33 that

$$x' + y' = \sum_{k=1}^{n} (\xi_k + \eta_k) e_k'.$$

But

$$\omega(x') = \sum_{k=1}^{n} \xi_k e_k'', \qquad \omega(y') = \sum_{k=1}^{n} \eta_k e_k''$$

by the definition of the mapping ω, and moreover

$$\omega(x' + y') = \sum_{k=1}^{n} (\xi_k + \eta_k) e_k'' = \sum_{k=1}^{n} \xi_k e_k'' + \sum_{k=1}^{n} \eta_k e_k'' = \omega(x') + \omega(y'),$$

so that condition a) of Sec. 2.71 is satisfied. Similarly,

$$\omega(\alpha x') = \omega\left(\alpha \sum_{k=1}^{n} \xi_k e_k'\right) = \omega\left(\sum_{k=1}^{n} \alpha \xi_k e_k'\right)$$

$$= \sum_{k=1}^{n} \alpha \xi_k e_k'' = \alpha \sum_{k=1}^{n} \xi_k e_k'' = \alpha \omega(x')$$

for every $\alpha \in K$, so that condition b) is also satisfied. Therefore ω is a morphism of $\mathbf{K}'$ into $\mathbf{K}''$, as asserted.

b. Obviously, the morphism ω just described is an *epimorphism* if and only if every vector $x'' \in \mathbf{K}''$ can be represented in the form

$$\sum_{k=1}^{n} \xi_k e_k'',$$

i.e., if and only if $\mathbf{K}''$ *coincides with the linear manifold spanned by the vectors* $e_1'', \ldots, e_n''$.

c. Similarly, our morphism ω is a *monomorphism* if and only if every pair of vectors

$$\sum_{k=1}^{n} \xi_k e_k'', \qquad \sum_{k=1}^{n} \eta_k e_k''$$

differing in at least one component (i.e., such that $\xi_k \neq \eta_k$ for at least one value of k) are distinct vectors of $\mathbf{K}''$. But this is equivalent to linear independence of the vectors $e_1'', \ldots, e_n''$. Therefore *the morphism ω is a monomorphism if and only if the vectors $e_1'', \ldots, e_n''$ are linearly independent.*

d. It follows that the morphism ω described above is an *isomorphism* if and only if the vectors $e_1'', \ldots, e_n''$ are linearly independent and the linear manifold spanned by them coincides with the whole space $\mathbf{K}''$. In other words, *the morphism ω is an isomorphism if and only if the vectors $e_1'', \ldots, e_n''$ form a basis in the space* $\mathbf{K}''$.

2.74. THEOREM. *Any two n-dimensional spaces* $\mathbf{K}'$ *and* $\mathbf{K}''$ (*over the same field K*) *are K-isomorphic.*

Proof. Let $e_1', \ldots, e_n'$ be a basis in the space $\mathbf{K}'$ and $e_1'', \ldots, e_n''$ a basis in the space $\mathbf{K}''$, and use these two systems of vectors to construct a morphism ω of $\mathbf{K}'$ into $\mathbf{K}''$ in the way described in Sec. 2.73a. Then ω is an isomorphism, by Sec. 2.73d. ∎

2.75. COROLLARY. *Every n-dimensional linear space over a field K is K-isomorphic to the space K_n of Sec. 2.15b. In particular, every n-dimensional complex space is C-isomorphic to the space C_n, and every n-dimensional real space is R-isomorphic to the space R_n.*

2.76. We now discuss further properties of epimorphisms and monomorphisms.

a. Given a morphism $\omega: \mathbf{K}' \to \mathbf{K}''$, consider the set $\mathbf{L}''$ of all vectors $\omega(x') \in \mathbf{K}''$ such that $x' \in \mathbf{K}'$. The set $\mathbf{L}''$, which is obviously a subspace of $\mathbf{K}''$, is called the *range* of the morphism ω. It is clear that the mapping ω

of $\mathbf{K}'$ into $\mathbf{L}''$ is an epimorphism. If the morphism $\omega:\mathbf{K}' \to \mathbf{K}''$ is a monomorphism, then the morphism $\omega:\mathbf{K}' \to \mathbf{L}''$ is an isomorphism.

b. Given a morphism $\omega:\mathbf{K}' \to \mathbf{K}''$, consider the set $\mathbf{L}'$ of all vectors $x' \in \mathbf{K}'$ such that $\omega(x') = 0$. The set $\mathbf{L}'$, which is obviously a subspace of $\mathbf{K}'$, is called the *null space* (or *kernel*) of the morphism ω.

We now construct the factor space $\mathbf{K}'/\mathbf{L}'$ (see Sec. 2.48). All the elements x' belonging to the same class $\mathbf{X}' \in \mathbf{K}'/\mathbf{L}'$ are carried by the morphism ω into the same element of the space $\mathbf{K}''$. In fact, given two such elements x' and y', we have $x' - y' = z' \in \mathbf{L}'$, and hence

$$\omega(x') - \omega(y') = \omega(z') = 0, \qquad \omega(x') = \omega(y').$$

Suppose that with every class $\mathbf{X}' \in \mathbf{K}'/\mathbf{L}'$ we associate the element $x'' = \omega(x') \in \mathbf{K}''$ where x' is an arbitrary element of $\mathbf{X}'$ (as just shown x'' is uniquely determined). Let $x'' = \Omega(\mathbf{X}')$. Then it is easy to see that Ω is a morphism of $\mathbf{K}'/\mathbf{L}'$ into $\mathbf{K}''$. Moreover Ω is a monomorphism, since it follows from $\mathbf{X}' \neq \mathbf{Y}'$, $x' \in \mathbf{X}'$, $y' \in \mathbf{Y}'$ that

$$\Omega(\mathbf{X}') - \Omega(\mathbf{Y}') = \omega(x') - \omega(y') = \omega(x' - y') \neq 0.$$

Thus any morphism $\omega:\mathbf{K}' \to \mathbf{K}''$ generates a monomorphism $\Omega:\mathbf{K}'/\mathbf{L}' \to \mathbf{K}''$. If the morphism ω is an epimorphism, then, obviously, the monomorphism Ω is also an epimorphism, so that the epimorphism $\omega:\mathbf{K}' \to \mathbf{K}''$ generates an isomorphism $\Omega:\mathbf{K}'/\mathbf{L}' \to \mathbf{K}''$.

We will continue the study of morphisms in Chapter 4.

PROBLEMS

1. Consider the set of vectors in the plane whose initial points are located at the origin of coordinates and whose final points lie within the first quadrant. Does this set form a linear space (with the usual operations)?

2. Consider the set of all vectors in the plane with the exception of the vectors which are parallel to a given straight line. Does this set form a linear space?

3. Consider the set P consisting of the *positive* real numbers only. We introduce operations according to the following rules: By the "sum" of two numbers we mean their product (in the usual sense), and by the "product" of an element $r \in P$ and a real number λ we mean r raised to the power λ (in the usual sense). If P a linear space (with these operations)?

4. Show that a criterion for the linear independence of n given vectors in the space K_n is that the determinant formed from the coordinates of the vectors does not vanish.

5. Show that the functions $t^{r_1}, t^{r_2}, \ldots, t^{r_k}$ are linearly independent in the space $K(a, b)$, where $0 < a < b$ and $r_1, r_2, \ldots, r_k$ are distinct real numbers.

6. The following is known about a system of vectors $e_1, e_2, \ldots, e_n$ in a linear space $\mathbf{K}$:

a) Every vector $x \in \mathbf{K}$ has an expansion of the form

$$x = \xi_1 e_1 + \xi_2 e_2 + \cdots + \xi_n e_n;$$

b) This expansion is *unique* for some fixed vector $x_0 \in \mathbf{K}$.

Show that the system $e_1, e_2, \ldots, e_n$ forms a basis in $\mathbf{K}$.

7. Does there exist a basis in the space P of Problem 3?

8. What is the dimension of the space P of Problem 3?

9. Find the intersection and sum of two distinct two-dimensional subspaces of the space V_3 (two distinct planes passing through the origin of coordinates).

10. Prove that if the dimension of the subspace $\mathbf{L} \subset \mathbf{K}$ is the same as that of the space $\mathbf{K}$, then $\mathbf{L} = \mathbf{K}$.

11. Is the shift vector x_0 figuring in the construction of a hyperplane uniquely determined by the hyperplane itself?

12. Show that every hyperplane $\mathbf{H} \subset \mathbf{K}$ has the following property: If $x \in \mathbf{H}$, $y \in \mathbf{H}$, then $\alpha x + (1 - \alpha)y \in \mathbf{H}$ for every element of the field K. Conversely, show that if a *subset* $\mathbf{H} \subset \mathbf{K}$ has this property, then $\mathbf{H}$ is a hyperplane. What geometric characteristic of a hyperplane is expressed by this property?

13. The hyperplanes $\mathbf{H}_1$ and $\mathbf{H}_2$ have dimensions p and q, respectively. What is the (smallest) dimension which the hyperplane $\mathbf{H}_3$ must have in order to be sure to contain both $\mathbf{H}_1$ and $\mathbf{H}_2$?

14. Solve the analogous problem for three hyperplanes $\mathbf{H}_1$, $\mathbf{H}_2$ and $\mathbf{H}_3$, with dimensions p, q and r, respectively.

15. According to Theorem 2.74, the one-dimensional spaces R_1 and P (see Problem 3) are isomorphic. How can one establish this isomorphism in practice?

chapter 3

SYSTEMS OF LINEAR EQUATIONS

3.1. More on the Rank of a Matrix

3.11. We have already touched upon the subject of matrices several times. In this section we will study in more detail those properties of matrices which are connected with the concept of rank (see Sec. 1.9). This will allow us to give a general solution of the basic problems of the theory of systems of linear equations, posed in Sec. 1.2.

We begin by recalling some basic definitions from Sec. 1.9. Suppose we have a matrix

$$\begin{Vmatrix} a_{11} & a_{12} & \cdots & a_{1k} \\ a_{21} & a_{22} & \cdots & a_{2k} \\ \cdot & \cdot & \cdots & \cdot \\ a_{n1} & a_{n2} & \cdots & a_{nk} \end{Vmatrix} \qquad (1)$$

with n rows and k columns, consisting of the numbers a_{ij} from the field K, where i is the row index ranging from 1 to n and j is the column index ranging from 1 to k.† If we choose any m rows and m columns of this matrix, then the elements which appear at the intersections of these rows and columns

† Sometimes the indices of an element of the matrix A will be written differently, i.e., sometimes we will denote the element appearing in the ith row and jth column of A by the symbol a_i^j.

form a square matrix of order m. The determinant of this matrix is called *a minor of order m of the matrix A*. The integer m is said to be the *rank* of the matrix A if A has a nonvanishing minor of order r and all its minors of order $r + 1$ and higher vanish. If the matrix A has rank $r > 0$, then each of its nonvanishing minors of order r is called a *basis minor*. The columns and rows of the matrix which intersect at the elements of the basis minor are called the *basis columns* and *basis rows*.

The considerations that follow are based on the possibility of regarding any column of numbers as a geometric object, i.e., as a vector in the n-dimensional space K_n of Sec. 2.15b. With this geometric interpretation, the matrix A itself corresponds to a certain set of k vectors of the space K_n. Let x_j $(j = 1, \ldots, k)$ denote the vector corresponding to the jth column of A. Then any linear relation between the columns of A can be interpreted as the same linear relation between the corresponding vectors (see Sec. 2.22b).

Let $L(x_1, x_2, \ldots, x_k)$ be the linear manifold spanned by the vectors $x_1, x_2, \ldots, x_k$ of K_n (see Sec. 2.51). We now prove that *the vectors corresponding to the basis columns of the matrix A form a basis for this linear manifold.* To be explicit, suppose that the first r columns of A are basis columns. Then, to prove our assertion, it suffices to show first that the vectors $x_1, x_2, \ldots, x_r$ are linearly independent, and secondly that any of the other vectors $x_{r+1}, \ldots, x_n$ is a linear combination of the first r vectors (see Sec. 2.54). To prove the first assertion, suppose that the vectors $x_1, x_2, \ldots, x_r$ are linearly dependent, or equivalently, that the first r columns of A are linearly dependent. Then, by Theorem 1.96, any determinant of order r constructed from these columns and any r rows of A would vanish. In particular, the basis minor of A would vanish, contrary to its very definition. This contradiction establishes the first assertion. The second assertion, as applied to columns of the matrix A, has already been proved in Sec. 1.93 under the guise of the "basis minor theorem." This completes the proof that the vectors $x_1, x_2, \ldots, x_r$ form a basis for the space $L(x_1, x_2, \ldots, x_k)$. According to Theorem 2.35, the dimension of this space equals the number r, i.e., the rank of the matrix A. Thus we have established the following important

THEOREM. *The dimension of the linear manifold spanned by the vectors corresponding to the columns of the matrix A equals the rank of A. Moreover, the vectors corresponding to the basis columns of A form a basis for this linear manifold.*

3.12. The following propositions are obvious consequences of conclusions a)–c) of Sec. 2.54:

a. THEOREM. *If the rank of the matrix A is less than the number of columns in A* $(r < k)$, *then the columns of A are linearly dependent. If the rank of A* *equals the number of columns in* A ($r = k$), *then the columns of* A *are linearly independent.*

b. THEOREM. *Any* $r + 1$ *columns of the matrix* A *are linearly dependent.*

c. THEOREM. *The rank of any matrix* A *equals the maximum number of linearly independent columns in* A.

This last theorem is of fundamental importance, since it constitutes a *new definition of the rank of a matrix.*

3.13. Suppose we transpose the matrix A, i.e., suppose we go over to the matrix A' whose rows are the columns of A (cf. Sec. 1.41). Clearly, the rank of the transposed matrix A' is the same as the rank of A. But according to Theorem 3.12c, the rank of A' equals the maximum number of linearly independent columns in A', or equivalently, the maximum number of linearly independent rows in A. Thus we arrive at the following somewhat unexpected conclusion:

THEOREM. *The maximum number of linearly independent rows in a matrix* A *is the same as the maximum number of linearly independent columns in* A.

We note that this theorem is not trivial. In fact, any direct proof of the theorem would require a chain of reasoning equivalent to the proof of Theorems 1.93 and 3.11.

3.14. Finally we note the following result, which is a consequence of Theorem 3.11 and Lemma 2.53b:

THEOREM. *Any column of the matrix* A *which is a linear combination of the other columns can be deleted without changing the rank of* A.

3.2. Nontrivial Compatibility of a Homogeneous Linear System

3.21. Suppose we have a homogeneous linear system

$$\begin{aligned} a_{11}x_1 + a_{12}x_2 + \cdots + a_{1n}x_n &= 0, \\ a_{21}x_1 + a_{22}x_2 + \cdots + a_{2n}x_n &= 0, \\ \cdots\cdots\cdots\cdots&\cdots\cdots \\ a_{k1}x_1 + a_{k2}x_2 + \cdots + a_{kn}x_n &= 0. \end{aligned} \tag{2}$$

As we know, this system is always compatible, since it has the trivial solution

$$x_1 = x_2 = \cdots = x_n = 0.$$

The basic problem encountered in studying homogeneous linear systems is the following: *Under what conditions is a homogeneous linear system "nontrivially compatible," i.e., under what conditions does such a system have solutions other than the trivial solution*? The results of Sec. 3.1 allow us to solve this problem immediately. In fact, as we have seen in Sec. 2.22b, the existence of a nontrivial solution of the system (2) is equivalent to the columns of the matrix

$$A = \begin{Vmatrix} a_{11} & a_{12} & \cdots & a_{1n} \\ a_{21} & a_{22} & \cdots & a_{2n} \\ \cdot & \cdot & \cdots & \cdot \\ a_{k1} & a_{k2} & \cdots & a_{kn} \end{Vmatrix}$$

being linearly dependent. But, according to Theorem 3.12a, this occurs if and only if the rank of the matrix A is less than the number of columns in A. Thus we obtain the following

THEOREM. *The system* (2) *is nontrivially compatible, i.e., has nontrivial solutions if and only if the rank of the matrix A is less than n. If the rank of the matrix A equals n, the system* (2) *has no nontrivial solutions.*

3.22. In particular, if the number of equations in the system (2) is less than the number of unknowns ($k < n$), the rank of the matrix A is certainly less than n, and in this case nontrivial solutions always exist. If $k = n$, the question of whether or not nontrivial solutions exist depends on the value of $\det A$. If $\det A \neq 0$, there are no nontrivial solutions ($r = n$), while if $\det A = 0$, there are nontrivial solutions ($r < n$). If $k > n$, we have to examine all possible determinants of order n which are obtained by fixing any n rows of the matrix A. If all these determinants vanish, then $r < n$ and nontrivial solutions exist. If at least one of these determinants is nonvanishing, then $r = n$ and there is only the trivial solution.

3.3. The Compatibility Condition for a General Linear System

3.31. Suppose we have a general (i.e., nonhomogeneous) system of linear equations

$$\begin{aligned} a_{11}x_1 + a_{12}x_2 + \cdots + a_{1n}x_n &= b_1, \\ a_{21}x_1 + a_{22}x_2 + \cdots + a_{2n}x_n &= b_2, \\ \cdots\cdots\cdots\cdots\cdots\cdots\cdots\cdots \\ a_{k1}x_1 + a_{k2}x_2 + \cdots + a_{kn}x_n &= b_k. \end{aligned} \tag{3}$$

With this system we associate two matrices, the matrix

$$A = \begin{Vmatrix} a_{11} & a_{12} & \cdots & a_{1n} \\ a_{21} & a_{22} & \cdots & a_{2n} \\ \cdot & \cdot & \cdots & \cdot \\ a_{k1} & a_{k2} & \cdots & a_{kn} \end{Vmatrix},$$

called the *coefficient matrix of the system* (3), and the matrix

$$A_1 = \begin{Vmatrix} a_{11} & a_{12} & \cdots & a_{1n} & b_1 \\ a_{21} & a_{22} & \cdots & a_{2n} & b_2 \\ \cdot & \cdot & \cdot & \cdots & \cdot \\ a_{k1} & a_{k2} & \cdots & a_{kn} & b_k \end{Vmatrix},$$

called the *augmented matrix of the system* (3). Regarding the compatibility of the system (3), we then have the following basic

THEOREM (***Kronecker-Capelli***). *The system* (3) *is compatible if and only if the rank of the augmented matrix of the system equals the rank of the coefficient matrix.*

Proof. Assume first that the system (3) is compatible. Then if $c_1, c_2, \ldots, c_n$ is a solution of the system, we have the equations

$$\begin{aligned} a_{11}c_1 + a_{12}c_2 + \cdots + a_{1n}c_n &= b_1, \\ a_{21}c_1 + a_{22}c_2 + \cdots + a_{2n}c_n &= b_2, \\ \cdots\cdots\cdots\cdots\cdots\cdots\cdots \\ a_{k1}c_1 + a_{k2}c_2 + \cdots + a_{kn}c_n &= b_k. \end{aligned}$$

These equations imply that the last column of A_1 is a linear combination of the other columns of A_1 (with coefficients $c_1, c_2, \ldots, c_n$). By Theorem 3.14, we can delete the last column of A_1 without changing its rank. But when the last column of A_1 is deleted, it becomes just A. Hence if the system (3) is compatible, the matrices A and A_1 have the same rank.

We now assume that the matrices A and A_1 have the same rank, and show that the system (3) is compatible. Let r be the rank of the matrix A (and consequently also of the matrix A_1). Consider r basis columns of A; they will also be basis columns of A_1. By Theorem 1.93, the last column of A_1 can be written as a linear combination of the basis columns, and hence it can be written as a linear combination of *all* the columns of A. If we denote the coefficients of this linear combination by $c_1, c_2, \ldots, c_n$, we find that the equations

$$\begin{aligned} a_{11}c_1 + a_{12}c_2 + \cdots + a_{1n}c_n &= b_1, \\ a_{21}c_1 + a_{22}c_2 + \cdots + a_{2n}c_n &= b_2, \\ \cdots\cdots\cdots\cdots\cdots\cdots \\ a_{k1}c_1 + a_{k2}c_2 + \cdots + a_{kn}c_n &= b_k \end{aligned}$$

are satisfied. Thus the values

$$x_1 = c_1,\ x_2 = c_2, \ldots, x_n = c_n$$

satisfy the system (3), which is therefore compatible. ∎

3.4. The General Solution of a Linear System

3.41. The Kronecker–Capelli theorem, which gives the general condition for the compatibility of a linear system, does not give a method for solving the system. We now derive a formula which constitutes a general solution of a linear system.

By a *general solution of the system* (3) we mean a set of expressions

$$x_j = f_j(a_{11}, \ldots, a_{kn}, b_1, \ldots, b_k, q_1, \ldots, q_s) \qquad (j = 1, \ldots, n),$$

where the right-hand sides are functions depending on the coefficients a_{ij} of the system (3), the constant terms b_j of (3) and certain undetermined parameters $q_1, \ldots, q_s$, such that

1) The quantities $x_j = c_j$ $(j = 1, \ldots, n)$ obtained for arbitrary fixed values of the parameters $q_1, \ldots, q_s$ (from the field K) constitute a solution of the system (3);

2) Any given solution of the system (3) can be obtained in this way by suitably choosing the values of the parameters $q_1, \ldots, q_s$ in K.

As shown in Sec. 2.62b, the set of all sums of the form $x_0 + y$, where x_0 is any ("particular") solution of the system (3) and y ranges over the set of all solutions of the corresponding homogeneous system, is just the set of all solutions of (3). This fact can now be expressed as follows: *The general solution of the nonhomogeneous system* (3) *is the sum of any particular solution of* (3) *and the general solution of the corresponding homogeneous system* (2).

Suppose we have a compatible linear system (3) with a coefficient matrix $A = \|a_{ij}\|$ of rank r. It can be assumed that the basis minor M of the matrix A appears in its upper left-hand corner; otherwise, we can achieve this configuration by interchanging rows and columns of A, which corresponds to renumbering some of the equations and unknowns in the system (3). We take the first r equations of the system (3) and rewrite them in the form

$$\begin{aligned} a_{11}x_1 + a_{12}x_2 + \cdots + a_{1r}x_r &= b_1 - a_{1,r+1}x_{r+1} - \cdots - a_{1n}x_n, \\ a_{21}x_1 + a_{22}x_2 + \cdots + a_{2r}x_r &= b_2 - a_{2,r+1}x_{r+1} - \cdots - a_{2n}x_n, \\ &\cdots\cdots\cdots\cdots\cdots\cdots \\ a_{r1}x_1 + a_{r2}x_2 + \cdots + a_{rr}x_r &= b_r - a_{r,r+1}x_{r+1} - \cdots - a_{rn}x_n. \end{aligned} \tag{4}$$

Next we assign the unknowns $x_{r+1}, \ldots, x_n$ completely arbitrary values $c_{r+1}, \ldots, c_n$. Then (4) becomes a system of r equations in the r unknowns $x_1, x_2, \ldots, x_r$, with a determinant M which is nonvanishing (a basis minor of the matrix A). This system can be solved by using Cramer's rule (see Sec. 1.73). Hence there exist numbers $c_1, c_2, \ldots, c_n$ which, when substituted for the unknowns $x_1, x_2, \ldots, x_n$ of the system (4), reduce all the equations of the system to identities. We now show that these values $c_1, c_2, \ldots, c_n$ satisfy all the other equations of the system (3) as well.

The first r rows of the augmented matrix A_1 of the system (3) are basis rows of this matrix, since by the compatibility condition, the rank of the augmented matrix is r, while by construction, the nonvanishing minor M appears in the first r rows of A_1. By Theorem 1.93 (applied to rows), each of the last $n - r$ rows of A_1 is a linear combination of the first r rows. This means that every equation of the system (3) beginning with the $(r + 1)$st equation is a linear combination of the first r equations of the system. Therefore, if the values

$$x_1 = c_1, \ldots, x_n = c_n$$

satisfy the first r equations of the system (3), they also satisfy all the other equations of (3).

3.42. To write an explicit formula for the solution of the system (3) just constructed, let $M_j(\alpha_j)$ denote the determinant obtained from the basis minor

$$M = \det \|a_{ij}\| \qquad (i, j = 1, 2, \ldots, r)$$

by replacing its jth column by the column consisting of the quantities $\alpha_1, \alpha_2, \ldots, \alpha_r$. Then, using Cramer's rule to write the solution of the system (4), we obtain

$$\begin{aligned} c_j &= \frac{1}{M} M_j(b_i - a_{i,r+1}c_{r+1} - \cdots - a_{in}c_n) \\ &= \frac{1}{M} [M_j(b_i) - c_{r+1}M_j(a_{i,r+1}) - \cdots - c_nM_j(a_{in})] \qquad (j = 1, 2, \ldots, r). \end{aligned} \tag{5}$$

These formulas express the values of the unknowns $x_j = c_j$ $(j = 1, 2, \ldots, r)$ in terms of the coefficients of the system, the constant terms and the arbitrary quantities (parameters)

$$c_{r+1}, c_{r+2}, \ldots, c_n.$$

Finally, we show that (5) comprises any solution of the system (3). In fact, let $c_1^{(0)}, c_2^{(0)}, \ldots, c_r^{(0)}, c_{r+1}^{(0)}, \ldots, c_n^{(0)}$ be an arbitrary solution of the system (3). Obviously, it is also a solution of the system (4). But, using Cramer's rule to solve the system (4), we obtain unique expressions for the quantities $c_1^{(0)}, c_2^{(0)}, \ldots, c_r^{(0)}$ in terms of the quantities $c_{r+1}^{(0)}, \ldots, c_n^{(0)}$, namely the formulas (5). Thus, choosing

$$c_{r+1} = c_{r+1}^{(0)}, \ldots, c_n = c_n^{(0)}$$

in (5), we get just the solution $c_1^{(0)}, c_2^{(0)}, \ldots, c_n^{(0)}$, as asserted. Thus (5) is the general solution of the system (3).

3.5. Geometric Properties of the Solution Space

3.51. Consider first the case of the homogeneous linear system (2). As we have already seen (Sec. 2.42e), the set of all solutions of this system forms a linear "solution space," which we denote by L. We now calculate the dimension of L and construct a basis for L.

For a homogeneous system, the equations (5) become

$$-Mc_j = c_{r+1}M_j(a_{i,r+1}) + \cdots + c_nM_j(a_{in}) \qquad (j = 1, 2, \ldots, r), \quad (6)$$

since $M_j(b_i) = M_j(0) = 0$. With every solution $c_1, c_2, \ldots, c_r, c_{r+1}, \ldots, c_n$ of the system (2) we associate a vector $(c_{r+1}, \ldots, c_n)$ of the space K_{n-r} (see Sec. 2.15b). Since the numbers $c_{r+1}, \ldots, c_n$ can be chosen arbitrarily and since they uniquely define a solution of the system (2), the correspondence between the space of solutions of the system (2) and the space K_{n-r} is one-to-one. This correspondence is an isomorphism, since it preserves linear operations, as is easily verified. Thus *the space L of solutions of a homogeneous system of linear equations in n unknowns with a coefficient matrix of rank r is isomorphic to the space K_{n-r}.* In particular, the dimension of the space L is $n - r$.

3.52. Any system of $n - r$ linearly independent solutions of a homogeneous linear system of equations (which, by Theorem 2.34, forms a basis in the space of all solutions) is called a *fundamental system of solutions.* To construct a fundamental system of solutions, we can use any basis of the space K_{n-r}. Then, because of the isomorphism, the corresponding solutions of the system (2) will form a basis in the space of all solutions of the system. The simplest basis of the space K_{n-r} consists of the vectors

$$\begin{aligned} e_1 &= (1, 0, \ldots, 0), \\ e_2 &= (0, 1, \ldots, 0), \\ &\ldots\ldots\ldots\ldots \\ e_{n-r} &= (0, 0, \ldots, 1) \end{aligned}$$

(see Sec. 2.32c). For example, to obtain the solution of the system (2) corresponding to the vector e_1, we set $c_{r+1} = 1$, $c_{r+2} = \cdots = c_n = 0$ in the formulas (6) and determine the corresponding values

$$c_i = c_i^{(1)} \qquad (i = 1, 2, \ldots, n).$$

Similarly, we construct the solution corresponding to any other basis vector e_j ($j = 2, \ldots, n - r$). The set of solutions of the system (2) constructed in this way is called a *normal fundamental system of solutions*. If we denote these solutions by $x^{(1)}, x^{(2)}, \ldots, x^{(n-r)}$, then by the definition of a basis, any solution x is given by the formula

$$x = \alpha_1 x^{(1)} + \alpha_2 x^{(2)} + \cdots + \alpha_{n-r} x^{(n-r)}. \tag{7}$$

Since any solution of the system (2) is a special case of (7), this formula gives the *general solution* of (2).

3.53. Consider now the general case of a nonhomogeneous system (3). As shown in Sec. 2.62b, the geometric object H corresponding to the set of all solutions of a nonhomogeneous system is a hyperplane in the n-dimensional space K_n. This hyperplane is obtained by shifting the subspace L of all solutions of the corresponding homogeneous system (L has been shown to be isomorphic to the space K_{n-r}) by a vector x_0 which is an arbitrary particular solution of the nonhomogeneous system. From this we conclude that the dimension of the hyperplane H is the same as the dimension of the subspace L. Moreover, if r is the rank of the coefficient matrix of the system (3), then any vector y of the subspace L can be represented as a sum

$$y = \alpha_1 y^{(1)} + \alpha_2 y^{(2)} + \cdots + \alpha_{n-r} y^{(n-r)},$$

where $y^{(1)}, y^{(2)}, \ldots, y^{(n-r)}$ are basis vectors of the space L (a fundamental system of solutions). Consequently, any vector x of the hyperplane H can be represented as a sum

$$x = x_0 + y = x_0 + \alpha_1 y^{(1)} + \alpha_2 y^{(2)} + \cdots + \alpha_{n-r} y^{(n-r)}.$$

In the language appropriate to solutions of the systems (2) and (3), this agrees with the prescription established in Sec. 3.41, i.e., *the general solution of the nonhomogeneous system* (3) *is the sum of any particular solution of* (3) *and the general solution of the corresponding homogeneous system* (2).

3.6. Methods for Calculating the Rank of a Matrix

3.61. To make practical use of the methods for solving systems of linear equations developed in the preceding sections, one must be able to calculate the rank of a matrix and find its basis minor. Obviously, the definition of the rank of a matrix given in Sec. 1.92 cannot serve *per se* as a reasonable practical means of calculating the rank. For example, a square matrix of order five contains one minor of order five, 25 minors of order four, 100 minors of order three, and 100 minors of order two. Clearly, it would be a very laborious task to find the rank of such a matrix by direct calculation of all its minors. In this section, we will give simple methods for calculating the rank of a matrix and determining its basis minor. These methods are based on a study of certain operations on rows and columns of a matrix which do not change its rank; these operations will be called *elementary operations*. Since, as already noted, the rank of a matrix does not change when it is transposed, we will define these operations only for the columns of a matrix. In keeping with this, our proofs will make use of the geometric interpretation of a matrix with n rows and k columns as the matrix formed from the components of a system of k vectors $x_1, x_2, \ldots, x_k$ in the n-dimensional (real) space R_n. We will also make use of Theorem 3.11, which asserts that the rank of this matrix equals the dimension of the linear manifold spanned by the vectors $x_1, x_2, \ldots, x_k$.

We now study the following elementary operations:

a. *Permutation of columns.* Suppose the columns of the matrix A are permuted in any way. This operation does not change the rank of A. In fact, the dimension of the linear manifold spanned by the vectors $x_1, x_2, \ldots, x_k$ does not depend on the order in which they are written, and hence the rank of the matrix does not depend on the order of its columns.

b. *Dividing out a nonzero common factor of the elements of a column.* Suppose the number $\lambda \neq 0$ being divided out is a common factor of the elements of the first column of the matrix A. This operation is equivalent to replacing the system of vectors $\lambda x_1, x_2, \ldots, x_k$ by the system $x_1, x_2, \ldots, x_k$. But obviously the linear manifolds spanned by these two systems have the same dimension (since the linear manifolds themselves are the same). Therefore the rank of the matrix A does not change as a result of this elementary operation.

c. *Adding an arbitrary multiple of one column to another column.* Suppose we multiply the mth column of the matrix A by the number λ and add it to the jth column. This means that the system of vectors $x_1, \ldots, x_j, \ldots, x_m, \ldots, x_k$ has been replaced by the system

$$x_1, \ldots, x_j + \lambda x_m, \ldots, x_m, \ldots, x_k.$$

We have to show that the linear manifolds L_1 and L_2 spanned by these two systems are the same. In the first place, all the vectors of the second system lie in the linear manifold spanned by the vectors of the first system. Hence, by Lemma 2.53a, we have $L_2 \subset L_1$. On the other hand, the equation

$$x_j = (x_j + \lambda x_m) - \lambda x_m$$

shows that the vector x_j lies in the linear manifold spanned by the vectors of the second system. Since all the other vectors of the first system obviously belong to this linear manifold, we have $L_1 \subset L_2$. It follows that $L_1 = L_2$. Therefore the rank of A does not change as a result of this elementary operation.

d. *Deletion of a column consisting entirely of zeros.* A column consisting entirely of zeros corresponds to the zero vector of the space R_n. Obviously, eliminating the zero vector from the system $x_1, x_2, \ldots, x_k$ does not change the linear manifold $L(x_1, x_2, \ldots, x_k)$ and hence does not change the rank of the matrix A.

e. *Deletion of a column which is a linear combination of the other columns.* The legitimacy of this elementary operation was proved in Theorem 3.14.

3.62. Calculation of the rank of a matrix and determination of a basis minor. We now show how to calculate the rank and find a basis minor of a given matrix A by using the elementary operations just enumerated. If the matrix A consists only of zeros, then its rank is obviously zero. Suppose A contains a nonzero element. Then, by suitably permuting the rows and columns, we can bring this element over to the upper left-hand corner of the matrix. Then, subtracting from every column the first column multiplied by a suitable coefficient, we can make all the other elements of the first row vanish. We shall make no further changes in the first row and first column (except for the rearrangements described below). If there are no nonzero elements among the remaining elements (i.e., the elements which do not belong to the first row and the first column), then the rank of the matrix A is obviously 1. If there is a nonzero element among the remaining elements, then by suitably rearranging rows and columns, we can bring this element over to the intersection of the second row and the second column and then make all the elements following it in the second row vanish, just as before. (We note that these operations do not affect the first row and the first column.)

Continuing in this fashion, and assuming that the number of columns in A does not exceed the number of rows in A (this can always be achieved by transposition), we reduce A to one of the following two forms:

$$A_1 = \left\| \begin{array}{ccccccccc} \alpha_1 & 0 & 0 & \cdots & 0 & 0 & \cdots & 0 \\ c_{21} & \alpha_2 & 0 & \cdots & 0 & 0 & \cdots & 0 \\ c_{31} & c_{32} & \alpha_3 & \cdots & 0 & 0 & \cdots & 0 \\ \cdot & \cdot & \cdot & \cdots & \cdot & \cdot & \cdots & \cdot \\ \cdot & \cdot & \cdot & \cdots & \cdot & \cdot & \cdots & \cdot \\ \cdot & \cdot & \cdot & \cdots & \cdot & \cdot & \cdots & \cdot \\ c_{k1} & c_{k2} & c_{k3} & \cdots & \alpha_k & 0 & \cdots & 0 \\ c_{k+1,1} & c_{k+1,2} & c_{k+1,3} & \cdots & c_{k+1,k} & 0 & \cdots & 0 \\ \cdot & \cdot & \cdot & \cdots & \cdot & \cdot & \cdots & \cdot \\ \cdot & \cdot & \cdot & \cdots & \cdot & \cdot & \cdots & \cdot \\ \cdot & \cdot & \cdot & \cdots & \cdot & \cdot & \cdots & \cdot \\ c_{n1} & c_{n2} & c_{n3} & \cdots & c_{nk} & 0 & \cdots & 0 \end{array} \right\|$$

or

$$A_2 = \left\| \begin{array}{cccccc} \alpha_1 & 0 & 0 & \cdots & 0 \\ c_{21} & \alpha_2 & 0 & \cdots & 0 \\ c_{31} & c_{32} & \alpha_3 & \cdots & 0 \\ \cdot & \cdot & \cdot & \cdots & \cdot \\ \cdot & \cdot & \cdot & \cdots & \cdot \\ \cdot & \cdot & \cdot & \cdots & \cdot \\ c_{m1} & c_{m2} & c_{m3} & \cdots & \alpha_m \\ \cdot & \cdot & \cdot & \cdots & \cdot \\ \cdot & \cdot & \cdot & \cdots & \cdot \\ \cdot & \cdot & \cdot & \cdots & \cdot \\ c_{n1} & c_{n2} & c_{n3} & \cdots & c_{nm} \end{array} \right\|.$$

Here the numbers α_1, α_2, etc. are nonzero. In the first case, the rank of A_1 equals k and its basis minor (in the transformed matrix) stands in the upper left-hand corner. In the second case, the rank of A_2 equals m (the number of columns) and its basis minor (in the transformed matrix) appears in the first m rows. This determines the rank of A. The location of the basis minor of A is easily found by following back in reverse order all the operations performed on A.

As an example, consider the following matrix with five columns and six rows:

$$A = \begin{Vmatrix} 1 & 2 & 6 & -2 & -1 \\ -2 & -1 & 0 & -5 & -1 \\ 3 & 1 & -1 & 8 & 1 \\ -1 & 0 & 2 & -4 & -1 \\ -1 & -2 & -7 & 3 & 2 \\ -2 & -2 & -5 & -1 & 1 \end{Vmatrix}.$$

There is one zero in the second row of A; by using the general method described above, we can produce three more zeros in this row. However, for convenience, we first interchange the first and second rows. Then, interchanging the first and second columns (so that an element -1 with the smallest nonzero absolute value again appears in the upper left-hand corner), we obtain†

$$A \sim \begin{Vmatrix} -2 & -1 & 0 & -5 & -1 \\ 1 & 2 & 6 & -2 & -1 \\ 3 & 1 & -1 & 8 & 1 \\ -1 & 0 & 2 & -4 & -1 \\ -1 & -2 & -7 & 3 & 2 \\ -2 & -2 & -5 & -1 & 1 \end{Vmatrix} \sim \begin{Vmatrix} -1 & -2 & 0 & -5 & -1 \\ 2 & 1 & 6 & -2 & -1 \\ 1 & 3 & -1 & 8 & 1 \\ 0 & -1 & 2 & -4 & -1 \\ -2 & -1 & -7 & 3 & 2 \\ -2 & -2 & -5 & -1 & 1 \end{Vmatrix}.$$

To obtain three more zeros in the first row, we multiply the first column by 2, 5, and 1, and subtract the results from the second, fourth, and fifth columns, respectively. This gives

$$A \sim \begin{Vmatrix} -1 & 0 & 0 & 0 & 0 \\ 2 & -3 & 6 & -12 & -3 \\ 1 & 1 & -1 & 3 & 0 \\ 0 & -1 & 2 & -4 & -1 \\ -2 & 3 & -7 & 13 & 4 \\ -2 & 2 & -5 & 9 & 3 \end{Vmatrix}.$$

The simplest thing to do next is to produce additional zeros in the third row. First we interchange this row with the second row. Then we multiply

† Here the symbol $\sim$ written between two matrices means that they have the same rank.

the second column by 1 and -3 and add the results to the third and fourth columns, respectively. Thus we have

$$A \sim \begin{Vmatrix} -1 & 0 & 0 & 0 & 0 \\ 1 & 1 & -1 & 3 & 0 \\ 2 & -3 & 6 & -12 & -3 \\ 0 & -1 & 2 & -4 & -1 \\ -2 & 3 & -7 & 13 & 4 \\ -2 & 2 & -5 & 9 & 3 \end{Vmatrix} \sim \begin{Vmatrix} -1 & 0 & 0 & 0 & 0 \\ 1 & 1 & 0 & 0 & 0 \\ 2 & -3 & 3 & -3 & -3 \\ 0 & -1 & 1 & -1 & -1 \\ -2 & 3 & -4 & 4 & 4 \\ -2 & 2 & -3 & 3 & 3 \end{Vmatrix} = A_1.$$

The fourth and fifth columns of the matrix A_1 are proportional to the third column and can be deleted. The matrix which is left obviously has rank 3, so that the original matrix A also has rank 3. Moreover, A_1 has a basis minor in its first three rows and first three columns. By reversing the successive transformations which led from A to A_1, we can easily verify that none of the transformations which were carried out has any effect on the absolute value of this minor. Therefore the minor appearing in the first three rows and the first three columns of the original matrix is also a basis minor.

PROBLEMS

1. Prove the following theorem: A necessary and sufficient condition for a matrix $\|a_{ij}\|$ of order m to have rank $r \leqslant 1$ is that there exist numbers $a_1, a_2, \ldots, a_m$ and $b_1, b_2, \ldots, b_m$ such that

$$a_{ij} = a_i b_j \qquad (i, j = 1, 2, \ldots, m).$$

2. Let $x_1, x_2, \ldots, x_k$ be k linearly independent vectors in an n-dimensional space $\mathbf{K}_n$, and let $A = \|a_i^{(j)}\|$ be the matrix made up of the components of the vectors $x_1, x_2, \ldots, x_k$ with respect to some basis $e_1, e_2, \ldots, e_n$. Show that the linear manifold $\mathbf{L}(x_1, x_2, \ldots, x_k)$ is uniquely determined, provided one knows the values of all the minors of A of order k.

3. Show that when $k = n$, the system (2), p. 60 has the solution

$$c_1 = A_{i1}, c_2 = A_{i2}, \ldots, c_n = A_{in} \qquad (1 \leqslant i \leqslant n),$$

where A_{ij} is the cofactor of the element a_{ij} (i fixed), provided that the rank of the matrix A is less than n.

4. Solve the system of equations

$$\begin{aligned} x_1 + x_2 + x_3 + x_4 + x_5 &= 7, \\ 3x_1 + 2x_2 + x_3 + x_4 - 3x_5 &= -2, \\ x_2 + 2x_3 + 2x_4 + 6x_5 &= 23, \\ 5x_1 + 4x_2 + 3x_3 + 3x_4 - x_5 &= 12. \end{aligned}$$

5. Study the solutions of the system

$$\begin{aligned} \lambda x + y + z &= 1, \\ x + \lambda y + z &= \lambda, \\ x + y + \lambda z &= \lambda^2 \end{aligned}$$

as a function of λ.

6. What is the condition for the three straight lines

$$a_1x + b_1y + c_1 = 0, \qquad a_2x + b_2y + c_2 = 0, \qquad a_3x + b_3y + c_3 = 0$$

to pass through one point?

7. What is the condition for the n straight lines

$$a_1x + b_1y + c_1 = 0, \qquad a_2x + b_2y + c_2 = 0, \ldots, a_nx + b_ny + c_n = 0$$

to pass through one point?

8. Find the normal fundamental system of solutions for the system of equations

$$\begin{aligned} x_1 + x_2 + x_3 + x_4 + x_5 &= 0, \\ 3x_1 + 2x_2 + x_3 + x_4 - 3x_5 &= 0, \\ x_2 + 2x_3 + 2x_4 + 6x_5 &= 0, \\ 5x_1 + 4x_2 + 3x_3 + 3x_4 - x_5 &= 0. \end{aligned}$$

9. Write down the general solution of the system given in Problem 4, using the normal fundamental system of solutions of the corresponding homogeneous system (found in Problem 8).

10. Determine the rank and basis minor of the following matrices:

$$A_1 = \begin{Vmatrix} 1 & -2 & 3 & -1 & -1 & -2 \\ 2 & -1 & 1 & 0 & -2 & -2 \\ -2 & -5 & 8 & -4 & 3 & -1 \\ 6 & 0 & -1 & 2 & -7 & -5 \\ -1 & -1 & 1 & -1 & 2 & 1 \end{Vmatrix}, \quad A_2 = \begin{Vmatrix} 1 & 0 & 1 & 0 & 0 \\ 1 & 1 & 0 & 0 & 0 \\ 0 & 1 & 1 & 0 & 0 \\ 0 & 0 & 1 & 1 & 0 \\ 0 & 1 & 0 & 1 & 1 \end{Vmatrix}.$$

11. Suppose the matrix A has a nonvanishing minor M of order r, while every minor of order $r + 1$ containing all the elements of M vanishes. Prove that A has rank r.

12. Construct a matrix

$$A = \begin{Vmatrix} a_{11} & a_{12} & a_{13} \\ a_{21} & a_{22} & a_{23} \end{Vmatrix}$$

such that the minors

$$\begin{vmatrix} a_{11} & a_{12} \\ a_{21} & a_{22} \end{vmatrix} = P, \qquad \begin{vmatrix} a_{11} & a_{13} \\ a_{21} & a_{23} \end{vmatrix} = Q, \qquad \begin{vmatrix} a_{12} & a_{13} \\ a_{22} & a_{23} \end{vmatrix} = R$$

have the indicated values P, Q and R.

13. For the system of equations

$$\sum_{k=1}^{n} a_{jk}x_k = b_j \qquad (j = 1, \ldots, n) \tag{8}$$

with a square coefficient matrix, prove "Fredholm's alternative," which asserts that (8) either has a unique solution for arbitrary $b_1, \ldots, b_n$ or else the corresponding homogeneous system

$$\sum_{k=1}^{n} a_{jk}x_k = 0 \qquad (j = 1, \ldots, n)$$

has a nontrivial solution.

14. Prove that the system of equations

$$\begin{aligned} a_{11}x_1 + \cdots + a_{1n}x_n &= b_1, \\ \cdots\cdots\cdots&\cdots\cdots \\ a_{n1}x_1 + \cdots + a_{nn}x_n &= b_n, \\ a_{n+1,1}x_1 + \cdots + a_{n+1,n}x_n &= b_{n+1}, \end{aligned}$$

subject to the condition

$$\begin{vmatrix} a_{11} & \cdots & a_{1n} \\ \cdot & \cdots & \cdot \\ a_{n1} & \cdots & a_{nn} \end{vmatrix} \neq 0,$$

is solvable if and only if

$$\begin{vmatrix} a_{11} & \cdots & a_{1n} & b_1 \\ \cdot & \cdots & \cdot & \cdot \\ a_{n1} & \cdots & a_{nn} & b_n \\ a_{n+1,1} & \cdots & a_{n+1,n} & b_{n+1} \end{vmatrix} = 0.$$

15 (*Elimination of unknowns*). Prove that the system

$$\begin{aligned} a_{11}x_1 + \cdots + a_{1n}x_n &= b_{11}y_1 + \cdots + b_{1k}y_k + c_1, \\ \cdots\cdots\cdots&\cdots\cdots\cdots\cdots\cdots \\ a_{n1}x_1 + \cdots + a_{nn}x_n &= b_{n1}y_1 + \cdots + b_{nk}y_k + c_n, \\ a_{n+1,1}x_1 + \cdots + a_{n+1,n}x_n &= b_{n+1,1}y_1 + \cdots + b_{n+1,k}y_k + c_{n+1} \end{aligned}$$

containing the parameters $y_1, \ldots, y_k$, subject to the condition

$$\begin{vmatrix} a_{11} & \cdots & a_{1n} \\ \cdot & \cdots & \cdot \\ a_{n1} & \cdots & a_{nn} \end{vmatrix} \neq 0$$

is solvable if and only if the parameters $y_1, \ldots, y_k$ satisfy the equation

$$y_1 \begin{vmatrix} a_{11} & \cdots & a_{1n} & b_{11} \\ \cdot & \cdots & \cdot & \cdot \\ a_{n1} & \cdots & a_{nn} & b_{1n} \\ a_{n+1,1} & \cdots & a_{n+1,1} & b_{n+1,1} \end{vmatrix} + \cdots + y_k \begin{vmatrix} a_{11} & \cdots & a_{1n} & b_{1k} \\ \cdot & \cdots & \cdot & \cdot \\ a_{n1} & \cdots & a_{nn} & b_{nk} \\ a_{n+1,1} & \cdots & a_{n+1,n} & b_{n+1,k} \end{vmatrix}$$

$$+ \begin{vmatrix} a_{11} & \cdots & a_{1n} & c_1 \\ \cdot & \cdots & \cdot & \cdot \\ a_{n1} & \cdots & a_{nn} & c_n \\ a_{n+1,1} & \cdots & a_{n+1,n} & c_{n+1} \end{vmatrix} = 0.$$

chapter 4

LINEAR FUNCTIONS OF A VECTOR ARGUMENT

In courses on mathematical analysis one studies functions of one or more real variables. Such functions can be regarded as functions of a *vector argument*. For example, a function of three variables can be regarded as a function whose argument is a vector of the space V_3. This suggests studying functions whose arguments are vectors from an *arbitrary* linear space. In making this study, we will for the time being restrict ourselves to the simplest functions of this kind, namely *linear* functions. We will study both linear *numerical* functions of a vector argument, i.e., functions whose values are numbers, and linear *vector* functions of a vector argument, i.e., functions whose values are vectors. Linear vector functions, otherwise known as *linear operators*, are of great importance in linear algebra and its applications.

4.1. Linear Forms

4.11. A numerical function $L(x)$ of a vector argument x, defined on a linear space $\mathbf{K}$ over a number field K, is called a *linear form* if it satisfies the following conditions:

a) $L(x + y) = L(x) + L(y)$ for every $x, y \in \mathbf{K}$;
b) $L(\alpha x) = \alpha L(x)$ for every $x \in \mathbf{K}$ and every $\alpha \in K$.

In other words, a linear form $L(x)$ is a morphism of the linear space $\mathbf{K}$ into the one-dimensional space $K_1 = K$ (cf. Sec. 2.71). By using induction, we easily verify that conditions a) and b) imply the formula

$$L(\alpha_1 x_1 + \cdots + \alpha_k x_k) = \alpha_1 L(x_1) + \cdots + \alpha_k L(x_k), \tag{1}$$

where $x_1, \ldots, x_k$ are arbitrary vectors in $\mathbf{K}$ and $\alpha_1, \ldots, \alpha_k$ are arbitrary numbers in K.

4.12. *Examples*

a. Suppose a basis is chosen in an n-dimensional space $\mathbf{K}$, so that every vector $x \in \mathbf{K}$ can be specified by its components $\xi_1, \xi_2, \ldots, \xi_n$. Then $L(x) = \xi_1$ (the first component) is obviously a linear form in x.

b. A more general linear form in the same space is given by the expression

$$L(x) = \sum_{k=1}^{n} l_k \xi_k,$$

with arbitrary fixed coefficients $l_1, l_2, \ldots, l_n$.

c. An example of a linear form in the space $K(a, b)$ (where K is R or C)† is the expression

$$L(x) = x(t_0),$$

where t_0 is a fixed point of the interval $a \leqslant t \leqslant b$.

d. In the same space we can study the linear form

$$L(x) = \int_a^b l(t)x(t)\,dt,$$

where $l(t)$ is a fixed continuous function.

e. In the space V_3 the scalar product (x, x_0) of the vector x with a fixed vector $x_0 \in V_3$ is a linear form in x.

Linear forms defined on infinite-dimensional spaces are usually called *linear functionals*.

4.13. We now find the general representation of a linear form $L(x)$ defined on an n-dimensional space $\mathbf{K}_n$. Let $e_1, e_2, \ldots, e_n$ be an arbitrary basis of the space $\mathbf{K}_n$, and denote the quantity $L(e_k)$ by l_k ($k = 1, 2, \ldots, n$). Then, by (1), given any

$$x = \sum_{k=1}^{n} l_k e_k,$$

we have

$$L(x) = L\left(\sum_{k=1}^{n} \xi_k e_k\right) = \sum_{k=1}^{n} \xi_k L(e_k) = \sum_{k=1}^{n} l_k \xi_k,$$

i.e., the value of the linear form $L(x)$ is a linear combination of the components of the vector x, with the fixed coefficients $l_1, l_2, \ldots, l_n$. Thus the

† Recall Secs. 2.15c and 2.15d.

most general representation of a linear form in an n-dimensional linear space has already been encountered in Example 4.12b.

4.14. In a complex linear space $\mathbf{C}$ we can also consider another type of linear form, called a *linear form of the second kind* (in this context, the linear form defined in Sec. 4.11 is called a *linear form of the first kind*). A numerical function $L(x)$ of a vector argument x, defined on a complex linear space $\mathbf{C}$, is called a linear form of the second kind if it satisfies the following two conditions:

a′) $L(x + y) = L(x) + L(y)$ for every $x, y \in \mathbf{C}$;
b′) $L(\alpha x) = \bar{\alpha} L(x)$ for every $x \in \mathbf{C}$ and every complex number $\alpha = \alpha_1 + i\alpha_2$ (here $\bar{\alpha} = \alpha_1 - i\alpha_2$ is the complex conjugate of α).

For a linear form of the second kind, the analogue of formula (1) becomes

$$L(\alpha_1 x_1 + \cdots + \alpha_k x_k) = \bar{\alpha}_1 L(x_1) + \cdots + \bar{\alpha}_k L(x_k), \tag{1'}$$

valid for arbitrary $x_1, \ldots, x_k$ in $\mathbf{C}$ and arbitrary complex numbers $\alpha_1, \ldots, \alpha_k$.

4.15. An example of a linear form of the second kind in an n-dimensional complex space $\mathbf{C}_n$ with basis $e_1, \ldots, e_n$ is given by the function

$$L(x) = \sum_{k=1}^{n} l_k \bar{\xi}_k,$$

where $l_1, \ldots, l_n$ are arbitrary fixed complex numbers and $\xi_1, \ldots, \xi_n$ are the components of the vector x with respect to the basis $e_1, \ldots, e_n$. Moreover, this formula gives the general representation of a linear form of the second kind defined on the space $\mathbf{C}_n$. In fact, let $L(x)$ be an arbitrary linear form of the second kind, and let $l_1 = L(e_1), \ldots, l_n = L(e_n)$. Then, given any $x \in \mathbf{C}_n$, it follows from (1′) that

$$L(x) = L\left(\sum_{k=1}^{n} \xi_k e_k\right) = \sum_{k=1}^{n} \bar{\xi}_k L(e_k) = \sum_{k=1}^{n} l_k \bar{\xi}_k,$$

as required.

4.2. Linear Operators

4.21. As just shown, a linear form $L(x)$ defined on a linear space $\mathbf{K}$ is just a morphism of $\mathbf{K}$ into the one-dimensional space K_1. More generally, we now consider a morphism $\mathbf{A} = \mathbf{A}(x)$ of a linear space $\mathbf{X}$ into another linear space $\mathbf{Y}$ over the same field K ($\mathbf{X}$ and $\mathbf{Y}$ may coincide). As already noted in Sec. 2.71, $\mathbf{A}(x)$ is also called a *linear operator*, mapping $\mathbf{X}$ into $\mathbf{Y}$. Instead of $\mathbf{A}(x)$, we will often write simply $\mathbf{A}x$. By the definition of a morphism, $\mathbf{A}(x)$ satisfies the following conditions:

a) $\mathbf{A}(x + y) = \mathbf{A}x + \mathbf{A}y$ for every $x, y \in \mathbf{X}$;
b) $\mathbf{A}(\alpha x) = \alpha \mathbf{A}x$ for every $x \in \mathbf{X}$ and every $\alpha \in K$.

Just as for linear forms, conditions a) and b) imply the more general formula

$$\mathbf{A}(\alpha_1 x_1 + \cdots + \alpha_k x_k) = \alpha_1 \mathbf{A} x_1 + \cdots + \alpha_k \mathbf{A} x_k$$

for arbitrary $x_1, \ldots, x_k$ in $\mathbf{X}$ and arbitrary $\alpha_1, \ldots, a_k$ in K.

4.22. *Examples*

a. The operator† associating the zero vector of the space $\mathbf{Y}$ with every vector x of the space $\mathbf{X}$ is obviously a linear operator. This operator is called the *zero operator*, denoted by $\mathbf{0}$.

b. Given any linear operator $\mathbf{A}$ mapping the space $\mathbf{X}$ into the space $\mathbf{Y}$, let

$$\mathbf{B}x = -\mathbf{A}x.$$

It is easy to see that the operator $\mathbf{B}$ so defined is also a linear operator mapping $\mathbf{X}$ into $\mathbf{Y}$. This operator is called the *negative* of the operator $\mathbf{A}$.

c. Let $e_1, \ldots, e_n$ be a basis in the space $\mathbf{X}$, and let vectors $f_1, \ldots, f_n$ in the space $\mathbf{Y}$ be associated with the vectors $e_1, \ldots, e_n$ in an arbitrary way. Then there exists a unique linear operator $\mathbf{A}$ mapping $\mathbf{X}$ into $\mathbf{Y}$ and carrying every vector e_k into the corresponding vector f_k $(k = 1, \ldots, n)$. In fact, if such an operator $\mathbf{A}$ exists, then, given any vector

$$x = \sum_{k=1}^{n} \xi_k e_k \in \mathbf{X}, \tag{2}$$

we have

$$\mathbf{A}x = \mathbf{A}\left(\sum_{k=1}^{n} \xi_k e_k\right) = \sum_{k=1}^{n} \xi_k \mathbf{A} e_k = \sum_{k=1}^{n} \xi_k f_k,$$

thereby proving the uniqueness of $\mathbf{A}$. On the other hand, given any vector (2), we can set

$$\mathbf{A}x = \sum_{k=1}^{n} \xi_k f_k,$$

by definition. The resulting operator, as is easily verified, is linear, maps $\mathbf{X}$ into $\mathbf{Y}$, and at the same time carries every vector e_k into the corresponding vector f_k $(k = 1, \ldots, n)$.

d. Suppose that with every vector x of the space $\mathbf{X}$ we associate the same vector x, thereby obtaining a linear operator $\mathbf{E}$, mapping $\mathbf{X}$ into itself. Then $\mathbf{E}$ is called the *identity operator* or *unit operator*.

4.23. Matrix representation of linear operators. Let $\mathbf{A}$ be a linear operator mapping a space $\mathbf{X}$ of dimension n into a space $\mathbf{Y}$ of dimension m. Let

† Here we use the term *operator* as a synonym for *function* (mapping one linear space into another).

$e_1, \ldots, e_n$ be a fixed basis in $\mathbf{X}$ and $f_1, \ldots, f_m$ a fixed basis in $\mathbf{Y}$. The vector e_1 is mapped by $\mathbf{A}$ into some vector $\mathbf{A}e_1$ of the space $\mathbf{Y}$, which, like every vector of $\mathbf{Y}$, has an expansion

$$\mathbf{A}e_1 = a_1^{(1)}f_1 + a_2^{(1)}f_2 + \cdots + a_m^{(1)}f_m$$

with respect to the basis vectors $f_1, \ldots, f_m$. The operator $\mathbf{A}$ has a similar effect on the other basis vectors:

$$\mathbf{A}e_2 = a_1^{(2)}f_1 + a_2^{(2)}f_2 + \cdots + a_m^{(2)}f_m,$$
$$\cdots\cdots\cdots\cdots\cdots\cdots\cdots\cdots$$
$$\mathbf{A}e_n = a_1^{(n)}f_1 + a_2^{(n)}f_2 + \cdots + a_m^{(n)}f_m.$$

These formulas can be written more concisely as

$$\mathbf{A}e_j = \sum_{i=1}^{m} a_i^{(j)}f_i \qquad (j = 1, 2, \ldots, n). \tag{3}$$

The coefficients $a_i^{(j)}$ $(i = 1, \ldots, m; j = 1, \ldots, n)$ define an $m \times n$ matrix†

$$A = A_{(e,f)} = \begin{Vmatrix} a_1^{(1)} & a_1^{(2)} & \cdots & a_1^{(n)} \\ a_2^{(1)} & a_2^{(2)} & \cdots & a_2^{(n)} \\ \cdot & \cdot & \cdots & \cdot \\ a_m^{(1)} & a_m^{(2)} & \cdots & a_m^{(n)} \end{Vmatrix},$$

called the *matrix of the operator* $\mathbf{A}$ *relative to the bases* $\{e\} = \{e_1, \ldots, e_n\}$ *and* $\{f\} = \{f_1, \ldots, f_m\}$. The components of the vectors $\mathbf{A}e_1, \mathbf{A}e_2, \ldots, \mathbf{A}e_n$ with respect to the basis $\{f\}$ serve as the columns of this matrix.‡

Now, given any vector

$$x = \sum_{j=1}^{n} \xi_j e_j \in \mathbf{X},$$

let

$$y = \mathbf{A}x = \sum_{i=1}^{m} \eta_i f_i.$$

With a view to expressing the components $\eta_1, \ldots, \eta_m$ of the vector y in terms of the components $\xi_1, \ldots, \xi_n$ of the vector x, we observe that

$$y = \sum_{i=1}^{m} \eta_i f_i = \mathbf{A}x = \mathbf{A}\left(\sum_{j=1}^{n} \xi_j e_j\right) = \sum_{j=1}^{n} \xi_j \mathbf{A}e_j$$
$$= \sum_{j=1}^{n} \xi_j \sum_{i=1}^{m} a_i^{(j)} f_i = \sum_{i=1}^{m} \left(\sum_{j=1}^{n} a_i^{(j)} \xi_j\right) f_i.$$

† I.e., a matrix with m rows and n columns.

‡ Note the distinction between the symbol $\mathbf{A}$ (boldface Roman) for an *operator* and the corresponding symbol A (lightface Italic) for the *matrix* of $\mathbf{A}$.

Comparing coefficients of the vector f_i, we find that

$$\eta_i = \sum_{j=1}^{n} a_i^{(j)}\xi_j \qquad (i = 1, 2, \ldots, m), \tag{4}$$

or, in expanded form

$$\begin{aligned} \eta_1 &= a_1^{(1)}\xi_1 + a_1^{(2)}\xi_2 + \cdots + a_1^{(n)}\xi_n, \\ \eta_2 &= a_2^{(1)}\xi_1 + a_2^{(2)}\xi_2 + \cdots + a_2^{(n)}\xi_n, \\ &\cdots\cdots\cdots\cdots\cdots\cdots \\ \eta_m &= a_m^{(1)}\xi_1 + a_m^{(2)}\xi_2 + \cdots + a_m^{(n)}\xi_n. \end{aligned} \tag{5}$$

Therefore from a knowledge of the matrix of the operator **A** relative to the basis $e_1, e_2, \ldots, e_n$ we can determine the result of applying **A** to any vector

$$x = \sum_{j=1}^{n} \xi_j e_j$$

of the space **X**. In fact, the equations (5) express the components of the vector $y = \mathbf{A}x$ as linear combinations of the components of x. Note that *the coefficient matrix of the system of the equations* (5) *is just the matrix* $A_{(e,f)}$.

Next let $\|a_i^{(j)}\|$ be an arbitrary $m \times n$ matrix, where the superscript is the column number and the subscript is the row number. Given any vector

$$x = \sum_{j=1}^{n} \xi_j e_j,$$

we construct the vector

$$y = \sum_{i=1}^{m} \eta_i f_i$$

with components $\eta_1, \eta_2, \ldots, \eta_m$ determined by (5). It is easy to see that the operator **A** effecting this mapping of the vector x into the vector y is a linear operator. We now construct the matrix of the operator **A** relative to the basis $e_1, e_2, \ldots, e_n$. Since the vector e_1 has components $\xi_1 = 1$, $\xi_2 = 0$, $\ldots$, $\xi_n = 0$, it follows from (5) that the components of the vector $\mathbf{A}e_1$ will be the numbers $a_1^{(1)}, a_2^{(1)}, \ldots, a_m^{(1)}$, so that

$$\mathbf{A}e_1 = a_1^{(1)}f_1 + a_2^{(1)}f_2 + \cdots + a_m^{(1)}f_m.$$

Similarly,

$$\mathbf{A}e_j = a_1^{(j)}f_1 + a_2^{(j)}f_2 + \cdots + a_m^{(j)}f_m \qquad (j = 1, 2, \ldots, n).$$

Therefore the matrix of the operator **A** coincides with the original matrix $\|a_i^{(j)}\|$. Thus *every* $m \times n$ *matrix is the matrix of a linear operator* **A** *mapping an n-dimensional space* **X** *into an m-dimensional space* **Y**, with fixed bases $e_1, \ldots, e_n$ in **X** and $f_1, \ldots, f_m$ in **Y**. Thus (3), or equivalently (4), establishes a *one-to-one correspondence* between linear operators mapping a space **X** (with basis $e_1, \ldots, e_n$) into a space $\mathbf{Y}$ (with basis $f_1, \ldots, f_m$) and $m \times n$ matrices made up of numbers from the field K. In particular, identical operators $\mathbf{A}$ and $\mathbf{B}$ (i.e., operators such that $\mathbf{A}x = \mathbf{B}x$ for *every* $x \in \mathbf{X}$) have identical matrices.

Finally we note that (5) can be used to construct the operator $\mathbf{A}$ directly (and uniquely) from the matrix $A = \|a_i^{(j)}\|$. In fact, A is just the coefficient matrix of the system (5).

4.24. *Examples*

a. Clearly, the matrix of the zero operator (see Example 4.22a) relative to any basis in the space $\mathbf{X}$ and any basis in the space $\mathbf{Y}$ consists entirely of zeros.

b. If $\|a_i^{(j)}\|$ is the matrix of $\mathbf{A}$, then the matrix of the negative operator (see Example 4.22b) is obviously just $-\|a_i^{(j)}\|$.

c. Let $m \geqslant n$ and suppose the operator $\mathbf{A}$ carries the vectors of the basis $e_1, \ldots, e_n$ of the space $\mathbf{X}$ into linearly independent vectors $f_1, \ldots, f_n$ of the space $\mathbf{Y}$. We augment the vectors $f_1, \ldots, f_n$ by the vectors $f_{n+1}, \ldots, f_m$ to make a basis for the whole space $\mathbf{Y}$. Then the matrix of the operator $\mathbf{A}$ relative to the bases $e_1, \ldots, e_n$ and $f_1, \ldots, f_m$ is clearly of the form

$$m\left\{\begin{matrix} n\left\{\begin{matrix} \\ \\ \\ \\ \end{matrix}\right. \\ \\ \\ \\ \end{matrix}\right. \left\|\begin{matrix} \overbrace{\begin{matrix}1 & 0 & \cdots & 0\end{matrix}}^{n} \\ \begin{matrix}0 & 1 & \cdots & 0\end{matrix} \\ \begin{matrix}\cdot & \cdot & \cdots & \cdot\end{matrix} \\ \begin{matrix}0 & 0 & \cdots & 1\end{matrix} \\ \begin{matrix}0 & 0 & \cdots & 0\end{matrix} \\ \begin{matrix}\cdot & \cdot & \cdots & \cdot\end{matrix} \\ \begin{matrix}0 & 0 & \cdots & 0\end{matrix} \end{matrix}\right\|$$

d. In particular, the matrix of the identity operator $\mathbf{E}$ (see Example 4.22d) relative to the basis $e_1, \ldots, e_n$ of the space $\mathbf{X}$ (the domain of $\mathbf{E}$) and the basis $e_1, \ldots, e_n$ of the same space (the range of $\mathbf{E}$) is just

$$\left\|\begin{matrix} 1 & 0 & \cdots & 0 \\ 0 & 1 & \cdots & 0 \\ \cdot & \cdot & \cdots & \cdot \\ 0 & 0 & \cdots & 1 \end{matrix}\right\|$$

A matrix of this form is called the *unit matrix* or *identity matrix* of order n.

4.3. Sums and Products of Operators

We now consider addition of operators and multiplication of operators both by numbers and by other operators. First we note that two operators $\mathbf{A}$ and $\mathbf{B}$ mapping a space $\mathbf{X}$ into a space $\mathbf{Y}$ are said to be *equal* (written $\mathbf{A} = \mathbf{B}$) if $\mathbf{A}x = \mathbf{B}x$ for every $x \in \mathbf{X}$.

4.31. Addition of operators. Given two linear operators $\mathbf{A}$ and $\mathbf{B}$ mapping a space $\mathbf{X}$ into a space $\mathbf{Y}$, the operator $\mathbf{C} = \mathbf{A} + \mathbf{B}$ is defined by the formula

$$\mathbf{C}x \equiv (\mathbf{A} + \mathbf{B})x = \mathbf{A}x + \mathbf{B}x. \tag{6}$$

Obviously, $\mathbf{C}$ also maps the space $\mathbf{X}$ into the space $\mathbf{Y}$. To verify that $\mathbf{C}$ is again a linear operator, let $x = \alpha_1 x_1 + \alpha_2 x_2$. Then

$$\begin{aligned}\mathbf{C}(\alpha_1 x_1 + \alpha_2 x_2) &= \mathbf{A}(\alpha_1 x_1 + \alpha_2 x_2) + \mathbf{B}(\alpha_1 x_1 + \alpha_2 x_2)\\ &= \alpha_1 \mathbf{A}x_1 + \alpha_2 \mathbf{A}x_2 + \alpha_1 \mathbf{B}x_1 + \alpha_2 \mathbf{B}x_2\\ &= \alpha_1(\mathbf{A}x_1 + \mathbf{B}x_1) + \alpha_2(\mathbf{A}x_2 + \mathbf{B}x_2) = \alpha_1 \mathbf{C}x_1 + \alpha_2 \mathbf{C}x_2,\end{aligned}$$

so that both conditions a) and b) of Sec. 4.21 are satisfied. The linear operator $\mathbf{C}$ defined by (6) is called the *sum* of the operators $\mathbf{A}$ and $\mathbf{B}$.

It is easily verified that

$$\begin{aligned}\mathbf{A} + \mathbf{B} &= \mathbf{B} + \mathbf{A},\\ (\mathbf{A} + \mathbf{B}) + \mathbf{C} &= \mathbf{A} + (\mathbf{B} + \mathbf{C}),\\ \mathbf{A} + \mathbf{0} &= \mathbf{A},\\ \mathbf{A} + (-\mathbf{A}) &= \mathbf{0},\end{aligned} \tag{7}$$

where $\mathbf{A}$, $\mathbf{B}$ and $\mathbf{C}$ are arbitrary linear operators, $\mathbf{0}$ is the zero operator (see Example 4.22a), and $-\mathbf{A}$ is the negative of the operator $\mathbf{A}$ (see Example 4.22b), i.e., the operator carrying the vector $x \in \mathbf{X}$ into the vector $-\mathbf{A}x$.

4.32. Multiplication of an operator by a number. Let $\mathbf{A}$ be a linear operator mapping a space $\mathbf{X}$ into a space $\mathbf{Y}$, and let λ be a number from the field K. Then the operator $\mathbf{B} = \lambda\mathbf{A}$, called the *product of the operator* $\mathbf{A}$ *and the number* λ, is defined by the formula

$$\mathbf{B}x = (\lambda\mathbf{A})x = \lambda(\mathbf{A}x).$$

It is easily verified (just as in Sec. 4.31) that this operator is linear, and moreover that

$$\begin{aligned}\lambda_1(\lambda_2\mathbf{A}) &= (\lambda_1\lambda_2)\mathbf{A},\\ 1 \cdot \mathbf{A} &= \mathbf{A},\\ (\lambda_1 + \lambda_2)\mathbf{A} &= \lambda_1\mathbf{A} + \lambda_2\mathbf{A},\\ \lambda(\mathbf{A} + \mathbf{B}) &= \lambda\mathbf{A} + \lambda\mathbf{B}.\end{aligned} \tag{7'}$$

The relations (7) and (7′) show that *the set of all linear operators mapping a linear space* **X** *into a linear space* **Y** *is itself a linear space.*

4.33. Multiplication of operators. Let **A** be a linear operator mapping the space **X** into the space **Y** and **B** a linear operator mapping the space **Y** into the space **Z** (where all the spaces are over the same number field K). Then the operator $\mathbf{P} = \mathbf{BA}$, called the *product of the operator* **B** *and the operator* **A** (in that order), is defined as the operator mapping **X** into **Z** such that

$$\mathbf{P}x = (\mathbf{BA})x = \mathbf{B}(\mathbf{A}x)$$

(note that first the operator **A** acts on the vector x and then the operator **B** acts on the resulting vector in the space **Y**). The operator **P** is again linear, since

$$\begin{aligned}\mathbf{P}(\alpha_1 x_1 + \alpha_2 x_2) &= \mathbf{B}[\mathbf{A}(\alpha_1 x_1 + \alpha_2 x_2)] = \mathbf{B}(\alpha_1 \mathbf{A}x_1 + \alpha_2 \mathbf{A}x_2)\\ &= \alpha_1 \mathbf{BA}x_1 + \alpha_2 \mathbf{BA}x_2 = \alpha_1 \mathbf{P}x_1 + \alpha_2 \mathbf{P}x_2.\end{aligned}$$

4.34. The following relations are easily verified:

a) $\lambda(\mathbf{BA}) = (\lambda\mathbf{B})\mathbf{A}$ for every number $\lambda \in K$ and arbitrary operators **A** mapping the space **X** into the space **Y** and **B** mapping the space **Y** into the space **Z**;

b) $(\mathbf{A} + \mathbf{B})\mathbf{C} = \mathbf{AC} + \mathbf{BC}$ for arbitrary operators **A** and **B** mapping the space **Y** into the space **Z** and **C** mapping the space **X** into the space **Y**;

c) $\mathbf{A}(\mathbf{B} + \mathbf{C}) = \mathbf{AB} + \mathbf{AC}$ for arbitrary operators **B** and **C** mapping the space **X** into the space **Y** and **A** mapping the space **Y** into the space **Z**;

d) $(\mathbf{AB})\mathbf{C} = \mathbf{A}(\mathbf{BC})$ for arbitrary operators **C** mapping the space **X** into the space **Y**, **B** mapping the space **Y** into the space **Z**, and **C** mapping the space **Z** into the space **W**.†

For example, to verify d), according to the definition of operator equality we must prove the identity

$$[\mathbf{A}(\mathbf{BC})x] = [(\mathbf{AB})\mathbf{C}]x$$

for every $x \in \mathbf{X}$. But by the very definition of the operator product, we have

$$\begin{aligned}[\mathbf{A}(\mathbf{BC})x] &= \mathbf{A}[(\mathbf{BC})x] = \mathbf{A}[\mathbf{B}(\mathbf{C}x)],\\ [(\mathbf{AB})\mathbf{C}]x &= (\mathbf{AB})(\mathbf{C}x) = \mathbf{A}[\mathbf{B}(\mathbf{C}x)],\end{aligned}$$

which implies the required formula. The other formulas are proved similarly.

---

† The associative law for operator multiplication is expressed by d), and the distributive law by b) and c).

4.4. Corresponding Operations on Matrices

We now study the matrix analogues of the algebraic operations on linear operators described in Sec. 4.4.

4.41. Addition of operators. Let **A** and **B** be two linear operators mapping a space **X** with basis $e_1, \ldots, e_n$ into a space **Y** with basis $f_1, \ldots, f_m$. Moreover, let $A = \|a_i^{(j)}\|$ be the matrix of the operator **A** and $B = \|b_i^{(j)}\|$ the matrix of the operator **B**, relative to these bases. Then

$$\mathbf{A}e_j = \sum_{i=1}^{m} a_i^{(j)} f_i, \qquad \mathbf{B}e_j = \sum_{i=1}^{m} b_i^{(j)} f_i \qquad (j = 1, 2, \ldots, n),$$

and hence

$$(\mathbf{A} + \mathbf{B})e_j = \mathbf{A}e_j + \mathbf{B}e_j = \sum_{i=1}^{m} (a_i^{(j)} + b_i^{(j)}) f_i.$$

It follows that the matrix corresponding to the operator $\mathbf{A} + \mathbf{B}$ is just $\|a_i^{(j)} + b_i^{(j)}\|$. This matrix is called the *sum of the matrices* $\|a_i^{(j)}\|$ *and* $\|b_i^{(j)}\|$. Thus the sum $A + B$ is defined for every pair of matrices A and B with the same number of rows and the same number of columns.

4.42. Multiplication of an operator by a number. With the same notation as before, we have

$$(\lambda\mathbf{A})e_j = \lambda(\mathbf{A}e_j) = \sum_{i=1}^{n} \lambda a_i^{(j)} f_i.$$

It follows that the matrix corresponding to the operator $\lambda\mathbf{A}$ is just the matrix $\|\lambda a_i^{(j)}\|$, obtained by multiplying all the elements of the matrix $\|a_i^{(j)}\|$ by the number λ. This matrix is called the *product of the matrix* $\|a_i^{(j)}\|$ *and the number* λ.

Since there is a one-to-one correspondence between $m \times n$ matrices and linear operators mapping an n-dimensional space into an m-dimensional space (see Sec. 4.22), there is a one-to-one correspondence between algebraic operations involving operators and the analogous operations involving matrices. Hence, since operators obey the rules (7) and (7′), the same is also true of matrices (of course, this can easily be verified directly). Thus we see that *the set of all* $m \times n$ *matrices is itself a linear space*, which, by its very construction, is isomorphic to the linear space of all linear operators mapping an n-dimensional space **X** into an m-dimensional space **Y**.

4.43. Multiplication of operators. Let **X**, **Y** and **Z** be linear spaces, and let $e_1, \ldots, e_n$ be a basis in **X**, $f_1, \ldots, f_m$ a basis in **Y**, and $g_1, \ldots, g_q$ a basis in **Z**. Let **B** be a linear operator mapping **X** into **Y** with $m \times n$ matrix $\|b_i^{(j)}\|$, so that

$$\mathbf{B}e_j = \sum_{i=1}^{m} b_i^{(j)} f_i \qquad (j = 1, \ldots, n),$$

and let **A** be a linear operator mapping **Y** into **Z** with $q \times m$ matrix $\|a_k^{(i)}\|$. so that

$$\mathbf{A}f_i = \sum_{k=1}^{q} a_k^{(i)} g_k \qquad (i = 1, \ldots, m).$$

Then for the product $\mathbf{P} = \mathbf{AB}$ we have

$$(\mathbf{AB})e_j = \mathbf{A}(\mathbf{B}e_j) = \mathbf{A} \sum_{i=1}^{m} b_i^{(j)} f_i = \sum_{i=1}^{m} b_i^{(j)} \mathbf{A} f_i$$

$$= \sum_{i=1}^{m} b_i^{(j)} \sum_{k=1}^{q} a_k^{(i)} g_k = \sum_{k=1}^{q} \left(\sum_{i=1}^{m} a_k^{(i)} b_i^{(j)} \right) g_k.$$

Hence the elements $p_k^{(j)}$ of the matrix P of the operator $\mathbf{P} = \mathbf{AB}$ are given by

$$p_k^{(j)} = \sum_{i=1}^{m} a_k^{(i)} b_i^{(j)} \qquad (j = 1, \ldots, n;\ k = 1, \ldots, q). \tag{8}$$

This is the desired result, which can be expressed as follows: *The element of the matrix P belonging to the kth row and jth column equals the sum of the products of the elements of the kth row of the matrix A with the corresponding elements of the jth column of the matrix B.* The matrix $P = \|p_k^{(j)}\|$ which is obtained from the matrices $A = \|a_k^{(i)}\|$ and $B = \|b_i^{(j)}\|$ in accordance with formula (8) is called the *product of the matrices A and B* (in that order).

It should be noted that for the product $P = AB$ to make sense, the number of columns in A must equal the number of rows in B. Then P will have the same number of rows as A and the same number of columns as B. This fact can be expressed more strikingly in the "$m \times n$ notation," namely, the product AB of a $q \times l$ matrix A and an $m \times n$ matrix B is defined if $l = m$, in which case AB is a $q \times n$ matrix. Both products AB and BA are defined if $l = m$ and $q = n$, in which case AB is a square $n \times n$ matrix while BA is a square $m \times m$ matrix. Moreover, if $l = m = q = n$, i.e., if both matrices A and B are square $n \times n$ matrices, then AB and BA are also $n \times n$ matrices. However, these products need not be equal. For example,

$$\begin{Vmatrix} 0 & 1 \\ 1 & 0 \end{Vmatrix} \begin{Vmatrix} 1 & 0 \\ 0 & 0 \end{Vmatrix} = \begin{Vmatrix} 0 & 0 \\ 1 & 0 \end{Vmatrix},$$

$$\begin{Vmatrix} 1 & 0 \\ 0 & 0 \end{Vmatrix} \begin{Vmatrix} 0 & 1 \\ 1 & 0 \end{Vmatrix} = \begin{Vmatrix} 0 & 1 \\ 0 & 0 \end{Vmatrix}.$$

Thus multiplication of square matrices is in general noncommutative. As for the associative and distributive laws, the situation is more favorable. In fact, as shown in Sec. 4.34, operator multiplication obeys the associative and distributive laws, and hence we can assert that the same is true of matrix multiplication, since there is a one-to-one correspondence between operators and matrices associating sums and products of operators with the sums and products of the corresponding matrices.

4.44. *Examples*

In the following examples, we write both indices of matrix elements as subscripts, so that the element a_{jk} of the matrix $A = \|a_{jk}\|$ belongs to the jth row and the kth column. In this notation, formula (8) for the matrix product $P = AB$ takes the form

$$p_{kj} = \sum_{i=1}^{m} a_{ki} b_{ij} \qquad (j = 1, \ldots, n;\ k = 1, \ldots, q). \tag{8'}$$

a. Suppose we multiply an $m \times n$ matrix $A = \|a_{jk}\|$ from the left by an $m \times m$ matrix $B_{rs} = \|b_{jk}\|$ with all its elements b_{jk} equal to zero except the single element $b_{rs} = 1$. Then by (8′) we get the $m \times n$ matrix

$$B_{rs}A = (r)\begin{Vmatrix} & \overset{(s)}{\cdot} & \\ & \cdot & \\ & \cdot & \\ \cdots & 1 & \cdots \\ & \cdot & \\ & \cdot & \\ & \cdot & \end{Vmatrix} \begin{Vmatrix} a_{11} & a_{12} & \cdots & a_{1n} \\ \cdot & \cdot & \cdots & \cdot \\ a_{s1} & a_{s2} & \cdots & a_{sn} \\ \cdot & \cdot & \cdots & \cdot \\ a_{m1} & a_{m2} & \cdots & a_{mn} \end{Vmatrix}$$

$$= (r)\begin{Vmatrix} 0 & 0 & \cdots & 0 \\ \cdot & \cdot & \cdots & \cdot \\ a_{s1} & a_{s2} & \cdots & a_{sn} \\ \cdot & \cdot & \cdots & \cdot \\ 0 & 0 & \cdots & 0 \end{Vmatrix}$$

so that the rth row of the matrix $B_{rs}A$ consists of the elements of the sth row of the matrix A while all other elements of $B_{rs}A$ vanish.

b. Suppose we multiply an $m \times n$ matrix $A = \|a_{jk}\|$ on the right by an $n \times n$ matrix $C_{pq} = \|c_{jk}\|$ with all its elements c_{jk} equal to zero except the single element $c_{pq} = 1$. Then by (8′) we get the $m \times n$ matrix

$$AC_{pq} = \left\| \begin{matrix} a_{11} & \cdots & a_{1p} & \cdots & a_{1n} \\ a_{21} & \cdots & a_{2p} & \cdots & a_{2n} \\ \cdot & \cdots & \cdot & \cdots & \cdot \\ a_{m1} & \cdots & a_{mp} & \cdots & a_{mn} \end{matrix} \right\| \begin{matrix} \\ \\ (p) \\ \\ \\ \end{matrix} \left\| \begin{matrix} & \overset{(q)}{\cdot} & \\ & \cdot & \\ & \cdot & \\ \cdots & 1 & \cdots \\ & \cdot & \\ & \cdot & \\ & \cdot & \end{matrix} \right\|$$

$$= \left\| \begin{matrix} 0 & \cdots & \overset{(q)}{a_{1p}} & \cdots & 0 \\ 0 & \cdots & a_{2p} & \cdots & 0 \\ \cdot & \cdots & \cdot & \cdots & \cdot \\ 0 & \cdots & a_{mp} & \cdots & 0 \end{matrix} \right\|,$$

so that the qth column of the matrix AC_{pq} consists of the elements of the pth column of the matrix A while all other elements of AC_{pq} vanish.

c. With the same matrices B_{rs}, A and C_{pq} we have

$$B_{rs}AC_{pq} = (r) \left\| \begin{matrix} 0 & \cdots & \overset{(q)}{0} & \cdots & 0 \\ \cdot & \cdots & \cdot & \cdots & \cdot \\ 0 & \cdots & a_{sp} & \cdots & 0 \\ \cdot & \cdots & \cdot & \cdots & \cdot \\ 0 & \cdots & 0 & \cdots & 0 \end{matrix} \right\|.$$

Thus $B_{rs}AC_{pq}$ is an $m \times n$ matrix all of whose elements vanish with the (possible) exception of the single element, equal to a_{sp}, appearing in the rth row and qth column.

d. By what $m \times m$ matrix D must we multiply an $m \times n$ matrix A from the left to make the matrix DA coincide with the matrix obtained from A by interchanging its rth and sth rows?

Solution. Example 4.44a shows that the matrix whose rth row is the sth row of the matrix A is obtained by multiplying A on the left by the $m \times m$ matrix B_{rs}. But the other rows of the resulting matrix vanish. It is now clear that to get the required matrix, we must multiply A from the left by the $m \times m$ matrix

$$D = B_{rs} + B_{sr} + \sum_{\substack{j \neq r \\ j \neq s}} B_{jj} = \left\| \begin{array}{ccccccccc} & & & (r) & & (s) & & & \\ 1 & & & \vdots & & \vdots & & & \\ & \ddots & & \vdots & & \vdots & & & \\ & & 1 & & & & & & \\ \cdots & & & 0 & \cdots & 1 & & \cdots & \\ & & & & 1 & & & & \\ & & & \vdots & \ddots & \vdots & & & \\ & & & & & 1 & & & \\ \cdots & & & 1 & \cdots & 0 & & \cdots & \\ & & & & & & 1 & & \\ & & & \vdots & & \vdots & & \ddots & \\ & & & & & & & & 1 \end{array} \right\|.$$

e. By what $n \times n$ matrix G must we multiply an $m \times n$ matrix A from the right to make the matrix AG coincide with the matrix obtained from A by interchanging its pth and qth columns?

Solution. By an argument like that in Example 4.44d, we have

$$G = C_{pq} + C_{qp} + \sum_{\substack{k \neq p \\ k \neq q}} C_{kk}.$$

f. By what $m \times m$ matrix F must we multiply an $m \times n$ matrix A from the left to make the matrix FA coincide with the matrix obtained from A by adding λ times its sth row to its rth row?

Solution. Using Example 4.44a, we obviously have $F = E + \lambda B_{rs}$ where E is the unit matrix of order m.

g. By what $n \times n$ matrix H must we multiply an $m \times n$ matrix A from the right to make the matrix AH coincide with the matrix obtained from A by adding λ times its pth column to its qth column?

Solution. Clearly, $H = E + \mu C_{pq}$ where E is the unit matrix of order n.

4.5. Further Properties of Matrix Multiplication

4.51. Multiplication of block matrices. In multiplying matrices, it is sometimes convenient to partition the matrices into blocks and afterwards deal with the blocks as separate entities. Suppose we are given an $m \times n$ matrix A and an $n \times p$ matrix B, partitioned into blocks as follows:

$$A = m\left\{\overbrace{\left\|\begin{array}{cc|c} A_{11} & A_{12} & \cdots \\ A_{21} & A_{22} & \cdots \\ \hline \cdot & \cdot & \cdots \\ \cdot & \cdot & \cdots \end{array}\right\|}^{n}\right., \qquad B = n\left\{\overbrace{\left\|\begin{array}{cc|c} B_{11} & B_{12} & \cdots \\ B_{21} & B_{22} & \cdots \\ \hline \cdot & \cdot & \cdots \\ \cdot & \cdot & \cdots \end{array}\right\|}^{p}\right..$$

Suppose further that every "block-row" of the matrix A contains the same number of blocks as every "block-column" of the matrix B, and that the "width" of every block A_{jk} of the matrix A coincides with the "height" of every block B_{ks} of the matrix B. Then the products $A_{jk}B_{ks}$ all make sense, and in fact are rectangular matrices of size depending on the indices j and s (but not on the index k). We then have the following multiplication rule: *The product matrix AB is made up of blocks constructed from the blocks of the matrices A and B in the same way as the elements of AB are constructed from the elements of A and B, i.e.,*

$$AB = \left\|\begin{array}{c|c|c} A_{11}B_{11} + A_{12}B_{21} + \cdots & A_{11}B_{12} + A_{12}B_{22} + \cdots & \cdots \\ A_{21}B_{11} + A_{22}B_{21} + \cdots & A_{21}B_{12} + A_{22}B_{22} + \cdots & \cdots \\ \hline \cdots\cdots & \cdots\cdots & \cdots \\ \cdots\cdots & \cdots\cdots & \cdots \end{array}\right\|. \tag{9}$$

To prove (9), let i be the index of a block-row of A containing the kth ordinary row of A, and let j be the index of a block-column of B containing the qth ordinary column of B. By the general rule of Sec. 4.43, the elements of the product matrix $P = AB$ are of the form

$$\begin{aligned} p_{kq} &= a_{k1}b_{1q} + \cdots + a_{kn}b_{nq} \\ &= (a_{k1}b_{1q} + \cdots + a_{kp}b_{pq}) + \cdots + (a_{kr}b_{rq} + \cdots + a_{kn}b_{nq}), \end{aligned}$$

where parentheses are inserted in keeping with the widths of blocks of A (and heights of blocks of B). But the first term in parentheses is the element in the kth row and qth column of the block $A_{i1}B_{1j}$, the second term in parentheses (not written) is the element in the kth row and qth column of the block $A_{i1}B_{2j}$, and so on. Thus p_{kq} is the element in the kth row and qth column of the block $A_{i1}B_{1j} + \cdots + A_{ir}B_{rj}$, itself the block in the ith row and jth column of the matrix $P = AB$ regarded as a block matrix. The proof of (9) is now complete.

4.52. Multiplication of quasi-diagonal matrices. A matrix is said to be *quasi-diagonal* if it is of the form

$$A = \left\| \begin{array}{cccc} \boxed{A_{11}} & & & \\ & \boxed{A_{22}} & & \\ & & \ddots & \\ & & & \boxed{A_{ss}} \end{array} \right\|,$$

where the "off-diagonal" blocks consist entirely of zeros. Suppose the block A_{kk} is an $m_k \times n_k$ matrix $(k = 1, \ldots, s)$, and consider the quasi-diagonal matrix

$$B = \left\| \begin{array}{cccc} \boxed{B_{11}} & & & \\ & \boxed{B_{22}} & & \\ & & \ddots & \\ & & & \boxed{B_{ss}} \end{array} \right\|,$$

where the block B_{kk} is an $n_k \times p_k$ matrix $(k = 1, \ldots, s)$. Then, using the rule of Sec. 4.51 to multiply the matrices A and B, we immediately get

$$AB = \left\| \begin{array}{cccc} \boxed{A_{11}B_{11}} & & & \\ & \boxed{A_{22}B_{22}} & & \\ & & \ddots & \\ & & & \boxed{A_{ss}B_{ss}} \end{array} \right\|.$$

Thus in this case the matrix AB is again a quasi-diagonal matrix, where the block $A_{kk}B_{kk}$ has m_k rows and p_k columns.

4.53. Multiplication of transposed matrices. Given an $m \times n$ matrix $A = \|a_{jk}\|$, by the *transpose* of A (cf. Sec. 1.41) is meant the $n \times m$ matrix $A' = \|a'_{jk}\|$ such that

$$a'_{jk} = a_{kj} \qquad (j = 1, \ldots, n;\ k = 1, \ldots, m).$$

Let A be an $m \times n$ matrix and B an $n \times p$ matrix. Then the product $P = AB$ is defined and is an $m \times p$ matrix. Moreover, the product $B'A'$ of the transposed matrices A' and B' is also defined and is a $p \times m$ matrix. We now show that

$$B'A' = (AB)'. \tag{10}$$

Let the elements of the matrices A, B, $P = AB$, A', B' and P' be denoted by a_{ij}, b_{ij}, p_{ij}, $a'_{ij} = a_{ji}$, $b'_{ij} = b_{ji}$, $p'_{ij} = p_{ji}$. Then, by the rule for matrix multiplication,

$$p_{ik} = p'_{ki} = \sum_{j=1}^{n} a_{ij}b_{jk} = \sum_{j=1}^{n} a'_{ji}b'_{kj} = \sum_{j=1}^{n} b'_{kj}a'_{ji},$$

where the summation is over the index j with the indices i and j held fixed. Thus to form the element p'_{ki} of the matrix P', the elements of the kth row of B' are multiplied by the corresponding elements of the ith column of A' and then added. In other words, using the rule for matrix multiplication once again, we see that P' is the product of B' and A' (in that order), thereby proving (10).

4.54. Minors of the product of two matrices. Given an $m \times n$ matrix $A = \|a_{jk}\|$ and an $n \times p$ matrix $B = \|b_{jk}\|$, we construct the $m \times p$ matrix $P = AB = \|p_{jk}\|$. Fixing the rows with indices $\alpha_1, \ldots, \alpha_k$ $(\alpha_1 < \cdots < \alpha_k)$ and the columns with indices $\beta_1, \ldots, \beta_k$ $(\beta_1 < \cdots < \beta_k)$, where $k \leqslant m$, $k \leqslant p$, we now consider the problem of calculating the minor

$$M_{\beta_1,\ldots,\beta_k}^{\alpha_1,\ldots,\alpha_k}(AB) = \begin{vmatrix} a_{\alpha_1 1}b_{1\beta_1} + \cdots + a_{\alpha_1 n}b_{n\beta_1} & \cdots & a_{\alpha_1 1}b_{1\beta_k} + \cdots + a_{\alpha_1 n}b_{n\beta_k} \\ a_{\alpha_2 1}b_{1\beta_1} + \cdots + a_{\alpha_2 n}b_{n\beta_1} & \cdots & a_{\alpha_2 1}b_{1\beta_k} + \cdots + a_{\alpha_2 n}b_{n\beta_k} \\ \cdots & \cdots & \cdots \\ a_{\alpha_k 1}b_{1\beta_1} + \cdots + a_{\alpha_k n}b_{n\beta_1} & \cdots & a_{\alpha_k 1}b_{1\beta_k} + \cdots + a_{\alpha_k n}b_{n\beta_k} \end{vmatrix} \qquad (11)$$

formed from these rows and columns. To make this calculation, we use the linear property of determinants (Sec. 1.44). The νth column of the minor (11) is the sum of k "elementary columns" with elements of the form $a_{\alpha_j i}b_{i\beta_\nu}$ (where the column indices i and ν are fixed, and the row index j varies from 1 to k). Hence the whole minor (11) is the sum of k^k "elementary determinants" consisting only of elementary columns. Since in each elementary column the factor $b_{i\beta_\nu}$ does not change as we go down the column, it can be factored out of the elementary determinant. After this, each elementary determinant takes the form

$$b_{i_1\beta_1}b_{i_2\beta_2}\cdots b_{i_k\beta_k}\begin{vmatrix} a_{\alpha_1 i_1} & a_{\alpha_1 i_2} & \cdots & a_{\alpha_1 i_k} \\ a_{\alpha_2 i_1} & a_{\alpha_2 i_2} & \cdots & a_{\alpha_2 i_k} \\ \cdot & \cdot & \cdots & \cdot \\ a_{\alpha_k i_1} & a_{\alpha_k i_2} & \cdots & a_{\alpha_k i_k} \end{vmatrix}, \qquad (12)$$

where $i_1, i_2, \ldots, i_k$ are certain numbers from 1 to n. If some of these numbers are the same, then clearly the corresponding elementary determinant vanishes. Moreover, this is always the case if $k > n$. Therefore *if the matrix AB has minors of order $k > n$, they must all vanish.*

Returning to the case $k \leqslant n$, we note that it is only necessary to consider elementary determinants for which the indices $i_1, i_2, \ldots, i_k$ are all different. In this case, the determinant

$$\begin{vmatrix} a_{\alpha_1 i_1} & a_{\alpha_1 i_2} & \cdots & a_{\alpha_1 i_k} \\ a_{\alpha_2 i_1} & a_{\alpha_2 i_2} & \cdots & a_{\alpha_2 i_k} \\ \cdot & \cdot & \cdots & \cdot \\ a_{\alpha_k i_1} & a_{\alpha_k i_2} & \cdots & a_{\alpha_k i_k} \end{vmatrix} \tag{13}$$

is the same (except possibly for sign) as the minor $M_{j_1,\ldots,j_k}^{\alpha_1,\ldots,\alpha_k}(A)$ where the indices $j_1, \ldots, j_k$ $(j_1 < \cdots < j_k)$ are the indices $i_1, \ldots, i_k$ rearranged in increasing order. To find the sign which must be ascribed to (13) to get $M_{j_1,\ldots,j_k}^{\alpha_1,\ldots,\alpha_k}(A)$, we successively interchange adjacent columns of (13) until we arrive at the normal arrangement of the columns, i.e., the arrangement they have in the matrix A itself. At each interchange of two adjacent columns, the determinant (13) changes sign and the number of inversions in the permutation $i_1, i_2, \ldots, i_k$ changes by unity. Since in the final arrangement of the columns, the subscripts are in natural order (i.e., without inversions), *the number of successive changes of sign is equal to the number of inversions in the permutation* $i_1, i_2, \ldots, i_k$.† Let $N(i)$ denote the number of sign changes. Then the expression (12) takes the form

$$(-1)^{N(i)} b_{i_1\beta_1} b_{i_2\beta_2} \cdots b_{i_k\beta_k} M_{j_1,\ldots,j_k}^{\alpha_1,\ldots,\alpha_k}(A). \tag{14}$$

To obtain (11), we must now add up all the expressions of the form (14).

First we add up all the expressions with the same set of indices $j_1, \ldots, j_k$, taking out the common factors $M_{j_1,\ldots,kj}^{\alpha_1,\ldots,\alpha_k}(A)$. The remaining expression is then

$$(-1)^{N(i)} b_{i_1\beta_1} b_{i_2\beta_2} \cdots b_{i_k\beta_k},$$

where the summation is over all distinct sets of indices $i_1, i_2, \ldots, i_k$ (these indices range from 1 to n). But this expression is just the minor $M_{\beta_1,\ldots,\beta_k}^{j_1,\ldots,j_k}(B)$ Thus finally we get the formula

$$M_{\beta_1,\ldots,\beta_k}^{\alpha_1,\ldots,\alpha_k}(AB) = \sum M_{j_1,\ldots,j_k}^{\alpha_1,\ldots,\alpha_k}(A) M_{\beta_1,\ldots,\beta_k}^{j_1,\ldots,j_k}(B), \tag{15}$$

where the summation is over all distinct sets of indices $j_1, j_2, \ldots, j_k$ $(1 \leqslant j_1 < j_2 < \cdots < j_k \leqslant n)$. The total number of terms in the sum (15) is just the binomial coefficient

$$C_k^n = \frac{n!}{k!\,(n-k)!}.$$

† It is assumed that the change in the indices $i_1, i_2, \ldots, i_k$ produced by every column interchange causes a smaller index to appear before a larger index, with the result that the total number of inversions changes by exactly one.

Our result can be summarized in the following

THEOREM. *Every minor of order* $k \leqslant n$ *of the matrix* AB *can be expressed in terms of the minors of the same order of the matrices* A *and* B, *in the way given by formula* (15).

4.6. The Range and Null Space of a Linear Operator

4.61. Let $\mathbf{A}$ be a linear operator mapping a linear space $\mathbf{X}$ into a linear space $\mathbf{Y}$ (in the notation of Sec. 2.71, this is expressed by writing $\mathbf{A}:\mathbf{X} \to \mathbf{Y}$). Let n be the dimension of $\mathbf{X}$ and m the dimension of $\mathbf{Y}$, and choose an arbitrary basis $e_1, \ldots, e_n$ in $\mathbf{X}$ and $f_1, \ldots, f_m$ in $\mathbf{Y}$. Then, by the method of Sec. 4.23, we can associate the operator $\mathbf{A}$ with an $m \times n$ matrix

$$A = \|a_i^{(j)}\| \qquad (i = 1, \ldots, m; j = 1, \ldots, n).$$

Let $\mathbf{T}(\mathbf{A})$ be the *range* of $\mathbf{A}$, i.e., the set of all vectors $y = \mathbf{A}x$, $x \in \mathbf{X}$. We now consider the problem of finding the dimension of the subspace $\mathbf{T}(\mathbf{A})$ from a knowledge of the matrix A.

Writing

$$x = \sum_{k=1}^{n} \xi_k e_k,$$

we have

$$y = \mathbf{A}x = \sum_{k=1}^{n} \xi_k \mathbf{A}e_k.$$

Hence the range of the operator $\mathbf{A}$ coincides with the linear manifold spanned by the vectors $\mathbf{A}e_1, \ldots, \mathbf{A}e_n$. As noted on p. 51, the dimension of this linear manifold $\mathbf{L}(\mathbf{A}e_1, \ldots, \mathbf{A}e_n)$ equals the maximum number of linearly independent vectors in the system $\mathbf{A}e_1, \ldots, \mathbf{A}e_n$. We know from Sec. 4.23 that the columns of the matrix of the operator $\mathbf{A}$ consist of the components of the vectors $\mathbf{A}e_1, \ldots, \mathbf{A}e_n$ with respect to the basis $e_1, \ldots, e_n$, and hence the problem of finding the maximum number of linearly independent vectors in the system $\mathbf{A}e_1, \ldots, \mathbf{A}e_n$ reduces at once to that of finding the maximum number of linearly independent columns of the matrix A. But by Theorem 3.12c, the latter quantity is just the rank of the matrix of the operator $\mathbf{A}$. Thus *the dimension of the range of a linear operator* $\mathbf{A}$ *mapping an n-dimensional space* $\mathbf{X}$ *into an m-dimensional space* $\mathbf{Y}$ *equals the rank of the matrix of* $\mathbf{A}$ *relative to any basis* $\{e\}$ *in* $\mathbf{X}$ *and any basis* $\{f\}$ *in* $\mathbf{Y}$.

We note that the choice of bases does not matter here. Therefore *the rank of the matrix of an operator* $\mathbf{A}$ *does not depend on the choice of bases, i.e., depends only on the operator* $\mathbf{A}$ *itself.* In what follows, the rank of the matrix of the operator $\mathbf{A}$ (relative to any bases) will simply be called the *rank of the operator* $\mathbf{A}$, denoted by $r_{\mathbf{A}}$.

4.62. Next let $\mathbf{N}(\mathbf{A})$ be the *null space* of the operator $\mathbf{A}$, i.e., the set of all vectors $x \in \mathbf{X}$ such that $\mathbf{A}x = 0$, and as before let $A = \|a_i^{(j)}\|$ be the matrix of $\mathbf{A}$. We now consider the problem of finding the dimension of the subspace $\mathbf{N}(\mathbf{A})$ from a knowledge of the matrix A. Let

$$x = \sum_{i=1}^{n} \xi_i e_i \in \mathbf{N}(\mathbf{A}).$$

Then the system (5), p. 80 takes the form

$$\begin{aligned} a_1^{(1)}\xi_1 + a_1^{(2)}\xi_2 + \cdots + a_1^{(n)}\xi_n &= 0, \\ a_2^{(1)}\xi_1 + a_2^{(2)}\xi_2 + \cdots + a_2^{(n)}\xi_n &= 0, \\ \cdots\cdots\cdots\cdots\cdots\cdots\cdots\cdots \\ a_m^{(1)}\xi_1 + a_m^{(2)}\xi_2 + \cdots + a_m^{(n)}\xi_n &= 0. \end{aligned} \tag{16}$$

Moreover, it is obvious that, conversely, every vector $x \in \mathbf{X}$ whose components satisfy (16) belongs to the null space of the operator $\mathbf{A}$. Thus the problem of finding the dimension of the null space of the operator $\mathbf{A}$ is equivalent to the problem of finding the dimension of the subspace of $\mathbf{X}$ consisting of all solutions of the system (16). But according to Sec. 3.51, the dimension $n_{\mathbf{A}}$ of this subspace equals $n - r$, where r is the rank of the coefficient matrix of the system, or equivalently, the rank of the operator $\mathbf{A}$. It follows that $n_{\mathbf{A}} = n - r_{\mathbf{A}}$. Thus *the dimension of the null space of the operator* $\mathbf{A}$ *equals the rank of the space* $\mathbf{X}$ *(on which* $\mathbf{A}$ *acts) minus the rank of the operator* $\mathbf{A}$.

4.63. In particular, if the morphism $\mathbf{A}:\mathbf{X} \to \mathbf{Y}$ is an epimorphism, then $\mathbf{T}(\mathbf{A}) = \mathbf{Y}$ and hence $r_{\mathbf{A}} = m$. If the morphism $\mathbf{A}:\mathbf{X} \to \mathbf{Y}$ is a monomorphism, then $\mathbf{N}(\mathbf{A}) = \{0\}$ and hence $r_{\mathbf{A}} = n$. The converse assertions are also true: If the rank of the matrix A equals the number m of its rows, then the dimension of $\mathbf{T}(\mathbf{A})$ coincides with the dimension of the whole space $\mathbf{Y}$ and hence $\mathbf{T}(\mathbf{A}) = \mathbf{Y}$. Therefore *the morphism* $\mathbf{A}$ *is an epimorphism if and only if* $r_{\mathbf{A}} = m$. If the rank of the matrix A equals the number of its columns, then the vectors $f_1 = \mathbf{A}e_1, \ldots, f_n = \mathbf{A}e_n$ are linearly independent and hence the operator $\mathbf{A}$ is a monomorphism (see Sec. 2.73c). Therefore *the morphism* $\mathbf{A}$ *is a monomorphism if and only if* $r_{\mathbf{A}} = n$.

4.64. The following proposition is the converse of the results of Secs. 4.61 and 4.62:

THEOREM. *Let* $\mathbf{X}$ *be an* n*-dimensional linear space and* $\mathbf{Y}$ *an arbitrary linear space. Then, given any subspaces* $\mathbf{N} \subset \mathbf{X}$ *and* $\mathbf{T} \subset \mathbf{Y}$ *the sum of whose dimensions equals* n, *there exists a linear operator* $\mathbf{A}:\mathbf{X} \to \mathbf{Y}$ *such that* $\mathbf{N}(\mathbf{A}) = \mathbf{N}$, $\mathbf{T}(\mathbf{A}) = \mathbf{T}$.

Proof. Let the dimensions of **N** and **T** be k and $m = n - k$, respectively. Moreover, let $f_1, f_2, \ldots, f_m$ be m linearly independent vectors in the subspace **T**, and let $e_1, e_2, \ldots, e_n$ be any basis in the space **X** whose first k vectors lie in the subspace **N** (see Sec. 2.43). Defining an operator **A** by the conditions

$$\begin{aligned} \mathbf{A}e_i &= 0 \qquad (i = 1, 2, \ldots, k), \\ \mathbf{A}e_{i+k} &= f_i \qquad (i = 1, 2, \ldots, m), \end{aligned} \tag{17}$$

we now show that **A** satisfies the requirements of the theorem. First of all, it is obvious that **T**(**A**) is the linear manifold spanned by the vectors $f_1, f_2, \ldots, f_m$ and hence coincides with the subspace **T**. Moreover, by (17), every vector of the subspace **N** belongs to **N**(**A**), and it remains to show only that every vector of **N**(**A**) belongs to **N**. Suppose $\mathbf{A}x = 0$ for some

$$x = \sum_{i=1}^{n} \xi_i e_i.$$

Then, by (17),

$$0 = \mathbf{A}x = \mathbf{A}(\xi_1 e_1 + \cdots + \xi_n e_n) = \xi_{k+1} f_1 + \cdots + \xi_n f_m,$$

and hence $\xi_{k+1} = \cdots = \xi_n = 0$ since $f_1, \ldots, f_m$ are linearly independent. But then

$$x = \xi_1 e_1 + \cdots + \xi_k e_k \in \mathbf{N}. \quad \blacksquare$$

4.65. The following theorem on the rank of the product of two matrices is a consequence of the geometric notions just introduced:

THEOREM. *The rank of the product AB of two matrices A and B does not exceed the rank of each of the factors.*

Proof. Naturally, we must assume that the number of columns of the matrix A coincides with the number of rows of the matrix B, since otherwise the product AB could not be formed. Thus let A be an $m \times n$ matrix and B a $n \times p$ matrix, and introduce linear spaces **X**, **Y** and **Z** with dimensions n, m and p, respectively. Choose a basis $e_1, \ldots, e_n$ in the space **X**, a basis $f_1, \ldots, f_m$ in the space **Y** and a basis $g_1, \ldots, g_p$ in the space **Z**. Using these bases, we associate a linear operator $\mathbf{A}: \mathbf{X} \to \mathbf{Y}$ with the matrix A and a linear operator $\mathbf{B}: \mathbf{Z} \to \mathbf{X}$ wiht the matrix B (see Sec. 4.23). Then the product operator $\mathbf{AB}: \mathbf{Z} \to \mathbf{Y}$ corresponds to the product matrix AB. The range of the operator **AB** is contained in the range of the operator **A**, by the very definition of **AB**. Since by Sec. 4.61 the dimension of the range of any operator equals the rank of its matrix, we find that *the rank of the product of two matrices does not exceed the rank of the first factor.* To prove that it also does not exceed the rank of the second factor, we go over to transposed matrices. Using equation (10), p. 90, we find that

$$\text{rank } AB = \text{rank } (AB)' = \text{rank } B'A' \leqslant \text{rank } B' = \text{rank } B. \quad \blacksquare$$

4.66. The rank of the product of two matrices can actually be less than the rank of each factor. For example, the matrices

$$A = \begin{Vmatrix} 0 & 1 \\ 0 & 0 \end{Vmatrix}, \qquad B = \begin{Vmatrix} 1 & 0 \\ 0 & 0 \end{Vmatrix}$$

both have rank one, but their product

$$AB = \begin{Vmatrix} 0 & 0 \\ 0 & 0 \end{Vmatrix}$$

has rank zero. Therefore the following theorem, which gives a lower bound rather than an upper bound for the rank of the product of two matrices, is of interest:

THEOREM. *Let A be an $m \times n$ matrix of rank r_A and B an $n \times p$ matrix of rank r_B. Then the rank of the $m \times p$ matrix AB is no less than $r_A + r_B - n$.*

Proof. First we show that any operator $\mathbf{A}:\mathbf{X} \to \mathbf{Y}$ of rank r carries every k-dimensional subspace $\mathbf{X}' \subset \mathbf{X}$ into a subspace $\mathbf{Y}' \subset \mathbf{Y}$ of dimension no less than $r - (n - k)$. Choose a basis $e_1, e_2, \ldots, e_n$ in the space $\mathbf{X}$ such that the first k basis vectors lie in the subspace $\mathbf{X}'$ (see Sec. 2.43). The components of the vectors $\mathbf{A}e_1, \mathbf{A}e_2, \ldots, \mathbf{A}e_k$ generating the space $\mathbf{Y}'$ occupy the first k columns of the matrix of the operator $\mathbf{A}$. By hypothesis, there are r linearly independent columns in the matrix of $\mathbf{A}$. We divide these columns into two groups, the first consisting of columns whose numbers lie in the range 1 to k, the second consisting of columns whose numbers lie in the range $k + 1$ to n. The second group contains no more than $n - k$ columns, and hence the first group contains no more than $r - (n - k)$ columns. Thus the subspace $\mathbf{Y}'$ has no more than $r - (n - k)$ linearly independent vectors, as asserted.

Now let $\mathbf{A}:\mathbf{X} \to \mathbf{Y}$ and $\mathbf{B}:\mathbf{Z} \to \mathbf{X}$ be linear operators corresponding to the matrices A and B. By Sec. 4.61, the rank of the matrix of the operator $\mathbf{AB}$ is just the dimension of the range of $\mathbf{AB}$. The operator $\mathbf{B}$ maps the whole space $\mathbf{Z}$ into the subspace $\mathbf{T}(\mathbf{B}) \subset \mathbf{X}$ of dimension r_B. But as shown above, the operator $\mathbf{A}$ maps the subspace $\mathbf{T}(\mathbf{B})$ into a subspace of dimension no less than $r_A - (n - r_B) = r_A + r_B - n$. Thus the range of the operator $\mathbf{AB}$, and hence the rank of the matrix of $\mathbf{AB}$, is no less than $r_A + r_B - n$. ∎

4.67. COROLLARY. *Let A be an $m \times n$ matrix and B an $n \times p$ matrix, and suppose the rank of one of these matrices equals n. Then the rank of AB equals the rank of the other matrix.*

Proof. In this case, the upper and lower bounds for the rank of AB, given by Theorems 4.65 and 4.66, have the same value, equal to the rank of the other matrix. ∎

4.68. Let **A** be a linear operator mapping a linear space **X** into a linear space **Y**. A linear operator **B** mapping **Y** into **X** is called a *left inverse* of the operator **A** if

$$\mathbf{BA} = \mathbf{E}$$

is the unit operator in the space **X**. The operator **A** is then called a *right inverse* of the operator **B**. The following theorem gives conditions under which the operator **A** (or **B**) has a left (or right) inverse:

THEOREM. *The operator* $\mathbf{A}:\mathbf{X} \to \mathbf{Y}$ *has a left inverse if and only if* **A** *is a monomorphism. The operator* $\mathbf{B}:\mathbf{Y} \to \mathbf{X}$ *has a right inverse if and only if* **B** *is an epimorphism.*

Proof. Let **A** be a monomorphism with range $\mathbf{T}(\mathbf{A}) \subset \mathbf{Y}$. Then for every $y \in \mathbf{T}(\mathbf{A})$ there is an $x \in \mathbf{X}$ such that $\mathbf{A}x = y$, where x is uniquely determined by y since **A** is a monomorphism by hypothesis. Let $\mathbf{Q} \subset \mathbf{Y}$ be the subspace whose direct sum with $\mathbf{T}(\mathbf{A})$ is the whole space **Y** (see Sec. 2.46). We now define an operator $\mathbf{B}:\mathbf{Y} \to \mathbf{X}$ by the following rule: For $y \in \mathbf{T}(\mathbf{A})$ we set $\mathbf{B}y$ equal to the (unique) vector x for which $\mathbf{A}x = y$, while otherwise we set

$$\begin{aligned} \mathbf{B}y &= 0 \quad \text{if } y \in \mathbf{Q}, \\ \mathbf{B}y &= \mathbf{B}y_1 \quad \text{if } y = y_1 + y_2, y_1 \in \mathbf{T}(\mathbf{A}), y_2 \in \mathbf{Q}. \end{aligned}$$

Then it is easy to see that the operator **B** is linear and that $\mathbf{BA}x = x$, for every $x \in \mathbf{X}$, so that **B** is the left inverse of **A**. However, if **A** is not a monomorphism, there exists a nonzero vector $x \in \mathbf{X}$ such that $\mathbf{A}x = 0$. Then for any $\mathbf{B}:\mathbf{Y} \to \mathbf{X}$ we have $(\mathbf{BA})x = \mathbf{B}(\mathbf{A}x) = \mathbf{B}(0) = 0$, so that **A** indeed fails to have a left inverse.

Next let $\mathbf{B}:\mathbf{Y} \to \mathbf{X}$ be an epimorphism and let $\mathbf{N}(\mathbf{B}) \subset \mathbf{Y}$ be the null space of **B**, while $\mathbf{Q} \subset \mathbf{Y}$ is the subspace whose direct sum with $\mathbf{N}(\mathbf{B})$, denoted by $\mathbf{N}(\mathbf{B}) + \mathbf{Q}$, is the whole space **Y**. Since

$$\mathbf{X} = \mathbf{B}(\mathbf{Y}) = \mathbf{B}(\mathbf{N}(\mathbf{B}) + \mathbf{Q}) = \mathbf{B}(\mathbf{Q}),$$

the mapping $\mathbf{B}:\mathbf{Q} \to \mathbf{X}$ is also an epimorphism and in fact an isomorphism, since no nonzero element $y \in \mathbf{Q}$ is mapped into zero by the operator **B**. We now define an operator $\mathbf{A}:\mathbf{X} \to \mathbf{Y}$ by the following rule: Given any $x \in \mathbf{X}$, we set $\mathbf{A}x$ equal to the (unique) vector $y \in \mathbf{Q}$ for which $\mathbf{B}y = x$. Then it is easy to see that the operator **A** is linear and that $\mathbf{BA}x = x$ for every $x \in \mathbf{X}$, so that **A** is the right inverse of **B**. However, if $\mathbf{B}:\mathbf{Y} \to \mathbf{X}$ is not an epimorphism, then $\mathbf{BA}x \neq x$ for any operator $\mathbf{A}:\mathbf{X} \to \mathbf{Y}$ and any vector $x \in \mathbf{X}$ such that $x \notin \mathbf{T}(\mathbf{B})$, so that **B** has no right inverse. ∎

4.69. a. As we know, the result of multiplying an $n \times m$ matrix P by an $m \times n$ matrix A is a square $n \times n$ matrix

$$S = PA.$$

If S is the unit $n \times n$ matrix (see Example 4.24d), we call P the *left inverse* of the matrix A. Similarly, the result of multiplying an $m \times n$ matrix A by an $n \times m$ matrix Q is a square $m \times m$ matrix

$$T = AQ,$$

and if T is the unit $m \times m$ matrix, we call Q the *right inverse* of the matrix A.

b. Using the results of Sec. 4.63, we can now formulate Theorem 4.68 in terms of the rank of a matrix:

THEOREM. *An $m \times n$ matrix A has a left inverse if and only if its rank equals n and a right inverse if and only if its rank equals m.*

4.7. Linear Operators Mapping a Space $\mathbf{K}_n$ into Itself

4.71. Let $\mathbf{A}$ be a linear operator mapping the space $\mathbf{X}$ into itself (this corresponds to setting $\mathbf{Y} = \mathbf{X}$ in Sec. 4.21). Such an operator is said to be an *operator (acting) in the space* $\mathbf{X}$.

Suppose the operator $\mathbf{A}$ acts in an n-dimensional space $\mathbf{X} = \mathbf{K}_n$. Choosing a basis $e_1, \ldots, e_n$ in the space $\mathbf{X}$, *we use the same basis in* $\mathbf{Y} = \mathbf{X}$ *to construct the matrix of the operator* $\mathbf{A}$. Then formula (3), p. 79 becomes

$$\mathbf{A}e_j = \sum_{i=1}^{n} a_i^{(j)} e_i \tag{18}$$

(after setting $f_i = e_i$), so that the coefficients $a_i^{(j)}$ now form a square $n \times n$ matrix A, called the *matrix of the operator* $\mathbf{A}$ *in (or relative to) the basis* $\{e\} = \{e_1, \ldots, e_n\}$. We will sometimes denote this matrix by $A_{(e)}$. The corresponding formula relating the components of the vectors x and y, where

$$y = \mathbf{A}x, \qquad x = \sum_{j=1}^{n} \xi_j e_j, \qquad y = \sum_{j=1}^{n} \eta_j e_j$$

is

$$\eta_i = \sum_{j=1}^{n} a_i^{(j)} \xi_j \tag{19}$$

(cf. formula (4), p. 80). For a fixed basis $\{e\} = \{e_1, \ldots, e_n\}$, we get a one-to-one correspondence between all linear operators acting in the space $\mathbf{K}_n$ (i.e., mapping $\mathbf{K}_n$ into itself) and all square $n \times n$ matrices made up of elements of the underlying field K.

4.72. *Examples*

a. The operator associating the zero vector with every vector of the space $\mathbf{X}$ is obviously linear. As in Example 4.22a, this operator is called the *zero operator*. It is clear that the matrix of the zero operator relative to any basis consists entirely of zeros.

b. The *identity* (or *unit*) *operator* **E**, associating the vector x itself with every vector $x \in \mathbf{X}$, has already been considered in Example 4.22d. Its matrix is the *unit* (or *identity*) *matrix* of the form

$$E = \left\| \begin{matrix} 1 & 0 & \cdots & 0 \\ 0 & 1 & \cdots & 0 \\ \cdot & \cdot & \cdots & \cdot \\ 0 & 0 & \cdots & 1 \end{matrix} \right\|$$

(cf. Example 4.24d).

c. The operator **A** which carries every vector $x \in \mathbf{X}$ into λx, where λ is a fixed number from the field K, is obviously linear. This operator is called the *similarity operator* (with *ratio of similitude* λ). As in the preceding example, the similarity operator has the matrix

$$\left\| \begin{matrix} \lambda & 0 & \cdots & 0 \\ 0 & \lambda & \cdots & 0 \\ \cdot & \cdot & \cdots & \cdot \\ 0 & 0 & \cdots & \lambda \end{matrix} \right\|$$

in any basis.

d. We can specify a vector in the Euclidean plane V_2 by giving its polar coordinates ρ and φ. The operator **A** carrying the vector $x = (\rho, \varphi)$ into $\mathbf{A}x = (\rho, \varphi + \varphi_0)$, where φ_0 is a fixed angle, is linear (as can easily be verified by drawing a figure). This operator is called the *rotation operator through the angle* φ_0.

To construct the matrix of **A**, we choose a basis in V_2 consisting of two orthogonal unit vectors e_1 and e_2. Drawing a figure, we easily see that after rotation through the angle φ_0 the vector e_1 goes into the vector $e_1 \cos \varphi_0 + e_2 \sin \varphi_0$, while the vector e_2 goes into $-e_1 \sin \varphi_0 + e_2 \cos \varphi_0$. Hence the matrix of the rotation operator **A** has the form

$$\left\| \begin{matrix} \cos \varphi_0 & -\sin \varphi_0 \\ \sin \varphi_0 & \cos \varphi_0 \end{matrix} \right\|$$

in the basis e_1, e_2.

e. Let $e_1, e_2, \ldots, e_n$ be a basis in an n-dimensional space $\mathbf{K}_n$, and suppose that with the vector

$$x = \sum_{k=1}^{n} \xi_k e_k$$

we associate the vector

$$\mathbf{P}x = \sum_{k=1}^{m} \xi_k e_k$$

where $m < n$. Then $\mathbf{P}$ is a linear operator, called the *projection operator* onto the subspace $\mathbf{K}_m$ spanned by the vectors $e_1, e_2, \ldots, e_m$.

To construct the matrix of $\mathbf{P}$, we note that it carries the vectors $e_1, e_2, \ldots, e_m$ into themselves and the vectors $e_{m+1}, \ldots, e_n$ into the zero vector. Hence the matrix of the projection operator $\mathbf{P}$ in the basis $e_1, e_2, \ldots, e_n$ is just

$$\begin{array}{r} \\ \\ \\ (m) \\ \\ \\ \\ \end{array}\left\|\begin{array}{cccccccc} 1 & 0 & \cdots & 0 & 0 & \cdots & 0 \\ 0 & 1 & \cdots & 0 & 0 & \cdots & 0 \\ \cdot & \cdot & \cdots & \cdot & \cdot & \cdots & \cdot \\ 0 & 0 & \cdots & 1 & 0 & \cdots & 0 \\ 0 & 0 & \cdots & 0 & 0 & \cdots & 0 \\ \cdot & \cdot & \cdots & \cdot & \cdot & \cdots & \cdot \\ 0 & 0 & \cdots & 0 & 0 & \cdots & 0 \end{array}\right\|.$$

f. Let $e_1, e_2, \ldots, e_n$ be a basis in an n-dimensional space $\mathbf{K}_n$, and let $\lambda_1, \lambda_2, \ldots, \lambda_n$ be n fixed numbers. Defining an operator $\mathbf{A}$ for the basis vectors by the conditions

$$\mathbf{A}e_1 = \lambda_1 e_1, \qquad \mathbf{A}e_2 = \lambda_2 e_2, \ldots, \qquad \mathbf{A}e_n = \lambda_n e_n,$$

we then of course use linearity to define $\mathbf{A}$ for any other vector

$$x = \sum_{k=1}^{n} \xi_k e_k$$

by the condition

$$\mathbf{A}x = \sum_{k=1}^{n} \lambda_k \xi_k e_k.$$

The resulting operator $\mathbf{A}$ is said to be *diagonal* relative to the basis $e_1, e_2, \ldots, e_n$; we also call $\mathbf{A}$ a *diagonalizable* operator.

The matrix of an operator which is diagonal relative to the basis $e_1, e_2, \ldots, e_n$ is of the form

$$\left\|\begin{array}{cccc} \lambda_1 & 0 & \cdots & 0 \\ 0 & \lambda_2 & \cdots & 0 \\ \cdot & \cdot & \cdots & \cdot \\ 0 & 0 & \cdots & \lambda_n \end{array}\right\|$$

in the same basis. Such a matrix, which can have nonzero elements only on its principal diagonal, is said to be *diagonal* (hence the corresponding terminology for the operator itself). It should be noted that the matrix of an operator which is diagonal relative to the basis $e_1, e_2, \ldots, e_n$ will in general not be diagonal in another basis $f_1, f_2, \ldots, f_n$.

4.73. a. Using the rules of Secs. 4.31 and 4.32 to add linear operators acting in a space $\mathbf{X}$ and multiply them by numbers, we again get linear operators acting in $\mathbf{X}$. The rules (7) and (7′), p. 82 show that the set of all linear operators acting in a space $\mathbf{X}$ (equipped with the indicated operations of addition and multiplication by numbers) is again a linear space over the same field K. Moreover, the operation of multiplication described in Sec. 4.33 can always be defined for operators acting in a space $\mathbf{X}$, and the result is again an operator acting in $\mathbf{X}$. In particular, we can define the *powers* of a given operator $\mathbf{A}$ by the rules

$$\mathbf{A}^1 = \mathbf{A},$$
$$\mathbf{A}^2 = \mathbf{AA},$$
$$\mathbf{A}^3 = \mathbf{A}^2\mathbf{A} = (\mathbf{AA})\mathbf{A} = \mathbf{A}(\mathbf{AA}) = \mathbf{A}(\mathbf{A}^2),$$
$$\cdots\cdots\cdots\cdots\cdots\cdots\cdots\cdots\cdots$$
$$\mathbf{A}^n = \mathbf{A}^{n-1}\mathbf{A} = \mathbf{AA}^{n-1}.$$

We then have the formula

$$\mathbf{A}^{m+n} = \mathbf{A}^m\mathbf{A}^n \qquad (m, n = 1, 2, \ldots), \tag{20}$$

which can easily be proved by induction. Next we define

$$\mathbf{A}^0 = \mathbf{E},$$

where $\mathbf{E}$ is the identity operator, and show that (20) remains valid in the case where one of the indices is zero. In fact, if $\mathbf{B}$ is any operator, we have

$$(\mathbf{BE})x = \mathbf{B}(\mathbf{E}x) = \mathbf{B}x = \mathbf{E}(\mathbf{B}x),$$

so that

$$\mathbf{BE} = \mathbf{EB} = \mathbf{B}.$$

Setting $\mathbf{B} = \mathbf{A}^n$, we obtain

$$\mathbf{A}^n\mathbf{E} = \mathbf{EA}^n = \mathbf{A}^n,$$

as required.

b. Let $\mathbf{X} = \mathbf{K}_n$ be a finite-dimensional space, and let $e_1, \ldots, e_n$ be an arbitrary basis in $\mathbf{X}$. Then with every linear operator $\mathbf{A}$ acting in the space $\mathbf{X}$ we can associate the matrix of $\mathbf{A}$ in the basis $e_1, \ldots, e_n$. Just like the operators themselves, the corresponding matrices can be added, multiplied and raised to powers in accordance with the rules of Secs. 4.41–4.43. The dimension of the linear space of all matrices of order n can easily be found. In fact, let E_{ij} be the matrix whose elements are all zero except for the element in the ith row and jth column, which, to be explicit, we choose to be 1. Then the matrices E_{ij} $(i, j = 1, \ldots, n)$ are obviously linearly independent. On the other hand, every matrix of order n is a linear combination of the matrices E_{ij}. Hence the matrices E_{ij} form a basis in the space of all matrices of order n. Since the number of matrices E_{ij} is n^2, the dimension of the space of all matrices of order n is just n^2 (see Sec. 2.35). The space of all linear operators acting in $\mathbf{X} = \mathbf{K}_n$ obviously has the same dimension n^2.

4.74. *Examples*

a. Multiplication by the complex number $\omega = \alpha + i\beta$ is a linear transformation in the xy-plane, which can be described by a real matrix of order two. It follows from the multiplication formula

$$(\alpha + i\beta)(x + iy) = (\alpha x - \beta y) + i(\beta x + \alpha y)$$

that this matrix is of the form

$$\varpi = \begin{Vmatrix} \alpha & -\beta \\ \beta & \alpha \end{Vmatrix}.$$

This rule establishes a one-to-one correspondence between complex numbers $\omega = \alpha + i\beta$ and real matrices ϖ of order two, where (as is easily verified) the sum (or product) of two numbers goes into the sum (or product) of the corresponding matrices. This is described by saying that the matrices ϖ form an *exact representation of the field of complex numbers* (see Sec. 11.21).

b. Let $\mathbf{B}_k$ $(k \geqslant 0)$ denote the operator which "lowers indices by k," i.e., the operator carrying each basis vector e_m $(m = 1, \ldots, n)$ into the basis vector e_{m-k} if $m - k > 0$ and into 0 if $m - k \leqslant 0$. Obviously

$$\mathbf{B}_0 = \mathbf{E}, \qquad \mathbf{B}_k\mathbf{B}_r = \mathbf{B}_{k+r},$$

and, in particular,

$$\mathbf{B}_1^k = \mathbf{B}_k.$$

The matrix of the operator $\mathbf{B}_1$ is

$$\begin{Vmatrix} 0 & 1 & 0 & \cdots & 0 \\ 0 & 0 & 1 & \cdots & 0 \\ \cdot & \cdot & \cdot & \cdots & \cdot \\ 0 & 0 & 0 & \cdots & 1 \\ 0 & 0 & 0 & \cdots & 0 \end{Vmatrix},$$

while that of the operator $\mathbf{B}_k$ $(k < n)$ is

$$\begin{array}{c} \quad\quad\quad (k+1) \\ \left\|\begin{array}{ccccccc} 0 & \cdots & 1 & 0 & \cdots & 0 \\ 0 & \cdots & 0 & 1 & \cdots & 0 \\ \cdot & \cdots & \cdot & \cdot & \cdots & \cdot \\ 0 & \cdots & 0 & 0 & \cdots & 1 \\ \cdot & \cdots & \cdot & \cdot & \cdots & \cdot \\ 0 & \cdots & 0 & 0 & \cdots & 0 \end{array}\right\| \end{array} (n-k)$$

4.75. The determinant of the product of two matrices. Let $A = \|a_{jk}\|$ and $B = \|b_{jk}\|$ be any two $n \times n$ matrices, and let $C = AB$ be their product. Applying Theorem 4.54 to the minor $M^{1,\ldots,n}_{1,\ldots,n}(AB)$, which is just the determinant of the matrix AB, we get

$$\det AB = \det A \det B. \tag{21}$$

Thus we have proved the following

THEOREM. *The determinant of the product of two $n \times n$ matrices equals the product of the determinants of the matrices.*

There also exist direct proofs of this theorem, i.e., proofs which do not rest on a proposition like Theorem 4.54. Here is one such proof. Consider the determinant

$$D = \begin{vmatrix} b_{11} & \cdots & b_{1n} & -1 & 0 & \cdots & 0 \\ b_{21} & \cdots & b_{2n} & 0 & -1 & \cdots & 0 \\ \cdot & \cdots & \cdot & \cdot & \cdot & \cdots & \cdot \\ b_{n1} & \cdots & b_{nn} & 0 & 0 & \cdots & -1 \\ 0 & \cdots & 0 & a_{11} & a_{12} & \cdots & a_{1n} \\ 0 & \cdots & 0 & a_{21} & a_{22} & \cdots & a_{2n} \\ \cdot & \cdots & \cdot & \cdot & \cdot & \cdots & \cdot \\ 0 & \cdots & 0 & a_{n1} & a_{n2} & \cdots & a_{nn} \end{vmatrix}$$

of order $2n$. By Sec. 1.32, the determinant D equals the product of the determinants of the matrices

$$A = \begin{vmatrix} a_{11} & \cdots & a_{1n} \\ \cdot & \cdots & \cdot \\ a_{n1} & \cdots & a_{nn} \end{vmatrix}, \qquad B = \begin{vmatrix} b_{11} & \cdots & b_{1n} \\ \cdot & \cdots & \cdot \\ b_{n1} & \cdots & b_{nn} \end{vmatrix},$$

so that

$$D = \det A \det B. \tag{22}$$

But there is another way of evaluating D. Using the elements -1 in the first n rows and last n columns of D, we can make all the elements in the last n rows and last n columns of D vanish. This is done by adding to the $(n + 1)$st row of D the first row multiplied by a_{11}, the second row multiplied by $a_{12}, \ldots,$ the nth row multiplied by a_{1n}, then adding to the $(n + 2)$nd row of D the first row multiplied by a_{21}, the second row multiplied by $a_{22}, \ldots,$ the nth row multiplied by a_{2n}, and so on, until we finally arrive at the last ($2n$th) row. This gives

$$D = \begin{vmatrix} b_{11} & \cdots & b_{1n} & -1 & 0 & \cdots & 0 \\ b_{21} & \cdots & b_{2n} & 0 & -1 & \cdots & 0 \\ \cdot & \cdots & \cdot & \cdot & \cdot & \cdots & \cdot \\ b_{n1} & \cdots & b_{nn} & 0 & 0 & \cdots & -1 \\ b_{11}a_{11} + b_{21}a_{12} + \cdots + b_{n1}a_{1n} & \cdots & b_{1n}a_{11} + \cdots + b_{nn}a_{1n} & 0 & 0 & \cdots & 0 \\ b_{11}a_{21} + b_{21}a_{22} + \cdots + b_{n1}a_{2n} & \cdots & b_{1n}a_{21} + \cdots + b_{nn}a_{2n} & 0 & 0 & \cdots & 0 \\ \cdots & \cdots & \cdots & \cdot & \cdot & \cdots & \cdot \\ b_{11}a_{n1} + b_{21}a_{n2} + \cdots + b_{n1}a_{nn} & \cdots & b_{1n}a_{n1} + \cdots + b_{nn}a_{nn} & 0 & 0 & \cdots & 0 \end{vmatrix},$$

and hence, by Laplace's theorem Sec. (1.81)

$$D = (-1)^{1+2+\cdots+2n} \begin{vmatrix} -1 & 0 & \cdots & 0 & b_{11}a_{11} + \cdots + b_{n1}a_{1n} & \cdots & b_{1n}a_{11} + \cdots + b_{nn}a_{1n} \\ 0 & -1 & \cdots & 0 & b_{11}a_{21} + \cdots + b_{n1}a_{2n} & \cdots & b_{1n}a_{21} + \cdots + b_{nn}a_{2n} \\ \cdot & \cdot & \cdots & \cdot & \cdots & \cdots & \cdots \\ 0 & 0 & \cdots & -1 & b_{11}a_{n1} + \cdots + b_{n1}a_{nn} & \cdots & b_{1n}a_{n1} + \cdots + b_{nn}a_{nn} \end{vmatrix}$$

$$= \begin{vmatrix} a_{11}b_{11} + \cdots + a_{1n}b_{n1} & \cdots & a_{11}b_{1n} + \cdots + a_{1n}b_{nn} \\ a_{21}b_{11} + \cdots + a_{2n}b_{n1} & \cdots & a_{21}b_{1n} + \cdots + a_{2n}b_{nn} \\ \cdots & \cdots & \cdots \\ a_{n1}b_{11} + \cdots + a_{nn}b_{n1} & \cdots & a_{n1}b_{1n} + \cdots + a_{nn}b_{nn} \end{vmatrix} = \det(AB). \quad (23)$$

Comparing (22) and (23), we get (21), thereby proving the theorem.

A square matrix A is said to be *nonsingular* if $\det A \neq 0$ and *singular* if $\det A = 0$. It follows from (21) that if the matrices A and B are nonsingular, then so is the product matrix AB, while if at least one of the matrices A and B is singular, then so is AB. These conclusions can also be deduced from Theorem 4.65 and Corollary 4.67.

4.76. The inverse operator. In keeping with the definition given in Sec. 4.68, an operator $\mathbf{B}$ acting in a space $\mathbf{X}$ is called a *left inverse* of the operator $\mathbf{A}$ acting in the same space $\mathbf{X}$ if

$$\mathbf{BA} = \mathbf{E},$$

where $\mathbf{E}$ is the identity operator. The operator $\mathbf{A}$ is then called a *right inverse* of the operator $\mathbf{B}$.

a. It is possible for an operator **A** to have many left inverses and no right inverses at all (see Problems 25 and 26) or, conversely, many right inverses and no left inverses at all. However, suppose **A** has both a left inverse **P** and a right inverse **Q**, so that

$$\mathbf{P} = \mathbf{PE} = \mathbf{P(AQ)} = \mathbf{(PA)Q} = \mathbf{EQ} = \mathbf{Q}.$$

Fixing **Q**, we see that every left inverse coincides with **P** and hence is uniquely determined. In just the same way, the right inverse **Q** is uniquely determined under these circumstances. The uniquely determined operator $\mathbf{P} = \mathbf{Q}$, which is simultaneously both a left and a right inverse of the operator **A**, is called the *inverse* of the operator **A** and is denoted by $\mathbf{A}^{-1}$. The operator **A** itself, with the inverse $\mathbf{A}^{-1}$, is said to be *invertible* (or *nonsingular*).

b. Let **A** be an operator acting in an n-dimensional space $\mathbf{X} = \mathbf{K}_n$, and let A be the matrix of **A** in some fixed basis $e_1, \ldots, e_n$. Then either $\det A \neq 0$ or $\det A = 0$. In the first case, the rank of the matrix A equals n and it follows from Theorem 4.69b that A has both a left and a right inverse. Correspondingly, the operator **A** then has both a left and a right inverse, and hence is invertible. However, if $\det A = 0$, then, by Theorem 4.69b again, the matrix A has neither a left nor a right inverse, and hence the operator **A** acting in $\mathbf{K}_n$ has neither a left nor a right inverse.

4.77. The matrix of the inverse operator. Let **A** be an invertible operator acting in an n-dimensional space **X**, and let $\mathbf{B} = \mathbf{A}^{-1}$ be its inverse. Choosing a basis $e_1, \ldots, e_n$, let $A = \|a_i^{(j)}\|$ and $B = \|b_i^{(j)}\|$ be the matrices of the operators **A** and **B** in this basis.

We now find an explicit formula for the elements $b_i^{(j)}$ in terms of the elements $a_i^{(j)}$. Fixing the row number i, we use formula (8), p. 85 to write down expressions for the elements of the ith row of the matrix $BA = E$:

$$\begin{aligned} b_i^{(1)}a_1^{(1)} + b_i^{(2)}a_2^{(1)} + \cdots + b_i^{(n)}a_n^{(1)} &= 0, \\ &\cdots\cdots \\ b_i^{(1)}a_1^{(i)} + b_i^{(2)}a_2^{(i)} + \cdots + b_i^{(n)}a_n^{(i)} &= 1, \\ &\cdots\cdots \\ b_i^{(1)}a_1^{(n)} + b_i^{(2)}a_2^{(n)} + \cdots + b_i^{(n)}a_n^{(n)} &= 0. \end{aligned}$$

The unknowns $b_i^{(1)}, \ldots, b_i^{(n)}$ can be determined from this system of equations by using Cramer's rule (Sec. 1.73), since $\det A \neq 0$ by hypothesis. Expanding the determinant in the numerator of the resulting expression for $b_i^{(j)}$ with respect to the jth column, we get

$$b_i^{(j)} = \frac{A_j^{(i)}}{\det A}, \tag{24}$$

where $A_j^{(i)}$ is the cofactor of the element $a_j^{(i)}$ in the matrix A. In words, *the element* $b_i^{(j)}$ *of the inverse matrix* A^{-1} *equals the ratio of the cofactor of the element* $a_j^{(i)}$ *of the original matrix* A *to the determinant of* A. Thus we have proved the following.

THEOREM. *Every nonsingular matrix* $A = \|a_i^{(j)}\|$ *has a unique inverse matrix* $B = \|b_i^{(j)}\|$ *such that*

$$AB = BA = E.$$

The elements of the matrix B *are given by formula* (24).

4.78. Let $\mathbf{A}^{-1}$ be the inverse of the operator $\mathbf{A}$, as in Sec. 4.76a. Then by $\mathbf{A}^{-k}$ we mean the operator $(\mathbf{A}^{-1})^k$. It is easily proved by induction that formula (20) continues to hold for negative powers. Powers of the inverse matrix are defined in just the same way, and then the validity of the formula

$$A^{m+n} = A^m A^n \qquad (m, n = 1, 2, \ldots)$$

for negative powers of matrices is an immediate consequence of the validity of (20) for negative powers of operators.

4.8. Invariant Subspaces

4.81. Given a linear operator $\mathbf{A}$ acting in a linear space $\mathbf{K}$, we say that a subspace $\mathbf{K}' \subset \mathbf{K}$ is *invariant with respect to* (or *under*) $\mathbf{A}$ if $x \in \mathbf{K}'$ implies $\mathbf{A}x \in \mathbf{K}'$. In particular, the trivial subspaces, i.e., the whole space and the space whose only element is the zero vector, are invariant with respect to every linear operator. Naturally, we will be interested only in nontrivial invariant subspaces.

4.82. The linear operators given in the examples of Sec. 4.72 will now be examined from this point of view.

a–c. *Every subspace is invariant* with respect to the operators of Examples 4.72a–c (the zero operator, the identity operator, and the similarity operator).

d. The rotation operator in the plane (Example 4.72d) *has no nontrivial invariant subspaces*, unless the angle of rotation equals $m\pi$ where m is an integer (in which case, every one-dimensional subspace is invariant).

e. The projection operator (Example 4.72e) has the following invariant subspaces (among others): The subspace $\mathbf{K}'$ of vectors

$$x = \sum_{k=1}^{m} \xi_k e_k$$

which remain unchanged and the subspace $\mathbf{K}''$ of vectors

$$y = \sum_{k=m+1}^{n} \xi_k e_k$$

which are carried into zero.

f. Every subspace spanned by some of the basis vectors $e_1, e_2, \ldots, e_n$ is invariant under a diagonal operator (Example 4.72f).

4.83. Suppose an operator $\mathbf{A}$ acting in an n-dimensional space $\mathbf{K}_n$ has an invariant m-dimensional subspace $\mathbf{K}_m$. Choose a basis $e_1, \ldots, e_n$ for $\mathbf{K}_n$ such that the first m vectors $e_1, \ldots, e_m$ lie in $\mathbf{K}_m$. Then

$$\begin{aligned} \mathbf{A}e_1 &= a_1^{(1)}e_1 + \cdots + a_m^{(1)}e_m, \\ &\cdots\cdots\cdots\cdots\cdots \\ \mathbf{A}e_m &= a_1^{(m)}e_1 + \cdots + a_m^{(m)}e_m, \end{aligned}$$

and hence the matrix of the operator $\mathbf{A}$ is of the form

$$A = \left\| \begin{matrix} a_1^{(1)} & \cdots & a_1^{(m)} & a_1^{(m+1)} & \cdots & a_1^{(n)} \\ \cdot & \cdots & \cdot & \cdot & \cdots & \cdot \\ a_m^{(1)} & \cdots & a_m^{(m)} & a_m^{(m+1)} & \cdots & a_m^{(n)} \\ 0 & \cdots & 0 & a_{m+1}^{(m+1)} & \cdots & a_{m+1}^{(n)} \\ \cdot & \cdots & \cdot & \cdot & \cdots & \cdot \\ 0 & \cdots & 0 & a_n^{(m+1)} & \cdots & a_n^{(n)} \end{matrix} \right\| \qquad (25)$$

in the given basis. Note that all the elements in the first m columns of this matrix vanish if they appear in rows $m + 1$ through n. Conversely, if the matrix of an operator $\mathbf{A}$ is of the form (25), then the subspace spanned by the vectors $e_1, \ldots, e_m$ is invariant under $\mathbf{A}$.

4.84. Suppose the space $\mathbf{K}_n$ can be represented as a direct sum of invariant subspaces $\mathbf{E}, \mathbf{F}, \ldots, \mathbf{H}$ (see Sec. 2.45), and choose a basis for $\mathbf{K}_n$ such that the vectors

$$\begin{aligned} &e_1, \ldots, e_r \text{ lie in } \mathbf{E}, \\ &f_1, \ldots, f_s \text{ lie in } \mathbf{F}, \\ &\cdots\cdots\cdots\cdots \\ &h_1, \ldots, h_u \text{ lie in } \mathbf{H}. \end{aligned}$$

Then the matrix of the operator $\mathbf{A}$ has the quasi-diagonal form

$$\left\| \begin{matrix} \boxed{A_{(e)}} & & & & \\ & \boxed{A_{(f)}} & & & \\ & & \cdot & & \\ & & & \cdot & \\ & & & & \cdot \\ & & & & & \boxed{A_{(h)}} \end{matrix} \right\|, \qquad (26)$$

where the square matrices $A_{(e)}, A_{(f)}, \ldots, A_{(h)}$ along the diagonal are made up of elements $a_i^{(j)}, b_i^{(j)}, \ldots, d_i^{(j)}$ in accordance with the formulas†

$$\mathbf{A}e_j = \sum_{i=1}^{r} a_i^{(j)} e_i,$$

$$\mathbf{A}f_j = \sum_{i=1}^{s} b_i^{(j)} f_i,$$

$$\cdots\cdots$$

$$\mathbf{A}h_j = \sum_{i=1}^{u} d_i^{(j)} h_i,$$

while all the elements outside the matrices $A_{(e)}, A_{(f)}, \ldots, A_{(h)}$ vanish. Conversely, if the matrix of an operator $\mathbf{A}$ is of the form (26) in some basis, then the space $\mathbf{K}_n$ can be represented as the direct sum of the invariant subspaces spanned by the corresponding groups of basis vectors.

4.9. Eigenvectors and Eigenvalues

4.91. A special role is played by the one-dimensional invariant subspaces of a given operator $\mathbf{A}$; they are also called *invariant directions* (or *eigenrays*). Every (nonzero) vector belonging to a one-dimensional invariant subspace of the operator $\mathbf{A}$ is called an *eigenvector* of $\mathbf{A}$. In other words, a vector $x \neq 0$ is called an *eigenvector* of the operator $\mathbf{A}$ if $\mathbf{A}$ carries x into a collinear vector, i.e., if

$$\mathbf{A}x = \lambda x.$$

The number λ appearing in (27) is called the *eigenvalue* (or *characteristic value*) of the operator $\mathbf{A}$, corresponding to the eigenvector x.

4.92. We now reexamine the examples of Sec. 4.72 from this standpoint.

a–c. In Examples 4.72a–c, every nonzero vector of the space is an eigenvector and the corresponding eigenvalues 0, 1, λ.

d. The rotation operator (Example 4.72d) has no eigenvectors unless the angle of rotation equals $m\pi$ where m is an integer.

e. The projection operator (Example 4.72e) has eigenvectors of the form

$$x = \sum_{k=1}^{m} \xi_k e_k$$

and

$$y = \sum_{k=m+1}^{n} \xi_k e_k,$$

† Cf. formula (18), p. 98.

with corresponding eigenvalues 1 and 0. It can be verified that the projection operator has no other eigenvectors.

f. The diagonal operator (Example 4.72f) by its very definition has the eigenvectors $e_1, e_2, \ldots, e_n$ with corresponding eigenvalues $\lambda_1, \lambda_2, \ldots, \lambda_n$.

4.93. Next we prove two simple properties of eigenvectors.

a. LEMMA. *Given an operator* $\mathbf{A}$ *with eigenvectors* $x_1, x_2, \ldots, x_m$ *and corresponding eigenvalues* $\lambda_1, \lambda_2, \ldots, \lambda_m$, *suppose* $\lambda_i \neq \lambda_j$ *whenever* $i \neq j$. *Then the eigenvectors* $x_1, x_2, \ldots, x_m$ *are linearly independent.*

Proof. We prove this assertion by induction on the integer m. Obviously, the lemma is true for $m = 1$. Assuming that the lemma is true for any $m - 1$ eigenvectors of the operator $\mathbf{A}$, we now show that it remains true for any m eigenvectors of $\mathbf{A}$. In fact, assume to the contrary that $x_1, x_2, \ldots, x_m$ are linearly dependent, so that there is a linear relation

$$\alpha_1 x_1 + \alpha_2 x_2 + \cdots + \alpha_m x_m = 0$$

between the eigenvectors $x_1, x_2, \ldots, x_m$, with $\alpha_1 \neq 0$, say. Applying the operator $\mathbf{A}$ to this relation, we get

$$\alpha_1 \lambda_1 x_1 + \alpha_2 \lambda_2 x_2 + \cdots + \alpha_m \lambda_m x_m = 0.$$

Multiplying the first equation by λ_m and then subtracting it from the second equation, we find that

$$\alpha_1(\lambda_1 - \lambda_m)x_1 + \alpha_2(\lambda_2 - \lambda_m)x_2 + \cdots + \alpha_{m-1}(\lambda_{m-1} - \lambda_m)x_{m-1} = 0,$$

which by the induction hypothesis implies that all the coefficients

$$\alpha_1(\lambda_1 - \lambda_m), \alpha_2(\lambda_2 - \lambda_m), \ldots, \alpha_{m-1}(\lambda_{m-1} - \lambda_m)$$

vanish, in particular that

$$\alpha_1(\lambda_1 - \lambda_m) = 0,$$

contrary to the assumption that $\alpha_1 \neq 0$, $\lambda_1 \neq \lambda_m$. This contradiction shows that the eigenvectors $x_1, x_2, \ldots, x_m$ must be linearly independent. ▌

In particular, *a linear operator* $\mathbf{A}$ *acting in an n-dimensional space cannot have more than n eigenvectors with distinct eigenvalues.*

b. LEMMA. *The eigenvectors of a linear operator* $\mathbf{A}$ *corresponding to a given eigenvalue* λ *span a subspace* $\mathbf{K}^{(\lambda)} \subset \mathbf{K}$.

Proof. If

$$\mathbf{A}x_1 = \lambda x_1, \qquad \mathbf{A}x_2 = \lambda x_2,$$

then

$$\mathbf{A}(\alpha x_1 + \beta x_2) \quad \alpha \mathbf{A}x_1 + \beta \mathbf{A}x_2 = \alpha\lambda x_1 + \beta\lambda x_2 = \lambda(\alpha x_1 + \beta x_2).$$ ▌

The subspace $\mathbf{K}^{(\lambda)}$ is called the *eigenspace* (or *characteristic space*) of the operator $\mathbf{A}$, corresponding to the eigenvalue λ.

4.94. Next we show how to calculate the components of the eigenvectors of an operator $\mathbf{A}$, where $\mathbf{A}$ is specified by its matrix in some basis $e_1, e_2, \ldots, e_n$ of the space $\mathbf{K}_n$. Suppose the vector

$$x = \sum_{k=1}^{n} \xi_k e_k$$

is an eigenvector of $\mathbf{A}$, so that

$$\mathbf{A}x = \lambda x \tag{27}$$

for some λ. Using (5), p. 80, we can write (27) in component form as

$$\begin{aligned} a_1^{(1)}\xi_1 + a_1^{(2)}\xi_2 + \cdots + a_1^{(n)}\xi_n &= \lambda\xi_1, \\ a_2^{(1)}\xi_1 + a_2^{(2)}\xi_2 + \cdots + a_2^{(n)}\xi_n &= \lambda\xi_2, \\ \cdots\cdots\cdots\cdots\cdots\cdots\cdots\cdots \\ a_n^{(1)}\xi_1 + a_n^{(2)}\xi_2 + \cdots + a_n^{(n)}\xi_n &= \lambda\xi_n \end{aligned}$$

or

$$\begin{aligned} (a_1^{(1)} - \lambda)\xi_1 + a_1^{(2)}\xi_2 + \cdots + a_1^{(n)}\xi_n &= 0, \\ a_2^{(1)}\xi_1 + (a_2^{(2)} - \lambda)\xi_2 + \cdots + a_2^{(n)}\xi_n &= 0, \\ \cdots\cdots\cdots\cdots\cdots\cdots\cdots\cdots \\ a_n^{(1)}\xi_1 + a_n^{(2)}\xi_2 + \cdots + (a_n^{(n)} - \lambda)\xi_n &= 0. \end{aligned} \tag{28}$$

This homogeneous system of equations in the unknowns $\xi_1, \xi_2, \ldots, \xi_n$ has a nontrivial solution if and only if its determinant vanishes (see Sec. 3.22):

$$\Delta(\lambda) \equiv \begin{vmatrix} a_1^{(1)} - \lambda & a_1^{(2)} & \cdots & a_1^{(n)} \\ a_2^{(1)} & a_2^{(2)} - \lambda & \cdots & a_2^{(n)} \\ \cdot & \cdot & \cdots & \cdot \\ a_n^{(1)} & a_n^{(2)} & \cdots & a_n^{(n)} - \lambda \end{vmatrix} = 0. \tag{29}$$

The polynomial $\Delta(\lambda)$ of degree n in λ is called the *characteristic polynomial of the matrix A*.† To each of its roots $\lambda_0 \in K$ there corresponds an eigenvector of the operator $\mathbf{A}$ obtained by substituting λ_0 for λ in (28) and then solving the resulting compatible system for the quantities $\xi_1, \xi_2, \ldots, \xi_n$. Moreover, λ_0 is obviously the eigenvalue corresponding to this eigenvector. In particular, it follows that although the matrix of the operator $\mathbf{A}$ depends on the choice of the basis $e_1, e_2, \ldots, e_n$, the roots of the characteristic polynomial of the

† Correspondingly, equation (29) itself is called the *characteristic equation* of A.

matrix no longer depend on the choice of basis. We will discuss this matter further in Sec. 5.53.

4.95. We now study the various possibilities which can occur in solving the characteristic equation (29).

a. *The case of no roots in the field K.* If equation (29) has no roots at all in the field K, then the linear operator $\mathbf{A}$ has no eigenvectors in the space $\mathbf{K}_n$. For example, as already noted, the rotation operator in the plane V_2 corresponding to rotation through an angle

$$\varphi_0 \neq m\pi \qquad (m = 0, \pm 1, \pm 2, \ldots) \tag{30}$$

has no eigenvectors. This fact, which is geometrically obvious, is easily proved algebraically. Indeed, for the rotation operator, equation (29) takes the form

$$\begin{vmatrix} \cos\varphi_0 - \lambda & -\sin\varphi_0 \\ \sin\varphi_0 & \cos\varphi_0 - \lambda \end{vmatrix} = 0$$

(see Example 4.72d), which becomes

$$1 - 2\lambda\cos\varphi_0 + \lambda^2 = 0$$

after calculating the determinant. But this equation has no real roots if (30) holds.

b. If $K = C$ is the field of complex numbers, then by the fundamental theorem of algebra, equation (29) *always* has a root $\lambda_0 \in K$. Thus *in the space* $\mathbf{C}_n$ *every linear operator has at least one eigenvector.*

c. *The case of n distinct roots.* If all n roots of equation (29) lie in the field K and are distinct, we can find n distinct eigenvectors of the operator $\mathbf{A}$ in the space $\mathbf{K}_n$ by solving the system (28) for $\lambda = \lambda_1, \lambda_2, \ldots, \lambda_n$ in turn. By Lemma 4.93a, the eigenvectors $f_1, f_2, \ldots, f_n$ so obtained are linearly independent. Choosing them as a new basis, we can construct the matrix of the operator $\mathbf{A}$ in this basis. Since

$$\begin{aligned} \mathbf{A}f_1 &= \lambda_1 f_1, \\ \mathbf{A}f_2 &= \qquad \lambda_2 f_2, \\ &\cdots\cdots\cdots\cdots \\ \mathbf{A}f_n &= \qquad\qquad \lambda_n f_n, \end{aligned}$$

the matrix $A_{(f)}$ has the form

$$\begin{Vmatrix} \lambda_1 & 0 & \cdots & 0 \\ 0 & \lambda_2 & \cdots & 0 \\ \cdot & \cdot & \cdots & \cdot \\ 0 & 0 & \cdots & \lambda_n \end{Vmatrix}. \tag{31}$$

Recalling the definition of a diagonalizable operator (see Example 4.72f), we can formulate this result as follows: *Let* **A** *be an operator in the space* $\mathbf{K}_n$, *whose matrix (in any basis) has a characteristic polynomial with n distinct roots in the field K. Then* **A** *is diagonalizable. The matrix of* **A** *in the basis consisting of its eigenvectors is diagonal, with diagonal elements equal to the eigenvalues of* **A**.

d. On the other hand, if the operator **A** has a diagonal matrix of the form (31) in some basis $f_1, f_2, \ldots, f_n$ of the space $\mathbf{K}_n$ with arbitrary, not necessarily distinct numbers $\lambda_1, \lambda_2, \ldots, \lambda_n$ along the diagonal, then the vectors $f_1, f_2, \ldots, f_n$ are eigenvectors of **A** and the numbers $\lambda_1, \lambda_2, \ldots, \lambda_n$ are the corresponding eigenvalues.

To see that **A** has no eigenvalues other than $\lambda_1, \lambda_2, \ldots, \lambda_n$, suppose λ is an eigenvalue of **A** corresponding to the eigenvector

$$f = \sum_{i=1}^{n} \beta_i f_i,$$

so that $\mathbf{A}f = \lambda f$. Then, comparing coefficients of f_i in the equations

$$\mathbf{A}f = \mathbf{A}\left(\sum_{i=1}^{n} \beta_i f_i\right) = \sum_{i=1}^{n} \beta_i \mathbf{A} f_i = \sum_{i=1}^{n} \beta_i \lambda_i f_i,$$

$$\lambda f = \lambda \sum_{i=1}^{n} \beta_i f_i = \sum_{i=1}^{n} \lambda \beta_i f_i,$$

we get

$$\lambda\beta_i = \lambda_i \beta_i \qquad (i = 1, 2, \ldots, n). \tag{32}$$

But at least one of the numbers $\beta_1, \beta_2, \ldots, \beta_n$ is nonzero, say $\beta_1 \neq 0$. Thus, choosing $i = 1$ in (32), we find that $\lambda = \lambda_1$, i.e., λ is already one of the numbers $\lambda_1, \lambda_2, \ldots, \lambda_n$.

e. *The case of multiple roots.* Let $\lambda = \lambda_0$ be a root of multiplicity $r \geqslant 1$ of the characteristic equation (29). The following question then arises: What is the dimension of the corresponding eigenspace $\mathbf{K}^{(\lambda_0)}$, or in other words, how many linearly independent solutions does the system (28) have for $\lambda = \lambda_0$? This question can be answered exactly from a knowledge of the rank of the matrix of the system (28), but we would like an answer which involves only the multiplicity r of the root λ_0.

In Examples 4.72a–c and 4.72e, it is easily verified that the dimension of each eigenspace $\mathbf{K}^{(\lambda_0)}$ is the same as the multiplicity of λ_0 as a root of the characteristic equation of the given operator. However, this is not true in general. For example, let **A** be the operator in $\mathbf{R}_2$ with matrix

$$A = \begin{Vmatrix} \lambda_0 & 0 \\ \mu & \lambda_0 \end{Vmatrix},$$

where $\mu \neq 0$ is arbitrary. Here the characteristic polynomial is $(\lambda_0 - \lambda)^2$ and has a double root $\lambda = \lambda_0$. Correspondingly, the system (28) takes the form

$$0 \cdot \xi_1 + 0 \cdot \xi_2 = 0,$$
$$\mu \cdot \xi_1 + 0 \cdot \xi_2 = 0,$$

which, to within a numerical factor, has the unique solution

$$\xi_1 = 0, \qquad \xi_2 = 1.$$

Thus the eigenspace of the operator **A** corresponding to the eigenvalue λ_0 has dimension 1, which is less than the multiplicity of the root λ_0.

It can be shown that in the general case the dimension of the eigenspace $\mathbf{K}^{(\lambda_0)}$ does not exceed the multiplicity of the root λ_0 (see Chapter 5, Problem 7). A complete solution to the problem of finding the dimension of the space $\mathbf{K}^{(\lambda_0)}$ for the case $K = C$ will be given in Chapter 6, after showing how to determine the "canonical form" of the matrix of the given operator.

PROBLEMS

1. After defining in a natural way addition of linear forms and multiplication of a linear form by a real number, construct a new linear space $\mathbf{K}^*$ consisting of all the linear forms defined on some linear space **K**. If the dimension of the space **K** is n, what is the dimension of the space $\mathbf{K}^*$?

2. Which of the following vector functions defined on the space V_3 are linear operators:

a) $\mathbf{A}x = x + a$ (a is a fixed nonzero vector);
b) $\mathbf{A}x = a$;
c) $\mathbf{A}x = (a, x)a$;†
d) $\mathbf{A}x = (a, x)x$;
e) $\mathbf{A}x = (\xi_1^2, \xi_2 + \xi_3, \xi_3^2)$, where $x = (\xi_1, \xi_2, \xi_3)$;
f) $\mathbf{A}x = (\sin \xi_1, \cos \xi_2, 0)$;
g) $\mathbf{A}x = (2\xi_1 - \xi_3, \xi_2 + \xi_3, \xi_1)$?

3. Consider the following operations in the space of all polynomials in t:

a) Multiplication by t;
b) Multiplication by t^2;
c) Differentiation.

Are these linear operators?

4. Suppose the operator **A** defined on V_3 carries the vectors

$$x_1 = (0, 0, 1), \qquad x_2 = (0, 1, 1), \qquad x_3 = (1, 1, 1)$$

---

† Here (a, x) denotes the usual scalar product of the vectors a and x, i.e., the number equal to the product of the lengths of the vectors and the cosine of the angle between them.

into the vectors

$$y_1 = (2, 3, 5), \qquad y_2 = (1, 0, 0), \qquad y_3 = (0, 1, -1).$$

Form the matrix of **A** in the following bases:

a) $e_1 = (1, 0, 0)$, $e_2 = (0, 1, 0)$, $e_3 = (0, 0, 1)$;

b) x_1, x_2, x_3.

5. In three-dimensional space let **A** denote the operator corresponding to rotation through 90° about the axis OX (taking OY into OZ), let **B** denote the operator corresponding to rotation through 90° about the axis OY (taking OZ into OX), and let **C** denote the operator corresponding to rotation through 90° about OZ (taking OX into OY). Show that

$$\mathbf{A}^4 = \mathbf{B}^4 = \mathbf{C}^4 = \mathbf{E}, \qquad \mathbf{AB} \neq \mathbf{BA}, \qquad \mathbf{A}^2\mathbf{B}^2 = \mathbf{B}^2\mathbf{A}^2.$$

Is the relation $\mathbf{ABAB} = \mathbf{A}^2\mathbf{B}^2$ valid?

6. In the space of all polynomials in t, let **A** denote the differentiation operator and let **B** denote the operator corresponding to multiplication by the independent variable t, so that

$$\mathbf{A}P(t) = P'(t), \qquad \mathbf{B}P(t) = tP(t).$$

Is the relation $\mathbf{AB} = \mathbf{BA}$ valid? Find the operator $\mathbf{AB} - \mathbf{BA}$.

7. Assuming that $\mathbf{AB} = \mathbf{BA}$, prove the formulas

$$(\mathbf{A} + \mathbf{B})^2 = \mathbf{A}^2 + 2\mathbf{AB} + \mathbf{B}^2,$$

$$(\mathbf{A} + \mathbf{B})^3 = \mathbf{A}^3 + 3\mathbf{A}^2\mathbf{B} + 3\mathbf{AB}^2 + \mathbf{B}^3.$$

How must these formulas be changed if $\mathbf{AB} \neq \mathbf{BA}$?

8. Assuming that $\mathbf{AB} - \mathbf{BA} = \mathbf{E}$, prove the formula

$$\mathbf{A}^m\mathbf{B} - \mathbf{BA}^m = m\mathbf{A}^{m-1} \qquad (m = 1, 2, \ldots).$$

9. Find the dimension of the linear space $\mathbf{K}_n^m$ of all linear operators mapping an n-dimensional space $\mathbf{K}_n$ into an m-dimensional space $\mathbf{K}_m$, and construct a basis for $\mathbf{K}_n^m$.

10. Find the product AB of the matrices A and B, where

$$A = \begin{Vmatrix} 1 & 2 & 3 \\ 2 & 4 & 6 \\ 3 & 6 & 9 \end{Vmatrix}, \qquad B = \begin{Vmatrix} -1 & -2 & -4 \\ -1 & -2 & -4 \\ 1 & 2 & 4 \end{Vmatrix}.$$

11. Raise the following matrices to the nth power:

$$A = \begin{Vmatrix} 1 & 1 \\ 0 & 1 \end{Vmatrix}, \qquad B = \begin{Vmatrix} \cos\varphi & -\sin\varphi \\ \sin\varphi & \cos\varphi \end{Vmatrix}.$$

12. Find all matrices A of order two satisfying the condition

$$A^2 = \begin{Vmatrix} 0 & 0 \\ 0 & 0 \end{Vmatrix}.$$

13. Calculate $AB - BA$ where

a) $$A = \begin{Vmatrix} 1 & 2 & 2 \\ 2 & 1 & 2 \\ 1 & 2 & 3 \end{Vmatrix}, \qquad B = \begin{Vmatrix} 4 & 1 & 1 \\ -4 & 2 & 0 \\ 1 & 2 & 1 \end{Vmatrix};$$

b) $$A = \begin{Vmatrix} 2 & 1 & 0 \\ 1 & 1 & 2 \\ -1 & 2 & 1 \end{Vmatrix}, \qquad B = \begin{Vmatrix} 3 & 1 & -2 \\ 3 & -2 & 4 \\ -3 & 5 & -1 \end{Vmatrix}.$$

14. The sum $a_{11} + \cdots + a_{nn}$ of the diagonal elements of a matrix $A = \|a_{jk}\|$ is called the *trace* of A, denoted by tr A. Prove that

$$\operatorname{tr}(A + B) = \operatorname{tr} A + \operatorname{tr} B,$$
$$\operatorname{tr}(AB) = \operatorname{tr}(BA).$$

15. Prove that the formula $\mathbf{AB} - \mathbf{BA} = \mathbf{E}$ is impossible for operators $\mathbf{A}$ and $\mathbf{B}$ acting on an n-dimensional space $\mathbf{K}_n$.

Comment. The result of Problem 6 shows that the assumption that the space $\mathbf{K}_n$ is finite-dimensional plays an essential role here.

16. Given a square matrix C of order two such that tr $C = 0$ (cf. Problem 14), show that C can be represented in the form

$$C = AB - BA$$

where A and B are (unknown) matrices of order two.

17. Let

$$x_j = \sum_{i=1}^{n} \xi_i^{(j)} e_i \qquad (j = 1, 2, \ldots, m)$$

be m linearly independent vectors in an n-dimensional space, and let $\mathbf{A}$ be the operator defined on the linear manifold $\mathbf{L}(x_1, x_2, \ldots, x_m)$ such that

$$y_j = \mathbf{A}x_j = \sum_{k=1}^{m} a_k^{(j)} x_k \qquad (j = 1, 2, \ldots, m).$$

Show that every minor of order m of the matrix made up of the components of y_j (with respect to the basis $e_1, e_2, \ldots, e_n$) equals the product of det $\|a_k^{(j)}\|$ with the corresponding minor of the matrix made up of the components of the vectors x_j.

18. Show that if the basis minor of a matrix of rank r appears in the upper left-hand corner, then the ratio of any minor M of order r to the minor appearing in the same columns as M but in the first r rows depends only on the column indices of the minor M.

19. Show that if A is a matrix of rank r, then any second-order determinant of the form

$$\begin{vmatrix} M^{i_1,i_2,\ldots,i_r}_{i_1,i_2,\ldots,i_r} & M^{i_1,i_2,\ldots,i_r}_{k_1,k_2,\ldots,k_r} \\ M^{k_1,k_2,\ldots,k_r}_{i_1,i_2,\ldots,i_r} & M^{k_1,k_2,\ldots,k_r}_{k_1,k_2,\ldots,k_r} \end{vmatrix},$$

consisting of minors of order r of the matrix A, vanishes.

20. Show that every minor of order k of the matrix ABC equals a sum of products of certain minors of order k of the matrices A, B and C.

21. Find the inverses of the following matrices:

$$A = \begin{Vmatrix} 1 & 2 \\ 2 & 5 \end{Vmatrix}, \qquad B = \begin{Vmatrix} 1 & 2 & -3 \\ 0 & 1 & 2 \\ 0 & 0 & 1 \end{Vmatrix}, \qquad C = \begin{Vmatrix} \frac{1}{2} & \frac{1}{2} & \frac{1}{2} & \frac{1}{2} \\ \frac{1}{2} & \frac{1}{2} & -\frac{1}{2} & -\frac{1}{2} \\ \frac{1}{2} & -\frac{1}{2} & \frac{1}{2} & -\frac{1}{2} \\ \frac{1}{2} & -\frac{1}{2} & -\frac{1}{2} & \frac{1}{2} \end{Vmatrix}.$$

22. Prove that

$$(A')^{-1} = (A^{-1})'$$

for any nonsingular matrix A.

23. Find all solutions of the equation $XA = 0$, where A is a given second-order matrix, X is an unknown second-order matrix and 0 is the zero matrix (the matrix all of whose elements vanish).

24. Let $A = \|a_i^{(j)}\|$ be any square matrix of order n, and let $A_i^{(j)}$ be the cofactor of the element $a_i^{(j)}$ in the determinant of A. The matrix $\tilde{A} = \|A_j^{(i)}\|$ is called the *adjugate* of the matrix A. Prove that

$$\tilde{A}A = A\tilde{A} = (\det A)E.$$

25. In the space of all polynomials in the variable t, consider the operators **A** and **B** defined by the relations

$$\mathbf{A}[a_0 + a_1 t + \cdots + a_n t^n] = a_1 + a_2 t + \cdots + a_n t^{n-1},$$
$$\mathbf{B}[a_0 + a_1 t + \cdots + a_n t^n] = a_0 t + a_1 t^2 + \cdots + a_n t^{n+1}.$$

Show that **A** and **B** are linear operators and that

$$\mathbf{AB} = \mathbf{E}, \qquad \mathbf{BA} \neq \mathbf{E}.$$

Does the operator **A** have an inverse?

26. Show that the operator **B** of Problem 25 has infinitely many left inverses.

27. Prove that if **A** is a nonsingular linear operator acting in an n-dimensional linear space, then every subspace invariant under **A** is also invariant under $\mathbf{A}^{-1}$.

28. Prove that if the linear operators **A** and **B** commute (i.e., if $\mathbf{AB} = \mathbf{BA}$), then every eigenspace of the operator **A** is an invariant subspace of the operator **B**.

29. Prove that if a direct sum (Sec. 2.45) of eigenspaces of an operator **A** coincides with the whole space **K** and if each eigenspace of the operator **A** is invariant under an operator **B**, then **A** and **B** commute.

30. Let x and y be eigenvectors of the operator **A** corresponding to *distinct* eigenvalues. Show that $\alpha x + \beta y$ $(\alpha \neq 0, \beta \neq 0)$ cannot be an eigenvector of **A**.

31. Prove that if every vector of the space **K** is an eigenvector of the operator **A**, then $\mathbf{A} = \lambda\mathbf{E}$ $(\lambda \in K)$.

32. Prove that if the linear operator **A** commutes with *all* linear operators acting in the given space, then $\mathbf{A} = \lambda\mathbf{E}$.

33. Let the linear operator **A** have the eigenvector e_0, with eigenvalue λ_0. Show that e_0 is also an eigenvector of the operator $\mathbf{A}^2$, with eigenvalue λ_0^2.

34. Even if a linear operator **A** has no eigenvectors, the operator $\mathbf{A}^2$ may have eigenvectors (e.g., the operator corresponding to rotation through 90° in the plane). Show that if the operator $\mathbf{A}^2$ has an eigenvector with a *nonnegative* eigenvalue $\lambda = \mu^2$, then the operator **A** also has an eigenvector.

35. Find the eigenvalues and eigenvectors of the operators given by the following matrices:

$$\text{a)} \begin{Vmatrix} 2 & -1 & -1 \\ 0 & -1 & 0 \\ 0 & 2 & 1 \end{Vmatrix}; \qquad \text{b)} \begin{Vmatrix} -1 & -2 & 2 \\ 0 & 1 & 0 \\ 0 & 0 & 1 \end{Vmatrix};$$

$$\text{c)} \begin{Vmatrix} 2 & -1 & 0 \\ 0 & 1 & -1 \\ 0 & 1 & 3 \end{Vmatrix}; \qquad \text{d)} \begin{Vmatrix} 0 & 0 & 1 & -1 \\ -1 & 0 & 1 & -1 \\ 0 & 0 & 0 & 0 \\ 0 & 0 & 0 & 1 \end{Vmatrix}.$$

36. Verify the following facts:

a) The relation $\mathbf{N}(\mathbf{A}) \supset \mathbf{T}(\mathbf{A})$ is necessary and sufficient for the equality $\mathbf{A}^2 = 0$ to hold;

b) $\mathbf{N}(\mathbf{A}) \subset \mathbf{N}(\mathbf{A}^2) \subset \mathbf{N}(\mathbf{A}^3) \subset \cdots$ for any operator **A**;

c) $\mathbf{T}(\mathbf{A}) \supset \mathbf{T}(\mathbf{A}^2) \supset \mathbf{T}(\mathbf{A}^3) \supset \cdots$ for any operator **A**;

d) If $\mathbf{T}(\mathbf{A}^k) \supset \mathbf{N}(\mathbf{A}^m)$, then

$$\mathbf{T}(\mathbf{A}) \subset \mathbf{N}(\mathbf{A}^{m+k-1}), \qquad \mathbf{T}(\mathbf{A}^{m+k-1}) \subset \mathbf{N}(\mathbf{A}).$$

37. Show that every linear operator **A** of rank r can be represented as the sum of r linear operators of rank one.

38. Find all the invariant subspaces of a diagonal operator with n distinct diagonal elements, and show that there are 2^n such subspaces.

chapter 5

COORDINATE TRANSFORMATIONS

As is well known, in solving geometric problems by the methods of analytic geometry a very important role is played by the proper choice of a coordinate system. Proper choice of a coordinate system also plays a very important role in a much wider class of problems connected with the geometry of n-dimensional linear spaces. This chapter is devoted to a study of the rules governing coordinate transformations in n-dimensional spaces. In particular, the results obtained here are fundamental for the classification of quadratic forms which will be made in Chapter 7.

5.1. Transformation to a New Basis

5.11. Let

$$\{e\} = \{e_1, e_2, \ldots, e_n\}$$

be a basis in an n-dimensional space $\mathbf{K}_n$, and let

$$\{f\} = \{f_1, f_2, \ldots, f_n\}$$

be another basis in the same space. The vectors of the system $\{f\}$ are uniquely determined by their expansions in terms of the vectors of the original basis:

$$\begin{aligned} f_1 &= p_1^{(1)} e_1 + p_2^{(1)} e_2 + \cdots + p_n^{(1)} e_n, \\ f_2 &= p_1^{(2)} e_1 + p_2^{(2)} e_2 + \cdots + p_n^{(2)} e_n, \\ &\cdots\cdots\cdots\cdots\cdots\cdots \\ f_n &= p_1^{(n)} e_1 + p_2^{(n)} e_2 + \cdots + p_n^{(n)} e_n, \end{aligned} \tag{1}$$

or, more concisely,

$$f_j = \sum_{i=1}^{n} p_i^{(j)} e_i \qquad (j = 1, 2, \ldots, n). \tag{2}$$

The coefficients $p_i^{(j)}$ $(i, j = 1, 2, \ldots, n)$ in (1) and (2) define a matrix

$$P = \|p_i^{(j)}\| = \begin{Vmatrix} p_1^{(1)} & p_1^{(2)} & \cdots & p_1^{(n)} \\ p_2^{(1)} & p_2^{(2)} & \cdots & p_2^{(n)} \\ \cdot & \cdot & \cdots & \cdot \\ p_n^{(1)} & p_n^{(2)} & \cdots & p_n^{(n)} \end{Vmatrix},$$

called the *matrix of the transformation from the basis* $\{e\}$ *to the basis* $\{f\}$. As was done previously in similar cases (Sec. 4.2 ff.), we write the components of the vectors f_j (with respect to the basis $\{e\}$) as the columns of the matrix P. By the same token, the formulas (1) together with the matrix P specify a corresponding linear operator $\mathbf{P}$, defined by the relations $f_i = \mathbf{P}e_i$ $(i = 1, 2, \ldots, n)$ and called the *operator of the transformation from the basis* $\{e\}$ *to the basis* $\{f\}$.

The determinant D of the matrix P is *nonvanishing*, since otherwise the columns of P, and hence the vectors $f_1, f_2, \ldots, f_n$, would be linearly dependent (Sec. 3.12a). A matrix with a nonvanishing determinant is said to be *nonsingular* (recall Sec. 4.75). Thus *the transformation from one basis of the n-dimensional space* $\mathbf{K}_n$ *to another basis is always accomplished by using a nonsingular matrix.*

5.12. Conversely, let $\{e\} = \{e_1, e_2, \ldots, e_n\}$ be a given basis of the space $\mathbf{K}_n$, and let $P = \|p_i^{(j)}\|$ be a nonsingular matrix of order n. Using the equations (1), construct the system of vectors $f_1, f_2, \ldots, f_n$. It is clear that these vectors are linearly independent, since the columns of every nonsingular matrix are linearly independent (Sec. 3.12a). Consequently, the vectors $f_1, f_2, \ldots, f_n$ form a new basis for the space $\mathbf{K}_n$. Thus *every nonsingular matrix* $P = \|p_i^{(j)}\|$ *determines via* (1) *a transformation from one basis of the n-dimensional space* $\mathbf{K}_n$ *to another basis.*

5.13. Next we note a particular case of a transformation to a new basis, i.e., the case where every vector f_k is just the corresponding vector e_k multiplied by a number $\lambda_k \neq 0$ $(k = 1, 2, \ldots, n)$. Then the equations (1) take the form

$$\begin{aligned} f_1 &= \lambda_1 e_1, \\ f_2 &= \qquad \lambda_2 e_2, \\ &\cdots\cdots\cdots\cdots \\ f_n &= \qquad\qquad \lambda_n e_n, \end{aligned}$$

and the matrix P has the diagonal form

$$P = \begin{Vmatrix} \lambda_1 & 0 & \cdots & 0 \\ 0 & \lambda_2 & \cdots & 0 \\ \cdot & \cdot & \cdots & \cdot \\ 0 & 0 & \cdots & \lambda_n \end{Vmatrix}. \tag{3}$$

In particular, for $\lambda_1 = \lambda_2 = \cdots = \lambda_n = 1$, we obtain the matrix of the *identity transformation*, namely the unit matrix

$$E = \begin{Vmatrix} 1 & 0 & \cdots & 0 \\ 0 & 1 & \cdots & 0 \\ \cdot & \cdot & \cdots & \cdot \\ 0 & 0 & \cdots & 1 \end{Vmatrix}$$

(the original basis is not changed by the identity transformation).

5.2. Consecutive Transformations

5.21. Let $P = \|p_i^{(j)}\|$ be the matrix of the transformation from the basis

$$\{e\} = \{e_1, e_2, \ldots, e_n\}$$

to the basis

$$\{f\} = \{f_1, f_2, \ldots, f_n\},$$

and let $Q = \|q_j^{(k)}\|$ be the matrix of the transformation from the basis $\{f\}$ to the basis

$$\{g\} = \{g_1, g_2, \ldots, g_n\}.$$

We now determine the matrix of the transformation from the basis $\{e\}$ directly to the basis $\{g\}$. By (2), the formula for transforming from the basis $\{e\}$ to the basis $\{f\}$ is

$$f_j = \sum_{i=1}^{n} p_i^{(j)} e_i \qquad (j = 1, 2, \ldots, n), \tag{4}$$

while that for transforming from the basis $\{f\}$ to the basis $\{g\}$ is

$$g_k = \sum_{j=1}^{n} q_j^{(k)} f_j \qquad (k = 1, 2, \ldots, n). \tag{5}$$

Substituting (4) into (5), we obtain

$$\begin{aligned} g_k &= \sum_{j=1}^{n} q_j^{(k)} \sum_{i=1}^{n} p_i^{(j)} e_i \\ &= \sum_{i=1}^{n} \left(\sum_{j=1}^{n} p_i^{(j)} q_j^{(k)} \right) e_i \qquad (k = 1, 2, \ldots, n). \end{aligned} \tag{6}$$

On the other hand, if $T = \|t_i^{(k)}\|$ denotes the matrix of the transformation from the basis $\{e\}$ to the basis $\{g\}$, we can write

$$g_k = \sum_{i=1}^{n} t_i^{(k)} e_i \qquad (k = 1, 2, \ldots, n). \tag{7}$$

Comparison of (6) and (7) gives

$$t_i^{(k)} = \sum_{j=1}^{n} p_i^{(j)} q_j^{(k)} \qquad (i, k = 1, 2, \ldots, n). \tag{8}$$

Recalling formula (8), p. 85 (where the choice of indices is somewhat different, but not their role), we find that *the desired matrix T is the product PQ of the matrices P and Q.*

5.22. Consider the following special case of consecutive transformations. Since the matrix P is nonsingular, the system of equations (1) can be solved for the vectors $e_1, e_2, \ldots, e_n$. The resulting system of equations

$$\begin{aligned} e_1 &= q_1^{(1)} f_1 + q_2^{(1)} f_2 + \cdots + q_n^{(1)} f_n, \\ e_2 &= q_1^{(2)} f_1 + q_2^{(2)} f_2 + \cdots + q_n^{(2)} f_n, \\ &\cdots\cdots\cdots\cdots\cdots\cdots \\ e_n &= q_1^{(n)} f_1 + q_2^{(n)} f_2 + \cdots + q_n^{(n)} f_n \end{aligned} \tag{9}$$

obviously determines the transformation from the basis $\{f\}$ to the basis $\{e\}$. The consecutive transformation from the basis $\{e\}$ to the basis $\{f\}$ by using the matrix P and then from the basis $\{f\}$ to the basis $\{e\}$ by using the matrix $Q = \|q_j^{(k)}\|$ is equivalent to the transformation from the basis $\{e\}$ to itself, i.e., to the identity transformation with unit matrix (3).

5.3. Transformation of the Components of a Vector

5.31. Let $\{e\} = \{e_1, e_2, \ldots, e_n\}$ and $\{f\} = \{f_1, f_2, \ldots, f_n\}$ be two bases in an n-dimensional linear space $\mathbf{K}_n$. Any vector $x \in \mathbf{K}_n$ has the expansions

$$x = \xi_1 e_1 + \xi_2 e_2 + \cdots + \xi_n e_n = \eta_1 f_1 + \eta_2 f_2 + \cdots + \eta_n f_n, \tag{10}$$

where $\xi_1, \xi_2, \ldots, \xi_n$ are the components of the vector x with respect to the basis $\{e\}$ and $\eta_1, \eta_2, \ldots, \eta_n$ are its components with respect to the basis $\{f\}$. We now show how to calculate the components of the vector x with respect to the basis $\{f\}$ in terms of its components with respect to the basis $\{e\}$.

Suppose we are given the matrix $P = \|p_i^{(j)}\|$ of the transformation from the basis $\{e\}$ to the basis $\{f\}$. Then the vectors $\{e\}$ are given in terms of the vectors $\{f\}$ by (9) or, more briefly, by

$$e_j = \sum_{k=1}^{n} q_k^{(j)} f_k \qquad (k = 1, 2, \ldots, n), \tag{11}$$

where the matrix $Q = \|q_k^{(j)}\|$ is the inverse of the matrix P. Substituting (11) into the expansion (10), we get

$$x = \sum_{j=1}^{n} \xi_j e_j = \sum_{k=1}^{n} \eta_k f_k = \sum_{j=1}^{n} \xi_j \left(\sum_{k=1}^{n} q_k^{(j)} f_k \right) = \sum_{k=1}^{n} \left(\sum_{j=1}^{n} q_k^{(j)} \xi_j \right) f_k.$$

It follows by the uniqueness of the expansion of the vector x with respect to the basis $\{f\}$ that

$$\eta_k = \sum_{j=1}^{n} q_k^{(j)} \xi_j \qquad (k = 1, 2, \ldots, n), \tag{12}$$

or, in expanded form

$$\begin{aligned} \eta_1 &= q_1^{(1)} \xi_1 + q_1^{(2)} \xi_2 + \cdots + q_1^{(n)} \xi_n, \\ \eta_2 &= q_2^{(1)} \xi_1 + q_2^{(2)} \xi_2 + \cdots + q_2^{(n)} \xi_n, \\ &\cdots\cdots\cdots\cdots\cdots\cdots\cdots\cdots \\ \eta_n &= q_n^{(1)} \xi_1 + q_n^{(2)} \xi_2 + \cdots + q_n^{(n)} \xi_n. \end{aligned}$$

Thus *the components of the vector x with respect to the basis $\{f\}$ are linear combinations of the components of the vector x with respect to the basis $\{e\}$; the coefficients of these linear combinations form a matrix which is the transpose of the matrix of the transformation from the basis $\{f\}$ to the basis $\{e\}$, i.e., the transpose of the inverse of the matrix P.* Denoting the inverse of the matrix P by P^{-1} and the transpose of a matrix by a prime, we find that the matrix S describing the transformation from the components $\xi_1, \xi_2, \ldots, \xi_n$ to the components $\eta_1, \eta_2, \ldots, \eta_n$ is given by

$$S = (P^{-1})'.$$

5.32. The converse proposition is also valid:

THEOREM. *Let $\xi_1, \xi_2, \ldots, \xi_n$ be the components of an arbitrary vector x with respect to the basis $\{e\} = \{e_1, e_2, \ldots, e_n\}$ of the n-dimensional space $\mathbf{K}_n$, and let the quantities $\eta_1, \eta_2, \ldots, \eta_n$ be defined by the formulas*

$$\begin{aligned} \eta_1 &= s_{11} \xi_1 + s_{12} \xi_2 + \cdots + s_{1n} \xi_n, \\ \eta_2 &= s_{21} \xi_1 + s_{22} \xi_2 + \cdots + s_{2n} \xi_n, \\ &\cdots\cdots\cdots\cdots\cdots\cdots\cdots\cdots \\ \eta_n &= s_{n1} \xi_1 + s_{n2} \xi_2 + \cdots + s_{nn} \xi_n, \end{aligned}$$

where $\det \|s_{jk}\| \neq 0$. *Then a new basis $\{f\} = \{f_1, f_2, \ldots, f_n\}$ can be found in the space $\mathbf{K}_n$ such that the numbers $\eta_1, \eta_2, \ldots, \eta_n$ are the components of the vector x with respect to the basis $\{f\}$.*

Proof. Introduce the matrix $S = \|s_{jk}\|$ and the matrix $P = (S')^{-1}$ with elements denoted by $p_i^{(j)}$. Substituting these elements into the formulas (1), we get a new basis $\{f\} = \{f_1, f_2, \ldots, f_n\}$. We assert that this is the desired basis. In fact, consider the transformation formulas (12), which give the components of the vector x with respect to the new basis. As we have seen, these formulas can be written in terms of the matrix $(P^{-1})'$. But in the present case, $(P^{-1})'$ coincides with S, since

$$(P^{-1})' = ([(S')^{-1}]^{-1})' = (S')' = S.$$

Hence, given any vector x, the quantities $\eta_1, \eta_2, \ldots, \eta_n$ are just the components of x with respect to the basis $\{f\}$. ∎

5.33. Just as in Sec. 5.21, we can construct the matrix corresponding to consecutive transformations of the components of a vector. Let $\xi_1, \xi_2, \ldots, \xi_n$ be the components of the vector x with respect to the basis $\{e\}$, and let the quantities $\eta_1, \eta_2, \ldots, \eta_n$ and $\tau_1, \tau_2, \ldots, \tau_n$ be defined by the equations

$$\eta_j = \sum_{i=1}^{n} p_{ji}\xi_i \qquad (j = 1, 2, \ldots, n),$$

$$\tau_k = \sum_{j=1}^{n} q_{kj}\eta_j \qquad (k = 1, 2, \ldots, n),$$

respectively, where the matrices $P = \|p_{ji}\|$ and $Q = \|q_{kj}\|$ are nonsingular. Then, just as before, we can express the quantities $\tau_1, \tau_2, \ldots, \tau_n$ directly in terms of the quantities $\xi_1, \xi_2, \ldots, \xi_n$ by the formulas

$$\tau_k = \sum_{i=1}^{n} \left(\sum_{j=1}^{n} q_{kj}p_{ji}\right)\xi_i = \sum_{i=1}^{n} t_{ki}\xi_i \qquad (k = 1, 2, \ldots, n),$$

where the quantities t_{ki} $(i, k = 1, 2, \ldots, n)$ form a matrix T equal to the product QP of the matrices Q and P.

5.4. Transformation of the Coefficients of a Linear Form

Let $L(x)$ be a linear form defined on a space $\mathbf{K}_n$. As we saw in Sec. 4.1, if a basis $\{e\} = \{e_1, e_2, \ldots, e_n\}$ is chosen in $\mathbf{K}_n$, then the values of $L(x)$ can be calculated from the formula

$$L(x) = \sum_{k=1}^{n} l_k\xi_k,$$

where ξ_k $(k = 1, 2, \ldots, n)$ are the components of the vector x with respect to the basis $\{e\}$, and the coefficients l_k are given by

$$l_k = L(e_k) \qquad (k = 1, 2, \ldots, n).$$

The coefficients l_k obviously depend on the choice of the basis $\{e\}$. We now derive the rule governing the transformation of the coefficients of a linear form when we go over to a new basis.

Suppose the formulas

$$f_j = \sum_{i=1}^{n} p_i^{(j)} e_i \qquad (j = 1, 2, \ldots, n) \tag{13}$$

give the transformation from the basis $\{e\}$ to the new basis $\{f\}$. We wish to find the coefficients of the linear form $L(x)$ in the basis $\{f\}$. These coefficients are the numbers $\lambda_j = L(f_j)$, which can easily be found by using (13):

$$\lambda_j = L(f_j) = \sum_{i=1}^{n} p_i^{(j)} L(e_i) = \sum_{i=1}^{n} p_i^{(j)} l_i.$$

Thus *the coefficients of a linear form transform in the same way as the basis vectors themselves.*

5.5. Transformation of the Matrix of a Linear Operator

5.51. Given a linear operator **A** in an n-dimensional space $\mathbf{K}_n$, let $A_{(e)} = \|a_i^{(j)}\|$ be the matrix of **A** in the basis $\{e\} = \{e_1, e_2, \ldots, e_n\}$, while $A_{(f)} = \|a_q^{(p)}\|$ is its matrix in the basis $\{f\} = \{f_1, f_2, \ldots, f_n\}$. Moreover, suppose the transformation formulas from the basis $\{e\}$ to the basis $\{f\}$ have the form

$$f_k = \sum_{j=1}^{n} p_j^{(k)} e_j \qquad (k = 1, 2, \ldots, n), \tag{14}$$

and let P denote the matrix $\|p_j^{(k)}\|$. We now find the relation between the matrices $A_{(e)}$, $A_{(f)}$ and P.

The matrix $A_{(e)}$ is defined by the system of equations

$$\mathbf{A}e_j = \sum_{i=1}^{n} a_i^{(j)} e_i \qquad (j = 1, 2, \ldots, n), \tag{15}$$

and the matrix $A_{(f)}$ by the system of equations

$$\mathbf{A}f_m = \sum_{k=1}^{n} \alpha_k^{(m)} f_k \qquad (m = 1, 2, \ldots, n).$$

In the last equation, we use (14) to replace the vectors f_k by their expressions in terms of the vectors e_j. The result is

$$\mathbf{A}f_m = \sum_{k=1}^{n} \alpha_k^{(m)} \sum_{i=1}^{n} p_i^{(k)} e_i = \sum_{i=1}^{n} \left(\sum_{k=1}^{n} p_i^{(k)} \alpha_k^{(m)} \right) e_i,$$

after changing the index of summation from j to i. Next we apply the operator

$\mathbf{A}$ to both sides of (14), changing k to m and using the expansion for $\mathbf{A}e_j$ given by (15):

$$\mathbf{A}f_m = \mathbf{A}\sum_{j=1}^{n} p_j^{(m)} e_j = \sum_{j=1}^{n} p_j^{(m)} \mathbf{A}e_j$$

$$= \sum_{j=1}^{n} p_j^{(m)} \sum_{i=1}^{n} a_i^{(j)} e_i = \sum_{i=1}^{n} \left(\sum_{j=1}^{n} a_i^{(j)} p_j^{(m)} \right) e_i.$$

Comparing coefficients of e_i in the last two expansions, we find that

$$\sum_{k=1}^{n} p_i^{(k)} \alpha_k^{(m)} = \sum_{j=1}^{n} a_i^{(j)} p_j^{(m)},$$

or

$$PA_{(f)} = A_{(e)}P \tag{16}$$

in matrix form. This is the desired relation between the matrices $A_{(e)}$, $A_{(f)}$ and P. Multiplying on the left by the matrix P^{-1}, we get the following expression for the matrix $A_{(f)}$:

$$A_{(f)} = P^{-1}A_{(e)}P.$$

5.52. It follows from (16) and the theorem on the determinant of a product of two matrices (Sec. 4.75) that

$$\det P \det A_{(f)} = \det A_{(e)} \det P,$$

or, since $\det P \neq 0$,

$$\det A_{(e)} = \det A_{(f)}.$$

Thus *the determinant of the matrix of an operator does not depend on the choice of a basis in the space*. Therefore we can talk about the determinant of an operator, meaning thereby the determinant of the matrix of the operator in any basis.

5.53. Besides the determinant, there exist other functions of the matrix elements of an operator which remain unchanged under transformation to a new basis. To construct such functions, consider the operator $\mathbf{A} - \lambda\mathbf{E}$, where λ is a parameter. This operator obviously has the matrices $A_{(e)} - \lambda E$ and $A_{(f)} - \lambda E$ in the bases $\{e\}$ and $\{f\}$. By what was just proved, we have

$$\det (A_{(e)} - \lambda E) = \det (A_{(f)} - \lambda E)$$

for any λ. Both sides of this equation are polynomials of degree n in λ. Since these polynomials are identically equal, they have the same coefficients for any power of λ. Hence these coefficients, which are functions of the matrix elements of the operator, are invariant under changes of basis.

We now examine the nature of these functions. The determinant of the matrix $A_{(e)} - \lambda E$ has the form

$$\begin{vmatrix} a_1^{(1)} - \lambda & a_1^{(2)} & \cdots & a_1^{(n)} \\ a_2^{(1)} & a_2^{(2)} - \lambda & \cdots & a_2^{(n)} \\ \cdot & \cdot & \cdots & \cdot \\ a_n^{(1)} & a_n^{(2)} & \cdots & a_n^{(n)} - \lambda \end{vmatrix}$$

$$= (-1)^n\lambda^n + \Delta_1\lambda^{n-1} + \cdots + \Delta_{n-1}\lambda + \Delta_n.$$

It is an easy consequence of the definition of a determinant that the coefficient Δ_1 of λ^{n-1} equals the sum

$$a_1^{(1)} + a_2^{(2)} + \cdots + a_n^{(n)}$$

of the diagonal elements, taken with the sign $(-1)^{n-1}$.† The coefficient Δ_2 of λ^{n-2} is the sum of all the principal minors of order 2, taken with the sign $(-1)^{n-2}$.‡ Similarly, the coefficient Δ_k of λ^{n-k} is the sum of all the principal minors of order k, taken with the sign $(-1)^{n-k}$. Finally, the coefficient Δ_n of λ^0, i.e., the constant term, is obviously equal to just the determinant of the operator. The polynomial det $(A_{(e)} - \lambda E)$, which, as we have just seen, is independent of the choice of basis, is called the *characteristic polynomial of the operator* **A**.

*5.6. Tensors

5.61. The components of a vector, the coefficients of a linear form, the elements of the matrix of a linear operator, these are all examples of a general class of geometric objects called *tensors*. Before giving the definition of a tensor, we first revise and "rationalize" our notation somewhat. The basis vectors of an n-dimensional space $\mathbf{K}_n$ will be denoted, as before, by the symbols $e_1, e_2, \ldots, e_n$ (with subscripts). The components of vectors, e.g., x and y, will be denoted by $\xi^1, \xi^2, \ldots, \xi^n$ and $\eta^1, \eta^2, \ldots, \eta^n$ (with superscripts). The coefficients of a linear form $L(x)$ will be denoted by $l_1, l_2, \ldots, l_n$ (with subscripts). The matrix elements of a linear operator will be denoted by a_i^j, where the superscript designates the *row number* and the subscript designates the *column number* (in contradistinction to the notation adopted in Sec. 4.23). The convenience of this arrangement of indices is determined by the following *summation convention:* If we have a sum of terms such that the summation index i (say) occurs twice in the general term, once as a superscript and once

† The sum $a_1^{(1)} + a_2^{(2)} + \cdots + a_n^{(n)}$ is called the *trace* of the operator **A** (cf. Problem 14, p. 115).

‡ The minor $M_{j_1, j_2, \ldots, j_k}^{i_1, i_2, \ldots, i_k}$ is said to be a *principal minor* if $i_1 = j_1, i_2 = j_2, \ldots, i_k = j_k$.

as a subscript, then *we will omit the summation sign.* For example, with our convention, the expansion of the vector x with respect to the basis $\{e_1, e_2, \ldots, e_n\}$ takes the form

$$x = \xi^i e_i$$

(although the summation sign is omitted, summation over i is implied). The expression for a linear form $L(x)$ in terms of the components of the vector x and the coefficients of the form becomes

$$L(x) = l_i \xi^i$$

(summation over i is implied). The result of applying the operator $\mathbf{A}$ to the basis vector e_i takes the form

$$\mathbf{A}e_i = a_i^j e_j$$

(summation over j is implied). The components η^j of the vector $\mathbf{A}x$ are expressed in terms of the components of the vector x as follows:

$$\eta^j = a_i^j \xi^i$$

(summation over i is implied).

We will denote quantities pertaining to a new coordinate system by the same symbols as in the old coordinate system but with primes on the indices. Thus we denote new basis vectors by $e_{1'}, e_{2'}, \ldots, e_{n'}$, new components of a vector x by $\xi^{1'}, \xi^{2'}, \ldots, \xi^{n'}$, etc. The elements of the matrix of a transformation from the basis e_i to the basis $e_{i'}$ will be denoted by $p_{i'}^i$, so that

$$e_{i'} = p_{i'}^i e_i \tag{17}$$

(summation over i is implied). The elements of the matrix of the inverse transformation will be denoted by $q_i^{i'}$, i.e.,

$$e_i = q_i^{i'} e_{i'} \tag{18}$$

(summation over i' is implied). The matrix $q_i^{i'}$ is the inverse of the matrix $p_{i'}^i$; this can be expressed by writing

$$p_{i'}^i q_j^{i'} = \begin{cases} 0 \text{ for } i \neq j, \\ 1 \text{ for } i = j, \end{cases} \tag{19}$$

or

$$p_{i'}^i q_i^{j'} = \begin{cases} 0 \text{ for } i' \neq j', \\ 1 \text{ for } i' = j'. \end{cases} \tag{20}$$

To make the notation more concise, let δ_i^j denote the quantity which depends on the indices i and j in such a way that it equals 0 when the indices are different and 1 when the indices are the same. Then we can write (19) in the form

$$p_{i'}^i q_j^{i'} = \delta_j^i \tag{21}$$

and (20) in the form

$$p^i_{i'}q^{j'}_i = \delta^{j'}_{i'}. \tag{22}$$

5.62. To show the advantages of using our new notation, we derive once again the formulas by which the components of a vector, the coefficients of a linear form and the matrix elements of an operator transform in going over to a new basis. Thus suppose we have a vector

$$x = \xi^i e_i = \xi^{i'} e_{i'}.$$

Using (18) to replace e_i by $q^{i'}_i e_{i'}$, we obtain

$$x = \xi^i q^{i'}_i e_{i'} = \xi^{i'} e_{i'},$$

which implies

$$\xi^{i'} = q^{i'}_i \xi^i, \tag{23}$$

since the $e_{i'}$ form a basis. This is just the transformation formula for the components of a vector.

Next suppose we have a linear form $L(x)$. The numbers l_i, are defined as usual by the relations $l_{i'} = L(e_{i'})$. Using (17) to substitute the expression $p^i_{i'}e_i$ for $e_{i'}$, we obtain

$$l_{i'} = L(p^i_{i'}e_i) = p^i_{i'}L(e_i) = p^i_{i'}l_i,$$

so that

$$l_{i'} = p^i_{i'}l_i, \tag{24}$$

which is the desired formula.

Finally suppose we have an operator **A**. The elements of its matrix in the new basis are defined by the relations

$$\mathbf{A}e_{i'} = a^{j'}_{i'}e_{j'}.$$

Using (17) to substitute $p^i_{i'}e_i$ and $p^j_{j'}e_j$ for the quantities $e_{i'}$ and $e_{j'}$, we get

$$p^i_{i'}\mathbf{A}e_i = a^{j'}_{i'}p^j_{j'}e_j.$$

But $\mathbf{A}e_i = a^j_i e_j$, so that the result is

$$p^i_{i'}a^j_i e_j = a^{j'}_{i'}p^j_{j'}e_j.$$

Since the e_j are basis vectors, we have

$$p^i_{i'}a^j_i = a^{j'}_{i'}p^j_{j'}.$$

To get $a^{k'}_{i'}$ on the right, we multiply both sides by $q^{k'}_j$ and sum over the index j. Using the relation (22), we obtain

$$p^i_{i'}a^j_i q^{k'}_j = a^{j'}_{i'}p^j_{j'}q^{k'}_j = a^{j'}_{i'}\delta^{k'}_{j'}.$$

By the definition of the quantity $\delta^{k'}_{j'}$, the sum over j' reduces to the single term corresponding to the value $j' = k'$. Then $\delta_{k'}^{k'} = 1$ (no summation implied) and we get

$$a_{i'}^{k'} = p_{i'}^{i} q_{j}^{k'} a_{i}^{j}, \tag{25}$$

which is the desired formula.

It is not hard to verify that the three transformation formulas just derived are the same as those derived earlier in the ordinary way (see Secs. 5.3–5.5). Formulas (23)–(25) have much in common. In the first place, these formulas are linear in the transformed quantities. Secondly, the coefficients in these formulas are elements of the matrix transforming the old basis into the new basis or elements of the matrix of the inverse transformation or, finally, elements of both matrices.

5.63. We are now in a position to give the definition of a tensor. Tensors are divided into three classes, *covariant*, *contravariant* and *mixed*. Moreover, every tensor has a definite *order*. We begin by defining a covariant tensor, which, to be explicit, we take to have order three. Suppose there is a rule which in every coordinate system of an n-dimensional space $\mathbf{K}_n$ allows us to construct n^3 numbers (components) T_{ijk}, each of which is specified by giving the indices i, j, k definite values from 1 to n. By definition, these numbers T_{ijk} form a *covariant tensor of order three* if in going to a new basis, the quantities T_{ijk} transform according to the formula

$$T_{i'j'k'} = p_{i'}^{i} p_{j'}^{j} p_{k'}^{k} T_{ijk}.$$

A covariant tensor of any other order is defined similarly; a tensor of order m has n^m components instead of n^3 components, and in the transformation formula there appear m factors of the form $p_{i'}^{i}$ instead of three factors. In particular, the coefficients of a linear form, which transform by formula (24), constitute a covariant tensor of order one.

Next we define a contravariant tensor of order three. Suppose we have a rule which in every coordinate system allows us to construct n^3 numbers T^{ijk}, each of which is specified by giving the indices i, j, k definite values from 1 to n. By definition, these numbers T^{ijk} form a *contravariant tensor of order three* if in going to a new basis, the quantities T^{ijk} transform according to the formula

$$T^{i'j'k'} = q_{i}^{i'} q_{j}^{j'} q_{k}^{k'} T^{ijk}.$$

A contravariant tensor of any other order is defined similarly. In particular, the components of a vector form a contravariant tensor of order one.

The terms "covariant" and "contravariant," which have just been introduced, are very simply explained. "Covariant" means "transforming in the same way" as the basis vectors, i.e., by using the coefficients $p_{i'}^{i}$. "Contravariant" means "transforming in the opposite direction," i.e., by using the coefficients $q_{i'}^{i}$.

There is still the case of mixed tensors to consider. For example, n^3 numbers T_{ij}^k, specified in every coordinate system, form a *mixed tensor of order three, with two covariant indices and one contravariant index*, if in going to a new basis, the quantities T_{ij}^k transform according to the formula

$$T_{i'j'}^{k'} = p_{i'}^i p_{j'}^j q_k^{k'} T_{ij}^k.$$

A mixed tensor with l covariant indices and m contravariant indices is defined similarly. In particular, the elements of the matrix of a linear operator form a mixed tensor of order two, with one covariant index and one contravariant index. Note the convenience of our arrangement of indices, which has been deliberately chosen to indicate the character of any tensor at a glance.

5.64. Operations on tensors. We can define the operation of *addition* for two tensors of the same structure, e.g., for two tensors T_{ij}^k and S_{ij}^k (with two covariant indices and one contravariant index). In this case, the sum is a tensor Q_{ij}^k of the same structure, defined as follows: In every coordinate system, the component of Q_{ij}^k with fixed indices i, j, k is the sum of the corresponding components of T_{ij}^k and S_{ij}^k. The fact that the quantities Q_{ij}^k actually form a tensor, and indeed one of the same structure as T_{ij}^k and S_{ij}^k, is implied by the following equality:

$$\begin{aligned} Q_{i'j'}^{k'} = T_{i'j'}^{k'} + S_{i'j'}^{k'} &= p_{i'}^i p_{j'}^j q_k^{k'} T_{ij}^k + p_{i'}^i p_{j'}^j q_k^{k'} S_{ij}^k \\ &= p_{i'}^i p_{j'}^j q_k^{k'} (T_{ij}^k + S_{ij}^k) = p_{i'}^i p_{j'}^j q_k^{k'} Q_{ij}^k. \end{aligned}$$

The operation of *multiplication* is applicable to tensors of any structure. For example, let us multiply a tensor T_{ij} by a tensor S_k^l. The result is a tensor Q_{ijk}^l of order four. In any coordinate system its component with fixed indices i, j, k, l is defined as equal to the product of the corresponding components of the factors T_{ij} and S_k^l. The tensor character of Q_{ijk}^l can be verified as follows:

$$Q_{i'j'k'}^{l'} = T_{i'j'} S_{k'}^{l'} = p_{i'}^i p_{j'}^j T_{ij} p_{k'}^k q_l^{l'} S_k^l = p_{i'}^i p_{j'}^j p_{k'}^k q_l^{l'} T_{ij} S_k^l = p_{i'}^i p_{j'}^j p_{k'}^k q_l^{l'} Q_{ijk}^l.$$

Next we consider still another operation called *contraction*. This operation can be applied to tensors which have at least one covariant index and one contravariant index. For example, suppose we have a tensor T_{ij}^k. To *contract* T_{ij}^k with respect to the superscript and the first subscript means to form the quantity

$$T_{ij}^i$$

in every coordinate system. Here summation over the index i is implied; as a result, the quantity $T_j = T_{ij}^k$ depends only on the index j. *Contraction of a tensor yields another tensor, whose order is two less than the order of the original tensor.* We verify this for the present example. We have

$$T_{j'} = T_{i'j'}^{i'} = p_{i'}^i p_{j'}^j q_k^{i'} T_{ij}^k = (p_{i'}^i q_k^{i'}) p_{j'}^j T_{ij}^k = \delta_k^i p_{j'}^j T_{ij}^k.$$

Here the summation over k reduces to only one term, corresponding to the value $k = i$. Since $\delta_i^i = 1$ (no summation implied), we obtain

$$T_{j'} = p_{j'}^j T_{ij}^i = p_{j'}^j T_j,$$

as required.

What is the result of contracting a mixed tensor T_i^j of order two with respect to its two indices? The quantity $T = T_i^i$ no longer has even a single index, i.e., in every coordinate system it consists of just one number. This number is the same in every coordinate system, since

$$T' = T_{i'}^{i'} = p_{i'}^i q_j^{i'} T_i^j = \delta_j^i T_i^j = T_i^i = T.$$

Such a *scalar quantity*, which does not depend on the coordinate system, is called an *invariant*. Thus, by contracting tensors, we can obtain invariants of the tensors.

For example, if we contract the tensor a_i^j corresponding to the linear operator **A**, the invariant a_i^i so obtained is the trace of the matrix of **A**, i.e., the sum of its diagonal elements. The invariance of this quantity has already been proved in a different way in Sec. 5.53. As another example, the matrix c_j^i of the product of two operators with matrices a_k^i and b_j^l, respectively, is the mixed second-order tensor obtained by contracting the fourth-order tensor $a_k^i b_j^l$ with respect to the indices k and l.

PROBLEMS

1. A vector $x \in \mathbf{K}_n$ has components $\xi_1, \xi_2, \ldots, \xi_n$ with respect to a basis $e_1, e_2, \ldots, e_n$. How does one construct a new basis in $\mathbf{K}_n$ such that the components of x with respect to this basis equal $1, 0, \ldots, 0$?

2. A basis $e_1, e_2, \ldots, e_n$ is chosen in an n-dimensional space $\mathbf{K}_n$. Show that every subspace $\mathbf{K}' \subset \mathbf{K}_n$ can be specified as the set of all vectors $x \in \mathbf{K}_n$ whose components (with respect to the basis $e_1, e_2, \ldots, e_n$) satisfy a system of equations of the form

$$\sum_{j=1}^{n} a_{ij}\xi_j = 0 \qquad (i = 1, 2, \ldots, k).$$

3 (*Continuation*). Show that every hyperplane $\mathbf{H} \subset \mathbf{K}_n$ can be specified as the set of all vectors $x \in \mathbf{K}_n$ whose components (with respect to the basis $e_1, e_2, \ldots, e_n$) satisfy a system of equations of the form

$$\sum_{j=1}^{n} a_{ij}\xi_j = b_i \qquad (i = 1, 2, \ldots, k).$$

4. Let the components of a vector in the plane be ξ_1, ξ_2 with respect to one basis, η_1, η_2 with respect to another basis, and τ_1, τ_2 with respect to a third basis.

Suppose that

$$\eta_1 = a_{11}\xi_1 + a_{12}\xi_2, \qquad \eta_2 = a_{21}\xi_1 + a_{22}\xi_2,$$
$$\tau_1 = b_{11}\xi_1 + b_{12}\xi_2, \qquad \tau_2 = b_{21}\xi_1 + b_{22}\xi_2,$$
$$A = \|a_{ij}\|, \qquad B = \|b_{ij}\|.$$

Express the components τ_1, τ_2 in terms of the components ξ_1, ξ_2.

5. Given a linear form $L(x) \not\equiv 0$ in the space $\mathbf{K}_n$, find a basis $f_1, f_2, \ldots, f_n$ such that the relation

$$L(x) = \eta_1$$

holds for every vector

$$x = \sum_{k=1}^{n} \eta_k f_k.$$

6. Let the operator **A** acting in an n-dimensional space **R** have a k-dimensional invariant subspace $\mathbf{R}'$. Then, temporarily regarding **A** as defined only in the subspace $\mathbf{R}'$, we can construct the characteristic polynomial of degree k for **A**. Show that this polynomial is a factor of the characteristic polynomial of the operator **A** acting in the whole space **R**.

7. Let $\lambda = \lambda_0$ be an r-fold root of the equation $\det \|A_{(e)} - \lambda E\| = 0$. Show that the dimension m of the eigenspace $\mathbf{R}^{(\lambda_0)}$ of **A** corresponding to the root λ_0 does not exceed r.

8. Show that the quantity δ_i^j is a second-order tensor, with one covariant index and one contravariant index.

9. A set of quantities S_{ij} is defined in every coordinate system as the solution of the system of equations

$$T^{ik}S_{ij} = \delta_j^k,$$

where T^{ik} is a contravariant tensor of order two and $\det \|T^{ik}\| \neq 0$. Show that S_{ij} is a covariant tensor of order two.

chapter 6

THE CANONICAL FORM OF THE MATRIX OF A LINEAR OPERATOR

Two operators $\mathbf{A}$ and $\mathbf{B}$ acting in an n-dimensional space $\mathbf{K}_n$ are said to be *equivalent* if there exist two bases in $\mathbf{K}_n$ such that the matrix of the operator $\mathbf{A}$ in the first basis coincides with the matrix of the operator $\mathbf{B}$ in the second basis. Clearly, the "linear transformations" in $\mathbf{K}_n$ corresponding to equivalent operators have identical properties. But how can we decide whether or not the operators $\mathbf{A}$ and $\mathbf{B}$ are equivalent by examining their matrices in the same basis?

In this chapter, starting from a given linear operator $\mathbf{A}$ in an n-dimensional (real or complex) space, we will find a basis in which the matrix A of the operator $\mathbf{A}$ has "canonical form," i.e., a form which is the simplest possible in a certain sense. This canonical form can be obtained directly from the elements of the matrix of the operator $\mathbf{A}$ in any basis. Moreover, it turns out that if the operators $\mathbf{A}$ and $\mathbf{B}$ are equivalent, then their matrices have the same canonical form. Thus a necessary and sufficient condition for two operators to be equivalent is that their canonical matrices coincide.

We begin our considerations by studying a special class of operators (Sec. 6.1). The general case will be studied in Sec. 6.3.

6.1. Canonical Form of the Matrix of a Nilpotent Operator

6.11. A linear operator $\mathbf{B}$ acting in an n-dimensional space $\mathbf{K}_n$ is said to be *nilpotent* if $\mathbf{B}^r = \mathbf{0}$ (i.e., if $\mathbf{B}^r x = 0$ for every $x \in \mathbf{K}_n$) for some positive integer r. Given a nilpotent operator $\mathbf{B}$ such that $\mathbf{B}^r = \mathbf{0}$, we will assume that $\mathbf{B}^{r-1} \neq \mathbf{0}$, i.e., that there are vectors $x \in \mathbf{K}_n$ such that $\mathbf{B}^{r-1}x \neq 0$. By the *height* of a vector $x \in \mathbf{K}_n$, we mean the smallest positive integer m for which $\mathbf{B}^m x = 0$. By hypothesis, every vector $x \in \mathbf{K}_n$ is of height $\leqslant r$, and there are vectors of height equal to r. Given any $k \leqslant r$, let $\mathbf{H}_k$ denote the set of all vectors of height $\leqslant k$. Obviously, $\mathbf{H}_k$ is a subspace of $\mathbf{K}_n$. In fact, if $x, y \in \mathbf{H}_k$, then $\mathbf{B}^k x = 0$, $\mathbf{B}^k y = 0$ and hence $\mathbf{B}^k(\alpha x + \beta y) = 0$ for arbitrary $\alpha, \beta \in K$, so that the height of the vector $\alpha x + \beta y$ does not exceed k, i.e., $\alpha x + \beta y \in \mathbf{H}_k$. Moreover, it is obvious that $\mathbf{H}_r = \mathbf{K}_n$ and that†

$$\{0\} = \mathbf{H}_0 \subset \mathbf{H}_1 \subset \cdots \subset \mathbf{H}_{r-1} \subset \mathbf{H}_r = \mathbf{K}_n.$$

Let m_k denote the dimension of $\mathbf{H}_k$, so that

$$0 = m_0 \leqslant m_1 \leqslant \cdots \leqslant m_r = n.$$

Next we construct a basis in the space $\mathbf{K}_n$ as follows: As we have seen, $\mathbf{H}_{r-1}$ does not coincide with the whole space $\mathbf{K}_n = \mathbf{H}_r$. Therefore we can find vectors $f_1, \ldots, f_{p_1}$ lying in $\mathbf{H}_r$ and linearly independent over $\mathbf{H}_{r-1}$, where $p_1 = m_r - m_{r-1}$ (see Sec. 2.44). The vectors $\mathbf{B}f_1, \ldots, \mathbf{B}f_{p-1}$ lie in $\mathbf{H}_{r-1}$ and are linearly independent over $\mathbf{H}_{r-2}$. In fact, if we had

$$\alpha_1 \mathbf{B}f_1 + \cdots + \alpha_{p_1}\mathbf{B}f_{p_1} = g \in \mathbf{H}_{r-2} \qquad (g \neq 0),$$

then application of the operator $\mathbf{B}^{r-2}$ would give

$$\alpha_1 \mathbf{B}^{r-1}f_1 + \cdots + \alpha_{p_1}\mathbf{B}^{r-1}f_{p_1} = 0,$$

or equivalently

$$\alpha_1 f_1 + \cdots + \alpha_{p_1} f_{p_1} \in \mathbf{H}_{r-1},$$

which is impossible, by construction. It follows that the dimension $m_{r-1} - m_{r-2}$ of the space $\mathbf{H}_{r-1}$ over $\mathbf{H}_{r-2}$ (again see Sec. 2.44) is equal to or greater than the dimension $m_r - m_{r-1}$ of the space $\mathbf{H}_r$ over $\mathbf{H}_{r-1}$. We now supplement the vectors $\mathbf{B}f_1, \ldots, \mathbf{B}f_{p_1}$ with vectors $f_{p_1+1}, \ldots, f_{p_2}$ in $\mathbf{H}_{r-1}$ to make the largest system which is linearly independent over $\mathbf{H}_{r-2}$ ($p_2 = m_{r-1} - m_{r-2}$). Applying the operator $\mathbf{B}$ to all these vectors, we get vectors

$$\mathbf{B}^2 f_1, \ldots, \mathbf{B}^2 f_{p_1}, \mathbf{B}f_{p_1+1}, \ldots, \mathbf{B}f_{p_2}$$

lying in $\mathbf{H}_{r-2}$ and linearly independent over $\mathbf{H}_{r-3}$ (this is proved in the same way as before). It follows that $m_{r-2} - m_{r-3} \geqslant m_{r-1} - m_{r-2}$, and we can construct vectors $f_{p_2+1}, \ldots, f_{p_3}$ in $\mathbf{H}_{r-2}$ which together with the preceding system form a "full system" of vectors linearly independent over $\mathbf{H}_{r-3}$. Continuing this construction in the subspaces $\mathbf{H}_{r-3}, \ldots, \mathbf{H}_0 = \{0\}$, we finally

† $\{0\}$ denotes the set whose only element is the zero vector.

get a full system of n linearly independent vectors. This system can be written in the form of a table

$$\begin{array}{l} f_1, \ldots, f_{p_1}, \\ \mathbf{B}f_1, \ldots, \mathbf{B}f_{p_1}, f_{p_1+1}, \ldots, f_{p_2} \\ \cdots\cdots\cdots\cdots\cdots\cdots\cdots\cdots\cdots\cdots \\ \mathbf{B}^{r-1}f_1, \ldots, \mathbf{B}^{r-1}f_{p_1}, \mathbf{B}^{r-2}f_{p_1+1}, \ldots, \mathbf{B}^{r-2}f_{p_2}, \ldots, f_{p_{r-1}+1}, \ldots, f_{p_r}, \end{array}$$

where the vectors in the first row are of height r, those in the second row are of height $r - 1$, and so on, with the vectors in the last row being of height 1 (so that the operator $\mathbf{B}$ carries them all into the zero vector).

6.12. Every column of the above table determines an invariant subspace of the operator $\mathbf{B}$. The first p_1 invariant subspaces all have dimension r, the next $p_2 - p_1$ invariant subspaces all have dimension $r - 1$, and so on, with the last $p_r - p_{r-1}$ single-element columns determining one-dimensional invariant subspaces. The whole space $\mathbf{K}_n$ is the direct sum of these p_r invariant subspaces.

6.13. Next we write the matrix of the operator $\mathbf{B}$ in the subspace determined by the vectors of the first column. For a basis we choose the vectors $\mathbf{B}^{r-1}f_1, \mathbf{B}^{r-2}f_1, \ldots, \mathbf{B}f_1, f_1$, arranged in order of increasing height. With this arrangement, the operator $\mathbf{B}$ carries the first vector of the basis into the zero vector, the second vector into the first vector, etc., and finally the rth vector into the $(r - 1)$st vector. Therefore, according to Sec. 4.23, the matrix of the operator $\mathbf{B}$ has r rows and r columns, and is of the form

$$\left\| \begin{array}{cccccc} 0 & 1 & 0 & \cdots & 0 & 0 \\ 0 & 0 & 1 & \cdots & 0 & 0 \\ \cdot & \cdot & \cdot & \cdots & \cdot & \cdot \\ 0 & 0 & 0 & \cdots & 0 & 1 \\ 0 & 0 & 0 & \cdots & 0 & 0 \end{array} \right\| \qquad (1)$$

with zeros everywhere except for the elements (equal to 1) along the diagonal just above the principal diagonal. The matrix of the operator $\mathbf{B}$ takes a similar form in the other invariant subspaces, corresponding to the remaining columns of the table, and in fact can differ from the matrix (1) only by having a different number of rows and columns.

6.14. Thus the matrix of the operator $\mathbf{B}$ in the whole space $\mathbf{K}_n$ is quasi-diagonal (see Sec. 4.84), with blocks of the form (1) along the principal diagonal:

$$B = \left\| \begin{array}{ccccccc} \boxed{\begin{matrix} 0 & 1 & 0 & \cdots & 0 & 0 \\ 0 & 0 & 1 & \cdots & 0 & 0 \\ \cdot & \cdot & \cdot & \cdots & \cdot & \cdot \\ 0 & 0 & 0 & \cdots & 0 & 1 \\ 0 & 0 & 0 & \cdots & 0 & 0 \end{matrix}} & & & & & & \\ & \boxed{\begin{matrix} 0 & 1 & 0 & \cdots & 0 & 0 \\ 0 & 0 & 1 & \cdots & 0 & 0 \\ \cdot & \cdot & \cdot & \cdots & \cdot & \cdot \\ 0 & 0 & 0 & \cdots & 0 & 1 \\ 0 & 0 & 0 & \cdots & 0 & 0 \end{matrix}} & & & & & \\ & & \ddots & & & & \\ & & & \boxed{\begin{matrix} 0 & 1 \\ 0 & 0 \end{matrix}} & & & \\ & & & & \boxed{0} & & \\ & & & & & \ddots & \\ & & & & & & \boxed{0} \end{array} \right\| \tag{2}$$

The number of blocks of size r equals p_1, the number of blocks of size $r - 1$ equals $p_2 - p_1, \ldots$, the number of blocks of size (2) equals $p_{r-1} - p_{r-2}$, and the number of blocks of size (1) equals $p_r - p_{r-1}$. Naturally, if $p_{r-j+1} = p_{r-j}$ for some j, then the matrix (2) contains no blocks of size j.

6.2. Algebras. The Algebra of Polynomials

6.21. We begin with some definitions. A linear space $\mathbf{K}$ over a number field K is called an *algebra* (more exactly, an *algebra over* K) if there is defined on the elements $x, y, \ldots$ of $\mathbf{K}$ an operation of *multiplication*, denoted by $x \cdot y$ (or xy), which satisfies the following conditions:

1) $\alpha(xy) = (\alpha x)y = x(\alpha y)$ for every x, y in $\mathbf{K}$ and every α in K;
2) $(xy)z = x(yz)$ for every x, y, z in $\mathbf{K}$ (the *associative law*);
3) $(x + y)z = xz + yz$ for every x, y, z in $\mathbf{K}$ (the *distributive law*).

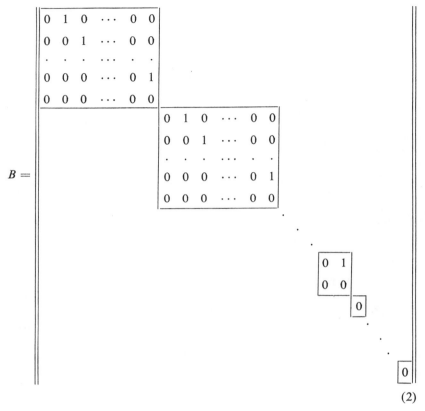

In general, multiplication may not be commutative, i.e., we may have $xy \neq yx$. If multiplication is commutative, i.e., if

4) $xy = yx$ for every x, y in **K**,

then the algebra **K** is said to be *commutative*.

An element $e \in \mathbf{K}$ is called a *left unit* if $ex = x$ for every $x \in \mathbf{K}$, a *right unit* if $xe = x$ for every $x \in \mathbf{K}$, and a *two-sided unit* or simply a *unit* (in **K**) if $ex = xe = x$ for every $x \in \mathbf{K}$.

An element $x \in \mathbf{K}$ is called a *left inverse* of the element $y \in \mathbf{K}$ if xy is the unit of the algebra **K**; in this case, y is called a *right inverse* of x. If an element z has both a left and a right inverse, then the two inverses are unique and in fact coincide (cf. Sec. 4.76a). The element z is then said to be *invertible*, and its inverse is denoted by z^{-1}.

The product zu of an invertible element z and an invertible element u is an invertible element with inverse $u^{-1}z^{-1}$. If the element u is invertible, then the equation $ux = v$ has the solution $x = u^{-1}v$. This solution is unique, being obtained by multiplying the equation $ux = v$ on the left by u^{-1}. In the commutative case, we write $x = v/u$ or $x = v:u$, calling the element x the *quotient* of the elements v and u.

The ordinary rules of arithmetic are valid for quotients, i.e.,

$$\frac{v_1}{u_1} + \frac{v_2}{u_2} = \frac{v_1u_2 + u_1v_2}{u_1u_2} \qquad \text{(if } u_1 \text{ and } u_2 \text{ are invertible)},$$

$$\frac{v_1}{u_1} \cdot \frac{v_2}{u_2} = \frac{v_1v_2}{u_1u_2} \qquad \text{(if } u_1 \text{ and } u_2 \text{ are invertible)},$$

$$\frac{v_1}{u_1} : \frac{v_2}{u_2} = \frac{v_1u_2}{u_1v_2} \qquad \text{(if } u_1, u_2, \text{ and } v_2 \text{ are invertible)}.$$

The proof of these facts is left to the reader.

An algebra **K** is said to have *dimension n* if **K** has dimension n regarded as a linear space.

6.22. *Examples*

a. Given any linear space **K**, suppose we set $x \cdot y = 0$ for every $x, y \in \mathbf{K}$. This gives an algebra, called the *trivial algebra*.

b. An example of a nontrivial commutative algebra over a field K is given by the set Π of all polynomials

$$P(\lambda) = \sum_{k=0}^{m} a_k \lambda^k$$

with coefficients in K, equipped with the usual operations of addition and multiplication. This "polynomial algebra" has a unit, namely the polynomial $e(\lambda)$ with $a_0 = 1$ and all other coefficients equal to 0.

c. The linear space $M(K_n)$ of all matrices of order n with elements in K, with the usual definition of matrix multiplication, is an example of a finite-dimensional noncommutative algebra of dimension n^2 (see Sec. 4.73b).

d. A more general example of a noncommutative algebra with a unit is the linear space of all linear operators acting in a linear space $\mathbf{K}$, with the usual definition of operator multiplication (see Sec. 4.33).

6.23. a. A subspace $\mathbf{L} \subset \mathbf{K}$ is called a *subalgebra* of the algebra $\mathbf{K}$ if $x \in \mathbf{L}$, $y \in \mathbf{L}$ implies $xy \in \mathbf{L}$. A subspace $\mathbf{L} \subset \mathbf{K}$ is called a *right ideal* in $\mathbf{K}$ if $x \in \mathbf{L}$, $y \in \mathbf{K}$ implies $xy \in \mathbf{L}$ and a *left ideal* in $\mathbf{K}$ if $x \in \mathbf{L}$, $y \in \mathbf{K}$ implies $yx \in \mathbf{L}$. An ideal which is both a left and a right ideal is called a *two-sided ideal*. In a commutative algebra there is no distinction between left, right and two-sided ideals. There are two obvious two-sided ideals in every algebra $\mathbf{K}$, i.e., the algebra $\mathbf{K}$ itself and the ideal $\{0\}$ consisting of the zero element alone.† All other one-sided and two-sided ideals are called *proper ideals*. Every ideal is a subalgebra, but the converse is in general false. Thus the set of all polynomials $P(\lambda)$ satisfying the condition $P(0) = P(1)$ is a subalgebra of the algebra Π which is not an ideal, while the set of all polynomials $P(\lambda)$ satisfying the condition $P(0) = 0$ is a proper ideal of the algebra Π.

b. Let $\mathbf{L} \subset \mathbf{K}$ be a subspace of the algebra $\mathbf{K}$, and consider the factor space $\mathbf{K}/\mathbf{L}$ (Sec. 2.48), i.e., the linear space consisting of the classes $\mathbf{X}$ of elements $x \in \mathbf{K}$ which are comparable relative to $\mathbf{L}$. If $\mathbf{L}$ is a two-sided ideal in $\mathbf{K}$, then, besides linear operations, we can introduce an operation of multiplication for the classes $\mathbf{X} \in \mathbf{K}/\mathbf{L}$. In fact, given two classes $\mathbf{X}$ and $\mathbf{Y}$, choose arbitrary elements $x \in \mathbf{X}$, $y \in \mathbf{Y}$ and interpret $\mathbf{X} \cdot \mathbf{Y}$ as the class containing the product xy. This uniquely defines $\mathbf{X} \cdot \mathbf{Y}$, since if $x' \in \mathbf{X}$, $y' \in \mathbf{Y}$, then

$$x'y' - xy = x'(y' - y) + (x' - x)y,$$

and hence $x'y' - xy$ belongs to $\mathbf{L}$ together with $y' - y$ and $x' - x$. Moreover, since conditions 1)–3), p. 136 hold in $\mathbf{K}$, the analogous conditions hold for the classes $\mathbf{X} \in \mathbf{K}/\mathbf{L}$. Therefore the factor space $\mathbf{K}/\mathbf{L}$ equipped with the above operation of multiplication, is also an algebra, called the *factor algebra of the algebra* $\mathbf{K}$ *with respect to the two-sided ideal* $\mathbf{L}$. If the algebra $\mathbf{K}$ is commutative, then obviously so is the factor algebra $\mathbf{K}/\mathbf{L}$.

6.24. Let $\mathbf{K}'$ and $\mathbf{K}''$ be two algebras over a field K. Then a morphism ω of the space $\mathbf{K}'$ into the space $\mathbf{K}''$ (Sec. 2.71) is called a *morphism of the algebra* $\mathbf{K}'$ *into the algebra* $\mathbf{K}''$ if besides satisfying the two conditions

a) $\omega(x' + y') = \omega(x') + \omega(y')$ for every $x', y' \in \mathbf{K}'$,
b) $\omega(\alpha x') = \alpha\omega(x') = \alpha\omega(x')$ for every $x' \in \mathbf{K}'$ and every $\alpha \in K$

---

† As in Theorem 2.14c, $0 \cdot x = 0$ for every $x \in \mathbf{K}$.

for the morphism of two spaces (see p. 53), it also satisfies the condition

c) $\omega(x'y') = \omega(x')\omega(y')$ for every $x', y' \in \mathbf{K}'$.

A morphism ω which is an epimorphism, monomorphism or isomorphism of the space $\mathbf{K}'$ into the space $\mathbf{K}''$, as defined in Sec. 2.71, is called an *epimorphism, monomorphism or isomorphism of the algebra* $\mathbf{K}'$ *into the algebra* $\mathbf{K}''$, provided condition c) is satisfied.

6.25. *Examples*

a. Let $\mathbf{L}$ be a subalgebra of an algebra $\mathbf{K}$. Then the mapping ω which assigns to every vector $x \in \mathbf{L}$ the same vector $x \in \mathbf{K}$ is a morphism of the algebra $\mathbf{L}$ into the algebra $\mathbf{K}$, and in fact a monomorphism. As in Example 2.72a, this monomorphism is said to *embed* $\mathbf{L}$ in $\mathbf{K}$.

b. Let $\mathbf{L}$ be a two-sided ideal of an algebra $\mathbf{K}$, and let $\mathbf{K}/\mathbf{L}$ be the corresponding factor algebra (Sec. 6.23b). Then the mapping ω which assigns to every vector $x \in \mathbf{K}$ the class $\mathbf{X} \in \mathbf{K}/\mathbf{L}$ containing x is a morphism of the algebra $\mathbf{K}$ into the algebra $\mathbf{K}/\mathbf{L}$, and in fact an epimorphism. As in Example 2.72b, this epimorphism is called the *canonical mapping of* $\mathbf{K}$ *onto* $\mathbf{K}/\mathbf{L}$.

c. Let ω be a monomorphism of an algebra $\mathbf{K}'$ into an algebra $\mathbf{K}''$. Then the set of all vectors $\omega(x') \in \mathbf{K}''$ is a subalgebra $\mathbf{L}'' \subset \mathbf{K}''$, and the monomorphism ω is an isomorphism of the algebra $\mathbf{K}'$ onto the algebra $\mathbf{L}''$.

d. Let ω be a morphism of an algebra $\mathbf{K}'$ into an algebra $\mathbf{K}''$. Then the set $\mathbf{L}'$ of all vectors $x' \in \mathbf{K}'$ such that $\omega(x') = 0$, which is obviously a subspace of $\mathbf{K}'$ (cf. Sec. 2.76b), is a two-sided ideal of the algebra $\mathbf{K}'$. In fact, if $x' \in \mathbf{L}'$, $y' \in \mathbf{K}$, then

$$\omega(x'y') = \omega(x')\omega(y') = 0,$$

so that $x'y' \in \mathbf{L}'$, and similarly $y'x' \in \mathbf{L}'$, i.e., $\mathbf{L}'$ is a two-sided ideal of $\mathbf{K}'$, as asserted. As in Sec. 2.76b, let Ω be the monomorphism of the space $\mathbf{K}'/\mathbf{L}'$ into the space $\mathbf{K}''$ which assigns to each class $\mathbf{X}' \in \mathbf{K}'/\mathbf{L}'$ the (unique) element $\omega(x')$, $x' \in \mathbf{X}'$. Then Ω is a monomorphism of the algebra $\mathbf{K}'/\mathbf{L}'$ into the algebra $\mathbf{K}''$. In fact, choosing $x' \in \mathbf{X}'$, $y' \in \mathbf{Y}'$, we have $x'y' \in \mathbf{X}'\mathbf{Y}'$ and

$$\Omega(\mathbf{X}'\mathbf{Y}') = \omega(x'y') = \omega(x')\omega(y') = \Omega(\mathbf{X}')\Omega(\mathbf{Y}').$$

If the morphism ω is an epimorphism of the algebra $\mathbf{K}'$ into the algebra $\mathbf{K}''$, then the morphism Ω is an isomorphism of the algebra $\mathbf{K}'/\mathbf{L}'$ onto the algebra $\mathbf{K}''$.

e. Let $\mathbf{A}$ be a linear operator acting in a space $\mathbf{K}$ over a field K. Since addition and multiplication by constants in K are defined for linear operators acting in $\mathbf{K}$, with every polynomial

$$P(\lambda) = \sum_{k=0}^{m} a_k \lambda^k$$

($a_k \in K$) we can associate an operator

$$P(\mathbf{A}) = \sum_{k=0}^{m} a_k \mathbf{A}^k$$

acting in the same space $\mathbf{K}$ as $\mathbf{A}$ itself. Then the rule associating $P(\lambda)$ with $P(\mathbf{A})$ has the three properties figuring in Sec. 6.24. In fact, if

$$P(\lambda) = P_1(\lambda) + P_2(\lambda) = \sum_{k=0}^{m} a_k \lambda^k + \sum_{k=0}^{m} b_k \lambda^k = \sum_{k=0}^{m} (a_k + b_k)\lambda^k,$$

then clearly

$$P(\mathbf{A}) = \sum_{k=0}^{m} (a_k + b_k)\mathbf{A}^k = \sum_{k=0}^{m} a_k \mathbf{A}^k + \sum_{k=0}^{m} b_k \mathbf{A}^k = P_1(\mathbf{A}) + P_2(\mathbf{A}),$$

and similarly for property b), while if

$$Q(\lambda) = P_1(\lambda)P_2(\lambda) = \sum_{j=0}^{m} a_j \lambda^j \sum_{k=0}^{m} b_k \lambda^k = \sum_{j=0}^{m} \sum_{k=0}^{m} a_j b_k \lambda^{j+k},$$

then

$$Q(\mathbf{A}) = \sum_{j=0}^{m} \sum_{k=0}^{m} a_j b_k \mathbf{A}^{j+k} = \sum_{j=0}^{m} a_j \mathbf{A}^j \sum_{k=0}^{m} b_k \mathbf{A}^k = P_1(\mathbf{A})P_2(\mathbf{A}),$$

by the distributive law for operators (Sec. 4.34). Note that the operators $P_1(\mathbf{A})$ and $P_2(\mathbf{A})$ always commute with each other, regardless of the choice of the polynomials $P_1(\lambda)$ and $P_2(\lambda)$. The resulting morphism of the algebra Π of polynomials (Example 6.22b) into the algebra $\mathbf{B}(\mathbf{K})$ of linear operators acting in $\mathbf{K}$ (Example 6.22d) is in general not an epimorphism, if only because operators of the form $P(\mathbf{A})$ commute with each other, while the whole algebra $\mathbf{B}(\mathbf{K})$ is noncommutative.†

f. There exists an isomorphism between the algebra $\mathbf{L}(K_n)$ of all linear operators acting in the n-dimensional space K_n and the algebra $M(K_n)$ of all matrices of order n with elements from the field K. This isomorphism is established by fixing a basis $e_1, \ldots, e_n$ in the space K_n and assigning every operator $\mathbf{A} \in \mathbf{L}(K_n)$ its matrix in this basis. Both algebras $\mathbf{L}(K_n)$ and $M(K_n)$ have the same dimension n^2.

6.26. The set of all polynomials of the form $P(\lambda)Q_0(\lambda)$, where $Q_0(\lambda)$ is a fixed polynomial and $P(\lambda)$ an arbitrary polynomial, is obviously an ideal in the commutative algebra Π of all polynomials $P(\lambda)$ with coefficients in a field K (Example 6.22b). Conversely, we now show that *every ideal* $I \neq \{0\}$ *of the algebra* Π *is of this structure*, i.e., is obtained from some polynomial $Q_0(\lambda)$ by multiplication by an arbitrary polynomial $P(\lambda)$. To this end, we

† Except in the trivial case where $\mathbf{K}$ is one-dimensional.

find the nonzero polynomial of lowest degree, say q, in the ideal I, and denote it by $Q_0(\lambda)$. We then assert that every polynomial $Q(\lambda) \in I$ is of the form $P(\lambda)Q_0(\lambda)$, where $P(\lambda) \in \Pi$. In fact, as is familiar from elementary algebra,

$$Q(\lambda) \equiv P(\lambda)Q_0(\lambda) + R(\lambda), \tag{3}$$

where $P(\lambda)$ is the quotient obtained by dividing $Q(\lambda)$ by $Q_0(\lambda)$ and $R(\lambda)$ is the remainder, of degree less than the divisor $Q_0(\lambda)$, i.e., less than the number q. But the polynomials $Q(\lambda)$ and $Q_0(\lambda)$ belong to the ideal I, and hence, as is apparent from (3), so does the remainder $R(\lambda)$. Since the degree of $R(\lambda)$ is less than q and since $Q_0(\lambda)$ has the lowest degree, namely q, of all nonzero polynomials in I, it follows that $R(\lambda) \equiv 0$, and the italicized assertion is proved.

The polynomial $Q_0(\lambda)$ is said to *generate* the ideal I.

6.27. *The polynomial $Q_0(\lambda)$ is uniquely determined by the ideal I to within a numerical factor.* In fact, if the polynomial $Q_1(\lambda)$ has the same property as the polynomial $Q_0(\lambda)$, then, as just shown,

$$Q_1(\lambda) = P_1(\lambda)Q_0(\lambda),$$

$$Q_0(\lambda) = P_0(\lambda)Q_1(\lambda).$$

It follows that the degrees of the polynomials $Q_1(\lambda)$ and $Q_0(\lambda)$ coincide and that $P_1(\lambda)$ and $P_0(\lambda)$ do not contain λ and hence are numbers, as asserted.

6.28. Given polynomials $Q_1(\lambda), \ldots, Q_m(\lambda)$ not all equal to zero and with no common divisors of degree $\geqslant 1$, we now show that there exist polynomials $P_1^0(\lambda), \ldots, P_m^0(\lambda)$ such that

$$P_1^0(\lambda)Q_1(\lambda) + \cdots + P_m^0(\lambda)Q_m(\lambda) \equiv 1. \tag{4}$$

In fact, let I be the set of all polynomials of the form

$$P_1(\lambda)Q_1(\lambda) + \cdots + P_m(\lambda)Q_m(\lambda)$$

with arbitrary $P_1(\lambda), \ldots, P_m(\lambda)$ in Π. Then I is obviously an ideal in Π. By Sec. 6.26, the ideal I is generated by some polynomial

$$Q_0(\lambda) = \sum_{k=1}^{m} \tilde{P}_k^0(\lambda)Q_k(\lambda). \tag{5}$$

In particular,

$$Q_1(\lambda) = S_1(\lambda)Q_0(\lambda), \ldots, Q_m(\lambda) = S_m(\lambda)Q_0(\lambda),$$

where $S_1(\lambda), \ldots, S_m(\lambda)$ are certain polynomials, from which it follows that $Q_0(\lambda)$ is a common divisor of the polynomials $Q_1(\lambda), \ldots, Q_m(\lambda)$. But, by hypothesis, the degree of $Q_0(\lambda)$ is zero, and hence $Q_0(\lambda)$ is a constant a_0, where $a_0 \neq 0$ since otherwise $I = \{0\}$. Multiplying (5) by $1/a_0$ and writing $P_k^0(\lambda) = \tilde{P}_k^0(\lambda)/a_0$, we get (4), as required.

6.3. Canonical Form of the Matrix of an Arbitrary Operator

6.31. Let $\mathbf{A}$ denote an arbitrary linear operator acting in an n-dimensional space $\mathbf{K}_n$. Since the operations of addition and multiplication are defined for such operators (Secs. 4.31–4.33), with every polynomial

$$P(\lambda) = \sum_{k=0}^{m} a_k \lambda^k$$

we can associate an operator

$$P(\mathbf{A}) = \sum_{k=0}^{m} a_k \mathbf{A}^k$$

acting in the same space $\mathbf{K}_n$ (cf. Example 6.25e), where addition and multiplication of polynomials corresponds to addition and multiplication of the associated operators in the sense of Sec. 4.4. In fact, if

$$P(\lambda) = P_1(\lambda) + P_2(\lambda) = \sum_{k=0}^{m} a_k \lambda^k + \sum_{k=0}^{m} b_k \lambda^k = \sum_{k=0}^{m} (a_k + b_k)\lambda^k,$$

then

$$P(\mathbf{A}) = \sum_{k=0}^{m} (a_k + b_k)\mathbf{A}^k = \sum_{k=0}^{m} a_k \mathbf{A}^k + \sum_{k=0}^{m} b_k \mathbf{A}^k = P_1(\mathbf{A}) + P_2(\mathbf{A}).$$

Similarly, if

$$Q(\lambda) = P_1(\lambda)P_2(\lambda) = \sum_{k=0}^{m} a_k \lambda^k \sum_{j=0}^{m} b_j \lambda^j = \sum_{k=0}^{m} \sum_{j=0}^{m} a_k b_j \lambda^{k+j},$$

then

$$Q(\mathbf{A}) = \sum_{k=0}^{m} \sum_{j=0}^{m} a_k b_j \mathbf{A}^{k+j} = \sum_{k=0}^{m} a_k \mathbf{A}^k \sum_{j=0}^{m} b_j \mathbf{A}^j = P_1(\mathbf{A})P_2(\mathbf{A}),$$

by the distributive law for operator multiplication (Sec. 4.34). In particular, the operators $P_1(\mathbf{A})$ and $P_2(\mathbf{A})$ always commute.

Thus the mapping $\omega(P(\lambda)) = P(\mathbf{A})$ is an epimorphism (Sec. 6.24) of the algebra Π of all polynomials with coefficients in the field K into the algebra $\Pi_{\mathbf{A}}$ of all linear operators of the form $P(\mathbf{A})$ acting in the space $\mathbf{K}_n$. By Sec. 6.25d, the algebra $\Pi_{\mathbf{A}}$ is isomorphic to the factor algebra $\Pi/I_{\mathbf{A}}$, where $I_{\mathbf{A}}$ is the ideal consisting of all polynomials $P(\lambda)$ such that

$$\omega(P(\lambda)) = P(\mathbf{A}) = \mathbf{0}.$$

We now analyze the structure of this ideal.

6.32. As noted in Example 6.25f, the set of all linear operators acting in a space $\mathbf{K}_n$ is an algebra of dimension n^2 over the field K. Hence, given any operator $\mathbf{A}$, it follows that the first $n^2 + 1$ terms of the sequence

$$\mathbf{A}^0 = \mathbf{E}, \mathbf{A}, \mathbf{A}^2, \ldots, \mathbf{A}^m, \ldots$$

must be linearly dependent. Suppose that

$$\sum_{k=0}^{m} a_k \mathbf{A}^k = 0 \qquad (m \leqslant n^2).$$

Then, by the correspondence between polynomials and operators established in Sec. 6.31, the polynomial

$$Q(\lambda) = \sum_{k=0}^{m} a_k \lambda^k$$

must correspond to the zero operator. Every polynomial $Q(\lambda)$ for which the operator $Q(\mathbf{A})$ is the zero operator is called an *annihilating polynomial of the operator* $\mathbf{A}$. Thus we have just shown that *every operator* $\mathbf{A}$ *has an annihilating polynomial of degree* $\leqslant n^2$.

6.33. The set of all annihilating polynomials of the operator $\mathbf{A}$ is an ideal in the algebra Π. By Secs. 6.26–6.27 there is a polynomial $Q_0(\lambda)$ uniquely determined to within a numerical factor such that all annihilating polynomials are of the form $P(\lambda)Q_0(\lambda)$ where $P(\lambda)$ is an arbitrary polynomial in Π. In particular, $Q_0(\lambda)$ is the annihilating polynomial of lowest degree among all annihilating polynomials of the operator $\mathbf{A}$. Hence $Q_0(\lambda)$ is called the *minimal annihilating polynomial of the operator* $\mathbf{A}$.

6.34. THEOREM. *Let $Q(\lambda)$ be an annihilating polynomial of the operator* $\mathbf{A}$, *and suppose that*

$$Q(\lambda) = Q_1(\lambda)Q_2(\lambda),$$

where the factors $Q_1(\lambda)$ and $Q_2(\lambda)$ are relatively prime. Then the space $\mathbf{K}_n$ *can be represented as the direct sum*

$$\mathbf{K}_n = \mathbf{T}_1 + \mathbf{T}_2$$

of two subspaces $\mathbf{T}_1$ *and* $\mathbf{T}_2$ *both invariant with respect to the operator* $\mathbf{A}$,† *where*

$$Q_1(\mathbf{A})x_2 = 0, \qquad Q_2(\mathbf{A})x_1 = 0$$

for arbitrary $x_1 \in \mathbf{T}_1$, $x_2 \in \mathbf{T}_2$, so that $Q_1(\lambda)$ and $Q_2(\lambda)$ are annihilating polynomials for the operator $\mathbf{A}$ *acting in the subspaces* $\mathbf{T}_2$ *and* $\mathbf{T}_1$, *respectively.*

† Thus $x_1 \in \mathbf{T}_1$ implies $\mathbf{A}x_1 \in \mathbf{T}_1$ and similarly $x_2 \in \mathbf{T}_2$ implies $\mathbf{A}x_2 \in \mathbf{T}_2$.

Proof. By Sec. 6.28 there exist polynomials $P_1(\lambda)$ and $P_2(\lambda)$ such that

$$P_1(\lambda)Q_1(\lambda) + P_2(\lambda)Q_2(\lambda) \equiv 1,$$

and hence

$$P_1(\mathbf{A})Q_1(\mathbf{A}) + P_2(\mathbf{A})Q_2(\mathbf{A}) = \mathbf{E}.$$

Let $\mathbf{T}_k$ $(k = 1, 2)$ denote the range of the operator $Q_k(\mathbf{A})$, i.e., the set of all vectors of the form $Q_k(\mathbf{A})x$, $x \in \mathbf{K}_n$ (see Sec. 4.61). Then obviously $y = Q_k(\mathbf{A})x \in \mathbf{T}_k$ implies $\mathbf{A}y = Q_k(\mathbf{A})\mathbf{A}x \in \mathbf{T}_k$, so that the subspace $\mathbf{T}_k$ is invariant with respect to the operator $\mathbf{A}$. Given any $x_1 \in \mathbf{T}_1$, there is a vector $y \in \mathbf{K}_n$ such that

$$Q_2(\mathbf{A})x_1 = Q_2(\mathbf{A})Q_1(\mathbf{A})y = Q(\mathbf{A})y = 0,$$

and similarly, given any $x_2 \in \mathbf{T}_2$, there is a vector $z \in \mathbf{K}_n$ such that

$$Q_1(\mathbf{A})x_2 = Q_1(\mathbf{A})Q_2(\mathbf{A})z = Q(\mathbf{A})z = 0.$$

Moreover, given any $x \in \mathbf{K}_n$, we have

$$x = Q_1(\mathbf{A})P_1(\mathbf{A})x + Q_2(\mathbf{A})P_2(\mathbf{A})x = x_1 + x_2,$$

where

$$x_k = Q_k(\mathbf{A})P_k(\mathbf{A})x \in \mathbf{T}_k \qquad (k = 1, 2).$$

It follows that $\mathbf{K}_n$ is the sum of the subspaces $\mathbf{T}_1$ and $\mathbf{T}_2$. If $x_0 \in \mathbf{T}_1 \cap \mathbf{T}_2$, then $Q_1(\mathbf{A})x_0 = Q_2(\mathbf{A})x_0 = 0$, and hence

$$x_0 = P_1(\mathbf{A})Q_1(\mathbf{A})x_0 + P_2(\mathbf{A})Q_2(\mathbf{A})x_0 = 0.$$

Therefore $\mathbf{T}_1 \cap \mathbf{T}_2 = \{0\}$, and the sum $\mathbf{K}_n = \mathbf{T}_1 + \mathbf{T}_2$ is direct.† ∎

6.35. *Remark*. By construction, the operator $Q_1(\mathbf{A})$ annihilates the subspace $\mathbf{T}_2$, while the operator $Q_2(\mathbf{A})$ annihilates the subspace $\mathbf{T}_1$. We now show that *every vector x annihilated by the operator $Q_1(\mathbf{A})$ belongs to $\mathbf{T}_2$, while every vector x annihilated by the operator $Q_2(\mathbf{A})$ belongs to $\mathbf{T}_1$*. In fact, suppose $Q_1(\mathbf{A})x = 0$. We have $x = x_1 + x_2$ where $x_1 \in \mathbf{T}_1$, $x_2 \in \mathbf{T}_2$, and hence $Q_1(\mathbf{A})x_1 = Q_1(\mathbf{A})x - Q_1(\mathbf{A})x_2 = 0$ since $Q_1(\mathbf{A})x_2 = 0$. But $Q_2(\mathbf{A})x_1 = 0$ as well, since $x_1 \in \mathbf{T}_1$. It follows that

$$x_1 = P_1(\mathbf{A})Q_1(\mathbf{A})x_1 + P_2(\mathbf{A})Q_2(\mathbf{A})x_1 = 0, \qquad x = x_2 \in \mathbf{T}_2.$$

Similarly, $Q_2(\mathbf{A})x = 0$ implies $x \in \mathbf{T}_1$, and our assertion is proved.

6.36. Representing the polynomials $Q_1(\lambda)$ and $Q_2(\lambda)$ themselves as products of further prime factors, we can decompose the space $\mathbf{K}_n$ into smaller subspaces invariant with respect to the operator $\mathbf{A}$ and annihilated by the

† Naturally, the possibility is not excluded that one of the subspaces $\mathbf{T}_1$ and $\mathbf{T}_2$ consists of the zero vector alone.

appropriate factors of $Q_1(\lambda)$ and $Q_2(\lambda)$. Suppose the annihilating polynomial $Q(\lambda)$ has a factorization of the form

$$Q(\lambda) = \prod_{k=1}^{m} (\lambda - \lambda_k)^{r_k} \tag{6}$$

where $\lambda_1, \ldots, \lambda_m$ are all the (distinct) roots of $Q(\lambda)$ and r_k is the multiplicity of λ_k. For example, such a factorization is always possible (to within a numerical factor) in the field C of complex numbers. Then we have the following

THEOREM. *Suppose the operator* $\mathbf{A}$ *has an annihilating polynomial of the form* (6). *Then the space* $\mathbf{K}_n$ *can be represented as the direct sum*

$$\mathbf{K}_n = \mathbf{T}_1 + \cdots + \mathbf{T}_m$$

of m subspaces $\mathbf{T}_1, \ldots, \mathbf{T}_m$, *all invariant with respect to* $\mathbf{A}$, *where the subspace* $\mathbf{T}_k$ *is annihilated by* $\mathbf{B}_k^{r_k}$, *the* r_k*th power of the operator*

$$\mathbf{B}_k = \mathbf{A} - \lambda_k \mathbf{E}.$$

Proof. Apply Theorem 6.34 repeatedly to the factorization (6) of $Q(\lambda)$ into m relatively prime factors of the form $(\lambda - \lambda_j)^{r_j}$. ∎

6.37. By construction, the operator $\mathbf{B}_k$ is nilpotent in the subspace $\mathbf{T}_k$. Hence, by Sec. 6.14, in every subspace $\mathbf{T}_k$ $(\neq\{0\})$ we can choose a basis in which the matrix of $\mathbf{B}_k$ takes the canonical form (2). In this basis, the matrix of the operator $\mathbf{A} = \mathbf{B}_k + \lambda_k \mathbf{E}$ takes the form

$$\left\| \begin{array}{ccc} \boxed{\begin{array}{cccccc} \lambda_k & 1 & 0 & \cdots & 0 & 0 \\ 0 & \lambda_k & 1 & \cdots & 0 & 0 \\ \cdot & \cdot & \cdot & \cdots & \cdot & \cdot \\ 0 & 0 & 0 & \cdots & \lambda_k & 1 \\ 0 & 0 & 0 & \cdots & 0 & \lambda_k \end{array}} & & \\ & \boxed{\begin{array}{cccccc} \lambda_k & 1 & 0 & \cdots & 0 & 0 \\ 0 & \lambda_k & 1 & \cdots & 0 & 0 \\ \cdot & \cdot & \cdot & \cdots & \cdot & \cdot \\ 0 & 0 & 0 & \cdots & \lambda_k & 1 \\ 0 & 0 & 0 & \cdots & 0 & \lambda_k \end{array}} & \\ & & \ddots \\ & & & \boxed{\lambda_k} \end{array} \right\|. \tag{7}$$

Hence the matrix of the operator **A** in the whole space $\mathbf{K}_n = \mathbf{T}_1 + \cdots + \mathbf{T}_m$ takes the form

$$J(\mathbf{A}) = \left\| \begin{array}{cccccccc} \boxed{\begin{matrix}\lambda_1 & 1 & \cdots & 0\\ 0 & \lambda_1 & \cdots & 0\\ \cdot & \cdot & \cdots & \cdot\\ 0 & 0 & \cdots & 1\\ 0 & 0 & \cdots & \lambda_1\end{matrix}} & & & & & & & \\ & \boxed{\begin{matrix}\lambda_1 & 1 & \cdots & 0\\ 0 & \lambda_1 & \cdots & 0\\ \cdot & \cdot & \cdots & \cdot\\ 0 & 0 & \cdots & 1\\ 0 & 0 & \cdots & \lambda_1\end{matrix}} & & & & & & \\ & & \ddots & & & & & \\ & & & \boxed{\lambda_1} & & & & \\ & & & & \ddots & & & \\ & & & & & \boxed{\begin{matrix}\lambda_m & 1 & \cdots & 0\\ 0 & \lambda_m & \cdots & 0\\ \cdot & \cdot & \cdots & \cdot\\ 0 & 0 & \cdots & 1\\ 0 & 0 & \cdots & \lambda_m\end{matrix}} & & \\ & & & & & & \ddots & \\ & & & & & & & \boxed{\lambda_m} \end{array} \right\| \qquad (8)$$

in the basis obtained by combining all the canonical bases constructed in the spaces $\mathbf{T}_1, \ldots, \mathbf{T}_m$. Thus finally we have the following

THEOREM. *Given any operator* **A** *in an n-dimensional space* $\mathbf{K}_n$ *with an annihilating polynomial of the form* (6) (*in particular, any operator* **A** *in an n-dimensional complex space* $\mathbf{C}_n$), *there exists a basis, called a* ***Jordan basis,*** *in which the matrix of* **A** *takes the form* (8), *called the* ***Jordan canonical form*** *of* **A**.†

In the case $\mathbf{K}_n = \mathbf{C}_n$ the complex numbers $\lambda_1, \ldots, \lambda_n$ can be arranged in

† Synonymously, the *Jordan normal form* of **A**.

accordance with any rule, e.g., in order of increasing absolute value.† The representation (8) is not always possible in the case of an operator **A** acting in a space $\mathbf{K}_n \neq \mathbf{C}_n$. In Sec. 6.6 we will consider the canonical form of the matrix of an operator **A** acting in a real space $\mathbf{K}_n = \mathbf{R}_n$.

6.4. Elementary Divisors

6.41. The matrix (8) can be specified by a table

$$\begin{matrix} \lambda_1 : n_1^{(1)}, \ldots, n_{r_1}^{(1)} \\ \lambda_2 : n_1^{(2)}, \ldots, n_{r_2}^{(2)} \\ \cdots\cdots\cdots \\ \lambda_m : n_1^{(m)}, \ldots, n_{r_m}^{(m)} \end{matrix} \qquad (n_1^{(k)} \geqslant n_2^{(k)} \geqslant \cdots \geqslant n_{r_k}^{(k)}), \tag{9}$$

which for each diagonal element λ_k indicates the sizes $n_1^{(k)}, \ldots, n_{r_k}^{(k)}$ of the corresponding "elementary Jordan blocks" of the form

$$n_j^{(k)} \left\{ \left\| \begin{matrix} \lambda_k & 1 & 0 & \cdots & 0 \\ 0 & \lambda_k & 1 & \cdots & 0 \\ \cdot & \cdot & \cdot & \cdots & \cdot \\ 0 & 0 & 0 & \cdots & 1 \\ 0 & 0 & 0 & \cdots & \lambda_k \end{matrix} \right\| \right. \tag{10}$$

appearing in the matrix (8). We now show how to construct the table (9) and thereby determine the form of the matrix $J(\mathbf{A})$ of the operator **A**, from a knowledge of the matrix A of the operator **A** in any basis of the space $\mathbf{K}_n$.

6.42. As shown in Sec. 5.53, the characteristic polynomial of the operator **A** does not depend on the choice of a basis. Forming this polynomial for the Jordan basis, we get

$$\det (A - \lambda E) = \det (J(\mathbf{A}) - \lambda E) = \prod_{k=1}^{m} (\lambda_k - \lambda)^{n_1^{(k)} + \cdots + n_{r_k}^{(k)}}, \tag{11}$$

since every element below the principal diagonal in (8) is zero. Thus the numbers λ_k $(k = 1, \ldots, m)$ are the roots of the characteristic polynomial, and the numbers $r_k = n_1^{(k)} + \cdots + n_{r_k}^{(k)}$ are the multiplicities of these roots.

† Or in order of increasing argument θ (varying in the interval $0 \leqslant \theta < 2\pi$), in the case of identical absolute values.

Hence, by calculating the characteristic polynomial (which can be done by using the matrix A) and finding its roots, we can determine the quantities λ_k and $r_k = n_1^{(k)} + \cdots + n_{r_k}^{(k)}$ in the table (9).

6.43. Next (here and in Sec. 6.44) we show how to use the matrix A of the operator $\mathbf{A}$ in the original basis to calculate the numbers $n_j^{(k)}$ themselves. Since $J(\mathbf{A})$ and A are matrices of the same operator $\mathbf{A}$ in different bases, it follows from Sec. 5.51 that

$$J(\mathbf{A}) = T^{-1}AT,$$

where T is a nonsingular matrix, and hence that

$$J(\mathbf{A}) - \lambda E = T^{-1}(A - \lambda E)T.$$

The minors of a fixed order, say p, of the matrix $A - \lambda E$ are certain polynomials in λ of degree $\leqslant p$. Let $I_p(A)$ be the ideal in the algebra Π generated by all these minors, and let $I_p(J(\mathbf{A}))$ have the analogous meaning. Then the two ideals $I_p(A)$ and $I_p(J(\mathbf{A}))$ coincide. In fact, according to Sec. 4.54, every minor of order p of the matrix $J(\mathbf{A}) - \lambda E$ is a sum of products of minors of order p of the matrices $A - \lambda E$, T and T^{-1}. But the elements of T and T^{-1} are numbers. Thus every minor of order p of the matrix $J(\mathbf{A}) - \lambda E$ is simply a linear combination of minors of order p of the matrix $A - \lambda E$, and hence belongs to the ideal $I_p(A)$. By symmetry, every minor of order p of the matrix $A - \lambda E$ belongs to the ideal $I_p(J(\mathbf{A}))$. It follows that the ideals $I_p(A)$ and $I_p(J(\mathbf{A}))$ coincide, as asserted.

Now let $D_p(\lambda)$ be the polynomial generating this ideal. According to Sec. 6.26, $D_p(\lambda)$ is just the greatest common divisor of the polynomials generating $I_p(A)$. Thus the greatest common divisor of the minors of order p of the matrix $J(\mathbf{A}) - \lambda E$ is the same as the greatest common divisor of the minors of order p of the matrix $A - \lambda E$, and hence can be regarded as known. The greatest common divisor of the minors of order p of the matrix $J(\mathbf{A}) - \lambda E$ can be calculated directly as follows: Instead of the matrix $J(\mathbf{A}) - \lambda E$, we can again consider a matrix of the form $S(J(\mathbf{A}) - \lambda E)T$, where S and T are invertible numerical matrices (not containing λ). The operations of interchanging rows (or columns) and adding an arbitrary multiple of one row (or column) to another lead to matrices of just this kind (see Examples 4.44d–4.44g). We now assert that the elementary block

$$\begin{Vmatrix} \lambda_k - \lambda & 1 & 0 & \cdots & 0 \\ 0 & \lambda_k - \lambda & 1 & \cdots & 0 \\ \cdot & \cdot & \cdot & \cdots & \cdot \\ 0 & 0 & 0 & \cdots & 1 \\ 0 & 0 & 0 & \cdots & \lambda_k - \lambda \end{Vmatrix}$$

can be reduced to the form

$$n_j^{(k)}\left\{\left\|\begin{matrix} 1 & 0 & \cdots & 0 \\ 0 & 1 & \cdots & 0 \\ \cdot & \cdot & \cdots & \cdot \\ 0 & 0 & \cdots & (\lambda_k - \lambda)^{n_j^{(k)}} \end{matrix}\right\|\right. \tag{12}$$

by operations of the indicated type. In fact, to get (12) we first subtract the first row multiplied by $\lambda_k - \lambda$ from the second row, then the second row multiplied by $\lambda_k - \lambda$ from the third row, and so on. This gives the matrix

$$\left\|\begin{matrix} \lambda_k - \lambda & 1 & 0 & \cdots & 0 \\ -(\lambda_k - \lambda)^2 & 0 & 1 & \cdots & 0 \\ \cdot & \cdot & \cdot & \cdots & \cdot \\ (-1)^{q-2}(\lambda_k - \lambda)^{q-1} & 0 & 0 & \cdots & 1 \\ (-1)^{q-1}(\lambda_k - \lambda)^{q} & 0 & 0 & \cdots & 0 \end{matrix}\right\|,$$

where $q = n_j^{(k)}$. Then from the first column we subtract the second column multiplied by $\lambda_k - \lambda$, the third column multiplied by $-(\lambda_k - \lambda)^2$, etc., and finally the $(q-1)$th column multiplied by $(-1)^{q-2}(\lambda_k - \lambda)^{q-1}$. This gives the matrix

$$\left\|\begin{matrix} 0 & 1 & 0 & \cdots & 0 \\ 0 & 0 & 1 & \cdots & 0 \\ \cdot & \cdot & \cdot & \cdots & \cdot \\ 0 & 0 & 0 & \cdots & 1 \\ (-1)^{q-1}(\lambda_k - \lambda)^{q} & 0 & 0 & \cdots & 0 \end{matrix}\right\|,$$

from which the matrix (12) can be obtained by interchanging columns.†

We now calculate the greatest common divisor $D_p(\lambda)$ of the minors of order p of the matrix $\hat{J}(\lambda)$ with blocks of the form (12) along its principal diagonal. Since all nondiagonal elements of $\hat{J}(\lambda)$ vanish, the only minors of $\hat{J}(\lambda)$ which can be nonzero are those with the same set of row and column indices, and such a minor is simply equal to the product of its diagonal elements. Among the elements along the principal diagonal of the matrix $\hat{J}(\lambda)$, a certain number, say N, are binomials of the form $(\lambda_k - \lambda)^{n_j^{(k)}}$, while the other $n - N$ elements are all equal to 1. The number N is just the total number of Jordan blocks in the matrix $J(\mathbf{A})$, i.e., $N = r_1 + \cdots + r_m$. Clearly $D_p(\lambda) \equiv 1$ if $p \leqslant n - N$, since some of the minors of $\hat{J}(\lambda)$ of order $p \leqslant n - N$ are certainly equal to 1. Suppose we replace the matrix $\hat{J}(\lambda)$ by

† Except possibly for the sign of the element $(\lambda_k - \lambda)^q$, which is irrelevant to the subsequent determination of $D_p(\lambda)$.

the diagonal matrix

$$J(\lambda) = \left\| \begin{matrix} (\lambda_1 - \lambda)^{n_1^{(1)}} & & & & & & & & & \\ & \ddots & & & & & & & & \\ & & (\lambda_1 - \lambda)^{n_{r_1}^{(1)}} & & & & & & & \\ & & & (\lambda_2 - \lambda)^{n_1^{(2)}} & & & & & & \\ & & & & \ddots & & & & & \\ & & & & & (\lambda_m - \lambda)^{n_{r_m}^{(m)}} & & & & \\ & & & & & & 1 & & & \\ & & & & & & & \ddots & & \\ & & & & & & & & 1 \end{matrix} \right\| \begin{matrix} \\ \\ \\ \\ \\ \\ \left.\begin{matrix} \\ \\ \\ \end{matrix}\right\} n - N \end{matrix}$$

which obviously has the same polynomial $D_p(\lambda)$ as $J(\lambda)$. The greatest common divisor of the minors of order p of the matrix $J(\lambda)$ are clearly of the form

$$D_p(\lambda) = \prod_{k=1}^{m} (\lambda_k - \lambda)^{\mu_k^{(p)}}, \tag{13}$$

with nonnegative exponents $\mu_k(p)$. The exponents in (13) are easily found. For example, to determine $\mu_1(p)$, we note that $\mu_1(p)$ is the smallest exponent with which $\lambda_1 - \lambda$ appears in all minors of $J(\lambda)$ of order p. If $p \leqslant n - r_1$, then there is a minor of order p which does not contain $\lambda_1 - \lambda$ at all, so that $\mu_1(p) = 0$. However, if $p = n - r_1 + 1$, then, bearing in mind that the exponents $n_1^{(1)}, \ldots, n_{r_1}^{(1)}$ are arranged in decreasing order, we have

$$\mu_1(p) = n_{r_1}^{(1)}.$$

Moreover, each time p is increased further by 1, the exponent $\mu_1(p)$ increases, first by $n_{r_1-1}^{(1)}$, then by $n_{r_1-2}^{(1)}$, and so on, until finally we get $\mu_1(p) = n_{r_1}^{(1)} + \cdots + n_1^{(1)}$ for $p = n$. Similarly,

$$\mu_k(p) = \begin{cases} 0 & \text{if } p \leqslant n - r_k, \\ n_{r_k}^{(k)} & \text{if } p = n - r_k + 1, \\ n_{r_k}^{(k)} + n_{r_k-1}^{(k)} & \text{if } p = n - r_k + 2, \\ \cdots\cdots\cdots\cdots\cdots\cdots\cdots\cdots & \\ n_{r_k}^{(k)} + \cdots + n_1^{(k)} & \text{if } p = n. \end{cases}$$

Note that

$$\mu_k(n) - \mu_k(n-1) = n_1^{(k)},$$
$$\mu_k(n-1) - \mu_k(n-2) = n_2^{(k)},$$
$$\cdots\cdots\cdots\cdots\cdots\cdots\cdots\cdots$$
$$\mu_k(n - r_k + 1) - \mu_k(n - r_k) = n_{r_k}^{(k)},$$

so that

$$\mu_k(n - j + 1) - \mu_k(n - j) = n_j^{(k)} \qquad (j = 1, 2, \ldots, n - 1) \tag{14}$$

(we set $n_j^{(k)} = 0$ if $j > r_k$).

6.44. The ratio

$$E_p(\lambda) = \frac{D_{p+1}(\lambda)}{D_p(\lambda)}$$

is called an *elementary divisor of the operator* **A**. The elementary divisors, like the polynomials $D_p(\lambda)$ themselves, do not depend on the choice of a basis and hence can be calculated from the matrix of **A** in any basis. It follows from (13) that

$$E_p(\lambda) = \frac{\prod_{k=1}^{m} (\lambda_k - \lambda)^{\mu_k(p+1)}}{\prod_{k=1}^{m} (\lambda_k - \lambda)^{\mu_k(p)}} = \prod_{k=1}^{m} (\lambda_k - \lambda)^{\mu_k(p+1)-\mu_k(p)}$$

$$(p = 1, 2, \ldots, n - 1)$$

or equivalently,

$$E_{n-j}(\lambda) = \prod_{k=1}^{m} (\lambda_k - \lambda)^{\mu_j(n-j+1)-\mu_j(n-j)} \qquad (j = 1, 2, \ldots, n - 1).$$

Using (14), we get

$$E_{n-j}(\lambda) = \prod_{k=1}^{m} (\lambda_k - \lambda)^{n_j^{(k)}} \qquad (j = 1, 2, \ldots, n - 1),$$

where the roots of $E_{n-j}(\lambda)$ have multiplicities equal to the sizes of certain Jordan blocks in the matrix $J(\mathbf{A})$. Thus by calculating the elementary divisors of **A**, we can find the numbers $n_j^{(k)}$, thereby finally solving the problem of constructing the table (9).

6.45. ***Examples***

a. The "Jordan matrix"

$$\left\| \begin{array}{ccc|cc|c|cc|cc}
0 & 1 & 0 & & & & & & & \\
0 & 1 & 1 & & & & & & & \\
0 & 0 & 1 & & & & & & & \\
\hline
 & & & 1 & 1 & & & & & \\
 & & & 0 & 1 & & & & & \\
\hline
 & & & & & 1 & & & & \\
\hline
 & & & & & & 2 & 1 & & \\
 & & & & & & 0 & 2 & & \\
\hline
 & & & & & & & & 2 & 1 \\
 & & & & & & & & 0 & 2
\end{array} \right\|$$

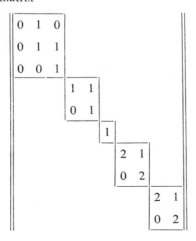

of order ten has three blocks of sizes 3, 2 and 1 corresponding to the root $\lambda_1 = 1$, and two blocks of sizes 2 and 2 corresponding to the root $\lambda_2 = 2$. Hence the elementary divisors are

$$\begin{aligned} E_9(\lambda) &= (1-\lambda)^3(2-\lambda)^2, \\ E_8(\lambda) &= (1-\lambda)^2(2-\lambda)^2, \\ E_7(\lambda) &= 1-\lambda, \\ E_6(\lambda) &= \cdots = E_1(\lambda) = 1. \end{aligned}$$

b. Suppose a given matrix $A = \|a_{ik}\|$ of order ten has elementary divisors

$$\begin{aligned} E_9(\lambda) &= (3-\lambda)^2(4-\lambda)^3, \\ E_8(\lambda) &= (3-\lambda)^2(4-\lambda), \\ E_7(\lambda) &= 4-\lambda, \\ E_6(\lambda) &= 4-\lambda, \\ E_5(\lambda) &= \cdots = E_1(\lambda) = 1 \end{aligned}$$

(calculated from the minors of the matrix $A - \lambda E$, as in Secs. 6.43–6.44). Then, according to Sec. 6.44, the Jordan matrix $J(\mathbf{A})$ has two blocks of sizes 2 and 2 corresponding to the root $\lambda_1 = 3$, and four blocks of sizes 3, 1, 1 and 1 corresponding to the root $\lambda_2 = 4$. It follows that

$$J(\mathbf{A}) = \left\| \begin{array}{cc|cc|ccc|c|c|c} 3 & 1 & & & & & & & & \\ 0 & 3 & & & & & & & & \\ \hline & & 3 & 1 & & & & & & \\ & & 0 & 3 & & & & & & \\ \hline & & & & 4 & 1 & 0 & & & \\ & & & & 0 & 4 & 1 & & & \\ & & & & 0 & 0 & 4 & & & \\ \hline & & & & & & & 4 & & \\ \hline & & & & & & & & 4 & \\ \hline & & & & & & & & & 4 \end{array} \right\| .$$

6.46. Thus from a knowledge of the elementary divisors of an operator **A**, we can determine all the numbers $n_j^{(k)}$ and hence the structure of the Jordan canonical form of **A**. In particular, we see that *the Jordan canonical form of an operator* **A** *is uniquely determined by* **A**.

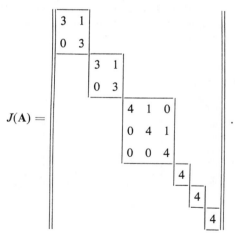

On the other hand, since the elementary divisors of an operator **A** are determined by the minors of the matrix $A - \lambda E$ in any basis, *two equivalent operators* **A** *and* **B**, i.e., *two operators with the same matrix in two (distinct) bases, have the same Jordan canonical form.* Conversely, it is obvious that *if two operators have the same Jordan canonical form, then they are equivalent.* This completely solves the problem of the equivalence of linear operators (in a complex space), posed at the beginning of the chapter.

6.5. Further Implications

6.51. If it is known that the operator **A** can be reduced to diagonal form, i.e., that its matrix has the form

$$A = \left\| \begin{matrix} \lambda_1 & & & & & & & & \\ & \ddots & & & & & & & \\ & & \lambda_1 & & & & & & \\ & & & \lambda_2 & & & & & \\ & & & & \ddots & & & & \\ & & & & & \lambda_2 & & & \\ & & & & & & \ddots & & \\ & & & & & & & & \lambda_m \end{matrix} \right\|$$

in some basis, then A is just the Jordan matrix of the operator **A** (all the Jordan blocks are of size 1). In particular, the elementary divisors all have simple roots. Conversely, if all the elementary divisors of an operator **A** have only simple roots, then the Jordan matrix $J(\mathbf{A})$ has blocks of size 1 only and hence is diagonal.

6.52. Given the Jordan canonical form of an operator **A**, we can easily find its minimal annihilating polynomial. Suppose the operator **B** has the matrix

$$\left\| \begin{matrix} 0 & 1 & 0 & \cdots & 0 \\ 0 & 0 & 1 & \cdots & 0 \\ \cdot & \cdot & \cdot & \cdots & \cdot \\ 0 & 0 & 0 & \cdots & 1 \\ 0 & 0 & 0 & \cdots & 0 \end{matrix} \right\|$$

in the basis $e_1, \ldots, e_p$, so that

$$\mathbf{B}e_1 = 0, \mathbf{B}e_2 = e_1, \ldots, \mathbf{B}e_p = e_{p-1}.$$

Then

$$\mathbf{B}^p x = 0$$

for every

$$x = \sum_{k=1}^{p} c_k e_k.$$

Thus λ^p is an annihilating polynomial of the operator **B**. The minimal annihilating polynomial is a divisor of λ^p (see Sec. 6.33), and hence must be of the form λ^m, $m \leqslant p$. But $\mathbf{B}^{p-1}e_p = e_1 \neq 0$, so that λ^p is in fact the minimal annihilating polynomial of **B**.

Now suppose the operator **A** has the matrix

$$\begin{Vmatrix} \lambda_0 & 1 & 0 & \cdots & 0 \\ 0 & \lambda_0 & 1 & \cdots & 0 \\ \cdot & \cdot & \cdot & \cdots & \cdot \\ 0 & 0 & 0 & \cdots & 1 \\ 0 & 0 & 0 & \cdots & \lambda_0 \end{Vmatrix}$$

in the same basis $e_1, \ldots, e_p$, so that $\mathbf{A} = \mathbf{B} + \lambda_0\mathbf{E}$. As just shown,

$$(\mathbf{A} - \lambda_0\mathbf{E})^p = \mathbf{B}^p = \mathbf{0},$$

and hence $(\lambda_0 - \lambda)^p$ is an annihilating polynomial of **A**, in fact the minimal annihilating polynomial, by the same argument as before.

Next suppose the operator **A** has the quasi-diagonal matrix

$$\begin{Vmatrix} \boxed{\begin{matrix} \lambda_0 & 1 & 0 & \cdots & 0 \\ 0 & \lambda_0 & 1 & \cdots & 0 \\ \cdot & \cdot & \cdot & \cdots & \cdot \\ 0 & 0 & 0 & \cdots & 1 \\ 0 & 0 & 0 & \cdots & \lambda_0 \end{matrix}} & & \\ & \ddots & \\ & & \boxed{\begin{matrix} \lambda_0 & 1 & 0 & \cdots & 0 \\ 0 & \lambda_0 & 1 & \cdots & 0 \\ \cdot & \cdot & \cdot & \cdots & \cdot \\ 0 & 0 & 0 & \cdots & 1 \\ 0 & 0 & 0 & \cdots & \lambda_0 \end{matrix}} \end{Vmatrix},$$

where the blocks along the diagonal have sizes $p_1 \geqslant p_2 \geqslant \cdots \geqslant p_r$. Then a polynomial $Q(\lambda)$ annihilating the operator **A** must annihilate each block separately. Clearly the polynomial $(\lambda_0 - \lambda)^{p_1}$ has this property (cf. Sec. 4.52), and in fact is the minimal annihilating polynomial, by the same argument as before.

Finally, in the general case where the operator **A** has the Jordan matrix described by the table (9), the polynomial

$$Q(\lambda) = \prod_{k=1}^{m} (\lambda_k - \lambda)^{n_1^{(k)}}$$

is clearly an annihilating polynomial of **A**, in fact the minimal annihilating polynomial, since none of the exponents $n_1^{(k)}$ can be lowered, for the reasons given above.

Thus *the polynomial $Q(\lambda)$ is the minimal annihilating polynomial of the operator* **A**. The degree of $Q(\lambda)$, equal to $n_1^{(1)} + \cdots + n_1^{(m)}$, is the sum of the sizes of the largest Jordan blocks, each corresponding to a root of the characteristic polynomial. Note that this number cannot exceed the order of the matrix A, i.e., the dimension n of the space in which the operator **A** acts. The characteristic polynomial

$$\det (A - \lambda E) = \prod_{k=1}^{m} (\lambda_k - \lambda)^{n_1^{(k)} + \cdots + n_{r_k}^{(k)}}$$

of the operator **A** (see Sec. 6.42) contains $Q(\lambda)$ as a factor, and hence *is also an annihilating polynomial* (a result known as the *Hamilton–Cayley theorem*). However, the characteristic polynomial is in general not the minimal annihilating polynomial of **A**. Clearly, the characteristic polynomial coincides with the minimal annihilating polynomial of **A** if and only if each root of the characteristic polynomial figures in only one Jordan block, of size equal to the multiplicity of the root.

6.6. The Real Jordan Canonical Form

6.61. Let **A** be a linear operator acting in a real n-dimensional space $\mathbf{R}_n$. Then in general there is no canonical basis in which the matrix of **A** takes the Jordan form (8), if only because the characteristic polynomial of **A** can have imaginary roots. Nevertheless, we can still find a modification of the Jordan matrix (8) suitable for the case of a real space.

Let $A = \|a_j^{(k)}\|$ be the matrix of the operator **A** in some basis $e_1, \ldots, e_n$ of the space $\mathbf{R}_n$, and consider the complex n-dimensional space $\mathbf{C}_n$ consisting of the vectors

$$x = \alpha_1 e_1 + \cdots + \alpha_n e_n,$$

where $\alpha_1, \ldots, \alpha_n$ are arbitrary *complex* numbers. The matrix A specifies a linear operator $\hat{\mathbf{A}}$ in the space $\mathbf{C}_n$ in accordance with the formula

$$\hat{\mathbf{A}}x = \sum_{k=1}^{n} \alpha_k \mathbf{A} e_k = \sum_{k=1}^{n} \alpha_k \left(\sum_{j=1}^{n} a_j^{(k)} e_j \right),$$

the same formula specifying the operator $\mathbf{A}$ itself for vectors x with real components α_k.

6.62. First we consider the case of an operator $\mathbf{A}$ with an annihilating polynomial of the special form

$$P(\lambda) = (\lambda^2 + \tau^2)^p,$$

where τ is a positive number. For the operator $\hat{\mathbf{A}}$ it makes sense to talk about polynomials $Q(\hat{\mathbf{A}})$ with complex coefficients, in particular, the polynomials $(\hat{\mathbf{A}} + i\tau)^p$ and $(\hat{\mathbf{A}} - i\tau)^p$. The polynomial $P(\lambda) = (\lambda^2 + \tau^2)^p$ is also an annihilating polynomial of the operator $\hat{\mathbf{A}}$. According to Theorem 6.34, the factorization

$$(\lambda^2 + \tau^2)^p = (\lambda - i\tau)^p(\lambda + i\tau)^p$$

corresponds to a decomposition of the space $\mathbf{C}_n$ into a direct sum of two subspaces $\mathbf{C}_n^1$ and $\mathbf{C}_n^2$, both invariant with respect to $\hat{\mathbf{A}}$, in which $\hat{\mathbf{A}}$ has annihilating polynomials $(\lambda - i\tau)^p$ and $(\lambda + i\tau)^p$, respectively. Moreover, if the subspace $\mathbf{C}_n^1$ consists of the vectors

$$x = \alpha_1 e_1 + \cdots + \alpha_m e_m$$

with arbitrary complex coefficients $\alpha_1, \ldots, \alpha_m$, then the subspace $\mathbf{C}_n^2$ consists of the vectors

$$\bar{x} = \bar{\alpha}_1 e_1 + \cdots + \bar{\alpha}_m e_m,$$

where $\bar{\alpha}_k$ is the complex conjugate of α_k ($k = 1, \ldots, m$). In fact, if

$$(\hat{\mathbf{A}} - i\tau\mathbf{E})^p x = 0, \tag{15}$$

then, taking complex conjugates in both factors of the left-hand side, we get

$$(\hat{\mathbf{A}} + i\tau\mathbf{E})^p \bar{x} = 0, \tag{15'}$$

and conversely.† In particular, it follows that n is even, i.e., $n = 2m$ where m is the dimension of each of the subspaces $\mathbf{C}_n^1$ and $\mathbf{C}_n^2$.

† The subspaces $\mathbf{C}_n^1$ and $\mathbf{C}_n^2$ are uniquely determined by (15) and (15′), respectively (see Sec. 6.35).

Now let f_j^k be the Jordan basis of the operator $\hat{\mathbf{A}}$ in the space $\mathbf{C}_n^1$, as in Sec. 6.37. According to (7), the matrix of $\hat{\mathbf{A}}$ in this basis is of the form

$$\left\|\begin{array}{ccc} n_1\left\{\begin{array}{|ccc|} \hline i\tau & 1 & 0 \\ 0 & i\tau & 1 \\ 0 & 0 & i\tau \\ \hline \end{array}\right. & & \\ & \ddots & \\ & & \left.\begin{array}{|ccc|} \hline i\tau & 1 & 0 \\ 0 & i\tau & 1 \\ 0 & 0 & i\tau \\ \hline \end{array}\right\} n_q \end{array}\right\|$$

Hence the action of $\hat{\mathbf{A}}$ on the basis vectors is described by the formulas

$$\begin{aligned} &\hat{\mathbf{A}}f_1^1 = i\tau f_1^1, && \ldots, \quad \hat{\mathbf{A}}f_1^q = i\tau f_1^q, \\ &\hat{\mathbf{A}}f_2^1 = f_1^1 + i\tau f_2^1, && \ldots, \quad \hat{\mathbf{A}}f_2^q = f_1^q + i\tau f_2^2, \\ &\cdots\cdots\cdots\cdots\cdots\cdots && \\ &\hat{\mathbf{A}}f_{n_1}^1 = f_{n_1-1}^1 + i\tau f_{n_1}^1, && \ldots, \quad \hat{\mathbf{A}}f_{n_q}^q = f_{n_q-1}^q + i\tau f_{n_q}^q. \end{aligned}$$

The action of $\hat{\mathbf{A}}$ on the complex conjugate vectors $\overline{f_j^k}$ in $\mathbf{C}_n^2$ is described by the complex conjugates of these formulas:

$$\begin{aligned} &\hat{\mathbf{A}}\overline{f_1^1} = -i\tau\overline{f_1^1}, && \ldots, \quad \hat{\mathbf{A}}\overline{f_1^q} = -i\tau\overline{f_1^q}, \\ &\hat{\mathbf{A}}\overline{f_2^1} = \overline{f_1^1} - i\tau\overline{f_2^1}, && \ldots, \quad \hat{\mathbf{A}}\overline{f_2^q} = \overline{f_1^q} - i\tau\overline{f_2^2}, \\ &\cdots\cdots\cdots\cdots\cdots\cdots && \\ &\hat{\mathbf{A}}\overline{f_{n_1}^1} = \overline{f_{n_1-1}^1} - i\tau\overline{f_{n_1}^1}, && \ldots, \quad \hat{\mathbf{A}}\overline{f_{n_q-1}^q} = \overline{f_{n_q-1}^q} - i\tau\overline{f_{n_q}^q}. \end{aligned}$$

Thus we see that the vectors $\overline{f_j^k}$ form a Jordan basis for the operator $\hat{\mathbf{A}}$ in the space $\mathbf{C}_n^2$. Hence all the vectors f_j^k, $\overline{f_j^k}$ taken together form a Jordan basis for the operator $\hat{\mathbf{A}}$ in the whole space $\mathbf{C}_n$.

We now construct a basis in the real space $\mathbf{R}_n$, by replacing each pair of complex vectors f_j^k and $\overline{f_j^k}$ by a pair of real vectors

$$g_j^k = \frac{1}{2}(f_j^k + \overline{f_j^k}), \qquad h_j^k = \frac{1}{2i}(f_j^k - \overline{f_j^k}). \tag{16}$$

It follows from the formulas

$$\begin{aligned} &\hat{\mathbf{A}}f_j^k = f_{j-1}^k + i\tau f_j^k, \\ &\hat{\mathbf{A}}\overline{f_j^k} = \overline{f_{j-1}^k} - i\tau\overline{f_j^k} \qquad (f_0^k = \overline{f_0^k} = 0) \end{aligned}$$

that

$$\hat{\mathbf{A}}\left\{\frac{1}{2}(f_j^k + \overline{f_j^k})\right\} \equiv \mathbf{A}g_j^k = g_{j-1}^k - \tau h_j^k,$$

$$\hat{\mathbf{A}}\left\{\frac{1}{2i}(f_j^k - \overline{f_j^k})\right\} \equiv \mathbf{A}h_j^k = h_{j-1}^k + \tau g_j^k \qquad (g_0^k = \overline{h_0^k} = 0).$$

Thus the action of the operator $\mathbf{A}$ on the vectors g_j^k and h_j^k is described by the formulas

$$\begin{aligned}
\mathbf{A}g_1^k &= \quad -\tau h_1^k,\\
\mathbf{A}h_1^k &= \tau g_1^k,\\
\mathbf{A}g_2^k &= g_1^k \quad -\tau h_2^k,\\
\mathbf{A}h_2^k &= h_1^k + \tau g_2^k,\\
&\cdots\cdots\cdots\cdots\cdots\cdots\\
\mathbf{A}g_{n_k}^k &= g_{n_k-1}^k \quad -\tau h_{n_k}^k,\\
\mathbf{A}h_{n_k}^k &= h_{n_k-1}^k + \tau g_{n_k}^k.
\end{aligned} \tag{17}$$

Moreover, (16) implies

$$f_j^k = g_j^k + ih_j^k, \qquad \overline{f_j^k} = g_j^k - ih_j^k.$$

Therefore the (complex) linear manifold spanned by all the vectors g_j^k, h_j^k is the same as the linear manifold spanned by all the vectors f_j^k, $\overline{f_j^k}$. But the number of vectors g_j^k, h_j^k is the same as the number of vectors f_j^k, $\overline{f_j^k}$. Hence the vectors g_j^k, h_j^k are linearly independent over the field C, just like the vectors f_j^k, $\overline{f_j^k}$. Thus, a fortiori, the vectors g_j^k, h_j^k are linearly independent over the field R, i.e., in the real space $\mathbf{R}_n$.

It follows from the formulas (17) that the matrix of the operator $\mathbf{A}$ in the basis g_j^k, h_j^k is a quasi-diagonal matrix, made up of blocks of the form

$$\left\|\begin{matrix}
0 & \tau & 1 & 0 & & & & & & \\
-\tau & 0 & 0 & 1 & & & & & & \\
 & & 0 & \tau & 1 & 0 & & & & \\
 & & -\tau & 0 & 0 & 1 & & & & \\
 & & & & 0 & \tau & & & & \\
 & & & & -\tau & 0 & & & & \\
 & & & & & & \ddots & & & \\
 & & & & & & & & & \\
 & & & & & & 0 & \tau & 1 & 0\\
 & & & & & & -\tau & 0 & 0 & 1\\
 & & & & & & & & 0 & \tau\\
 & & & & & & & & -\tau & 0
\end{matrix}\right\|, \tag{18}$$

of sizes $2n_1, \ldots, 2n_q$, respectively.

6.63. We now consider the general case. Let **A** be a linear operator in a real n-dimensional space $\mathbf{R}_n$, and let $P(\lambda)$ be an annihilating polynomial of $P(\lambda)$. Then $P(\lambda)$ has a factorization of the form

$$P(\lambda) = \prod_{k=1}^{m} (\lambda_k - \lambda)^{r_k} \prod_{l=1}^{s} [(\lambda - \sigma_l)^2 + \tau_l^2]^{p_l}$$

(to within a numerical factor) in the real domain, where λ_k $(k = 1, \ldots, n)$ are the distinct real roots of $P(\lambda)$ and $\sigma_l + i\tau_l = \mu_l$, $\sigma_l - i\tau_l = \bar{\mu}_l$ are the distinct imaginary roots of $P(\lambda)$. According to the general theory (Sec. 6.36), the space $\mathbf{R}_n$ can be represented as a direct sum

$$\mathbf{R}_n = \sum_{k=1}^{n} \mathbf{E}_k + \sum_{l=1}^{s} \mathbf{F}_l$$

of subspaces invariant with respect to **A**, where $(\lambda_k - \lambda)^{r_k}$ is an annihilating polynomial of the operator **A** in the subspace $\mathbf{E}_k$, while $(\sigma_l - \lambda)^2 + \tau_l^2$ is an annihilating polynomial of **A** in the subspace $\mathbf{F}_l$. In the subspace $\mathbf{E}_k$ the operator **A** can be reduced to the Jordan canonical form (7). As for the subspace $\mathbf{F}_l$, let $\mathbf{B}_l = \mathbf{A} - \sigma_l \mathbf{E}$. Then $(\lambda^2 + \tau_l^2)^{p_l}$ is an annihilating polynomial for the operator $\mathbf{B}_l$ in $\mathbf{F}_l$, and hence, by Sec. 6.62, there is a basis in which the matrix of $\mathbf{B}_l$ is of the form (18), with τ replaced by τ_l. In this same basis the matrix of the operator $\mathbf{A} = \mathbf{B}_l + \sigma_l \mathbf{E}$ is quasi-diagonal, made up of blocks of the form

$$\left\| \begin{array}{cccccccccc} \sigma_l & \tau_l & 1 & 0 & & & & & & \\ -\tau_l & \sigma_l & 0 & 1 & & & & & & \\ & & \sigma_l & \tau_l & 1 & 0 & & & & \\ & & -\tau_l & \sigma_l & 0 & 1 & & & & \\ & & & & \sigma_l & \tau_l & & & & \\ & & & & -\tau_l & \sigma_l & & & & \\ & & & & & & \cdot & & & \\ & & & & & & & \cdot & & \\ & & & & & & & & \cdot & \\ & & & & & & \sigma_l & \tau_l & 1 & 0 \\ & & & & & & -\tau_l & \sigma_l & 0 & 1 \\ & & & & & & & & \sigma_l & \tau_l \\ & & & & & & & & -\tau_l & \sigma_l \end{array} \right\| \qquad (19)$$

of sizes $2n_1, \ldots, 2n_q$, respectively. Thus we can choose a basis in the space $\mathbf{R}_n$ in which the matrix of the operator **A** consists of diagonal blocks of the form (10) and (19). This "real Jordan matrix" will be denoted by $J_R(\mathbf{A})$.

6.64. As in Sec. 6.4, the structure of the matrix $J_R(\mathbf{A})$ can be deduced from the elementary divisors of the operator **A**, which in turn can be calculated from the minors of the matrix $A - \lambda E$ in the original basis. Since the polynomials $D_p(\lambda)$ and $E_p(\lambda)$ are obtained from the minors $A - \lambda E$ by rational operations, the polynomials $E_p(\lambda)$ have real coefficients and hence are of the form

$$E_{n-j}(\lambda) = \prod_{k=1}^{m} (\lambda_k - \lambda)^{n_j^{(k)}} \prod_{l=1}^{s} [(\lambda - \sigma_l)^2 + \tau_l^2]^{p_j^{(l)}} \qquad (j = 1, 2, \ldots, n - 1)$$

(cf. Sec. 6.44). To every exponent $n_j^{(k)}$ there corresponds a Jordan block of size $n_j^{(k)}$, and to every exponent $p_j^{(l)}$ a block of the form (19) of size $2p_j^{(l)}$.

6.65. The above results can be summarized in the form of the following

THEOREM. *Given any operator* **A** *in a real n-dimensional space* $\mathbf{R}_n$, *there exists a basis in which the matrix of* **A** *is quasi-diagonal, made up of blocks of the form* (10) *and* (19), *where* λ_k $(k = 1, \ldots, m)$ *are the real roots and* $\sigma_l \pm i\tau_l$ $(l = 1, \ldots, s)$ *the complex roots of the characteristic polynomial of* **A**. *The sizes of the blocks are uniquely determined by the elementary divisors of* **A** *in the way indicated in Sec. 6.64.*

6.66. COROLLARY. *Every linear operator* **A** *in a real n-dimensional space* $\mathbf{R}_n$ *has an invariant subspace of dimension* 2.

Proof. The basis vectors g_1^k and h_1^k obviously generate a two-dimensional invariant subspace of **A** (see (17)). ▮

The number of distinct two-dimensional subspaces of **A** can always be estimated (from below). In fact, there are at least as many such subspaces as there are distinct diagonal blocks (19) of size $\geqslant 2$ in $J_R(\mathbf{A})$.

*6.7. Spectra, Jets and Polynomials

In many problems of algebra and analysis, the need arises to calculate various functions (in particular, polynomials) of given linear operators acting in a finite-dimensional space. Such functions, which have a number of special properties, will be investigated in the next two sections. A natural arithmetic model for functions of a single operator is the algebra of jets, with which we begin our discussion.

6.71. By a *spectrum*, denoted by S, we mean any set of points $\lambda_1, \ldots, \lambda_k$, where it is assumed that each point λ_k is assigned a "multiplicity," i.e., a positive integer r_k $(k = 1, \ldots, m)$, a fact indicated by writing

$$S = \{\lambda_1^{r_1}, \ldots, \lambda_m^{r_m}\}.$$

Moreover, we assume that each point λ_k is assigned a set of r_k numbers from the field K, denoted by

$$f(\lambda_k) = f^{(0)}(\lambda_k), f'(\lambda_k), \ldots, f^{(r_k-1)}(\lambda_k).$$

Such a set of numbers will be called a *jet* f, defined on S.

We now introduce the following algebraic operations in $\mathscr{J}(S)$, the set of all jets on a given spectrum S:

a. *Addition of jets.* By the *sum* $f+g$ of two jets $f = \{f^{(j)}(\lambda_k)\}$ and $g = \{g^{(j)}(\lambda_k)\}$ we mean the jet defined by the set of numbers

$$(f+g)^{(j)}(\lambda_k) = f^{(j)}(\lambda_k) + g^{(j)}(\lambda_k)$$

$(k = 1, \ldots, m; j = 0, 1, \ldots, r_k - 1)$.

b. *Multiplication of a jet by a number.* By the *product* αf *of a jet* $f = \{f^{(j)}(\lambda_k)\}$ *and a number* $\alpha \in K$ we mean the jet defined by the set of numbers

$$(\alpha f)^{(j)}(\lambda_k) = \alpha f^{(j)}(\lambda_k).$$

These two operations obviously convert the set $\mathscr{J}(S)$ into a linear space, whose zero element is the jet 0 whose "components" are all zero.

c. *Multiplication of jets.* By the *product* fg *of two jets* $f = \{f^{(j)}(\lambda_k)\}$ *and* $g = \{g^{(j)}(\lambda_k)\}$ we mean the jet defined by†

$$\begin{aligned}(fg)(\lambda_k) &= f(\lambda_k)g(\lambda_k),\\ (fg)'(\lambda_k) &= f(\lambda_k)g'(\lambda_k) + f'(\lambda_k)g(\lambda_k),\\ &\cdots\cdots\cdots\cdots\cdots\\ (fg)^{(j)}(\lambda_k) &= \sum_{l=0}^{j} C_l^j f^{(l)}(\lambda_k)g^{(j-l)}(\lambda_k)\end{aligned}$$

$(k = 1, \ldots, m; j = 0, 1, \ldots, r_k - 1)$, where C_l^j is the binomial coefficient

$$C_l^j = \frac{j!}{l!\,(j-l)!}.$$

It is easily verified that this operation is commutative and satisfies conditions 1)–3) of Sec. 6.21. Therefore $\mathscr{J}(S)$ is a commutative *algebra* over the field K. This algebra has a *unit*, i.e., a jet e such that $ef = f$ for every $f \in \mathscr{J}(S)$. In fact, we need only choose

$$e^{(j)}(\lambda_k) = \begin{cases} 1 & \text{if } j = 0, \\ 0 & \text{if } 0 < j \leqslant r_k \end{cases}$$

$(k = 1, \ldots, m)$.

---

† These formulas are formally identical with Leibniz's rule for repeated differentiation of the product of two functions f and g.

In what follows, *we will set up a correspondence between the algebra* $\mathscr{J}(S)$ *and the algebra of all polynomials with coefficients in the field K*, for the case where the points $\lambda_1, \ldots, \lambda_m$ all belong to K.

6.72. It will be assumed that the field K has infinitely many distinct elements. Making this assumption, we first show how to "reconstruct" the coefficients of a polynomial from a knowledge of its values.

a. Let

$$P(\lambda) = \sum_{k=0}^{p} a_k \lambda^k$$

be a polynomial with coefficients in the field K, whose argument λ also takes values in K. Then the coefficients $a_0, a_1, \ldots, a_p$ of $P(\lambda)$ are uniquely determined by the values of $P(\lambda)$. In fact, let $\lambda_0, \lambda_1, \ldots, \lambda_p$ be distinct elements of K, and consider the equations

$$\begin{aligned} a_0 + a_1\lambda_0 + \cdots + a_p\lambda_0^p &= P(\lambda_0), \\ a_0 + a_1\lambda_1 + \cdots + a_p\lambda_1^p &= P(\lambda_1), \\ &\cdots\cdots\cdots \\ a_0 + a_1\lambda_p + \cdots + a_p\lambda_p^p &= P(\lambda_p), \end{aligned}$$

which can be regarded as a system of $p + 1$ equations in the unknowns $a_0, a_1, \ldots, a_p$. The system has a nonvanishing determinant (see Example 1.55c), and hence, as asserted, has a unique solution by Cramer's rule (Sec. 1.73).

b. In particular, it follows that if two polynomials

$$P(\lambda) = \sum_{k=0}^{p} a_k \lambda^k, \qquad Q(\lambda) = \sum_{k=0}^{p} b_k \lambda^k$$

coincide for every value $\lambda \in K$, then

$$a_k = b_k \qquad (k = 0, 1, \ldots, p).$$

6.73. We will subsequently need the concept of the derivative of a polynomial $P(\lambda)$, and the notions of higher derivatives and Taylor's formula as well. In analysis these concepts are introduced for the case of polynomials which are functions of a real (or complex) argument, but here we are concerned with polynomials $P(\lambda)$ whose argument λ varies in an arbitrary field K. We must therefore introduce the corresponding definitions *independently*, i.e., without recourse to the notion of a limit which may not exist in the field K.

a. Fixing a point $\mu \in K$, we write the formula

$$\sum_{k=0}^{p} a_k \lambda^k = \sum_{k=0}^{p} a_k [\mu + (\lambda - \mu)]^k = \sum_{k=0}^{p} \frac{b_k(\mu)}{k!} (\lambda - \mu)^k, \tag{20}$$

where the quantities

$$\frac{b_k(\mu)}{k!} \qquad (k = 0, 1, \ldots, p)$$

are the polynomials in μ obtained after expanding $[\mu + (\lambda - \mu)]^k$ in powers of μ and $\lambda - \mu$ and collecting similar terms. The polynomials $b_k(\mu)$ are then given the following names:

$$b_0(\mu) = \sum_{k=0}^{p} a_k\mu^k \equiv P(\mu), \text{ the polynomial } P(\mu) \text{ itself},$$

$$b_1(\mu) = \sum_{k=1}^{p} ka_k\mu^{k-1} \equiv P'(\mu), \text{ the } \textit{first derivative} \text{ of } P(\mu),$$

$$b_2(\mu) = \sum_{k=2}^{p} k(k-1)a_k\mu^{k-2} \equiv P''(\mu), \text{ the } \textit{second derivative} \text{ of } P(\mu),$$

$$\cdots\cdots\cdots\cdots\cdots\cdots\cdots\cdots\cdots\cdots$$

$$b_p(\mu) = p(p-1)\cdots 1 \cdot a_p \equiv P^{(p)}(\mu), \text{ the } \textit{pth derivative} \text{ of } P(\mu).$$

For a polynomial of degree p, we set $P^{(q)}(\mu) \equiv 0$ if $q > p$.

In the new notation, formula (20) takes the form

$$P(\lambda) = \sum_{k=0}^{p} \frac{1}{k!} P^{(k)}(\mu)(\lambda - \mu)^k, \tag{20'}$$

known as *Taylor's formula for the polynomial* $P(\lambda)$.

b. In particular, for the polynomial

$$P(\lambda) = (\lambda - a)^p \qquad (a \in K),$$

we have

$$P(a) = P'(a) = \cdots = P^{(p-1)}(a) = 0,$$
$$P^{(p)}(\lambda) = p!,$$
$$P^{(q)}(\lambda) = 0 \qquad (q > p).$$

c. More generally, if

$$P(\lambda) = (\lambda - a)^p Q(\lambda),$$

we have

$$Q(\lambda) = \sum_{k=0}^{r} b_k(\lambda - a)^k, \qquad P(\lambda) = \sum_{k=0}^{r} b_k(\lambda - a)^{k+p},$$

and hence

$$P(a) = P'(a) = \cdots = P^{(p-1)}(a) = 0. \tag{21}$$

d. Conversely, if it is known that (21) holds, then

$$\begin{aligned} P(\lambda) &= \sum_{k=0}^{s} \frac{1}{k!} P^{(k)}(a)(\lambda - a)^k \\ &= (\lambda - a)^p \sum_{k=p}^{s} \frac{1}{k!} P^{(k)}(a)(\lambda - a)^{k-p} = (\lambda - a)^p Q(\lambda), \end{aligned}$$

where $Q(\lambda)$ is a new polynomial.

6.74. It should be noted that the representation of the polynomial $P(\lambda)$ in the form

$$\sum_{k=0}^{p} a_k\lambda^k \equiv P(\lambda) = \sum_{k=0}^{p} b_k(\mu)(\lambda - \mu)^k,$$

where the $b_k(\mu)$ are polynomials in μ, is necessarily unique. In fact, suppose we fix $\mu = \mu_0$ and give λ the distinct values $\lambda_0, \lambda_1, \ldots, \lambda_p$ in turn. Then $\tau = \lambda - \mu$ takes the distinct values $\lambda_0 - \mu_0, \lambda_1 - \mu_0, \ldots, \lambda_p - \mu_0$, and the values of the polynomial

$$\sum_{k=0}^{p} b_k(\mu_0)\tau^k$$

are known for these values of τ, being equal to $P(\lambda_0), P(\lambda_1), \ldots, P(\lambda_p)$. But then the quantities $b_k(\mu_0)$ are uniquely determined, by Sec. 6.72a. Since this is true for arbitrary $\mu = \mu_0 \in K$, the polynomials $b_k(\mu)$ $(k = 0, 1, \ldots, p)$ are themselves uniquely determined.

6.75. a. Given two polynomials $P(\lambda)$ and $Q(\lambda)$, we now verify the formulas

$$(P + Q)^{(k)}(\mu) = P^{(k)}(\mu) + Q^{(k)}(\mu), \tag{22}$$

$$(PQ)^{(k)}(\mu) = \sum_{j=0}^{k} C_j^k P^{(j)}(\mu)Q^{(k-j)}(\mu) \tag{23}$$

$(k = 0, 1, 2, \ldots)$, where

$$C_j^k = \frac{k!}{j!\,(k - j)!}.$$

In fact, by definition,

$$(P + Q)(\lambda) = \sum_{k=0}^{p} \frac{1}{k!} (P + Q)^{(k)}(\mu)(\lambda - \mu)^k,$$

$$P(\lambda) = \sum_{k=0}^{p} \frac{1}{k!} P^{(k)}(\mu)(\lambda - \mu)^k,$$

$$Q(\lambda) = \sum_{k=0}^{p} \frac{1}{k!} Q^{(k)}(\mu)(\lambda - \mu)^k,$$

$$P(\lambda) + Q(\lambda) = \sum_{k=0}^{p} \frac{1}{k!} [P^{(k)}(\mu) + Q^{(k)}(\mu)](\lambda - \mu)^k,$$

so that (22) follows from the uniqueness theorem of Sec. 6.74. Similarly,

$$(PQ)(\lambda) = \sum_{k=0}^{p} \frac{1}{k!} (PQ)^{(k)}(\mu)(\lambda - \mu)^k,$$

while on the other hand,

$$P(\lambda) = \sum_{j=0}^{p} \frac{1}{j!} P^{(j)}(\mu)(\lambda - \mu)^j, \qquad Q(\lambda) = \sum_{l=0}^{p} \frac{1}{l!} Q^{(l)}(\mu)(\lambda - \mu)^l,$$

$$P(\lambda)Q(\lambda) = \sum_{j=0}^{p} \sum_{l=0}^{p} \frac{1}{j!\, l!} P^{(j)}(\mu)Q^{(l)}(\mu)(\lambda - \mu)^{j+l}$$

$$= \sum_{k=0}^{2p} \left\{ \sum_{j=0}^{k} \frac{1}{j!\,(k-j)!} P^{(j)}(\mu)Q^{(k-j)}(\mu) \right\} (\lambda - \mu)^k.$$

Thus the uniqueness theorem of Sec. 6.74 implies

$$\frac{1}{k!} (PQ)^{(k)}(\mu) = \sum_{j=0}^{k} \frac{1}{j!\,(k-j)!} P^{(j)}(\mu)Q^{(k-j)}(\mu),$$

which is equivalent to (23).

b. In particular, formula (23) implies the following important

THEOREM. *If*

$$P^{(k)}(\mu) = 0 \qquad (k = 0, 1, \ldots, m),$$

then

$$(PQ)^{(k)}(\mu) = 0 \qquad (k = 0, 1, \ldots, m)$$

for any polynomial $Q(\lambda)$.

6.76. Now suppose we are given a spectrum

$$S = \{\lambda_1^{r_1}, \ldots, \lambda_m^{r_m}\} \qquad (\lambda_j \in K)$$

and the corresponding algebra $\mathscr{J}(S)$ of jets on S (see Sec. 6.71). Then with every polynomial $P(\lambda)$ we associate the jet $P \in \mathscr{J}(S)$ which assigns to λ_j the numbers

$$P(\lambda_k), P'(\lambda_k), \ldots, P^{(r_k-1)}(\lambda_k),$$

where the $P^{(j)}(\lambda_k)$ are the derivatives of the polynomial $P(\lambda)$, as defined in Sec. 6.73. It follows from formulas (22) and (23) that the operations on jets defined in Sec. 6.71 correspond to the usual operations of addition and multiplication of polynomials. Thus the mapping $P(\lambda) \to P$ is a morphism (Sec. 6.24) of the algebra of polynomials Π into the algebra of jets $\mathscr{J}(S)$. As we now show, this morphism is an *epimorphism*, i.e., given any jet f, we can find a polynomial $P(\lambda)$ such that

$$P(\lambda_k) = f(\lambda_k),\ P'(\lambda_k) = f'(\lambda_k), \ldots, P^{(r_k-1)}(\lambda_k) = f^{(r_k-1)}(\lambda_k)$$
$$(k = 1, \ldots, m).$$

To prove the assertion, it is enough to consider the case where all the numbers $f^{(j)}(\lambda_k)$ vanish except one, corresponding to any given value $k = k_1$. In fact, having solved the problem for this case, we need only construct a polynomial $P_k(\lambda)$ for each $k = 1, \ldots, m$ satisfying the conditions

$$P_k(\lambda_k) = f(\lambda_k), \ldots, P_k^{(r_k-1)}(\lambda_k) = f^{(r_k-1)}(\lambda_k), \tag{24}$$

$$P_k^{(j)}(\lambda_s) = 0 \qquad (s \neq k; j = 0, 1, \ldots, r_s - 1), \tag{25}$$

and the solution will then be given by the formula

$$P(\lambda) = P_1(\lambda) + \cdots + P_m(\lambda).$$

Thus we must find a polynomial $P_k(\lambda)$ satisfying the conditions (24) and (25). To this end, we look for $P_k(\lambda)$ in the form

$$P_k(\lambda) = Q_k(\lambda)R_k(\lambda), \tag{26}$$

where $Q_k(\lambda)$ is a new polynomial and

$$R_k(\lambda) = \prod_{s \neq k} (\lambda - \lambda_s)^{r_s}. \tag{27}$$

By Sec. 6.73c, we have

$$R_k^{(j)}(\lambda_s) = 0 \qquad (s \neq k; j = 0, 1, \ldots, r_s - 1),$$

and hence, by Theorem 6.75b,

$$P_k^{(j)}(\lambda_s) = 0 \qquad (s \neq k; j = 0, 1, \ldots, r_s - 1)$$

for any polynomial $Q_k(\lambda)$. Hence the condition (25) is clearly satisfied. We must still subject the polynomial $P_k(\lambda)$ to the condition (24). Since

$$R_k(\lambda_k) = \prod_{s \neq k} (\lambda_k - \lambda_s)^{r_s} \neq 0,$$

the condition

$$f(\lambda_k) = P_k(\lambda_k) = Q_k(\lambda_k)R_k(\lambda_k)$$

uniquely determines $Q_k(\lambda_k)$. Moreover, once $Q_k(\lambda_k)$ is known, the condition

$$f'(\lambda_k) = P_k'(\lambda_k) = Q_k'(\lambda_k)R_k(\lambda_k) + Q_k(\lambda_k)R_k'(\lambda_k)$$

uniquely determines $Q_k'(\lambda_k)$. Continuing in this way, we are able to uniquely determine all the numbers $Q_k(\lambda_k), Q_k'(\lambda_k), \ldots, Q_k^{(r_k-1)}(\lambda_k)$. But once these numbers are known, we can determine the desired polynomial $Q_k(\lambda)$ by using Taylor's formula

$$Q_k(\lambda) = \sum_{j=0}^{r_k-1} \frac{1}{j!} Q_k^{(j)}(\lambda_k)(\lambda - \lambda_k)^j. \tag{28}$$

Reasoning backwards, we see that the polynomial $P_k(\lambda)$ defined by formulas (26)–(28) satisfies the stipulated conditions (24) and (25).

6.77. Next, applying Sec. 6.52d, we find that the algebra $\mathscr{J}(S)$ of all jets defined on the given spectrum S is isomorphic to the factor algebra Π/I, where I is the ideal in Π consisting of all polynomials for which

$$P^{(j)}(\lambda_k) = 0 \qquad (k = 1, \ldots, m; j = 0, 1, \ldots, r_k - 1).$$

It follows from Sec. 6.73d that every polynomial $P(\lambda) \in I$ is divisible by the polynomial

$$T(\lambda) = \prod_{k=1}^{m} (\lambda - \lambda_k)^{r_k}, \tag{29}$$

and from Sec. 6.73c that every polynomial divisible by $T(\lambda)$ belongs to I. The ideal I, like every ideal in the algebra Π, is generated by the polynomial in I of lowest degree (see Sec. 6.26), and this polynomial is just $T(\lambda)$ itself. Hence *the algebra $\mathcal{J}(S)$ is isomorphic to the factor algebra Π/I, where I is the ideal generated by the polynomial $T(\lambda)$.*

6.78. We now use the result of Sec. 6.77 to solve the problem of describing all invertible elements (Sec. 6.21) of the algebra $\mathcal{J}(S)$.

Obviously, a jet f for which $f(\lambda_k) = 0$ for at least one value of k cannot be invertible, since then

$$(fg)(\lambda_k) = f(\lambda_k)g(\lambda_k) = 0 \neq 1 = e(\lambda_k)$$

for every jet g. Thus let f be a jet such that

$$f(\lambda_k) \neq 0 \qquad (k = 1, \ldots, m),$$

and let $P(\lambda)$ be the polynomial for which

$$P(\lambda_k) = f(\lambda_k), \ldots, P^{(r_k-1)}(\lambda_k) = f^{(r_k-1)}(\lambda_k) \qquad (k = 1, \ldots, m)$$

(see Sec. 6.76). This polynomial clearly has no factors in common with the polynomial $T(\lambda)$ defined by (29), and hence, by Sec. 6.28, there are polynomials $Q(\lambda)$ and $S(\lambda)$ such that

$$P(\lambda)Q(\lambda) + T(\lambda)S(\lambda) \equiv 1. \tag{30}$$

Let q be the jet corresponding to the polynomial $Q(\lambda)$. Applying the epimorphism $\Pi \to \mathcal{J}(S)$ constructed in Sec. 6.76 to equation (30), and using the fact that this epimorphism carries the polynomial $T(\lambda)$ into 0, we find that

$$fq = 1,$$

i.e., *the jet $f \in \mathcal{J}(S)$ is invertible.*

Let u be any invertible jet. Then, as we know from Sec. 6.21, the equation

$$ux = v$$

where x is an unknown jet and v any given jet, has the unique solution $x = v/u$. We can find an explicit expression for the ratio v/u by successively solving the equations

$$u(\lambda_k)x(\lambda_k) = v(\lambda_k),$$
$$u(\lambda_k)x'(\lambda_k) + u'(\lambda_k)x(\lambda_k) = v'(\lambda_k),$$
$$\cdots\cdots\cdots\cdots$$
$$\sum_{l=0}^{j} C_l^j u^{(l)}(\lambda_k)x^{(j-l)}(\lambda_k) = v^{(j)}(\lambda_k)$$

$(k = 1, \ldots, m; j = 0, 1, \ldots, r_k - 1)$.

6.79. a. A spectrum $S = \{\lambda_1^{r_1}, \ldots, \lambda_m^{r_m}\}$ with complex $\lambda_1, \ldots, \lambda_m$ is said to be *symmetric* if whenever S contains an imaginary number $\lambda_k = \sigma_k + i\tau_k$, it also contains the complex conjugate number $\bar{\lambda}_k = \sigma_k - i\tau_k$ with the same multiplicity r_k. A jet $f = \{f^{(j)}(\lambda_k)\}$ defined on a symmetric spectrum S is said to be *symmetric* if the numbers $f^{(j)}(\lambda_k)$ and $f^{(j)}(\bar{\lambda}_k)$ are complex conjugates $(j = 0, 1, \ldots, r_k - 1)$. If $P(\lambda)$ is a polynomial with real coefficients, then the jet defined on a symmetric spectrum by the numbers

$$P^{(j)}(\lambda_k) \qquad (k = 1, \ldots, m; j = 0, 1, \ldots, r_k - 1)$$

is symmetric, since the derivatives $P^{(j)}(\lambda)$ also have real coefficients and hence

$$P^{(j)}(\lambda_k) = \overline{P^{(j)}(\bar{\lambda}_k)}. \tag{31}$$

Conversely, given a symmetric jet $f = \{f^{(j)}(\lambda_k)\}$ on a symmetric spectrum $S = \{\lambda_1^{r_1}, \ldots, \lambda_m^{r_m}\}$, we can always find a polynomial $P_0(\lambda)$ with real coefficients such that

$$P_0^{(j)}(\lambda_k) = f^{(j)}(\lambda_k) \qquad (k = 1, \ldots, m; j = 0, 1, \ldots, r_k - 1).$$

In fact, by Sec. 6.76, we can construct a polynomial $P(\lambda)$ with complex coefficients satisfying the same conditions. Let $\bar{P}(\lambda)$ denote the polynomial whose coefficients are the complex conjugates of those of $P(\lambda)$. Then it follows from (31) that

$$\frac{1}{2}[P^{(j)}(\lambda_k) + \bar{P}^{(j)}(\lambda_k)] = \frac{1}{2}[P^{(j)}(\lambda_k) + \overline{P^{(j)}(\bar{\lambda}_k)}] = P^{(j)}(\lambda_k) = f^{(j)}(\lambda_k),$$

i.e., the polynomial

$$P_0(\lambda) = \frac{1}{2}[P(\lambda) + \bar{P}(\lambda)]$$

with real coefficients satisfies the required conditions.

b. The set of all symmetric jets f on a symmetric spectrum S obviously forms an algebra over the field of real numbers. According to Sec. 6.25d, *this algebra is isomorphic to the factor algebra Π/I, where Π is the algebra of all polynomials with real coefficients and $I \subset \Pi$ is the ideal consisting of all* *polynomials* $P(\lambda) \in \Pi$ *for which*

$$P^{(j)}(\lambda_k) = 0 \qquad (k = 1, \ldots, m; j = 0, 1, \ldots, r_k - 1),$$

i.e., the ideal generated by the (real) polynomial

$$T(\lambda) = \prod_{k=1}^{m} (\lambda - \lambda_k)^{r_k}.$$

*6.8. Operator Functions and Their Matrices

In this section we investigate functions of operators, finding matrices (and corresponding rules of operation) for polynomials of the form $P(\mathbf{A})$ and rational functions of the form $P(\mathbf{A})/Q(\mathbf{A})$, where $\mathbf{A}$ is any linear operator acting in an n-dimensional space $\mathbf{C}_n$ (or $\mathbf{R}_n$). In Sec. 6.89 we will extend the "calculus of operators" to the case of analytic functions of operators.

6.81. Given an operator $\mathbf{A}$ acting in an n-dimensional space $\mathbf{K}_n$, let $\Pi_{\mathbf{A}}$ be the algebra of all operators of the form $P(\mathbf{A})$, where $P(\lambda)$ is some polynomial. Then $\Pi_{\mathbf{A}}$ is isomorphic to the factor algebra $\Pi/I_{\mathbf{A}}$, where Π is the algebra of all polynomials and $I_{\mathbf{A}}$ is the ideal generated by the minimal annihilating polynomial $T(\lambda)$ of the operator $\mathbf{A}$ (see Secs. 6.31–6.33). Suppose it is known that the polynomial $T(\lambda)$ has the factorization

$$T(\lambda) = \prod_{k=1}^{m} (\lambda - \lambda_k)^{r_k} \tag{32}$$

in the field K. Then, by Sec. 6.77, the factor algebra $\Pi/I_{\mathbf{A}}$ is isomorphic to the algebra $\mathscr{J}(S)$ of all jets defined on the spectrum

$$S = S_{\mathbf{A}} = \{\lambda_1^{r_1}, \ldots, \lambda_m^{r_m}\}$$

(called the *spectrum of the operator* $\mathbf{A}$). Hence *the algebra* $\Pi_{\mathbf{A}}$ *is itself isomorphic to the algebra* $\mathscr{J}(S)$. The explicit form of this isomorphism is the following: To every jet $f \in \mathscr{J}(S)$ there corresponds the class of polynomials $P(\lambda) \in \Pi$ such that

$$P^{(j)}(\lambda_k) = f^{(j)}(\lambda_k) \qquad (k = 1, \ldots, m; j = 0, 1, \ldots, r_{k-1}), \tag{33}$$

and to each of these polynomials there corresponds the same uniquely defined polynomial operator $P(\mathbf{A})$, which we denote by $f(\mathbf{A})$.

Below we will investigate the explicit form of the matrix of the operator $P(\mathbf{A})$ for a given minimal annihilating polynomial (32), in the case where the matrix of $\mathbf{A}$ is in Jordan canonical form.

6.82. First suppose the operator **A** has a matrix (of order n) of the special form

$$\begin{Vmatrix} \lambda_0 & 1 & \cdots & 0 \\ 0 & \lambda_0 & \cdots & 0 \\ \cdot & \cdot & \cdots & \cdot \\ 0 & 0 & \cdots & 1 \\ 0 & 0 & \cdots & \lambda_0 \end{Vmatrix} \tag{34}$$

in some basis of the space $\mathbf{K}_n$. Then **A** is of the form $\lambda_0\mathbf{E} + \mathbf{B}$, where the operator **B** has the matrix

$$\begin{Vmatrix} 0 & 1 & \cdots & 0 \\ 0 & 1 & \cdots & 0 \\ \cdot & \cdot & \cdots & \cdot \\ 0 & 0 & \cdots & 1 \\ 0 & 0 & \cdots & 0 \end{Vmatrix}.$$

According to Example 4.74b, the matrix of $\mathbf{B}^k$ is

$$\begin{matrix} & & (k+1) & & & \\ \end{matrix}$$

$$\begin{Vmatrix} 0 & \cdots & 1 & 0 & \cdots & 0 \\ 0 & \cdots & 0 & 1 & \cdots & 0 \\ \cdot & \cdots & \cdot & \cdot & \cdots & \cdot \\ 0 & \cdots & 0 & 0 & \cdots & 1 \\ \cdot & \cdots & \cdot & \cdot & \cdots & \cdot \\ 0 & \cdots & \cdot & \cdot & \cdots & \cdot \end{Vmatrix} \begin{matrix} \\ \\ \\ (n-k), \\ \\ \\ \end{matrix} \tag{35}$$

where the diagonal consisting entirely of ones has moved over k steps to the right from the principal diagonal. If $P(\lambda)$ is an arbitrary polynomial of degree p, then

$$P(\lambda) = \sum_{k=0}^{p} \frac{1}{k!} P^{(k)}(\lambda_0)(\lambda - \lambda_0)^k,$$

by Taylor's formula (20′). Replacing λ by the operator **A**, we get the identity

$$P(\mathbf{A}) = \sum_{k=0}^{p} \frac{1}{k!} P^{(k)}(\lambda_0)(\mathbf{A} - \lambda_0\mathbf{E})^k = \sum_{k=0}^{p} \frac{1}{k!} P^{(k)}(\lambda_0)\mathbf{B}^k.$$

Then, taking account of the expression (35) for the matrix of $\mathbf{B}^k$, we find that

$P(\mathbf{A})$ has the matrix

$$\left\|\begin{matrix} P(\lambda_0) & P'(\lambda_0) & \frac{1}{2}P''(\lambda_0) & \cdots & \frac{1}{(n-1)!}P^{(n-1)}(\lambda_0) \\ 0 & P(\lambda_0) & P'(\lambda_0) & \cdots & \frac{1}{(n-2)!}P^{(n-2)}(\lambda_0) \\ \cdot & \cdot & \cdot & \cdots & \cdot \\ 0 & 0 & 0 & \cdots & P(\lambda_0) \end{matrix}\right\|. \tag{36}$$

Note that to construct the matrix of $P(\mathbf{A})$ from the polynomial $P(\lambda)$, we only need the n values $P(\lambda_0), P'(\lambda_0), \ldots, P^{(n-1)}(\lambda_0)$, where n is the order of the matrix of $\mathbf{A}$.

6.83. Next suppose the operator $\mathbf{A}$ has a quasi-diagonal matrix of order n, made up of blocks of the form (34), where λ_0 takes the values $\lambda_1, \ldots, \lambda_m$ with corresponding block sizes $n_1, \ldots, n_m$. By the rules for operating on quasidiagonal matrices (Sec. 4.52), each block of the matrix of the operator $P(\mathbf{A})$ can be calculated independently. Applying Sec. 6.82, we find that *the matrix of* $P(\mathbf{A})$ *is obtained by replacing each block* (34) *of the matrix of* $\mathbf{A}$ *by the block* (36). Thus to construct the matrix of $P(\mathbf{A})$, we now need the values

$$P^{(j)}(\lambda_k) \qquad (k = 1, \ldots, m; j = 0, 1, \ldots, n_k - 1).$$

6.84. Let $\mathbf{A}$ be any operator acting in an n-dimensional complex space $\mathbf{C}_n$. Then, as on pp. 146–147, there exists a basis in which the matrix of $\mathbf{A}$ is quasi-diagonal, made up of blocks of the form

$$n_j^{(k)}\left\{\left\|\begin{matrix} \lambda_k & 1 & 0 & \cdots & 0 \\ 0 & \lambda_k & 1 & \cdots & 0 \\ \cdot & \cdot & \cdot & \cdots & \cdot \\ 0 & 0 & 0 & \cdots & 1 \\ 0 & 0 & 0 & \cdots & \lambda_k \end{matrix}\right\|\right. \qquad (k = 1, \ldots, m; j = 1, \ldots, r_k), \tag{37}$$

where the numbers r_k and $n_j^{(k)}$ are those figuring in table (9). Correspondingly, the spectrum of the operator $\mathbf{A}$ is

$$S = S_{\mathbf{A}} = \{\lambda_1^{r_1}, \ldots, \lambda_m^{r_m}\}.$$

If

$$f = \{f^{(j)}(\lambda_k)\} \qquad (k = 1, \ldots, m; j = 0, 1, \ldots, r_k - 1)$$

is any jet defined on S, then, by Secs. 6.81–6.83, the corresponding operator $f(\mathbf{A})$ has a quasi-diagonal matrix, in which each block of the form (37) is replaced by the block

$$\left\|\begin{matrix} f(\lambda_k) & f'(\lambda_k) & \frac{1}{2} f''(\lambda_k) & \cdots & \frac{1}{(n_j^{(k)}-1)!} f^{(n_j^{(k)}-1)}(\lambda_k) \\ 0 & f(\lambda_k) & f'(\lambda_k) & \cdots & \frac{1}{(n_j^{(k)}-2)!} f^{(n_j^{(k)}-2)}(\lambda_k) \\ \cdot & \cdot & \cdot & \cdots & \cdot \\ 0 & 0 & 0 & \cdots & f(\lambda_k) \end{matrix}\right\|. \tag{38}$$

The isomorphism between the algebras $\Pi_{\mathbf{A}}$ and $\mathscr{J}(S)$ has now been made perfectly explicit.

6.85. a. Next we consider functions of an operator **A** which has a matrix of order $2m$ of the form

$$\left\|\begin{matrix} \sigma & \tau & 1 & 0 & & & & \\ -\tau & \sigma & 0 & 1 & & & & \\ & & \sigma & \tau & & & & \\ & & -\tau & \sigma & & & & \\ & & & & \cdot & & & \\ & & & & & \cdot & & \\ & & & & & & \cdot & \\ & & & & & & \sigma & \tau \\ & & & & & & -\tau & \sigma \end{matrix}\right\| \tag{39}$$

where σ and τ are elements of the field K. Introducing the 2×2 matrices

$$E = \left\|\begin{matrix} 1 & 0 \\ 0 & 1 \end{matrix}\right\|, \qquad \Lambda = \left\|\begin{matrix} \sigma & \tau \\ -\tau & \sigma \end{matrix}\right\|,$$

we can write the matrix of **A** as the following block matrix of order m:

$$\left\|\begin{matrix} \Lambda & E & 0 & \cdots & 0 & 0 \\ 0 & \Lambda & E & \cdots & 0 & 0 \\ \cdot & \cdot & \cdot & \cdots & \cdot & \cdot \\ 0 & 0 & 0 & \cdots & \Lambda & E \\ 0 & 0 & 0 & \cdots & 0 & \Lambda \end{matrix}\right\|$$

$$= \left\|\begin{matrix} \Lambda & 0 & 0 & \cdots & 0 & 0 \\ 0 & \Lambda & 0 & \cdots & 0 & 0 \\ \cdot & \cdot & \cdot & \cdots & \cdot & \cdot \\ 0 & 0 & 0 & \cdots & \Lambda & 0 \\ 0 & 0 & 0 & \cdots & 0 & \Lambda \end{matrix}\right\| + \left\|\begin{matrix} 0 & E & 0 & \cdots & 0 & 0 \\ 0 & 0 & E & \cdots & 0 & 0 \\ \cdot & \cdot & \cdot & \cdots & \cdot & \cdot \\ 0 & 0 & 0 & \cdots & 0 & E \\ 0 & 0 & 0 & \cdots & 0 & 0 \end{matrix}\right\|.$$

Therefore it follows from Sec. 6.82 and the rule for multiplication of block matrices (Sec. 4.51) that the matrix of $P(\mathbf{A})$ can be written in the form of the block matrix

$$\begin{Vmatrix} P(\Lambda) & P'(\Lambda) & \frac{1}{2}P''(\Lambda) & \cdots & \frac{1}{(m-1)!}P^{(m-1)}(\Lambda) \\ 0 & P(\Lambda) & P'(\Lambda) & \cdots & \frac{1}{(m-2)!}P^{(m-2)}(\Lambda) \\ \cdot & \cdot & \cdot & \cdots & \cdot \\ 0 & 0 & 0 & \cdots & P(\Lambda) \end{Vmatrix}. \qquad (40)$$

b. If the matrix of $\mathbf{A}$ is quasi-diagonal, made up of blocks of the form (34) and (39), then, just as in Sec. 6.83, we deduce that the matrix of $P(\mathbf{A})$ is obtained by replacing each block by the corresponding block of the form (36) or (40).

c. In the case where $K = R$, so that the numbers σ, τ and the polynomial $P(\lambda)$ are real, we can find the explicit form of the matrices $P^{(k)}(\Lambda)$ figuring in (40). In fact, introducing the matrix

$$I = \begin{Vmatrix} 0 & 1 \\ -1 & 0 \end{Vmatrix},$$

we easily verify that $I^2 = -E$, so that the algebra of real matrices

$$\Lambda = \sigma E + \tau I = \begin{Vmatrix} \sigma & \tau \\ -\tau & \sigma \end{Vmatrix} = \begin{Vmatrix} \operatorname{Re}\lambda & \operatorname{Im}\lambda \\ -\operatorname{Im}\lambda & \operatorname{Re}\lambda \end{Vmatrix} \qquad (\lambda = \sigma + i\tau)$$

is isomorphic to the ordinary algebra of complex numbers (cf. Example 4.74a). Hence for any polynomial $P(\lambda)$ with real coefficients we have

$$P(\Lambda) = P(\sigma E + \tau I) = \begin{Vmatrix} \operatorname{Re} P(\lambda) & \operatorname{Im} P(\lambda) \\ -\operatorname{Im} P(\lambda) & \operatorname{Re} P(\lambda) \end{Vmatrix} \qquad (\lambda = \sigma + i\tau),$$

and correspondingly

$$P^{(k)}(\Lambda) = P^{(k)}(\sigma E + \tau I) = \begin{Vmatrix} \operatorname{Re} P^{(k)}(\lambda) & \operatorname{Im} P^{(k)}(\lambda) \\ -\operatorname{Im} P^{(k)}(\lambda) & \operatorname{Re} P^{(k)}(\lambda) \end{Vmatrix}.$$

6.86. Let $K = R$ and $\mathbf{K}_n = \mathbf{R}_n$. Then, given any operator $\mathbf{A}$ acting in $\mathbf{K}_n$, the minimal annihilating polynomial $T(\lambda)$ has real coefficients and hence has a symmetric spectrum $S_{\mathbf{A}}$ (see Sec. 6.79a). The algebra $\Pi_{\mathbf{A}}$ of operators of the form $P(\mathbf{A})$ is isomorphic to the factor algebra $\Pi/I_{\mathbf{A}}$, where Π is the algebra of polynomials with real coefficients and $I_{\mathbf{A}}$ is the ideal generated by the minimal annihilating polynomial of the operator $\mathbf{A}$. According to Sec. 6.79b, this factor algebra is isomorphic to the algebra of symmetric jets on the spectrum $S_{\mathbf{A}}$. On the other hand, there is a basis in which the matrix of **A** is quasi-diagonal, with diagonal blocks of the form (34) and (39). Let f be any symmetric jet on the spectrum $S_{\mathbf{A}}$. Then it follows from the above considerations that the corresponding matrix $f(\mathbf{A})$ is obtained by replacing every block (34) by a block (38) and every block (39) of size $2m$ by the block matrix

$$\begin{Vmatrix} f(\Lambda) & f'(\Lambda) & \cdots & \dfrac{1}{(m-1)!}f^{(m-1)}(\Lambda) \\ \cdot & f(\Lambda) & \cdots & \dfrac{1}{(m-2)!}f^{(m-2)}(\Lambda) \\ \cdot & \cdot & \cdots & \cdot \\ 0 & 0 & \cdots & f(\Lambda) \end{Vmatrix}$$

of order m, where the $f^{(k)}(\Lambda)$ are 2×2 matrices of the form

$$f^{(k)}(\Lambda) = \begin{Vmatrix} \operatorname{Re} f^{(k)}(\lambda) & \operatorname{Im} f^{(k)}(\lambda) \\ -\operatorname{Im} f^{(k)}(\lambda) & \operatorname{Re} f^{(k)}(\lambda) \end{Vmatrix}.$$

6.87. Given a linear operator **A** acting in a space $\mathbf{C}_n$, suppose **A** has the Jordan canonical form (8) specified by the table (9), as on pp. 146–147. We now look for all *invertible* operators of the form $P(\mathbf{A})$, where $P(\lambda)$ is a polynomial. It is clear from the form of the operator of the matrix of $P(\mathbf{A})$ in the Jordan basis of the operator **A** that the determinant of this matrix is just

$$\prod_{k=1}^{m} [P(\lambda_k)]^{r_k}, \qquad \sum_{k=1}^{m} r_k = n$$

(cf. Example 1.55b). Therefore the operator $P(\mathbf{A})$ is invertible in the algebra $\mathbf{L}(\mathbf{C}_n)$ of all linear operators acting in the space $\mathbf{C}_n$ if and only if

$$P(\lambda_k) \neq 0 \qquad (k = 1, \ldots, m). \tag{41}$$

Moreover, if the condition (41) is satisfied, then the inverse operator $[P(\mathbf{A})]^{-1}$ already belongs to the algebra $\Pi_{\mathbf{A}}$. In fact, in this case the jet p corresponding to the polynomial $P(\lambda)$ in the algebra of jets $\mathscr{J}(S_{\mathbf{A}})$, i.e., the jet consisting of the numbers

$$P^{(j)}(\lambda_k) \qquad (k = 1, \ldots, m; j = 0, 1, \ldots, r_k - 1),$$

is invertible in the algebra $\mathscr{J}(S_{\mathbf{A}})$, by Sec. 6.78. But then the operator $P(\mathbf{A})$ is invertible in the algebra $\Pi_{\mathbf{A}}$, by the isomorphism between the algebras $\mathscr{J}(S_{\mathbf{A}})$ and $\Pi_{\mathbf{A}}$.

Again using the isomorphism between the algebras $\mathscr{J}(S_{\mathbf{A}})$ and $\Pi_{\mathbf{A}}$, we see that if $P(\mathbf{A})$ is invertible, then the equation

$$P(\mathbf{A})X(\mathbf{A}) = Q(\mathbf{A}),$$

where $X(\lambda)$ is an unknown polynomial and $Q(\lambda)$ any given polynomial, has the unique solution $X(\mathbf{A}) = Q(\mathbf{A})/P(\mathbf{A})$. Let x, p and q be the jets corresponding to the polynomials $X(\lambda)$, $P(\lambda)$ and $Q(\lambda)$, respectively, so that in particular $px = q, x = q/p$. Then, according to Secs. 6.78 and 6.84, the matrix of the operator $X(\mathbf{A})$ in the Jordan basis of the operator $\mathbf{A}$ is obtained by replacing every block of the form (36) by a block of the form

$$\left\| \begin{matrix} \dfrac{q(\lambda_k)}{p(\lambda_k)} & \left(\dfrac{q(\lambda)}{p(\lambda)}\right)'_{\lambda=\lambda_k} & \dfrac{1}{2}\left(\dfrac{q(\lambda)}{p(\lambda)}\right)''_{\lambda=\lambda_k} & \cdots \\ 0 & \dfrac{q(\lambda_k)}{p(\lambda_k)} & \left(\dfrac{q(\lambda)}{p(\lambda)}\right)'_{\lambda=\lambda_k} & \cdots \\ 0 & 0 & \dfrac{q(\lambda_k)}{p(\lambda_k)} & \cdots \\ \cdot & \cdot & \cdot & \cdots \end{matrix} \right\|. \tag{42}$$

6.88. The above result can be interpreted somewhat differently. Given a spectrum $S = \{\lambda_1^{r_1}, \ldots, \lambda_m^{r_m}\}$ in the complex plane, let $\mathscr{R}(S)$ denote the set of all complex rational functions

$$f(\lambda) = \frac{Q(\lambda)}{P(\lambda)},$$

where $P(\lambda)$ and $Q(\lambda)$ are polynomials, and $P(\lambda)$ has no roots at the points of the set S. In the set $\mathscr{R}(S)$ we define the operations of addition of two functions, multiplication of a function by a complex number, and multiplication of two functions in accordance with the usual rules, thereby making $\mathscr{R}(S)$ into an algebra over the field C. Moreover, we note that every function $f(\lambda) \in \mathscr{R}(S)$ has derivatives $f'(\lambda), f''(\lambda), \ldots$ in the usual sense of analysis. Assigning to each function $f(\lambda) \in \mathscr{R}(S)$ the jet

$$f = \{f^{(j)}(\lambda_k)\} \qquad (k = 1, \ldots, m; j = 0, 1, \ldots, r_k - 1),$$

where $f^{(j)}(\lambda_k)$ denotes the usual jth derivative of $f(\lambda)$, we get a morphism of the algebra $\mathscr{R}(S)$ of rational functions into the algebra $\mathscr{J}(S)$ of jets on the spectrum S, in fact an epimorphism, since by Sec. 6.76 the jets corresponding to just the polynomials $Q(\lambda)$ already fill the whole algebra $\mathscr{J}(S)$.

Now let $S = S_{\mathbf{A}}$ be the spectrum of some operator $\mathbf{A}$ acting in the space $\mathbf{C}_n$. Then the algebra $\Pi_{\mathbf{A}}$ of operators $P(\mathbf{A})$ is isomorphic to the algebra of jets $\mathscr{J}(S_{\mathbf{A}})$, and we can extend the given epimorphism $\mathscr{R}(S_{\mathbf{A}}) \to \mathscr{J}(S_{\mathbf{A}})$ to an epimorphism $\mathscr{R}(S_{\mathbf{A}}) \to \Pi_{\mathbf{A}}$. In other words, we can assign to each rational function $f(\lambda) \in \mathcal{R}(S)$ a linear operator $f(\mathbf{A}) \in \Pi_{\mathbf{A}}$ such that the correspondence $f(\lambda) \rightarrow f(\mathbf{A})$ is again an epimorphism, where the matrix of the operator $f(\mathbf{A})$ is given by the rule (42).

6.89. Instead of the algebra of rational functions, we can consider the algebra of analytic functions. Thus let $\mathcal{F}(S)$ be the set of all functions $f(\lambda)$ analytic at the points $\lambda_1, \ldots, \lambda_m$, i.e., analytic in a neighborhood of each of the points $\lambda_1, \ldots, \lambda_m$. Then the set $\mathcal{F}(S)$ equipped with the usual operations of addition and multiplication is again an algebra over the field C, in fact an algebra containing the algebra $\mathcal{R}(S)$. Analytic functions also have derivatives of all orders (in the usual sense of analysis), and using them, we can extend the epimorphism $\mathcal{R}(S_{\mathbf{A}}) \rightarrow \Pi_{\mathbf{A}}$ constructed in Sec. 6.88 to an epimorphism $\mathcal{F}(S_{\mathbf{A}}) \rightarrow \Pi_{\mathbf{A}}$. An important feature of this new epimorphism is that it now involves many transcendental functions of analysis, like $e^{t\lambda}$, $\cos t\lambda$, $\sin t\lambda$, etc. If $f(\mathbf{A})$ denotes the operator corresponding to the function $f(\lambda) \in \mathcal{F}(S_{\mathbf{A}})$, then its matrix in the Jordan basis of the operator $\mathbf{A}$ is calculated by the same rule (38) as before. We note in particular that the operator formula

$$e^{(t_1+t_2)\mathbf{A}} = e^{t_1\mathbf{A}}e^{t_2\mathbf{A}}$$

is an immediate consequence of the identity

$$e^{(t_1+t_2)\lambda} = e^{t_1\lambda}e^{t_2\lambda}$$

and the fact that the mapping $\mathcal{F}(S_{\mathbf{A}}) \rightarrow \Pi_{\mathbf{A}}$ is an epimorphism.

The results of Secs. 6.87–6.89, pertaining to linear operators in a complex space, can be carried over to linear operators in a real space, by using the real Jordan canonical form and the method of Secs. 6.85–6.86. We leave the details of this extension to the reader, since no new ideas are involved.

PROBLEMS

1. The matrix of an operator $\mathbf{A}$ is of the form

$$\left\|\begin{matrix} \lambda & 0 & 0 & \cdots & 0 & 0 \\ 1 & \lambda & 0 & \cdots & 0 & 0 \\ 0 & 1 & \lambda & \cdots & 0 & 0 \\ \cdot & \cdot & \cdot & \cdots & \cdot & \cdot \\ 0 & 0 & 0 & \cdots & \lambda & 0 \\ 0 & 0 & 0 & \cdots & 1 & \lambda \end{matrix}\right\|$$

in a basis $e_1, e_2, \ldots, e_n$. In what basis does it have Jordan canonical form?

2. Prove that the matrix A and the matrix A' (obtained by transposing A) are equivalent.

3. Find the Jordan canonical form of the matrix

$$\begin{Vmatrix} -2 & -1 & -1 & 3 & 2 \\ -4 & 1 & -1 & 3 & 2 \\ 1 & 1 & 0 & -3 & -2 \\ -4 & -2 & -1 & 5 & 1 \\ 1 & 1 & 1 & -3 & 0 \end{Vmatrix}.$$

4. Are the operators specified by the matrices

$$A = \begin{Vmatrix} 1 & 1 & 0 \\ 0 & 1 & 0 \\ 0 & 0 & 2 \end{Vmatrix}, \qquad B = \begin{Vmatrix} 4 & 1 & -1 \\ -6 & -1 & 2 \\ 2 & 1 & 1 \end{Vmatrix}$$

equivalent?

5. Find the elementary divisors of the following matrices of order n:

$$A_1 = \begin{Vmatrix} 1 & 1 & \cdots & 1 \\ 0 & 1 & \cdots & 1 \\ \cdot & \cdot & \cdots & \cdot \\ 0 & 0 & \cdots & 1 \end{Vmatrix}, \qquad A_2 = \begin{Vmatrix} 1 & 2 & 3 & \cdots & n \\ 0 & 1 & 2 & \cdots & n-1 \\ 0 & 0 & 1 & \cdots & 0 \\ \cdot & \cdot & \cdot & \cdots & \cdot \\ 0 & 0 & 0 & \cdots & 1 \end{Vmatrix},$$

$$A_3 = \begin{Vmatrix} n & n-1 & n-2 & \cdots & 1 \\ 0 & n & n-1 & \cdots & 2 \\ 0 & 0 & n & \cdots & 3 \\ \cdot & \cdot & \cdot & \cdots & \cdot \\ 0 & 0 & 0 & \cdots & n \end{Vmatrix}, \qquad A_4 = \begin{Vmatrix} 1 & 1 & 1 & \cdots & 1 \\ 0 & 2 & 2 & \cdots & 2 \\ 0 & 0 & 3 & \cdots & 3 \\ \cdot & \cdot & \cdot & \cdots & \cdot \\ 0 & 0 & 0 & \cdots & n \end{Vmatrix}.$$

6. Show that all matrices of the form

$$A = \begin{Vmatrix} \alpha & a_{12} & a_{13} & \cdots & a_{1n} \\ 0 & \alpha & a_{23} & \cdots & a_{2n} \\ 0 & 0 & \alpha & \cdots & a_{3n} \\ \cdot & \cdot & \cdot & \cdots & \cdot \\ 0 & 0 & 0 & \cdots & \alpha \end{Vmatrix}$$

with arbitrary elements $a_{12}, a_{13}, \ldots$ are equivalent if the elements $a_{12}, a_{23}, \ldots,$ $a_{n-1,n}$ are nonzero.

7. Find the Jordan canonical form of the matrix A satisfying the equation $P(A) = 0$, where the polynomial $P(\lambda)$ has no multiple roots.

8. Find the Jordan canonical form of the matrix A satisfying the equation $P(A) = 0$, where the polynomial $P(\lambda)$ is an arbitrary polynomial.

9. Prove that if the annihilating polynomial of an operator $\mathbf{A}$ acting in the space R_n is of degree 2, then every vector x lies in a plane or line invariant with respect to $\mathbf{A}$.

10. Find all matrices commuting with the $m \times m$ matrix

$$A_m(a) = \left\| \begin{matrix} a & 1 & 0 & \cdots & 0 & 0 \\ 0 & a & 1 & \cdots & 0 & 0 \\ \cdot & \cdot & \cdot & \cdots & \cdot & \cdot \\ 0 & 0 & 0 & \cdots & a & 1 \\ 0 & 0 & 0 & \cdots & 0 & a \end{matrix} \right\|.$$

11. Find all $m \times n$ matrices B satisfying the condition

$$BA_n(a) = A_m(a)B.$$

12. Find all matrices commuting with quasi-diagonal matrices of the form

$$\left\| \begin{matrix} A_{m_1}(a) & 0 & \cdots & 0 \\ 0 & A_{m_2}(a) & \cdots & 0 \\ \cdot & \cdot & \cdots & \cdot \\ 0 & 0 & \cdots & A_{m_k}(a) \end{matrix} \right\|.$$

13. Find all matrices commuting with quasi-diagonal matrices of the form

$$\left\| \begin{matrix} A_{m_1}(a_1) & 0 & \cdots & 0 \\ \cdot & A_{m_2}(a_2) & \cdots & 0 \\ \cdot & \cdot & \cdots & \cdot \\ 0 & 0 & \cdots & A_{m_k}(a_k) \end{matrix} \right\|,$$

where the numbers $a_1, a_2, \ldots, a_k$ are all distinct.

14. Find all matrices commuting with the general Jordan matrix (8).

15. Under what conditions is every matrix commuting with a given matrix A a polynomial in A?

chapter 7

BILINEAR AND QUADRATIC FORMS

In this chapter, we shall study linear numerical functions of two vector arguments. Unlike the theory of linear numerical functions of *one* vector argument, the theory of linear numerical functions of *two* vector arguments (such functions are called *bilinear forms*) has rich geometric content. Setting the second argument equal to the first in the expression for a bilinear form, we get an important new kind of function of one variable, called a *quadratic form*, which is no longer linear.

The considerations of Secs. 7.1–7.8 pertain to a linear space **K** over an arbitrary number field K, while those of Sec. 7.9 pertain to a real linear space.

7.1. Bilinear Forms

7.11. A numerical function $\mathbf{A}(x, y)$ of two vector arguments x and y in a linear space **K** is called a *bilinear form* (or a *bilinear function*) if it is a linear function of x for every fixed value of y and a linear function of y for every fixed value of x. In other words, $\mathbf{A}(x, y)$ is a bilinear form in x and y if and only if the following relations hold for any x, y and z:

$$\begin{aligned} \mathbf{A}(x + z, y) &= \mathbf{A}(x, y) + \mathbf{A}(z, y), \\ \mathbf{A}(\alpha x, y) &= \alpha \mathbf{A}(x, y), \\ \mathbf{A}(x, y + z) &= \mathbf{A}(x, y) + \mathbf{A}(x, z), \\ \mathbf{A}(x, \alpha y) &= \alpha \mathbf{A}(x, y). \end{aligned} \tag{1}$$

The first two equations mean that $\mathbf{A}(x, y)$ is linear in its first argument, and the last two equations that $\mathbf{A}(x, y)$ is linear in its second argument. Using induction and the relations (1), we easily obtain the general formula

$$\mathbf{A}\left(\sum_{i=1}^{k} \alpha_i x_i, \sum_{j=1}^{m} \beta_j y_j\right) = \sum_{i=1}^{k} \sum_{j=1}^{m} \alpha_i \beta_j \mathbf{A}(x_i, y_j), \tag{2}$$

where $x_1, \ldots, x_k, y_1, \ldots, y_m$ are arbitrary vectors of the space $\mathbf{K}$ and $\alpha_1, \ldots, \alpha_k, \beta_1, \ldots, \beta_m$ are arbitrary numbers from the field K.

Bilinear forms defined on infinite-dimensional spaces are usually called *bilinear functionals*.

7.12. *Examples*

a. If $L_1(x)$ and $L_2(x)$ are linear forms, then $\mathbf{A}(x, y) = L_1(x)L_2(y)$ is obviously a bilinear form in x and y.

b. An example of a bilinear form in an n-dimensional linear space with a fixed basis $e_1, e_2, \ldots, e_n$ is the function

$$\mathbf{A}(x, y) = \sum_{i=1}^{n} \sum_{k=1}^{n} a_{ik} \xi_i \eta_k,$$

where

$$x = \sum_{i=1}^{n} \xi_i e_i, \qquad y = \sum_{i=1}^{n} \eta_k e_k$$

are arbitrary vectors and the a_{ik} $(i, k = 1, 2, \ldots, n)$ are fixed numbers.

7.13. The general representation of a bilinear form in an n-dimensional linear space. Suppose we have a bilinear form $\mathbf{A}(x, y)$ *in* an n-dimensional linear space $\mathbf{K}_n$. Choose an arbitrary basis $e_1, e_2, \ldots, e_n$ in $\mathbf{K}_n$, and write

$$\mathbf{A}(e_i, e_k) = a_{ik} \qquad (i, k = 1, 2, \ldots, n).$$

Then for any two vectors

$$x = \sum_{i=1}^{n} \xi_i e_i, \qquad y = \sum_{k=1}^{n} \eta_k e_k,$$

it follows from (2) that

$$\mathbf{A}(x, y) = \mathbf{A}\left(\sum_{i=1}^{n} \xi_i e_i, \sum_{k=1}^{n} \eta_k e_k\right) = \sum_{i=1}^{n} \sum_{k=1}^{n} \xi_i \eta_k \mathbf{A}(e_i, e_k)$$

$$= \sum_{i=1}^{n} \sum_{k=1}^{n} a_{ik} \xi_i \eta_k. \tag{3}$$

Thus the most general representation of a bilinear form in an n-dimensional linear space has already been encountered in Example 7.12b.

The coefficients a_{ik} form a square matrix

$$A = A_{(e)} = \begin{Vmatrix} a_{11} & a_{12} & \cdots & a_{1n} \\ a_{21} & a_{22} & \cdots & a_{2n} \\ \cdot & \cdot & \cdots & \cdot \\ a_{n1} & a_{n2} & \cdots & a_{nn} \end{Vmatrix} = \|a_{ik}\|$$

which we will call the *matrix of the bilinear form* $\mathbf{A}(x, y)$ *in* (*or relative to*) *the basis* $\{e\} = \{e_1, e_2, \ldots, e_n\}$.

7.14. Symmetric bilinear forms. A bilinear form is called *symmetric* if

$$\mathbf{A}(x, y) = \mathbf{A}(y, x)$$

for arbitrary vectors x and y. If the bilinear form $\mathbf{A}(x, y)$ is symmetric, then

$$a_{ik} = \mathbf{A}(e_i, e_k) = \mathbf{A}(e_k, e_i) = a_{ki},$$

so that the matrix $A_{(e)}$ of a symmetric bilinear form in any basis $e_1, e_2, \ldots, e_n$ of the space $\mathbf{K}_n$ equals its own transpose $A'_{(e)}$. It is easily verified that the converse is also true, i.e., if $A'_{(e)} = A_{(e)}$ in any basis $e_1, e_2, \ldots, e_n$, then the form $\mathbf{A}(x, y)$ is symmetric. In fact, we have

$$\mathbf{A}(y, x) = \sum_{i,k=1}^{n} a_{ik}\eta_i\xi_k = \sum_{i,k=1}^{n} a_{ki}\eta_i\xi_k = \sum_{i,k=1}^{n} a_{ik}\xi_i\eta_k = \mathbf{A}(x, y),$$

as required. In particular, we have the following result: If the matrix of the bilinear form $\mathbf{A}(x, y)$ calculated in any basis equals its own transpose, then the matrix of the form calculated in any other basis also equals its own transpose. A matrix which equals its own transpose will henceforth be called *symmetric*.

7.15. Transformation of the matrix of a bilinear form when the basis is changed.

a. Of course, if we transform to a new basis, the matrix of a bilinear form changes according to a certain transformation law. We now find this law. Let $A_{(e)} = \|a_{ik}\|$ be the matrix of the bilinear form $\mathbf{A}(x, y)$ in the basis

$$\{e\} = \{e_1, e_2, \ldots, e_n\},$$

and let $A_{(f)} = \|b_{ik}\|$ be the matrix of the same form in the basis

$$\{f\} = \{f_1, f_2, \ldots, f_n\}$$

$(i, k = 1, 2, \ldots, n)$. Assuming that the transformation from one basis to the other is described by the formula

$$f_i = \sum_{j=1}^{n} p_j^{(i)} e_j \qquad (i = 1, 2, \ldots, n)$$

with the transformation matrix $P = \|p_j^{(i)}\|$, we have

$$b_{ik} = \mathbf{A}(f_i, f_k) = \mathbf{A}\left(\sum_{j=1}^{n} p_j^{(i)} e_j, \sum_{l=1}^{n} p_l^{(k)} e_l\right)$$

$$= \sum_{j,l=1}^{n} p_j^{(i)} p_l^{(k)} \mathbf{A}(e_j, e_l) = \sum_{j,l=1}^{n} p_j^{(i)} p_l^{(k)} a_{jl}.$$

This formula can be written in the form

$$b_{ik} = \sum_{j=1}^{n} \sum_{l=1}^{n} p_i^{(j)'} a_{jl} p_l^{(k)}, \tag{4}$$

where $p_i^{(j)'} = p_j^{(i)}$ is an element of the matrix P' which is the transpose of P. Equation (4) corresponds to the following relation between matrices (see Sec. 4.43):

$$A_{(f)} = P' A_{(e)} P. \tag{5}$$

b. Since the matrices P and P' are nonsingular, it follows from Corollary 4.67 that the rank of the matrix $A_{(f)}$ equals the rank of the matrix $A_{(e)}$, i.e., *the rank of the matrix of a bilinear form is independent of the choice of a basis.* Hence it makes sense to talk about the *rank of a bilinear form.* A bilinear form $\mathbf{A}(x, y)$ is said to be *nonsingular* if its rank equals the dimension n of the space $\mathbf{K}_n$.

c. Let $\mathbf{A}(x, y)$ be a nonsingular bilinear form. Then, as we now show, given any vector $x_0 \neq 0$, there exists a vector $y_0 \in \mathbf{K}_n$ such that $\mathbf{A}(x_0, y_0) \neq 0$. Suppose to the contrary that $\mathbf{A}(x_0, y) = 0$ for every $y \in \mathbf{K}_n$, and construct a basis $e_1, e_2, \ldots, e_n$ in the space $\mathbf{K}_n$ such that $e_1 = x_0$. Then the matrix of the form $\mathbf{A}(x, y)$ in this basis is such that

$$a_{1m} = \mathbf{A}(e_1, e_m) = \mathbf{A}(x_0, e_m) = 0,$$

so that the whole first row of the matrix consists of zeros. But then the rank of the matrix is less than n, contrary to the hypothesis that $\mathbf{A}(x, y)$ is nonsingular. This contradiction proves our assertion.

d. Note that a form $\mathbf{A}(x, y)$ which is nonsingular in the whole space $\mathbf{K}$ may be singular in a subspace $\mathbf{K}' \subset \mathbf{K}$. For example, the form

$$\mathbf{A}(x, y) = \xi_1 \eta_1 - \xi_2 \eta_2$$

is nonsingular in the space $\mathbf{R}_2$, where $x = (\xi_1, \xi_2)$, $y = (\eta_1, \eta_2)$. However, it vanishes identically in the subspace $\mathbf{R}_2' \subset \mathbf{R}_2$ where $\xi_1 = \xi_2$ (and $\eta_1 = \eta_2$).

e. It follows from (5) and Theorem 4.75 on the determinant of the product of two matrices that

$$\det A_{(f)} = \det A_{(e)} (\det P)^2. \tag{6}$$

7.2. Quadratic Forms

One of the basic problems of plane analytic geometry is to reduce the general equation of a second-degree curve to canonical form by transforming to a new coordinate system. The equation of a second-degree curve with center at the origin $x = 0$, $y = 0$, has the familiar form

$$Ax^2 + 2Bxy + Cy^2 = D. \tag{7}$$

A coordinate transformation is described by the formulas

$$x = a_{11}x' + a_{12}y',$$

$$y = a_{21}x' + a_{22}y',$$

where $a_{11}, a_{12}, a_{21}, a_{22}$ are certain numbers (usually sines and cosines of the angle through which the axes are rotated). As a result of this coordinate transformation, (7) takes the simpler form

$$A'x'^2 + B'y'^2 = D.$$

An analogous problem can be stated for a space with any number of dimensions. The solution of this and related problems is the fundamental aim of the theory of quadratic forms, which we now present.

7.21. We begin with the following definition:

A *quadratic form* defined on a linear space $\mathbf{K}$ is a function $\mathbf{A}(x, x)$ of one vector argument $x \in \mathbf{K}$ obtained by changing y to x in any bilinear form $\mathbf{A}(x, y)$ defined on $\mathbf{K}$.

According to (3), in an n-dimensional space $\mathbf{K}_n$ with a fixed basis $\{e\} = \{e_1, e_2, \ldots, e_n\}$, every quadratic form can be written as

$$\mathbf{A}(x, x) = \sum_{i=1}^{n} \sum_{k=1}^{n} a_{ik}\xi_i\xi_k, \tag{8}$$

where $\xi_1, \xi_2, \ldots, \xi_n$ are components of the vector x with respect to the basis $\{e\}$. Conversely, *every function* $\mathbf{A}(x, x)$ *of the vector* x *defined in the basis* $\{e\}$ *by formula* (8) *is a quadratic form in* x. In fact, we need only introduce the bilinear form

$$\mathbf{B}(x, y) = \sum_{i=1}^{n} \sum_{k=1}^{n} a_{ik}\xi_i\eta_k,$$

where $\eta_1, \eta_2, \ldots, \eta_n$ are the components of the vector y with respect to the basis $\{e\}$. Then the function $\mathbf{A}(x, x)$ is obviously just the quadratic form $\mathbf{B}(x, x)$.

7.22. We can write the double sum (8) somewhat differently by combining similar terms. Let $b_{ii} = a_{ii}$ and $b_{ik} = a_{ik} + a_{ki}$ $(i \neq k)$. Then, since

$$a_{ik}\xi_i\xi_k + a_{ki}\xi_k\xi_i = (a_{ik} + a_{ki})\xi_i\xi_k = b_{ik}\xi_i\xi_k,$$

the double sum (8) can be written as

$$\mathbf{A}(x, x) = \sum_{k=1}^{n} \sum_{i \leqslant k} b_{ik}\xi_i\xi_k,$$

and has fewer terms. It follows that two different bilinear forms

$$\mathbf{A}(x, y) = \sum_{i,k=1}^{n} a_{ik}\xi_i\eta_k, \qquad \mathbf{C}(x, y) = \sum_{i,k=1}^{n} c_{ik}\xi_i\eta_k$$

can reduce to the same quadratic form after y is replaced by x. All that is necessary is that $a_{ik} + a_{ki} = c_{ik} + c_{ki}$ for arbitrary i and k.

Thus, in general, we cannot reconstruct uniquely the bilinear form generating a given quadratic form. However, in the case where it is known that the original bilinear form is *symmetric*, it *can* be reconstructed. In fact, if $a_{ik} = a_{ki}$, then the relation $a_{ik} + a_{ki} = b_{ik}$ $(i \neq k)$ uniquely determines the coefficients a_{ik}, i.e.,

$$a_{ik} = a_{ki} = \frac{1}{2} b_{ik} \qquad (i \neq k), \tag{9}$$

while for $i = k$ we have

$$a_{ii} = b_{ii}, \tag{9'}$$

so that the bilinear form itself is uniquely determined. This assertion can be proved without recourse to bases and components. In fact, we have

$$\mathbf{A}(x + y, x + y) = \mathbf{A}(x, x) + \mathbf{A}(x, y) + \mathbf{A}(y, x) + \mathbf{A}(y, y)$$

by the definition of a bilinear form, and

$$\mathbf{A}(x, y) = \frac{1}{2}[\mathbf{A}(x, y) + \mathbf{A}(y, x)] = \frac{1}{2}[\mathbf{A}(x + y, x + y) - \mathbf{A}(x, x) - \mathbf{A}(y, y)]$$

by the assumption that $\mathbf{A}(x, y)$ is symmetric. Hence the value of the bilinear form $\mathbf{A}(x, y)$ for any pair of vectors x, y is uniquely determined by the values of the corresponding quadratic form for the vectors x, y and $x + y$.

On the other hand, to obtain all possible quadratic forms, we need only use symmetric bilinear forms. In fact, if $\mathbf{A}(x, y)$ is an arbitrary bilinear form, then

$$\mathbf{A}_1(x, y) = \frac{1}{2} [\mathbf{A}(x, y) + \mathbf{A}(y, x)]$$

is a symmetric bilinear form, and

$$\mathbf{A}_1(x, x) = \frac{1}{2} [\mathbf{A}(x, x) + \mathbf{A}(x, x)] = \mathbf{A}(x, x),$$

i.e., the quadratic forms $\mathbf{A}_1(x, x)$ and $\mathbf{A}(x, x)$ coincide.

7.23. These considerations show that in using bilinear forms to study the properties of quadratic forms, we need only consider symmetric bilinear forms, with corresponding symmetric matrices $\|a_{ik}\|$, $a_{ik} = a_{ki}$. By the *matrix of the quadratic form* $\mathbf{A}(x, x)$, we mean the symmetric matrix $A = \|a_{ik}\|$ of the symmetric bilinear form $\mathbf{A}(x, y)$ corresponding to $\mathbf{A}(x, x)$. When the basis is changed, the matrix A of the quadratic form $\mathbf{A}(x, x)$ transforms just like the matrix of the corresponding symmetric bilinear form $\mathbf{A}(x, y)$, i.e.,

$$A_{(f)} = P' A_{(e)} P,$$

where P is the matrix of the transformation from the basis $\{e\}$ to the basis $\{f\}$. In particular, *the rank of the matrix of a quadratic form does not depend on the choice of a basis.* Therefore we can talk about the *rank of a quadratic form* $\mathbf{A}(x, x)$, understanding it to mean the rank of the matrix of $\mathbf{A}(x, x)$ in any basis of the space $\mathbf{K}_n$. A quadratic form whose rank equals the dimension n of the space $\mathbf{K}_n$ is said to be *nonsingular.*

7.3. Reduction of a Quadratic Form to Canonical Form

7.31. Suppose we are given an arbitrary quadratic form $\mathbf{A}(x, x)$ defined on an n-dimensional linear space $\mathbf{K}_n$. We now show that *there exists a basis* $\{f\} = \{f_1, f_2, \ldots, f_n\}$ *in* $\mathbf{K}_n$ *such that given any vector*

$$x = \sum_{k=1}^{n} \eta_k f_k,$$

the value of the quadratic form $\mathbf{A}(x, x)$ *is given by*

$$\mathbf{A}(x, x) = \lambda_1 \eta_1^2 + \lambda_2 \eta_2^2 + \cdots + \lambda_n \eta_n^2, \tag{10}$$

where $\lambda_1, \lambda_2, \ldots, \lambda_n$ *are certain fixed numbers.* Every basis with this property will be called a *canonical basis* of $\mathbf{A}(x, x)$, and the expression (10) will be called a *canonical form* of $\mathbf{A}(x, x)$. In particular, the numbers $\lambda_1, \lambda_2, \ldots, \lambda_n$ will be called *canonical coefficients* of $\mathbf{A}(x, x)$.

Let $\{e_1, e_2, \ldots, e_n\}$ be an arbitrary basis of the space $\mathbf{K}_n$. If

$$x = \sum_{k=1}^{n} \xi_k e_k,$$

then, as we have already seen, $\mathbf{A}(x, x)$ can be written in the form

$$\mathbf{A}(x, x) = \sum_{k=1}^{n} \sum_{i \leqslant k} b_{ik} \xi_i \xi_k. \tag{11}$$

According to Sec. 5.22, our assertion will be proved if we can write a system

$$\begin{aligned} \eta_1 &= p_{11}\xi_1 + p_{12}\xi_2 + \cdots + p_{1n}\xi_n, \\ \eta_2 &= p_{21}\xi_1 + p_{22}\xi_2 + \cdots + p_{2n}\xi_n, \\ &\cdots\cdots\cdots\cdots\cdots\cdots \\ \eta_n &= p_{n1}\xi_1 + p_{n2}\xi_2 + \cdots + p_{nn}\xi_n \end{aligned} \tag{12}$$

with a nonsingular matrix $P = \|p_{ik}\|$ such that expressing the variables $\eta_1, \eta_2, \ldots, \eta_n$ appearing in (11) in terms of $\xi_1, \xi_2, \ldots, \xi_n$ has the effect of transforming (11) into the form (10). We will carry out the proof by induction on the number of variables ξ_i actually appearing in (11), i.e., those which have nonzero coefficients, assuming that every form containing $m - 1$ variables $\xi_1, \xi_2, \ldots, \xi_{m-1}$, say, can be reduced to the canonical form (10) with $n = m - 1$, by making a transformation (12) also with $n = m - 1$.

If (11) actually contains only one variable ξ_1, say, i.e., if (11) has the form

$$\mathbf{A}(x, x) = b_{11}\xi_1^2,$$

then the induction hypothesis is satisfied for any choice of $p_{11} \neq 0$. Consider a form (11) which actually contains m variables $\xi_1, \xi_2, \ldots, \xi_m$. First we assume that one of the numbers $b_{11}, b_{22}, \ldots, b_{mm}$, say b_{mm}, is nonzero, and we group together all the terms in (11) which contain the variable ξ_m. This group of terms can be written in the form

$$\begin{aligned} &b_{1m}\xi_1\xi_m + b_{2m}\xi_2\xi_m + \cdots + b_{m-1,m}\xi_{m-1}\xi_m + b_{mm}\xi_m^2 \\ &= b_{mm}\left(\frac{b_{1m}}{2b_{mm}}\xi_1 + \frac{b_{2m}}{2b_{mm}}\xi_2 + \cdots + \frac{b_{m-1,m}}{2b_{mm}}\xi_{m-1} + \xi_m\right)^2 + \mathbf{A}_1(x, x), \end{aligned} \tag{13}$$

where $\mathbf{A}_1(x, x)$ denotes a quadratic form which depends only on the variables $\xi_1, \xi_2, \ldots, \xi_{m-1}$. Now consider the coordinate transformation

$$\begin{aligned} \tau_1 &= \xi_1, \\ \tau_2 &= \qquad \xi_2, \\ &\cdots\cdots\cdots\cdots\cdots\cdots \\ \tau_{m-1} &= \qquad\qquad \xi_{m-1}, \\ \tau_m &= \frac{b_{1m}}{2b_{mm}}\xi_1 + \frac{b_{2m}}{2b_{mm}}\xi_2 + \cdots + \frac{b_{m-1,m}}{2b_{mm}}\xi_{m-1} + \xi_m. \end{aligned}$$

The matrix of this transformation is nonsingular (its determinant is actually 1). In the new coordinate system, $\mathbf{A}(x, x)$ clearly has the form

$$\mathbf{A}(x, x) = \mathbf{B}(x, x) + b_{mm}\tau_m^2,$$

where the quadratic form $\mathbf{B}(x, x)$ depends only on the variables $\tau_1, \tau_2, \ldots, \tau_{m-1}$. By the induction hypothesis, there exists a new transformation

$$\begin{aligned} \eta_1 &= p_{11}\tau_1 + p_{12}\tau_2 + \cdots + p_{1,m-1}\tau_{m-1}, \\ \eta_2 &= p_{21}\tau_1 + p_{22}\tau_2 + \cdots + p_{2,m-1}\tau_{m-1}, \\ &\cdots\cdots\cdots\cdots\cdots\cdots\cdots\cdots \\ \eta_{m-1} &= p_{m-1,1}\tau_1 + p_{m-1,2}\tau_2 + \cdots + p_{m-1,m-1}\tau_{m-1}, \end{aligned} \tag{12'}$$

with a nonsingular matrix $P = \|p_{ik}\|$, which carries $\mathbf{B}(x, x)$ into the canonical form

$$\mathbf{B}(x, x) = \lambda_1\eta_1^2 + \lambda_2\eta_2^2 + \cdots + \lambda_{m-1}\eta_{m-1}^2.$$

If we supplement the system of equations (12′) with the additional equation $\eta_m = \tau_m$, we obtain a nonsingular transformation of the variables $\tau_1, \tau_2, \ldots, \tau_m$ into the variables $\eta_1, \eta_2, \ldots, \eta_m$, which carries $\mathbf{A}(x, x)$ into the canonical form

$$\mathbf{A}(x, x) = \mathbf{B}(x, x) + b_{mm}\tau_m^2 = \lambda_1\eta_1^2 + \lambda_2\eta_2^2 + \cdots + \lambda_{m-1}\eta_{m-1}^2 + b_{mm}\eta_m^2.$$

According to Sec. 5.33, the direct transformation from the variables $\{\xi\}$ to the variables $\{\eta\}$ is accomplished by using the matrix equal to the product of the matrix of the transformation from $\{\tau\}$ to $\{\eta\}$ and the matrix of the transformation from $\{\xi\}$ to $\{\tau\}$.† Since both of these matrices are nonsingular, the product of the matrices is also nonsingular.

We must still consider the case of a quadratic form $\mathbf{A}(x, x)$ in m variables $\xi_1, \xi_2, \ldots, \xi_m$ which has all the numbers $b_{11}, b_{22}, \ldots, b_{mm}$ equal to zero. Consider one of the terms $b_{ik}\xi_i\xi_k$ with a nonzero coefficient, say $b_{12} \neq 0$. Then carry out the following coordinate transformation, where for convenience we write the transformation from the new variables to the old variables:

$$\begin{aligned} \xi_1 &= \xi_1' + \xi_2', \\ \xi_2 &= \xi_1' - \xi_2', \\ \xi_3 &= \xi_3', \\ &\cdots\cdots \\ \xi_m &= \xi_m'. \end{aligned} \tag{14}$$

The determinant of the matrix of the transformation (14) equals -2, and hence this transformation is again nonsingular. The term $b_{12}\xi_1\xi_2$ is transformed into

$$b_{12}\xi_1\xi_2 = b_{12}\xi_1'^2 - b_{12}\xi_2'^2,$$

so that two squared terms with nonzero coefficients are produced simultaneously in the new form. (Clearly these terms cannot cancel any of the other

† $\{\xi\}$ is shorthand for the set $\{\xi_1, \xi_2, \ldots, \xi_m\}$, $\{\eta\}$ for the set $\{\eta_1, \eta_2, \ldots, \eta_m\}$, etc.

terms, since all the other terms contain a variable ξ'_i with $i > 2$.) We can now apply our inductive method to the quadratic form (11) written in the new variables ξ'_i.

Thus, finally, we have proved our theorem for any integer $m = 1, 2, \ldots$. In particular, the case $m = n$ suffices to prove the theorem for an arbitrary quadratic form in an n-dimensional space.

The idea of our proof, i.e., consecutive splitting off of complete squares, can be used as a practical method for reducing a given quadratic form to canonical form. However, in Sec. 7.5 we will describe another method, which permits us to obtain directly both the canonical form and the vectors of the canonical basis.

7.32. *Example.* To reduce the quadratic form

$$\mathbf{A}(x, x) = \xi_1^2 + 6\xi_1\xi_2 + 5\xi_2^2 - 4\xi_1\xi_3 - 12\xi_2\xi_3 + 4\xi_3^2 - 4\xi_2\xi_4 - 8\xi_3\xi_4 - \xi_4^2$$

to canonical form, we first complete the square in the group of terms containing ξ_1, writing

$$\eta_1 = \xi_1 + 3\xi_2 - 2\xi_3.$$

Then the form is transformed into

$$\mathbf{A}(x, x) = \eta_1^2 - 4\xi_2^2 - 4\xi_2\xi_4 - 8\xi_3\xi_4 - \xi_4^2.$$

Next we complete the square in the group of terms containing ξ_2, writing

$$\eta_2 = 2\xi_2 + \xi_4.$$

This reduces the form to

$$\mathbf{A}(x, x) = \eta_1^2 - \eta_2^2 - 8\xi_3\xi_4.$$

There are no squares of the variables ξ_3 and ξ_4. Hence we write

$$\xi_3 = \eta_3 - \eta_4,$$
$$\xi_4 = \eta_3 + \eta_4,$$

so that $\xi_3\xi_4 = \eta_3^2 - \eta_4^2$. Thus the form $\mathbf{A}(x, x)$ is reduced to the canonical form

$$\mathbf{A}(x, x) = \eta_1^2 - \eta_2^2 - 8\eta_3^2 + 8\eta_4^2$$

by the transformation

$$\eta_1 = \xi_1 + 3\xi_2 - 2\xi_3,$$
$$\eta_2 = 2\xi_2 + \xi_4,$$
$$\eta_3 = \tfrac{1}{2}\xi_3 + \tfrac{1}{2}\xi_4,$$
$$\eta_4 = -\tfrac{1}{2}\xi_3 + \tfrac{1}{2}\xi_4.$$

It is apparent from the construction that this transformation is nonsingular, a fact which is easily verified directly.

7.33. a. Neither the canonical basis nor the canonical form of a quadratic form is uniquely determined. For example, any permutation of the vectors of a canonical basis gives another canonical basis. In Sec. 7.5 it will be shown, among other things, that with a few rare exceptions a canonical basis for a given quadratic form can be constructed by choosing an *arbitrary* vector of the space as the first vector of the basis. Moreover, if $\mathbf{A}(x, x)$ is written in the canonical form

$$\mathbf{A}(x, x) = \lambda_1\eta_1^2 + \lambda_2\eta_2^2 + \cdots + \lambda_n\eta_n^2,$$

where $\eta_1, \eta_2, \ldots, \eta_n$ are the components of the vector x, then the transformation

$$\eta_1 = \alpha_1\tau_1,$$

$$\eta_2 = \alpha_2\tau_2,$$

$$\cdots\cdot$$

$$\eta_n = \alpha_n\tau_n$$

(where $\alpha_1, \alpha_2, \ldots, \alpha_n$ are fixed numbers all different from zero and $\tau_1, \tau_2, \ldots, \tau_n$ are new components) carries $\mathbf{A}(x, x)$ into the new form

$$\mathbf{A}(x, x) = (\lambda_1\alpha_1^2)\tau_1^2 + (\lambda_2\alpha_2^2)\tau_2^2 + \cdots + (\lambda_n\alpha_n^2)\tau_n^2,$$

which is also canonical but has different coefficients. Hence there still remains the problem of describing all the canonical forms to which a given quadratic form can be reduced. This problem can be made more precise if we restrict the definition of a canonical form (as for example will be done in Sec. 7.93 for the case of a real space) or if we restrict the class of admissible coordinate transformations (as for example will be done in Sec. 10.1 for the case of a Euclidean space).

b. It should be noted that in the preceding example the number of nonzero coefficients remains unchanged when we transform from the variables $\{\eta\}$ to the variables $\{\tau\}$. In general, the number of nonzero canonical coefficients is obviously the rank of the matrix of the quadratic form in the corresponding canonical basis. Since the rank of the matrix of a quadratic form does not depend on the choice of a basis (Sec. 7.23), *the number of nonzero canonical coefficients of a quadratic form does not depend on the choice of a canonical basis.* Moreover, this number obviously coincides with the *rank* of the quadratic form (Sec. 7.23). Thus from a knowledge of a quadratic form $\mathbf{A}(x, x)$ in any basis $\{e\}$, we can predict the number of nonzero canonical coefficients of $\mathbf{A}(x, x)$ in any canonical basis, namely the rank of $\mathbf{A}(x, x)$. In particular, the canonical coefficients of a nonsingular quadratic form are all nonzero in any canonical basis.

7.4. The Canonical Basis of a Bilinear Form

7.41. a. The vector x_1 is said to be *conjugate to the vector* y_1 *with respect to the bilinear form* $\mathbf{A}(x, y)$ if

$$\mathbf{A}(x_1, y_1) = 0.$$

In this case, y_1 is also said to be conjugate to x_1.

b. Let $\|a_{ik}\|$ be the matrix of the form $\mathbf{A}(x, y)$ in any basis $e_1, e_2, \ldots, e_n$. Then, if

$$x_1 = \sum_{i=1}^{n} \xi_i e_i, \qquad y_1 = \sum_{k=1}^{n} \eta_k e_k,$$

the condition for x_1 and y_1 to be conjugate can be written in the form

$$\mathbf{A}(x_1, y_1) = \sum_{i,k=1}^{n} a_{ik}\xi_i\eta_k = 0.$$

c. *If the vectors* $x_1, x_2, \ldots, x_k$ *are all conjugate to the vector* y_1, *then every vector of the linear manifold* $\mathbf{L}(x_1, x_2, \ldots, x_k)$ *spanned by* $x_1, x_2, \ldots, x_k$ *is also conjugate to* y_1. In fact, it follows from the properties of a bilinear form that

$$\mathbf{A}(\alpha_1 x_1 + \alpha_2 x_2 + \cdots + \alpha_k x_k, y_1)$$
$$= \alpha_1 \mathbf{A}(x_1, y_1) + \alpha_2 \mathbf{A}(x_2, y_1) + \cdots + \alpha_k \mathbf{A}(x_k, y_1) = 0.$$

A vector y_1 conjugate to every vector of a subspace $\mathbf{K}' \subset \mathbf{K}$ is said to be *conjugate to the subspace* $\mathbf{K}'$.

d. The set $\mathbf{K}''$ of all vectors $y_1 \in \mathbf{K}$ conjugate to the subspace $\mathbf{K}'$ is obviously a subspace of the space $\mathbf{K}$. This subspace $\mathbf{K}''$ is said to be *conjugate* to $\mathbf{K}'$.

7.42. A basis $e_1, e_2, \ldots, e_n$ of the n-dimensional space $\mathbf{K}_n$ is called a *canonical basis of the bilinear form* $\mathbf{A}(x, y)$ if the basis vectors are conjugate to each other, i.e., if

$$\mathbf{A}(e_i, e_k) = 0 \quad \text{for} \quad i \neq k.$$

For example, in the space V_3 let the bilinear form $\mathbf{A}(x, y)$ be the scalar product of the vectors x and y. Then to say that x and y are conjugate with respect to $\mathbf{A}(x, y)$ means that x and y are orthogonal. In this case, any orthogonal basis of the space V_3 is a canonical basis.

7.43. The matrix of a bilinear form relative to a canonical basis is diagonal, since

$$a_{ik} = \mathbf{A}(e_i, e_k) = 0 \quad \text{for} \quad i \neq k.$$

Since a diagonal matrix coincides with its own transpose, a bilinear form which has a canonical basis must be symmetric. (We recall from Sec. 7.14 that whether or not the matrix of a bilinear form is symmetric does not depend on the choice of a basis.) Conversely, we now prove that *every symmetric bilinear form* $\mathbf{A}(x, y)$ *has a canonical basis.* To see this, consider the quadratic form $\mathbf{A}(x, x)$ corresponding to the given bilinear form $\mathbf{A}(x, y)$. We know that there exists a basis $e_1, e_2, \ldots, e_n$ in the space $\mathbf{K}_n$ in which $\mathbf{A}(x, x)$ can be written in the canonical form

$$\mathbf{A}(x, x) = \sum_{i=1}^{n} \lambda_i \xi_i^2.$$

It follows from formulas (9) and (9′), p. 184 that the corresponding symmetric bilinear form $\mathbf{A}(x, y)$ takes the canonical form

$$\mathbf{A}(x, y) = \sum_{i=1}^{n} \lambda_i \xi_i \eta_i, \tag{15}$$

in this basis, where

$$y = \sum_{i=1}^{n} \eta_i e_i,$$

and hence its matrix is diagonal. But this just means that the basis $e_1, e_2, \ldots, e_n$ is canonical for the form $\mathbf{A}(x, y)$, and our assertion is proved.

7.44. In analytic geometry it is shown that the locus of the midpoints of the chords of a second-degree curve which are parallel to a given vector is a straight line. We now prove this theorem. A second-degree curve in the x_1x_2-plane has an equation of the form

$$a_{11}x_1^2 + 2a_{12}x_1x_2 + a_{22}x_2^2 + b_1x_1 + b_2x_2 + c = 0$$

or

$$\mathbf{A}(x, x) + L(x) + c = 0,$$

where

$$\mathbf{A}(x, x) = a_{11}x_1^2 + 2a_{12}x_1x_2 + a_{22}x_2^2$$

is a quadratic form and

$$L(x) = b_1x_1 + b_2x_2$$

is a linear form in the vector $x = (x_1, x_2)$. Let x be the vector giving the position of the midpoint of a chord parallel to a fixed vector e. This means that the equations

$$\begin{aligned} &\mathbf{A}(x + te, x + te) + L(x + te) + c = 0, \\ &\mathbf{A}(x - te, x - te) + L(x - te) + c = 0 \end{aligned} \tag{16}$$

are satisfied for some $t \neq 0$. Let $\mathbf{A}(x, y)$ be the symmetric bilinear form corresponding to the quadratic form $\mathbf{A}(x, x)$. Then we can write (16) as

$$\mathbf{A}(x, x) + 2t\mathbf{A}(x, e) + t^2\mathbf{A}(e, e) + L(x) + tL(e) + c = 0,$$
$$\mathbf{A}(x, x) - 2t\mathbf{A}(x, e) + t^2\mathbf{A}(e, e) + L(x) - tL(e) + c = 0.$$

Subtracting the second equation from the first and dividing by $2t$, we get

$$2\mathbf{A}(x, e) + L(e) = 0. \tag{17}$$

This equation is linear in x and hence determines a straight line in the x_1x_2-plane, thereby proving the theorem.

Let x' be another point of the same line, so that

$$2\mathbf{A}(x', e) + L(e) = 0. \tag{18}$$

Then subtracting (18) from (17), we get

$$\mathbf{A}(x - x', e) = 0,$$

i.e., the vector e and the vector $x - x'$ determining the direction of the straight line in question are conjugate with respect to the bilinear form $\mathbf{A}(x, y)$, in the sense of Sec. 7.41.

7.45. Let $e_1, \ldots, e_k$ be a canonical basis of the form $\mathbf{A}(x, y)$ in a k-dimensional subspace $\mathbf{K}' \subset \mathbf{K}$, and let $\varepsilon_1, \ldots, \varepsilon_k$ be the corresponding canonical coefficients. Expressing the numbers $\mathbf{A}(x, e_i)$ in terms of the components of a vector $x \in \mathbf{K}'$, we get

$$\mathbf{A}(x, e_i) = \mathbf{A}\left(\sum_{j=1}^{k} \xi_j e_j, e_i\right) = \sum_{j=1}^{k} \xi_j \mathbf{A}(e_j, e_i) = \xi_i \mathbf{A}(e_i, e_i) = \varepsilon_i \xi_i,$$

so that the numbers $\mathbf{A}(x, e_i)$ are uniquely determined by the components of the vector x. If the form $\mathbf{A}(x, y)$ is nonsingular in the subspace $\mathbf{K}'$, then the numbers ε_i are all nonzero. In this case, the converse is also true, i.e., the values $\mathbf{A}(x, e_i)$ of the form $\mathbf{A}(x, y)$ uniquely determine the components of the vector x.

7.5. Construction of a Canonical Basis by Jacobi's Method

7.51. The construction of a canonical basis given in Sec. 7.31 has the drawback that the components of the vectors of a canonical basis and the corresponding canonical coefficients λ_i cannot be determined directly from a knowledge of the elements of the matrix $A_{(f)}$ of the symmetric bilinear form $\mathbf{A}(x, y)$ in a given basis $\{f\} = \{f_1, f_2, \ldots, f_n\}$. Jacobi's method, which will now be presented, does allow us to do just this. However, we must now impose the following supplementary condition on the matrix $A_{(f)}$: The *descending principal minors* of $A_{(f)}$ of order up to and including $n-1$, i.e., the principal minors of the form

$$\delta_1 = a_{11}, \qquad \delta_2 = \begin{vmatrix} a_{11} & a_{12} \\ a_{21} & a_{22} \end{vmatrix}, \ldots,$$

$$\delta_{n-1} = \begin{vmatrix} a_{11} & a_{12} & \cdots & a_{1,n-1} \\ a_{21} & a_{22} & \cdots & a_{2,n-1} \\ \cdot & \cdot & \cdots & \cdot \\ a_{n-1,1} & a_{n-1,2} & \cdots & a_{n-1,n-1} \end{vmatrix}, \tag{19}$$

must all be *nonvanishing*.

7.52. The vectors $e_1, e_2, \ldots, e_n$ are constructed by the formulas

$$\begin{aligned} e_1 &= f_1, \\ e_2 &= \alpha_1^{(1)} f_1 + f_2, \\ e_3 &= \alpha_1^{(2)} f_1 + \alpha_2^{(2)} f_2 + f_3, \\ &\cdots\cdots\cdots\cdots\cdots \\ e_{k+1} &= \alpha_1^{(k)} f_1 + \alpha_2^{(k)} f_2 + \alpha_3^{(k)} f_3 + \cdots + \alpha_k^{(k)} f_k + f_{k+1}, \\ &\cdots\cdots\cdots\cdots\cdots\cdots\cdots\cdots \\ e_n &= \alpha_1^{(n-1)} f_1 + \alpha_2^{(n-1)} f_2 + \alpha_3^{(n-1)} f_3 + \cdots + \alpha_{n-1}^{(n-1)} f_{n-1} + f_n, \end{aligned} \tag{20}$$

where the coefficients $\alpha_i^{(k)}$ $(i = 1, 2, \ldots, k;\ k = 1, 2, \ldots, n-1)$ are still to be determined. First of all, we note that the transformation from the vectors $f_1, f_2, \ldots, f_k$ to the vectors $e_1, e_2, \ldots, e_k$ is accomplished by using the matrix

$$\left\| \begin{matrix} 1 & 0 & 0 & \cdots & 0 & 0 \\ \alpha_1^{(1)} & 1 & 0 & \cdots & 0 & 0 \\ \cdot & \cdot & \cdot & \cdots & \cdot & \cdot \\ \cdot & \cdot & \cdot & \cdots & \cdot & \cdot \\ \alpha_1^{(k-1)} & \alpha_2^{(k-1)} & \alpha_3^{(k-1)} & \cdots & \alpha_{k-1}^{(k-1)} & 1 \end{matrix} \right\|,$$

whose determinant is unity. Hence for $k = 1, 2, \ldots, n$ the vectors $f_1, f_2, \ldots, f_k$ can be expressed as linear combinations of $e_1, e_2, \ldots, e_k$, so that the linear manifold $\mathbf{L}(f_1, f_2, \ldots, f_k)$ coincides with the linear manifold $\mathbf{L}(e_1, e_2, \ldots, e_k)$.

We now subject the coefficients $\alpha_i^{(k)}$ $(i = 1, 2, \ldots, k)$ to the condition that the vector e_{k+1} be conjugate to the subspace $L(e_1, e_2, \ldots, e_k)$. A necessary and sufficient condition for this is that the relations

$$\mathbf{A}(e_{k+1}, f_1) = 0,\ \mathbf{A}(e_{k+1}, f_2) = 0, \ldots, \mathbf{A}(e_{k+1}, f_k) = 0 \tag{21}$$

be satisfied. In fact, it follows from (21) that the vector e_{k+1} is conjugate to the linear manifold spanned by the vectors $f_1, f_2, \ldots, f_k$, which, as we have just proved, coincides with the linear manifold spanned by the vectors $e_1, e_2, \ldots, e_k$. Conversely, if the vector e_{k+1} is conjugate to the subspace $L(e_1, e_2, \ldots, e_k)$, it is conjugate to every vector in the subspace, in particular, to the vectors $f_1, f_2, \ldots, f_k$, so that the conditions (21) are satisfied.

Substituting the expression (20) for e_{k+1} into (21) and using the definition of a bilinear form, we obtain the following system of equations in the quantities $\alpha_i^{(k)}$ $(i = 1, 2, \ldots, k)$:

$$\begin{aligned}
\mathbf{A}(e_{k+1}, f_1) &= \alpha_1^{(k)}\mathbf{A}(f_1, f_1) + \alpha_2^{(k)}\mathbf{A}(f_2, f_1) + \cdots + \alpha_k^{(k)}\mathbf{A}(f_k, f_1) + \mathbf{A}(f_{k+1}, f_1) = 0,\\
\mathbf{A}(e_{k+1}, f_2) &= \alpha_1^{(k)}\mathbf{A}(f_1, f_2) + \alpha_2^{(k)}\mathbf{A}(f_2, f_2) + \cdots + \alpha_k^{(k)}\mathbf{A}(f_k, f_2) + \mathbf{A}(f_{k+1}, f_2) = 0,\\
&\cdots\cdots\cdots\cdots\cdots\cdots\cdots\cdots\cdots\cdots\cdots\cdots\cdots\cdots\\
\mathbf{A}(e_{k+1}, f_k) &= \alpha_1^{(k)}\mathbf{A}(f_1, f_k) + \alpha_2^{(k)}\mathbf{A}(f_2, f_k) + \cdots + \alpha_k^{(k)}\mathbf{A}(f_k, f_k) + \mathbf{A}(f_{k+1}, f_k) = 0.
\end{aligned} \tag{22}$$

By hypothesis, this nonhomogeneous system of equations with coefficients

$$\mathbf{A}(f_i, f_j) = a_{ij} \qquad (i, j = 1, 2, \ldots, k)$$

has a nonvanishing determinant, and hence can be solved uniquely. Therefore we can determine the quantities $\alpha_i^{(k)}$ and thereby construct the desired vector e_{k+1}. To determine all the coefficients $\alpha_i^{(k)}$ and all the vectors e_k, we must solve the appropriate system (22) for every k. Thus, in all, we must solve $n - 1$ systems of linear equations.

Let $\xi_1, \xi_2, \ldots, \xi_n$ denote the components of the vector x and $\eta_1, \eta_2, \ldots, \eta_n$ the components of the vector y with respect to the basis $e_1, e_2, \ldots, e_n$ just constructed. Then the bilinear form $\mathbf{A}(x, y)$ becomes

$$\mathbf{A}(x, y) = \sum_{i=1}^{n} \lambda_i \xi_i \eta_i \tag{23}$$

in this basis.

7.53. To calculate the coefficients λ_i, we argue as follows: Consider the bilinear form $\mathbf{A}(x, y)$ only in the subspace $\mathbf{L}_m = \mathbf{L}(e_1, e_2, \ldots, e_m)$ where $m \leqslant n$. The form $\mathbf{A}(x, y)$ clearly has the matrix

$$\begin{Vmatrix} a_{11} & a_{12} & \cdots & a_{1m} \\ a_{21} & a_{22} & \cdots & a_{2m} \\ \cdot & \cdot & \cdots & \cdot \\ \cdot & \cdot & \cdots & \cdot \\ a_{m1} & a_{m2} & \cdots & a_{mm} \end{Vmatrix}$$

in the basis $f_1, f_2, \ldots, f_m$ of the subspace $\mathbf{L}_m$ and the matrix

$$\left\|\begin{matrix} \lambda_1 & 0 & \cdots & 0 \\ 0 & \lambda_2 & \cdots & 0 \\ \cdot & \cdot & \cdots & \cdot \\ 0 & 0 & \cdots & \lambda_m \end{matrix}\right\|$$

in the basis $e_1, e_2, \ldots, e_m$. As we have seen, the matrix of the transformation (20) from the basis $f_1, f_2, \ldots, f_m$ to the basis $e_1, e_2, \ldots, e_m$ has determinant 1. Hence by equation (6), p. 182 we must have

$$\det \left\|\begin{matrix} a_{11} & a_{12} & \cdots & a_{1m} \\ a_{21} & a_{22} & \cdots & a_{2m} \\ \cdot & \cdot & \cdots & \cdot \\ \cdot & \cdot & \cdots & \cdot \\ a_{m1} & a_{m2} & \cdots & a_{mm} \end{matrix}\right\| = \det \left\|\begin{matrix} \lambda_1 & 0 & \cdots & 0 \\ 0 & \lambda_2 & \cdots & 0 \\ \cdot & \cdot & \cdots & \cdot \\ \cdot & \cdot & \cdots & \cdot \\ 0 & 0 & \cdots & \lambda_m \end{matrix}\right\|,$$

or, in the notation (19),

$$\delta_m = \lambda_1\lambda_2 \cdots \lambda_m \qquad (m = 1, 2, \ldots, n)$$

($\delta_n = \det A_{(f)}$). It follows immediately that

$$\lambda_1 = \delta_1 = a_{11}, \quad \lambda_2 = \frac{\delta_2}{\delta_1}, \quad \lambda_3 = \frac{\delta_3}{\delta_2}, \quad \ldots, \quad \lambda_n = \frac{\delta_n}{\delta_{n-1}}. \tag{24}$$

Using (24), we can find the coefficients of the bilinear form $\mathbf{A}(x, y)$ in a canonical basis without calculating the basis itself.

7.54. Consider once again the kth equation in the system (20), which we write in the form

$$f_{k+1} = -\alpha_1^{(k)} f_1 - \cdots - \alpha_k^{(k)} f_k + e_{k+1} = g_k + e_{k+1},$$

where g_k lies in the subspace $\mathbf{L}(f_1, \ldots, f_k)$ and e_{k+1} is conjugate to this subspace. The coefficients $\alpha_1^{(k)}, \ldots, \alpha_k^{(k)}$ are uniquely determined by the system (22) subject to the condition that $\det \|\mathbf{A}(f_i, f_j)\| \neq 0$ or, equivalently, that the form $\mathbf{A}(x, y)$ be nonsingular in the subspace $\mathbf{L}(f_1, \ldots, f_k)$. Since the vector f_{k+1} is arbitrary in this construction, then, writing

$$f = f_{k+1}, \qquad g = g_k, \qquad h = e_{k+1}, \qquad \mathbf{L}(f_1, \ldots, f_k) = \mathbf{K}' \subset \mathbf{K},$$

we arrive at the following

THEOREM. *Suppose the bilinear form* $\mathbf{A}(x, y)$ *is nonsingular in a subspace* $\mathbf{K}' \subset \mathbf{K}$, *and suppose the vector* f *does not belong to* $\mathbf{K}'$. *Then there exists a unique expansion*

$$f = g + h, \tag{25}$$

where $g \in \mathbf{K}'$ and h is conjugate to the space $\mathbf{K}'$.

7.55. Let $\mathbf{K}''$ denote the subspace conjugate to the subspace $\mathbf{K}'$ with respect to the form $\mathbf{A}(x, y)$. Then the existence and uniqueness of the expansion (25) shows that the whole space $\mathbf{K}$ is the direct sum of the subspaces $\mathbf{K}'$ and $\mathbf{K}''$ (see Sec. 2.45). Thus, given a subspace $\mathbf{K}' \subset \mathbf{K}$ in which a bilinear form $\mathbf{A}(x, y)$ defined on the whole space $\mathbf{K}$ is nonsingular, $\mathbf{K}$ can be written as the direct sum

$$\mathbf{K} = \mathbf{K}' + \mathbf{K}'',$$

where $\mathbf{K}''$ is conjugate to $\mathbf{K}'$ with respect to the form $\mathbf{A}(x, y)$.

7.6. Adjoint Linear Operators

7.61. Let (x, y) denote a fixed nonsingular symmetric bilinear form in the space $\mathbf{K}_n$. Let $\mathbf{A}$ and $\mathbf{B}$ be linear operators acting in $\mathbf{K}_n$, and use the formulas

$$\mathbf{A}(x, y) = (\mathbf{A}x, y), \qquad \mathbf{B}(x, y) = (x, \mathbf{B}y)$$

to define functions $\mathbf{A}(x, y)$ and $\mathbf{B}(x, y)$ of two vector arguments x and y. Then $\mathbf{A}(x, y)$ and $\mathbf{B}(x, y)$ are bilinear forms. In fact, it follows from the definition of a linear operator (Sec. 4.21) and the definition of a bilinear form (Sec. 7.11) that

$$\begin{aligned} \mathbf{A}(x_1 + x_2, y) &= (\mathbf{A}(x_1 + x_2), y) = (\mathbf{A}x_1 + \mathbf{A}x_2, y) \\ &= (\mathbf{A}x_1, y) + (\mathbf{A}x_2, y) = \mathbf{A}(x_1, y) + \mathbf{A}(x_2, y), \\ \mathbf{A}(\alpha x, y) &= (\mathbf{A}(\alpha x), y) = (\alpha\mathbf{A}x, y) = \alpha(\mathbf{A}x, y) = \alpha\mathbf{A}(x, y), \end{aligned}$$

which shows that $\mathbf{A}(x, y)$ is linear in its first argument. Similarly, the linearity of $\mathbf{A}(x, y)$ in its second argument is a consequence of the linearity of (x, y) in y. Then $\mathbf{A}(x, y)$ is a bilinear form, and similarly so is $\mathbf{B}(x, y)$.

Next let $e_1, \ldots, e_n$ be a canonical basis of the form (x,y), so that

$$(e_j, e_k) = 0 \quad \text{if} \quad j \neq k,$$

$$(e_m, e_m) = \varepsilon_m \in K, \qquad \varepsilon_m \neq 0.$$

We now compare the matrix of the operator $\mathbf{A}$ with that of the form $\mathbf{A}(x, y)$ in this basis. The matrix $\|a_k^{(j)}\|$ of the operator $\mathbf{A}$ is defined by the formula

$$\mathbf{A}e_j = \sum_{k=1}^{n} a_k^{(j)} e_k \qquad (j = 1, \ldots, n),$$

where here (in contradistinction to the notation adopted in Sec. 4.23) the superscript indicates the row number and the subscript the column number. The matrix $\|a_{jk}\|$ of the form $\mathbf{A}(x, y)$, where the first subscript indicates the row number and the second the column number, is defined by the formula

$$a_{jm} = \mathbf{A}(e_j, e_m) = (\mathbf{A}e_j, e_m) = \left(\sum_{k=1}^{n} a_k^{(j)} e_k, e_m\right) = a_m^{(j)}(e_m, e_m) = \varepsilon_m a_m^{(j)}. \quad (26)$$

Hence the mth column of the matrix $\|a_{jm}\|$ is obtained (for every $m = 1, \ldots, n$) by multiplying the mth column of the matrix $\|a_m^{(j)}\|$ by the canonical coefficient ε_m of the form (x, y). Similarly, for the matrix $\|b_k^{(j)}\|$ of the operator $\mathbf{B}$ (in the same basis $e_1, \ldots, e_n$) and the matrix $\|b_{jk}\|$ of the form $\mathbf{B}(x, y)$, we get

$$b_{jm} = \mathbf{B}(e_j, e_m) = (e_j, \mathbf{B}e_m) = \left(e_j, \sum_{k=1}^{n} b_k^{(m)} e_k\right) = b_j^{(m)}(e_j, e_j) = \varepsilon_j b_j^{(m)}, \quad (27)$$

i.e., the jth row of the matrix $\|b_{jm}\|$ is obtained (for every $j = 1, \ldots, n$) by multiplying the jth column of the matrix of the operator $\mathbf{B}$ by the corresponding canonical coefficient ε_j.

7.62. Conversely, given two bilinear forms $\mathbf{A}(x, y)$ and $\mathbf{B}(x, y)$ in the space $\mathbf{K}_n$, we assert that there exist unique linear operators $\mathbf{A}$ and $\mathbf{B}$ such that

$$\mathbf{A}(x, y) = (\mathbf{A}x, y), \qquad \mathbf{B}(x, y) = (x, \mathbf{B}y). \quad (28)$$

To show this, we specify $\mathbf{A}$ and $\mathbf{B}$ in the same basis $e_1, \ldots, e_n$ by the matrices with elements

$$a_j^{(m)} = \frac{1}{\varepsilon_m}\mathbf{A}(e_j, e_m), \qquad b_j^{(m)} = \frac{1}{\varepsilon_j}\mathbf{B}(e_j, e_m),$$

respectively. We then use these operators to construct the forms $\mathbf{A}_1(x, y) = (\mathbf{A}x, y)$ and $\mathbf{B}_1(x, y) = (x, \mathbf{B}y)$. It follows from Sec. 7.61 that the matrix of the form $\mathbf{A}_1(x, y)$ coincides with the matrix of the form $\mathbf{A}(x, y)$ in the basis $e_1, \ldots, e_n$, while the matrix of the form $\mathbf{B}_1(x, y)$ coincides with the matrix of the form $\mathbf{B}(x, y)$. But then

$$(\mathbf{A}x, y) = \mathbf{A}_1(x, y) = \mathbf{A}(x, y), \qquad (x, \mathbf{B}y) = \mathbf{B}_1(x, y) = \mathbf{B}(x, y)$$

for arbitrary $x, y \in \mathbf{K}_n$ (recall Sec. 7.13), so that the operators $\mathbf{A}$ and $\mathbf{B}$ satisfy (28). To prove the uniqueness, we need only verify that if an operator $\mathbf{A}$ satisfies the condition

$$(\mathbf{A}x, y) = 0 \text{ for arbitrary } x, y \in \mathbf{K}_n, \quad (29)$$

then $\mathbf{A}x = 0$ for every $x \in \mathbf{K}_n$, so that $\mathbf{A}$ is the zero operator. Suppose $\mathbf{A}x_0 \neq 0$ for some $x_0 \in \mathbf{K}_n$. Then, since the form (x, y) is nonsingular, it follows from Sec. 7.15c that there is a vector $y_0 \in \mathbf{K}_n$ such that $(\mathbf{A}x_0, y_0) \neq 0$.

This contradicts (29) and establishes the required uniqueness of $\mathbf{A}$. The uniqueness of $\mathbf{B}$ is proved similarly.

7.63. We now prove the following important

THEOREM. *Let (x, y) be a nonsingular symmetric bilinear form in the space $\mathbf{K}_n$. Then, given any linear operator $\mathbf{A}$ acting in $\mathbf{K}_n$, there exists a unique linear operator $\mathbf{A}'$ acting in $\mathbf{K}_n$ such that*

$$(\mathbf{A}x, y) = (x, \mathbf{A}'y)$$

for arbitrary $x, y \in \mathbf{K}_n$. The matrix of the operator $\mathbf{A}'$ in any canonical basis of the form (x, y) is obtained from the matrix of $\mathbf{A}$ by transposition, followed by multiplication of the mth row by the canonical coefficient ε_m and division of the jth column by the canonical coefficient ε_j $(j, m = 1, \ldots, n)$.

Proof. We use the given operator $\mathbf{A}$ to construct the form $\mathbf{A}(x, y) = (\mathbf{A}x, y)$, and then we define the operator $\mathbf{A}'$ by the formula

$$(\mathbf{A}x, y) \equiv \mathbf{A}(x, y) = (x, \mathbf{A}'y).$$

The existence and uniqueness of $\mathbf{A}'$ follow from Sec. 7.62. In any canonical basis of the form (x, y), the matrix $\|a_m^{(j)}\|$ of the operator $\mathbf{A}$, the matrix $\|a_{jm}\|$ of the form $\mathbf{A}(x, y)$ and the matrix $\|a_j'^{(m)}\|$ of the operator $\mathbf{A}'$ are related by formulas (26) and (27):

$$a_m^{(j)} = \frac{a_{jm}}{\varepsilon_m}, \qquad a_j'^{(m)} = \frac{a_{jm}}{\varepsilon_j}.$$

It follows that

$$a_j'^{(m)} = \frac{a_{jm}}{\varepsilon_j} = \frac{\varepsilon_m}{\varepsilon_j} a_m^{(j)}. \quad \blacksquare \tag{30}$$

The operator $\mathbf{A}'$ is called the *adjoint* (*or conjugate*) *of the operator* $\mathbf{A}$ *with respect to the form* (x, y).

7.64. The operation leading from an operator $\mathbf{A}$ to its adjoint $\mathbf{A}'$ has the following properties:

1) $(\mathbf{A}')' = \mathbf{A}$ for every operator $\mathbf{A}$;
2) $(\mathbf{A} + \mathbf{B})' = \mathbf{A}' + \mathbf{B}'$ for every pair of operators $\mathbf{A}$ and $\mathbf{B}$;
3) $(\lambda\mathbf{A})' = \lambda\mathbf{A}'$ for every operator $\mathbf{A}$ and every number $\lambda \in K$;
4) $(\mathbf{AB})' = \mathbf{B}'\mathbf{A}'$ for every pair of operators $\mathbf{A}$ and $\mathbf{B}$.

To prove property 1), we use the formula

$$(x, (\mathbf{A}')'y) = (\mathbf{A}'x, y) = (x, \mathbf{A}y)$$

implied by the definition of $(\mathbf{A}')'$, together with the uniqueness of the operator defined by a bilinear form (Sec. 7.62). The remaining properties are proved similarly. Thus

$$\begin{aligned}(x, (\mathbf{A} + \mathbf{B})'y) &= ((\mathbf{A} + \mathbf{B})x, y) = (\mathbf{A}x, y) + (\mathbf{B}x, y) \\ &= (x, \mathbf{A}'y) + (x, \mathbf{B}'y) = (x, (\mathbf{A}' + \mathbf{B}')y)\end{aligned}$$

implies property 2).

$$(x, (\lambda\mathbf{A})'y) = (\lambda\mathbf{A}x, y) = \lambda(\mathbf{A}x, y) = \lambda(x, \mathbf{A}'y) = (x, \lambda\mathbf{A}'y)$$

implies property 3), and

$$(x, (\mathbf{AB})'y) = (\mathbf{AB}x, y) = (\mathbf{B}x, \mathbf{A}'y) = (x, \mathbf{B}'\mathbf{A}'y)$$

implies property 4).

7.65. We point out another connection between the operators $\mathbf{A}$ and $\mathbf{A}'$. Suppose the subspace $\mathbf{K}' \subset \mathbf{K}_n$ is invariant under the operator $\mathbf{A}$. According to Sec. 4.81, this means that the operator $\mathbf{A}$ carries every vector $x \in \mathbf{K}'$ into another vector of the same subspace $\mathbf{K}'$. Let $\mathbf{K}''$ be the subspace conjugate to $\mathbf{K}'$ (Sec. 7.55). Then $\mathbf{K}''$ is invariant under the adjoint operator $\mathbf{A}'$. In fact, suppose $y \in \mathbf{K}''$, so that $(y, x) = 0$ for every $x \in \mathbf{K}'$. Then $(\mathbf{A}'y, x) = (y, \mathbf{A}x) = 0$, since $x \in \mathbf{K}'$ implies $\mathbf{A}x \in \mathbf{K}'$. But this means that the vector $\mathbf{A}'y$ is conjugate to every vector $x \in \mathbf{K}'$ and hence belongs to $\mathbf{K}''$, as required.

7.7. Isomorphism of Spaces Equipped with a Bilinear Form

7.71. Definition. Let $\mathbf{K}'$ and $\mathbf{K}''$ be two linear spaces over the same number field K. Suppose $\mathbf{K}'$ is equipped with a nonsingular symmetric bilinear form $\mathbf{A}(x', y')$, while $\mathbf{K}''$ is equipped with a nonsingular symmetric bilinear form $\mathbf{A}(x'', y'')$. Then $\mathbf{K}'$ and $\mathbf{K}''$ are said to be $\mathbf{A}$-*isomorphic* if

1) They are isomorphic regarded as linear spaces over the field K (see Sec. 2.71), i.e., there exists a one-to-one mapping (morphism) $\omega x' = x''$ preserving linear operations;

2) The values of the forms $\mathbf{A}(x', y')$ and $\mathbf{A}(x'', y'')$ coincide for all corresponding pairs of elements x', y' and $x'' = \omega x'$, $y'' = \omega y'$, i.e.,

$$\mathbf{A}(x', y') = \mathbf{A}(x'', y'').$$

7.72. THEOREM. *Given two finite-dimensional linear spaces* $\mathbf{K}'$ *and* $\mathbf{K}''$, *suppose* $\mathbf{K}'$ *is equipped with a nonsingular symmetric bilinear form* $\mathbf{A}(x', y')$, *while* $\mathbf{K}''$ *is equipped with a nonsingular symmetric bilinear form* $\mathbf{A}(x'', y'')$. *Then* $\mathbf{K}'$ *and* $\mathbf{K}''$ *are* $\mathbf{A}$-*isomorphic if and only if*

a) *They have the same dimension* n;

b) *There exists a canonical basis for* $\mathbf{A}(x', y')$ *in* $\mathbf{K}'$ *and a canonical basis for* $\mathbf{A}(x'', y'')$ *in* $\mathbf{K}''$ *relative to which the two forms have the same set of canonical coefficients* $\varepsilon_1, \ldots, \varepsilon_n$.

Proof. Suppose $\mathbf{K}'$ and $\mathbf{K}''$ are $\mathbf{A}$-isomorphic. Then they are isomorphic as linear spaces and hence have the same dimension, say n (see Sec. 2.73d). If $e_1', \ldots, e_n'$ is a canonical basis for the form $\mathbf{A}(x', y')$ in the space $\mathbf{K}'$, then

$$\mathbf{A}(e_i', e_j') = \begin{cases} 0 & \text{if } i \neq j, \\ \varepsilon_i & \text{if } i = j. \end{cases}$$

Let $e_1'', \ldots, e_n''$ be the vectors in $\mathbf{K}''$ corresponding to the vectors $e_1', \ldots, e_n'$ in $\mathbf{K}'$ under the given $\mathbf{A}$-isomorphism. By hypothesis,

$$\mathbf{A}(\omega e_i', \omega e_j') = \mathbf{A}(e_i'', e_j'') = \begin{cases} 0 & \text{if } i \neq j, \\ \varepsilon_i & \text{if } i = j. \end{cases}$$

Thus $e_1'', \ldots, e_n''$ is a canonical basis for $\mathbf{A}(x'', y'')$ in the space $\mathbf{K}''$. Moreover, $\mathbf{A}(x'', y'')$ has the same canonical coefficients $\varepsilon_1, \ldots, \varepsilon_n$ in the basis $e_1'', \ldots, e_n''$ as $\mathbf{A}(x', y')$ has in the basis $e_1', \ldots, e_n'$.

Conversely, suppose $\mathbf{K}'$ and $\mathbf{K}''$ have the same dimension n, and let $e_1', \ldots, e_n' \in \mathbf{K}'$ and $e_1'', \ldots, e_n'' \in \mathbf{K}''$ be canonical bases with the same canonical coefficients $\varepsilon_1, \ldots, \varepsilon_n$, so that

$$\mathbf{A}(e_i', e_j') = \mathbf{A}(e_i'', e_j'') = \begin{cases} 0 & \text{if } i \neq j, \\ \varepsilon_i & \text{if } i = j. \end{cases}$$

Given any vector

$$x' = \sum_{i=1}^{n} \xi_i e_i'$$

in $\mathbf{K}'$, let

$$x'' = \omega(x') = \sum_{i=1}^{n} \xi_i e_i''$$

(with the same components $\xi_1, \ldots, \xi_n$) be the corresponding vector in $\mathbf{K}''$. This correspondence defines an isomorphism ω of the spaces $\mathbf{K}'$ and $\mathbf{K}''$ (see Sec. 2.73d). Moreover, if

$$y' = \sum_{i=1}^{n} \eta_i e_i', \qquad y'' = \omega(y') = \sum_{i=1}^{n} \eta_i e_i'',$$

then

$$\mathbf{A}(x', y') = \sum_{i=1}^{n} \varepsilon_i \xi_i \eta_i = \mathbf{A}(x'', y''),$$

so that the isomorphism ω is an $\mathbf{A}$-isomorphism. ▌

7.73. Given an n-dimensional space $\mathbf{K}_n$ equipped with a nonsingular symmetric bilinear form $\mathbf{A}(x, y)$, consider an $\mathbf{A}$-isomorphism of $\mathbf{K}_n$, i.e., an invertible linear mapping $y = \mathbf{Q}x$ which does not change the form $\mathbf{A}(x, y)$ in the sense that

$$\mathbf{A}(\mathbf{Q}x, \mathbf{Q}y) = \mathbf{A}(x, y). \tag{31}$$

We will henceforth denote $\mathbf{A}(x, y)$ simply by (x, y). If $\mathbf{Q}'$ is the adjoint of the operator $\mathbf{Q}$ with respect to the form (x, y), then

$$(\mathbf{Q}x, \mathbf{Q}y) = (\mathbf{Q}'\mathbf{Q}x, y). \tag{32}$$

It follows from (31) and (32) that

$$\mathbf{Q}'\mathbf{Q} = \mathbf{E}, \tag{33}$$

and hence that $\mathbf{Q}'$ is the inverse of the operator $\mathbf{Q}$ (since $\mathbf{Q}$ is nonsingular, so is $\mathbf{Q}'$).

Conversely, (33) implies (32) and then (31), so that the condition (33) completely specifies the class of operators which do not change the form (x, y). These operators are said to be *invariant with respect to the form* (x, y).

7.74. If $\mathbf{Q}$ is invariant, then so is the inverse operator $\mathbf{Q}^{-1} = \mathbf{Q}'$, since

$$(\mathbf{Q}'x, \mathbf{Q}'y) = (\mathbf{Q}\mathbf{Q}'x, y) = (x, y)$$

for every x and y. The product of two invariant operators $\mathbf{Q}$ and $\mathbf{T}$ is also an invariant operator, since

$$(\mathbf{QT}x, \mathbf{QT}y) = (\mathbf{T}x, \mathbf{T}y) = (x, y)$$

for every x and y.

7.75. Let $e_1, \ldots, e_n$ be a canonical basis of the form (x, y), with canonical coefficients $\varepsilon_1, \ldots, \varepsilon_n$. Then, applying an invariant operator $\mathbf{Q}$ to the vectors $e_1, \ldots, e_n$, we get the vectors

$$f_1 = \mathbf{Q}e_1, \ldots, f_n = \mathbf{Q}e_n, \tag{34}$$

where

$$(f_j, f_k) = (\mathbf{Q}e_j, \mathbf{Q}e_k) = (e_j, e_k) = \begin{cases} \varepsilon_j & \text{if } j = k, \\ 0 & \text{if } j \neq k. \end{cases}$$

Thus $f_1, \ldots, f_n$ is also a canonical basis of the form (x, y), with the same canonical coefficients $\varepsilon_1, \ldots, \varepsilon_n$.

Conversely, if $f_1, \ldots, f_n$ is a canonical basis of the form (x, y) with the same canonical coefficients $\varepsilon_1, \ldots, \varepsilon_n$ as the basis $e_1, \ldots, e_n$, then the operator $\mathbf{Q}$ defined by (34) is invariant. In fact,

$$(\mathbf{Q}e_j, \mathbf{Q}e_k) = (f_j, f_k) = (e_j, e_k) = \begin{cases} \varepsilon_j & \text{if } j = k, \\ 0 & \text{if } j \neq k, \end{cases}$$

and hence (31) holds for any pair of basis vectors. But then, by the linearity, (31) holds for *arbitrary* vectors $x, y \in \mathbf{K}_n$, as required.

Thus an invariant operator $\mathbf{Q}$ is characterized by the fact that *it carries every canonical basis of the space* $\mathbf{K}_n$ *(with respect to the form* (x, y)*) into another canonical basis with the same canonical coefficients.*

7.76. We now find conditions characterizing the matrix of an invariant operator **Q** in a canonical basis of the form (x, y). Let $e_1, \ldots, e_n$ be such a basis, and let $\varepsilon_1, \ldots, \varepsilon_n$ be the corresponding canonical coefficients. Moreover, let $Q = \|q_i^{(j)}\|$ be the matrix of **Q** in the basis $e_1, \ldots, e_n$. Then, according to Sec. 7.63, the matrix of the adjoint operator has the form

$$Q' = \|q_i'^{(j)}\|, \qquad q_i'^{(j)} = \frac{\varepsilon_j}{\varepsilon_i} q_j^{(i)}.$$

In terms of matrix elements, we can write equation (33) as

$$\sum_{i=1}^{n} q_i'^{(j)} q_k^{(i)} = \sum_{i=1}^{n} \frac{\varepsilon_j}{\varepsilon_i} q_j^{(i)} q_k^{(i)} = \delta_j^{(k)} = \begin{cases} 1 & \text{if } j = k, \\ 0 & \text{if } j \neq k. \end{cases}$$

In other words,

$$\sum_{i=1}^{n} \frac{1}{\varepsilon_i} q_j^{(i)} q_k^{(i)} = \begin{cases} \dfrac{1}{\varepsilon_j} & \text{if } j = k, \\ 0 & \text{if } j \neq k. \end{cases} \tag{35}$$

Equation (35) is equivalent to (33), and can also serve as the definition of an invariant operator **Q**.

Thus an *invariant matrix*, i.e., the matrix of an invariant operator in any canonical basis of the form (x, y), is characterized by the fact that the sum of the squares of the elements of its jth column taken with coefficients $\varepsilon_1^{-1}, \ldots, \varepsilon_n^{-1}$ equals the number ε_j^{-1} $(j = 1, \ldots, n)$, while the sum of the products of the corresponding elements of two different columns also taken with the coefficients $\varepsilon_1^{-1}, \ldots, \varepsilon_n^{-1}$ equals zero. Since (33) also implies $\mathbf{QQ}' = \mathbf{E}$, we also have the relations

$$\sum_{k=1}^{n} q_k^{(j)} q_m'^{(k)} = \sum_{k=1}^{n} \frac{\varepsilon_k}{\varepsilon_m} q_k^{(j)} q_k^{(m)} = \delta_m^{(j)},$$

or

$$\sum_{k=1}^{n} \varepsilon_k q_k^{(j)} q_k^{(m)} = \begin{cases} \varepsilon_j & \text{if } j = m, \\ 0 & \text{if } j \neq m. \end{cases} \tag{35$'$}$$

This gives another characterization of an invariant matrix, namely the sum of the squares of the elements of its jth row taken with coefficients $\varepsilon_1, \ldots, \varepsilon_n$ equals the number ε_j $(j = 1, \ldots, n)$, while the sum of the products of the corresponding elements of two different rows also taken with the coefficients $\varepsilon_1, \ldots, \varepsilon_n$ equals zero.

*7.8. Multilinear Forms

7.81. By analogy with bilinear forms we can consider linear functions of a larger number of vectors (three, four or more). All such functions are called *multilinear forms*.

Definition. A function $\mathbf{A}(x_1, \ldots, x_k)$ of k vector arguments $x_1, \ldots, x_k$ varying in a linear space $\mathbf{K}$ is called a *multilinear* (more exactly, a *k-linear*) *form* if it is linear in each argument x_j ($j = 1, \ldots, k$) for fixed values of the remaining arguments $x_1, \ldots, x_{j-1}, x_{j+1}, \ldots, x_k$. A multilinear form $\mathbf{A}(x_1, \ldots, x_k)$ is called *symmetric* if it does not change when any two of its arguments are interchanged, and *antisymmetric* if it changes sign when any two of its arguments are interchanged.

An example of an antisymmetric multilinear form in three vectors x, y and z (a trilinear form) of the space V_3 is the mixed triple product of x, y and z.† An example of an antisymmetric multilinear form in n vectors

$$x_1 = (a_{11}, a_{12}, \ldots, a_{1n}),$$
$$x_2 = (a_{21}, a_{22}, \ldots, a_{2n}),$$
$$\cdots\cdots\cdots\cdots\cdots$$
$$x_n = (a_{n1}, a_{n2}, \ldots, a_{nn})$$

of an n-dimensional linear space $\mathbf{K}_n$‡ is the determinant

$$\mathbf{A}(x_1, x_2, \ldots, x_n) = \begin{vmatrix} a_{11} & a_{12} & \cdots & a_{1n} \\ a_{21} & a_{22} & \cdots & a_{2n} \\ \cdot & \cdot & \cdots & \cdot \\ a_{n1} & a_{n2} & \cdots & a_{nn} \end{vmatrix}. \tag{36}$$

A somewhat more general example is the product of the determinant (36) with a fixed number $\lambda \in K$.

7.82. We now show that ***every*** *antisymmetric multilinear form*

$$\mathbf{A}(x_1, x_2, \ldots, x_n)$$

in n vectors $x_1, x_2, \ldots, x_n$ of an n-dimensional linear space $\mathbf{K}_n$ with a fixed basis $e_1, e_2, \ldots, e_n$ equals the determinant (36) *multiplied by some constant* $\lambda \in K$.

Let λ denote the quantity $\mathbf{A}(e_1, e_2, \ldots, e_n)$. Then we can easily calculate the quantity $\mathbf{A}(e_{i_1}, e_{i_2}, \ldots, e_{i_n})$ where $i_1, i_2, \ldots, i_n$ are arbitrary integers from 1 to n. If two of these numbers are equal, then $\mathbf{A}(e_{i_1}, e_{i_2}, \ldots, e_{i_n})$ vanishes, since on the one hand it does not change when the arguments corresponding to these numbers are interchanged, while on the other hand it must change sign because of the antisymmetry property. If all the numbers $i_1, i_2, \ldots, i_n$ are different, then by making the same number of interchanges

† I.e., $(x, y \times z)$ where (,) denotes the scalar product and $\times$ the vector product.
‡ By $x_1 = (a_{11}, a_{12}, \ldots, a_{1n})$ we mean $x = a_{11}e_1 + a_{12}e_2 + \cdots + a_{1n}e_n$, where $e_1, e_2, \ldots, e_n$ is a fixed basis in $\mathbf{K}_n$, and so on.

of adjacent arguments as there are inversions in the sequence of indices $i_1, i_2, \ldots, i_n$, we can cause the arguments to be arranged in normal order†; let the required number of interchanges be N. Then we have

$$\mathbf{A}(e_{i_1}, e_{i_2}, \ldots, e_{i_n}) = (-1)^N \lambda.$$

Now let

$$x_i = \sum_{j=1}^{n} a_{ij} e_j \qquad (i = 1, 2, \ldots, n)$$

be an arbitrary system of n vectors of the space $\mathbf{K}_n$, and consider the multilinear form

$$\begin{aligned}\mathbf{A}(x_1, x_2, \ldots, x_n) &= \mathbf{A}\left(\sum_{i_1=1}^{n} a_{1i_1} e_{i_1}, \sum_{i_2=1}^{n} a_{2i_2} e_{i_2}, \ldots, \sum_{i_n=1}^{n} a_{ni_n} e_{i_n}\right) \\ &= \sum_{i_1, i_2, \ldots, i_n=1}^{n} a_{1i_1} a_{2i_2} \cdots a_{ni_n} \mathbf{A}(e_{i_1}, e_{i_2}, \ldots, e_{i_n}) \\ &= \lambda \sum_{i_1, i_2, \ldots, i_n=1}^{n} (-1)^N a_{1i_1} a_{2i_2} \cdots a_{ni_n}.\end{aligned}$$

Since in each term of the last sum, N denotes the number of inversions in the arrangement of the second subscripts of the elements a_{ij} when the first subscripts are in normal order, it follows that each term is one of the terms in the determinant (36) with the appropriate sign. Hence the sum of all the terms is just the determinant (36), and our assertion is proved.

In particular, this shows that the mixed triple product of three vectors x, y and z of the space V_3 in any basis can be written as the third-order determinant made up of the components of x, y and z, taken with a coefficient equal to the triple product of the basis vectors.

7.9. Bilinear and Quadratic Forms in a Real Space

7.91. Every real number has a definite sign (+ or −), and hence the theory of bilinear and quadratic forms in a real space can be carried somewhat further than in a space over an arbitrary field K. According to the general theory of Sec. 7.31, a quadratic form $\mathbf{A}(x, x)$ can be reduced in some basis to the canonical form

$$\mathbf{A}(x, x) = \lambda_1 \eta_1^2 + \lambda_2 \eta_2^2 + \cdots + \lambda_n \eta_n^2,$$

where the number of nonzero coefficients $\lambda_1, \lambda_2, \ldots, \lambda_n$, equal to the rank of the form $\mathbf{A}(x, x)$ (Sec. 7.33b), does not change when the canonical basis is changed. These coefficients are either positive or negative. It turns out

† Cf. the proof of Theorem 4.54.

that changing the canonical basis also has no effect on the total number of positive coefficients and the total number of negative coefficients:

THEOREM (***Law of inertia for quadratic forms***). *If a quadratic form* $\mathbf{A}(x, x)$ *in a real space is written in canonical form, the total number of positive coefficients and the total number of negative coefficients are invariants of the form, i.e., do not depend on the choice of the canonical basis.*

Proof. Suppose $\mathbf{A}(x, x)$ has the form

$$\mathbf{A}(x, x) = \sum_{i,k=1}^{n} a_{ik}\xi_i\xi_k$$

in the basis $\{e\} = \{e_1, e_2, \ldots, e_n\}$, where $\xi_1, \xi_2, \ldots, \xi_n$ are the components of the vector x with respect to $\{e\}$. Suppose $\mathbf{A}(x, x)$ has two canonical bases $\{f\} = \{f_1, f_2, \ldots, f_n\}$ and $\{g\} = \{g_1, g_2, \ldots, g_n\}$. Let $\eta_1, \eta_2, \ldots, \eta_n$ denote the components of x with respect to the basis $\{f\}$, and let $\tau_1, \tau_2, \ldots, \tau_n$ denote the components of x with respect to the basis $\{g\}$. Let the corresponding transformation formulas be

$$\begin{aligned} \eta_1 &= b_{11}\xi_1 + b_{12}\xi_2 + \cdots + b_{1n}\xi_n, \\ \eta_2 &= b_{21}\xi_1 + b_{22}\xi_2 + \cdots + b_{2n}\xi_n, \\ &\cdots\cdots\cdots\cdots\cdots\cdots \\ \eta_n &= b_{n1}\xi_1 + b_{n2}\xi_2 + \cdots + b_{nn}\xi_n \end{aligned} \tag{37}$$

and

$$\begin{aligned} \tau_1 &= c_{11}\xi_1 + c_{12}\xi_2 + \cdots + c_{1n}\xi_n, \\ \tau_2 &= c_{21}\xi_1 + c_{22}\xi_2 + \cdots + c_{2n}\xi_n, \\ &\cdots\cdots\cdots\cdots\cdots\cdots \\ \tau_n &= c_{n1}\xi_1 + c_{n2}\xi_2 + \cdots + c_{nn}\xi_n, \end{aligned} \tag{37'}$$

where the matrices $\|b_{ik}\|$ and $\|c_{ik}\|$ are nonsingular. In the basis $\{f\}$, $\mathbf{A}(x, x)$ has the form

$$\mathbf{A}(x, x) = \alpha_1\eta_1^2 + \cdots + \alpha_k\eta_k^2 - \alpha_{k+1}\eta_{k+1}^2 - \cdots - \alpha_m\eta_m^2, \tag{38}$$

while in the basis $\{g\}$ it has the form

$$\mathbf{A}(x, x) = \beta_1\tau_1^2 + \cdots + \beta_p\tau_p^2 - \beta_{p+1}\tau_{p+1}^2 - \cdots - \beta_q\tau_q^2, \tag{39}$$

where the numbers $\alpha_1, \ldots, \alpha_m, \beta_1, \ldots, \beta_q$ are assumed to be *positive.* We wish to show that $k = p$, $m = q$. Equating the right-hand sides of (38) and (39), and transposing negative terms to opposite sides of the equation, we obtain

$$\begin{aligned} &\alpha_1\eta_1^2 + \cdots + \alpha_k\eta_k^2 + \beta_{p+1}\tau_{p+1}^2 + \cdots + \beta_q\tau_q^2 \\ &\qquad = \alpha_{k+1}\eta_{k+1}^2 + \cdots + \alpha_m\eta_m^2 + \beta_1\tau_1^2 + \cdots + \beta_p\tau_p^2. \end{aligned} \tag{40}$$

Now suppose $k < p$, and consider the vectors x which satisfy the conditions

$$\begin{gathered}\eta_1 = 0,\ \eta_2 = 0, \ldots, \eta_k = 0, \\ \tau_{p+1} = 0, \ldots, \tau_q = 0,\ \tau_{q+1} = 0, \ldots, \tau_n = 0.\end{gathered} \tag{41}$$

There are clearly less than n of these conditions, since $k < p$. Using (37) and (37′) to express $\eta_1, \ldots, \eta_k, \tau_{p+1}, \ldots, \tau_n$ in terms of the variables $\xi_1, \xi_2, \ldots, \xi_n$, we obtain a homogeneous system of linear equations in the unknowns $\xi_1, \xi_2, \ldots, \xi_n$. The number of equations is less than the number of unknowns, and therefore this homogeneous system has a nontrivial solution $x = (\xi_1, \xi_2, \ldots, \xi_n)$. On the other hand, because of (40), every vector x satisfying the conditions (41) also satisfies the conditions

$$\tau_1 = \tau_2 = \cdots = \tau_p = 0.$$

However, since $\det \|c_{ik}\| \neq 0$, any vector x for which

$$\tau_1 = \tau_2 = \cdots = \tau_p = \tau_{p+1} = \cdots = \tau_n = 0$$

must be the zero vector, with all its components $\xi_1, \xi_2, \ldots, \xi_n$ equal to zero. Thus the assumption that $k < p$ leads to a contradiction. Because of the complete symmetry of the role played by the numbers k and p in this problem, the assertion $p < k$ also leads to a contradiction. It follows that $k = p$. Moreover, examining the conditions

$$\begin{gathered}\tau_1 = 0,\ \tau_2 = 0, \ldots, \tau_p = 0, \\ \eta_{k+1} = 0, \ldots, \eta_m = 0,\ \tau_{q+1} = 0, \ldots, \tau_n = 0,\end{gathered}$$

we can use the same argument to show that $m < q$ is impossible and hence, by symmetry, that $q < m$. Thus we finally find that $k = p$, $m = q$. ∎

7.92. The total number of terms appearing in the canonical form of a quadratic form $\mathbf{A}(x, x)$, i.e., its *rank* (see Sec. 7.33b), is also called its *index of inertia*. The total number of positive terms is called the *positive index of inertia*, and the total number of negative terms is called the *negative index of inertia*. If the positive index of inertia equals the dimension of the space, the form is said to be *positive definite*. In other words, a quadratic form $\mathbf{A}(x, x)$ is positive definite if and only if all n of its canonical coefficients are positive. It follows that *a positive definite quadratic form takes a positive value at every point of the space except the origin of coordinates.*

Conversely, if a quadratic form defined on an n-dimensional real space takes positive values everywhere except at the origin, then its rank is n and its positive index of inertia is also n, i.e., the form is positive definite. In fact, for a form of rank less than n or with less than n positive canonical coefficients, it is easy to find points in the space other than the origin where the form takes either the value 0 or negative values. For example, the quadratic form

$$\mathbf{A}(x, x) = \xi_1^2 + \xi_3^2$$

of rank 2 in a three-dimensional space takes the value 0 for any nonzero vector with components $\xi_1 = 0$, $\xi_2 \neq 0$, $\xi_3 = 0$. For these vectors the form

$$\mathbf{A}(x, x) = \xi_1^2 - \xi_2^2 + \xi_3^2$$

of rank 3 in a three-dimensional space takes negative values. Clearly, these examples illustrate the full generality of the situation.

7.93. The law of inertia just proved for quadratic forms generalizes immediately to the case of symmetric bilinear forms, i.e., *the total number of positive coefficients and the total number of negative coefficients in the canonical form* (22) *of a symmetric bilinear form* $\mathbf{A}(x, y)$ *is independent of the choice of a canonical basis.* Thus the positive and negative indices of inertia are well-defined concepts for a symmetric bilinear form. The values of the positive and negative indices of inertia of the bilinear form $\mathbf{A}(x, y)$ and hence of the quadratic form $\mathbf{A}(x, x)$ can be determined from the signs of the descending principal minors of the matrix of the form in any basis (provided only that the minors are nonzero) by using the formulas (24), p. 195.

It should be noted that *given any quadratic form* $\mathbf{A}(x, x)$ *in a real space* $\mathbf{R}_n$, *a canonical basis can always be found such that the corresponding canonical coefficients can only take the values* ± 1. In fact, having reduced $\mathbf{A}(x, x)$ to the form

$$\mathbf{A}(x, x) = \lambda_1\eta_1^2 + \cdots + \lambda_p\eta_p^2 - \mu_1\eta_{p+1}^2 - \cdots - \mu_q\eta_{p+q}^2$$

where the numbers $\lambda_1, \ldots, \lambda_p, \mu_1, \ldots, \mu_q$ are all positive, we make another coordinate transformation

$$\tau_1 = \sqrt{\lambda_1}\,\eta_1, \ldots, \tau_p = \sqrt{\lambda_p}\,\eta_p, \tau_{p+1} = \sqrt{\mu_1}\,\eta_{p+1}, \ldots, \tau_{p+q} = \sqrt{\mu_q}\,\eta_{p+q},$$

thereby reducing $\mathbf{A}(x, x)$ to the form

$$\mathbf{A}(x, x) = \tau_1^2 + \cdots + \tau_p^2 - \tau_{p+1}^2 - \cdots - \tau_{p+q}^2.$$

This shows that *in a real space the numbers p and q are the only invariants*† *of the quadratic form* $\mathbf{A}(x, x)$ *and the corresponding symmetric bilinear form* $\mathbf{A}(x, y)$.

THEOREM. *Two finite-dimensional real spaces* $\mathbf{R}'$ *and* $\mathbf{R}''$, *equipped with nonsingular symmetric bilinear forms* $\mathbf{A}(x', y')$ *and* $\mathbf{A}(x'', y'')$, *respectively, are* $\mathbf{A}$*-isomorphic if and only if they have the same dimension and the indices of*

† Apart from any function of p and q (like the rank $r = p + q$), which is obviously an invariant of $\mathbf{A}(x, x)$ and $\mathbf{A}(x, y)$.

inertia p', q' of the form $\mathbf{A}(x', y')$ coincide with the corresponding indices of inertia p'', q'' of the form $\mathbf{A}(x'', y'')$.

Proof. An immediate consequence of the above considerations and Theorem 7.72. ▮

7.94. Let $\mathbf{A}(x, y)$ be a symmetric bilinear form in a real space $\mathbf{R}_n$. Then, as in Sec. 7.15b, $\mathbf{A}(x, y)$ is said to be *nonsingular* if its rank equals the dimension of the space, i.e., if all the coefficients $\lambda_1, \lambda_2, \ldots, \lambda_n$ in the canonical form

$$\mathbf{A}(x, y) = \lambda_1\xi_1\eta_1 + \lambda_2\xi_2\eta_2 + \cdots + \lambda_n\xi_n\eta_n$$

(see Sec. 7.43) are nonzero. Suppose that in addition all the coefficients $\lambda_1, \lambda_2, \ldots, \lambda_n$ are *positive*, so that the corresponding quadratic form $\mathbf{A}(x, x)$ is positive definite (see Sec. 7.92). Then the bilinear form $\mathbf{A}(x, y)$ is said to be *positive definite*. Thus, according to Sec. 7.92, $\mathbf{A}(x, y)$ is positive definite if and only if the corresponding quadratic form $\mathbf{A}(x, x)$ takes a *positive* value for every nonzero vector x.

By its very definition, a positive definite form $\mathbf{A}(x, y)$ in a space $\mathbf{R}_n$ is nonsingular. But, because of the fact that $\mathbf{A}(x, x) > 0$, a positive definite form $\mathbf{A}(x, y)$ remains positive definite in any subspace $\mathbf{R}' \subset \mathbf{R}_n$. Hence a positive definite bilinear form, unlike the general bilinear form (see Sec. 7.15d), remains nonsingular in any subspace $\mathbf{R}' \subset \mathbf{R}_n$. Thus, given any k linearly independent vectors $f_1, \ldots, f_k$, the determinant

$$D = \begin{vmatrix} \mathbf{A}(f_1, f_1) & \cdots & \mathbf{A}(f_1, f_k) \\ \cdot & \cdots & \cdot \\ \mathbf{A}(f_k, f_1) & \cdots & \mathbf{A}(f_k, f_k) \end{vmatrix}$$

must be nonzero. We will see in a moment that D must in fact be positive.

7.95. An important example of a symmetric positive definite bilinear form in the space V_3 is given by the scalar product (x, y) of the vectors x and y. In fact, it follows at once from the definition of the scalar product that

$$(x, y) = (y, x),$$
$$(x, x) = |x|^2 > 0 \quad \text{for} \quad x \neq 0.$$

The first of these relations shows that the bilinear form (x, y) is symmetric, while the second shows that the corresponding quadratic form takes a positive value for every vector $x \neq 0$. Thus the bilinear form (x, y) is positive definite.

Positive definite bilinear forms will play a particularly important role below. In fact, by using such forms we will be able to introduce the concepts of the length of a vector and the angle between two vectors in a general linear space (Chap. 8).

7.96. The problem now arises of how to use the matrix of a symmetric bilinear form $\mathbf{A}(x, y)$ to determine whether or not $\mathbf{A}(x, y)$ is positive definite. The answer to this problem is given by the following

THEOREM. *A necessary and sufficient condition for the symmetric matrix* $A = \|a_{ik}\|$ *to define a positive definite bilinear form* $\mathbf{A}(x, y)$ *is that the descending principal minors*

$$a_{11}, \quad \begin{vmatrix} a_{11} & a_{12} \\ a_{21} & a_{22} \end{vmatrix}, \quad \begin{vmatrix} a_{11} & a_{12} & a_{13} \\ a_{21} & a_{22} & a_{23} \\ a_{31} & a_{32} & a_{33} \end{vmatrix}, \quad \ldots, \quad \det \|a_{ik}\| \tag{42}$$

of the matrix $\|a_{ik}\|$ *all be positive.*

Proof. If the principal minors (42) of the matrix A are all positive, then by the formulas (24), p. 195, all the canonical coefficients λ_k of the form $\mathbf{A}(x, y)$ are also positive in some basis, i.e., $\mathbf{A}(x, y)$ is positive definite.

Conversely, suppose the form $\mathbf{A}(x, y)$ is positive definite. Then the descending principal minors (42) of the matrix $\|a_{ik}\|$ are positive. In fact, the principal minor

$$M = \begin{vmatrix} a_{11} & a_{12} & \cdots & a_{1m} \\ a_{21} & a_{22} & \cdots & a_{2m} \\ \cdot & \cdot & \cdots & \cdot \\ a_{m1} & a_{m2} & \cdots & a_{mm} \end{vmatrix}$$

corresponds to the matrix $\|a_{ik}\|$ $(i, k = 1, 2, \ldots, m)$ of the bilinear form $\mathbf{A}(x, y)$ in the subspace $\mathbf{L}_m$ spanned by the first m basis vectors. Since $\mathbf{A}(x, y)$ is positive definite in the subspace $\mathbf{L}_m$ ($\mathbf{A}(x, x) > 0$ for $x \neq 0$), there exists a canonical basis in $\mathbf{L}_m$ in which $\mathbf{A}(x, y)$ can be written in canonical form with positive coefficients. In particular, the determinant of $\mathbf{A}(x, y)$ in this basis is positive, being equal to the product of the canonical coefficients. Bearing in mind the relation between determinants of a bilinear form in different bases (equation (6), p. 182), we see that the determinant of $\mathbf{A}(x, y)$ in the original basis of the subspace $\mathbf{L}_m$ is also positive. But the determinant of $\mathbf{A}(x, y)$ in the original basis of $\mathbf{L}_m$ is just the minor M. It follows that $M > 0$. ∎

Remark. In the second part of the proof, we could have taken M to be any principal minor instead of a descending principal minor, without changing the argument in any essential way. Thus *every principal minor of the matrix of a positive definite bilinear form is positive.*

7.97. For a positive definite form $\mathbf{A}(x, y)$ there always exists a canonical basis $e_1, \ldots, e_n$ in which all the canonical coefficients equal $+1$ (see Sec. 7.93). Hence two n-dimensional real spaces $\mathbf{R}_n'$ and $\mathbf{R}_n''$ equipped with positive definite forms $\mathbf{A}(x', y')$ and $\mathbf{A}(x'', y'')$, respectively, are $\mathbf{A}$-isomorphic, by Theorem 7.72.

7.98. The solution of the following problem is often needed in applications of linear algebra to analysis (i.e., in the theory of conditional extrema): Given the matrix $A = \|a_{ik}\|$ of a symmetric bilinear form $\mathbf{A}(x, y)$, determine whether the form is positive definite in the subspace specified by the system of k independent linear equations

$$\sum_{j=1}^{n} b_{ij}\xi_j = 0 \qquad (i = 1, 2, \ldots, k;\ k < n).$$

It turns out that a necessary and sufficient condition for this to be the case is that the descending principal minors of orders $2k + 1, 2k + 2, \ldots, k + n$ of the matrix

$$\Delta = (-1)^k \begin{Vmatrix} 0 & 0 & \cdots & 0 & b_{11} & b_{12} & \cdots & b_{1n} \\ 0 & 0 & \cdots & 0 & b_{21} & b_{22} & \cdots & b_{2n} \\ \cdot & \cdot & \cdots & \cdot & \cdot & \cdot & \cdots & \cdot \\ 0 & 0 & \cdots & 0 & b_{k1} & b_{k2} & \cdots & b_{kn} \\ b_{11} & b_{21} & \cdots & b_{k1} & a_{11} & a_{12} & \cdots & a_{1n} \\ b_{12} & b_{22} & \cdots & b_{k2} & a_{21} & a_{22} & \cdots & a_{2n} \\ \cdot & \cdot & \cdots & \cdot & \cdot & \cdot & \cdots & \cdot \\ b_{1n} & b_{2n} & \cdots & b_{kn} & a_{n1} & a_{n2} & \cdots & a_{nn} \end{Vmatrix}$$

be positive, under the assumption that the rank of the matrix $\|b_{ij}\|$ equals k and that the determinant made up of the first k columns of $\|b_{ij}\|$ is non-vanishing.†

PROBLEMS

1. Do the elements of the matrix of a bilinear form constitute a tensor (Sec. 5.61), and if so, of what type?

2. Reduce the quadratic form

$$\xi_1\xi_2 + \xi_2\xi_3 + \xi_3\xi_1$$

to canonical form.

3. Let p be the positive index of inertia of a quadratic form $\mathbf{A}(x, x)$ (defined on the space $\mathbf{R}_n$), and let q be its negative index of inertia. Moreover, let $\lambda_1, \lambda_2, \ldots, \lambda_p$ be any p positive numbers and $\mu_1, \mu_2, \ldots, \mu_q$ any q negative numbers. Show that there exists a basis in which the form $\mathbf{A}(x, x)$ takes the form

$$\mathbf{A}(x, x) = \lambda_1\tau_1^2 + \cdots + \lambda_p\tau_p^2 + \mu_1\tau_{p+1}^2 + \cdots + \mu_q\tau_{p+q}^2.$$

† See the note by R. Y. Shostak, Uspekhi Mat. Nauk, vol. 9, no. 2 (1954), pp. 199–206.

4. Show that the matrix of a quadratic form of rank r always has at least one nonvanishing principal minor of order r.

5. Reduce the bilinear form

$$\mathbf{A}(x, y) = \xi_1\eta_1 + \xi_1\eta_2 + \xi_2\eta_1 + 2\xi_2\eta_2 + 2\xi_2\eta_3 + 2\xi_3\eta_2 + 5\xi_3\eta_3$$

to canonical form.

6. Apply Jacobi's method to reduce the bilinear form

$$\mathbf{A}(x, y) = \xi_1\eta_1 - \xi_1\eta_2 - \xi_2\eta_1 + \xi_1\eta_3 + \xi_3\eta_1 + 2\xi_2\eta_3 + 2\xi_3\eta_2 + \xi_3\eta_3 + + \xi_2\eta_2$$

to canonical form.

7. State the conditions under which a symmetric matrix $\|a_{ik}\|$ defines a negative definite bilinear form.

8. Given a symmetric matrix $A = \|a_{ik}\|$ with the properties

$$a_{11} > 0, \quad \begin{vmatrix} a_{11} & a_{12} \\ a_{21} & a_{22} \end{vmatrix} > 0, \ldots, \det \|a_{ik}\| > 0,$$

show that $a_{nn} > 0$.

9. Prove that an antisymmetric multilinear form in $n + 1$ vectors of an n-dimensional space $\mathbf{K}_n$ vanishes identically.

10. Let $\mathbf{A}(x_1, \ldots, x_{n-1})$ be an antisymmetric multilinear form in $n - 1$ vectors of an n-dimensional space. Prove that $\mathbf{A}(x_1, \ldots, x_{n-1})$ can be written in any basis as a determinant whose first $n - 1$ rows consist of the components of the vector arguments and whose last (nth) row is fixed.

11. Prove that every antisymmetric bilinear form $\mathbf{A}(x, y) \not\equiv 0$ can be reduced to the canonical form

$$\mathbf{A}(x, y) = \sigma_1\tau_2 - \sigma_2\tau_1 + \sigma_3\tau_4 - \sigma_4\tau_3 + \cdots + \sigma_{2k-1}\tau_{2k} - \sigma_{2k}\tau_{2k-1}.$$

12. Prove that a real quadratic form

$$A(x, x) = \sum_{i,k=1}^{n} a_{ik}\xi_i\xi_k$$

is nonnegative for all $x \in \mathbf{R}_n$ if and only if *all* principal minors of the matrix $A = \|a_{ik}\|$ are nonnegative.

Comment. The descending principal minors δ_1 and δ_2 vanish for the matrix

$$\begin{Vmatrix} 0 & 0 \\ 0 & -1 \end{Vmatrix},$$

but the corresponding form fails to be nonnegative. Thus the conditions $\delta_1 \geqslant 0$, $\delta_2 \geqslant 0$ are not sufficient for nonnegativity of the form.

13. Let $\mathbf{A}(x, y)$ be a nonsingular symmetric bilinear form in an n-dimensional space $\mathbf{K}_n$, and let $\mathbf{K}' \subset \mathbf{K}_n$ be a subspace of dimension r. Prove that the space $\mathbf{K}'' \subset \mathbf{K}$ conjugate to $\mathbf{K}'$ with respect to $\mathbf{A}(x, y)$ is of dimension $n - r$.

14. Consider the symmetric bilinear form

$$(x, y) = \xi_1\eta_1 - \xi_2\eta_2$$

in the space $\mathbf{R}_2$. Find the operator which is the adjoint with respect to this form of the rotation operator with matrix

$$A = \begin{Vmatrix} \cos\alpha & \sin\alpha \\ -\sin\alpha & \cos\alpha \end{Vmatrix}.$$

15. Let (x, y) be a nonsingular quadratic form in the space $\mathbf{K}_n$. For the system

$$\sum_{k=1}^{n} a_{jk}\xi_k = b_j \qquad (j = 1, 2, \ldots, n) \tag{43}$$

of n linear equations in n unknowns, prove *Fredholm's theorem* which asserts that the system (43) has a solution for precisely those vectors $b = (b_1, \ldots, b_n)$ which are conjugate to all the solutions of the homogeneous system

$$\sum_{k=1}^{n} a'_{jk}\eta_k = 0, \tag{44}$$

where $\|a'_{jk}\|$ is the matrix conjugate to $\|a_{jk}\|$ with respect to the form (x, y). From this deduce that the number of independent linear conditions on the vector b which are necessary and sufficient for the system (43) to have a solution equals the dimension of the space of solutions of the homogeneous system

$$\sum_{k=1}^{n} a_{jk}\xi_k = 0 \qquad (j = 1, 2, \ldots, n).$$

Comment. For a general system

$$\sum_{k=1}^{n} a_{jk}\xi_k = b_j \qquad (j = 1, 2, \ldots, m \neq n), \tag{43'}$$

the two quantities in question no longer coincide, and their difference, equal to $m - n$, is called the *index* of the system (43′).

16. Prove that every nonnegative bilinear form of rank r in the space $\mathbf{R}_n$ can be represented as a sum of r nonnegative bilinear forms of rank 1.

17. Prove that every bilinear form of rank 1 in the space $\mathbf{K}_n$ is of the form

$$\mathbf{A}(x, y) = f(x)g(y),$$

where $f(x)$ and $g(y)$ are linear forms.

18. Prove that if

$$\mathbf{A}(x, y) = \sum_{j,k=1}^{n} a_{jk}\xi_j\eta_k$$

and

$$\mathbf{B}(x, y) = \sum_{j,k=1}^{n} b_{jk}\xi_j\eta_k$$

are nonnegative bilinear forms in the space $\mathbf{R}_n$, then the form

$$\mathbf{C}(x, y) = \sum_{j,k=1}^{n} a_{jk}b_{jk}\xi_j\eta_k$$

is also nonnegative.

chapter 8

EUCLIDEAN SPACES

8.1. Introduction

The explanation of a large variety of geometric facts rests to a great extent on the possibility of making measurements, basically measurements of the lengths of straight line segments and the angles between them. So far, we are not in a position to make such measurements in a general linear space; of course, this has the effect of narrowing the scope of our investigations. A natural way to extend these "metric" methods to the case of general linear spaces is to begin with the definition of the scalar product of two vectors which is adopted in analytic geometry (and which is suitable as of now only for ordinary vectors, i.e., elements of the space V_3 introduced in Sec. 2.15a). This definition reads as follows: *The scalar product of two vectors is the product of the lengths of the vectors and the cosine of the angle between them.* Thus the definition already rests on the possibility of measuring the lengths of vectors and the angles between them. On the other hand, if we know the scalar product for an arbitrary pair of vectors, we can deduce the lengths of vectors and the angles between them. In fact, the square of the length of a vector equals the scalar product of the vector with itself, while the cosine of the angle between two vectors is just the ratio of their scalar product to the product of their lengths. Therefore the possibility of measuring lengths and angles (and with it, the whole field of geometry associated with measurements, so-called "metric geometry"), is already implicit in the concept of the scalar product. In the case of a general linear space, the simplest approach is to introduce the concept of the scalar product of two vectors, and then use the scalar product (once it is available) to define lengths of vectors and angles between them.

We now look for properties of the ordinary scalar product which can be used to construct a similar quantity in a general linear space. For the time being, we restrict ourselves to the case of real spaces.

As already noted in Sec. 7.95, in the space V_3 the scalar product (x, y) is a symmetric positive definite bilinear form in the vectors x and y. Quite generally, we can define such a form in any real linear space. Thus we are led to consider a fixed but arbitrary symmetric positive definite bilinear form $\mathbf{A}(x, y)$ defined on a given real linear space, which we call the "scalar product" of the vectors x and y. We then use the scalar product to define the length of every vector and the angle between every pair of vectors by the same formulas as those used in the space V_3. Of course, only further study will show how successful this definition is; however, in the course of this and subsequent chapters, it will become apparent that with this definition we can in fact extend the methods of metric geometry to general linear spaces, thereby greatly enhancing our technique for investigating various mathematical objects encountered in algebra and analysis.

At this point, it is important to note that the initial positive definite bilinear form can be chosen in a variety of different ways in the given linear space. The length of a vector x calculated by using one such form will be different from the length of the same vector calculated by using another form; a similar remark applies to the angle between two vectors. Thus the lengths of vectors and the angles between them are not uniquely defined. However, this lack of uniqueness should not disturb us, for there is certainly nothing very surprising about the fact that different numbers will be assigned as the length of the same line segment if we measure the segment in different units. In fact, we can say that the choice of the original symmetric positive definite bilinear form is analogous to the choice of a "unit" for measuring lengths of vectors and angles between them.

A real linear space equipped with a "unit" symmetric positive definite bilinear form will henceforth be called a *Euclidean space*, while a linear space without a "unit" form will be called an *affine space*. The case of complex linear spaces will be considered in Chapter 9.

8.2. Definition of a Euclidean Space

8.21. A real linear space $\mathbf{R}$ is said to be *Euclidean* if there is a rule assigning to every pair of vectors $x, y \in \mathbf{R}$ a real number called the *scalar product* of the vectors x and y, denoted by (x, y), such that

a) $(x, y) = (y, x)$ for every $x, y \in \mathbf{R}$ (the *commutative law*);

b) $(x, y + z) = (x, y) + (x, z)$ for every $x, y, z \in \mathbf{R}$ (the *distributive law*);

c) $(\lambda x, y) = \lambda(x, y)$ for every $x, y \in \mathbf{R}$ and every real number λ;
d) $(x, x) > 0$ for every $x \neq 0$ and $(x, x) = 0$ for $x = 0$.

Taken together, these axioms imply that the scalar product of the vectors x and y is a *bilinear form* (axioms b) and c)), which is *symmetric* (axiom a)) and *positive definite* (axiom d)). *Conversely, any bilinear form which is symmetric and positive definite can be chosen as the scalar product.*

Since the scalar product of the vectors x and y is a bilinear form, equation (2) of Sec. 7.1 holds, and in the present case becomes

$$\left(\sum_{i=1}^{k} \alpha_i x_i, \sum_{j=1}^{m} \beta_j y_j\right) = \sum_{i=1}^{k} \sum_{j=1}^{m} \alpha_i \beta_j (x_i, y_j), \tag{1}$$

where $x_1, \ldots, x_k, y_1, \ldots, y_m$ are arbitrary vectors of the Euclidean space $\mathbf{R}$, and $\alpha_1, \ldots, \alpha_k, \beta_1, \ldots, \beta_m$ are arbitrary real numbers.

8.22. ***Examples***

a. In the space V_3 of free vectors (Sec. 2.15a), the scalar product is defined as in the beginning of Sec. 8.1, and axioms a)–d) express the familiar properties of the scalar product, proved in vector algebra.

b. In the n-dimensional space R_n (Sec. 2.15b) we define the scalar product of the vectors $x = (\xi_1, \xi_2, \ldots, \xi_n)$ and $y = (\eta_1, \eta_2, \ldots, \eta_n)$ by the formula

$$(x, y) = \xi_1\eta_1 + \xi_2\eta_2 + \cdots + \xi_n\eta_n. \tag{2}$$

This definition generalizes the familiar expression for the scalar product of three-dimensional vectors in terms of the components of the vectors with respect to an orthogonal coordinate system. The reader can easily verify that axioms a)–d) are satisfied in this case.

We note that formula (2) is not the only way of introducing a scalar product in R_n. A description of all possible ways of introducing a scalar product (i.e., a symmetric positive definite bilinear form) in the space R_n has essentially already been given in Sec. 7.96.

c. In the space $R(a, b)$ of continuous real functions on the interval $a \leqslant t \leqslant b$ (Sec. 2.15c), we define the scalar product of the functions $x = x(t)$ and $y = y(t)$ by the formula

$$(x, y) = \int_a^b x(t)y(t)\,dt. \tag{3}$$

Axioms a)–d) are then immediate consequences of the basic properties of the integral. Henceforth the space $R(a, b)$, with the scalar product defined by (3), will be denoted by $R_2(a, b)$.

8.3. Basic Metric Concepts

Equipped with the scalar product, we now proceed to define the basic metric concepts, i.e., the length of a vector and the angle between two vectors.

8.31. The length of a vector. By the *length* (or *norm*) of a vector x in a Euclidean space $\mathbf{R}$ we mean the quantity

$$|x| = +\sqrt{(x, x)}. \tag{4}$$

Examples

a. In the space V_3 our definition reduces to the usual definition of the length of a vector.

b. In the space R_n the length of the vector $x = (\xi_1, \xi_2, \ldots, \xi_n)$ is given by

$$|x| = +\sqrt{\xi_1^2 + \xi_2^2 + \cdots + \xi_n^2}.$$

c. In the space $R_2(a, b)$, the length of the vector $x(t)$ turns out to be

$$|x| = +\sqrt{(x, x)} = +\sqrt{\int_a^b x^2(t)\, dt}.$$

This quantity is sometimes written $\|x(t)\|$ and is best called the *norm* of the function $x(t)$ (in order to avoid misleading connotations connected with the phrase "length of a function").

8.32. It follows from axiom d) that every vector x of a Euclidean space $\mathbf{R}$ has a length; this length is positive if $x \neq 0$ and zero if $x = 0$ (i.e., if x is the zero vector). The formula

$$|\lambda x| = \sqrt{(\lambda x, \lambda x)} = \sqrt{\lambda^2(x, x)} = |\lambda| \sqrt{(x, x)} = |\lambda|\, |x| \tag{5}$$

shows that *the length of a vector multiplied by a numerical factor λ equals the absolute value of λ times the length of x.*

A vector x of length 1 is said to be a *unit vector*. Every nonzero vector x can be *normalized*, i.e., multiplied by a number λ such that the result is a unit vector. In fact, solving the equation $|\lambda x| = 1$ for λ, we see that λ need only be such that

$$|\lambda| = \frac{1}{|x|}.$$

A set $\mathbf{F} \subset \mathbf{R}$ is said to be *bounded* if the lengths of all the vectors $x \in \mathbf{F}$ are bounded by a fixed constant. The set of all vectors $x \in \mathbf{R}$ such that $|x| \leqslant 1$ is a bounded set called the *unit ball*, while the set of all $x \in \mathbf{R}$ such that $|x| = 1$ is a bounded set called the *unit sphere*.

8.33. The angle between two vectors. By *the angle between two vectors x and y* we mean the angle (lying between 0 and 180 degrees) whose cosine is the ratio

$$\frac{(x, y)}{|x|\, |y|}.$$

For ordinary vectors (in the space V_3) our definition agrees with the usual way of writing the angle between two vectors in terms of the scalar product. To apply this definition in a general Euclidean space, we must first prove that the ratio has an absolute value no greater than unity for any vectors x and y. To prove this, consider the vector $\lambda x - y$, where λ is a real number. By axiom d), we have

$$(\lambda x - y, \lambda x - y) \geqslant 0 \tag{6}$$

for any λ. Using (1), we can write this inequality in the form

$$\lambda^2(x, x) - 2\lambda(x, y) + (y, y) \geqslant 0. \tag{7}$$

The left-hand side of the inequality is a quadratic trinomial in λ with positive coefficients, which cannot have distinct real roots, since then it would not have the same sign for all λ. Therefore the discriminant $(x, y)^2 - (x, x)(y, y)$ of the trinomial cannot be positive, i.e.,

$$(x, y)^2 \leqslant (x, x)(y, y).$$

Taking the square root, we obtain

$$|(x, y)| \leqslant |x|\,|y|, \tag{8}$$

as required. The inequality (8) is called the *Schwarz inequality*.†

8.34. We now examine when the inequality (8) reduces to an inequality. Suppose the vectors x and y are collinear, so that $y = \lambda x$, $\lambda \in R$, say. Then obviously

$$|(x, y)| = |(x, \lambda x)| = |\lambda|\,(x, x) = |\lambda|\,|x|^2 = |x|\,|y|,$$

and (8) reduces to an equality.

Conversely, if the inequality (8) reduces to an equality for some pair of vectors x and y, then x and y are collinear. In fact, if

$$|(x, y)| = |x|\,|y|,$$

then the discriminant of (7) vanishes and hence (7) has a unique real root λ_0 (of multiplicity two). Therefore

$$\lambda_0^2(x, x) - 2\lambda_0(x, y) + (y, y) = (\lambda_0 x - y, \lambda_0 x - y) = 0,$$

whence it follows by axiom d) that $\lambda_0 x - y = 0$ or $y = \lambda_0 x$, i.e., the vectors x and y are collinear. Thus *the absolute value of the scalar product of two vectors equals the product of their lengths if and only if the vectors are collinear.*

Examples

a. In the space V_3 the Schwarz inequality is an obvious consequence of the definition of the scalar product as the product of the lengths of two vectors and the cosine of the angle between them.

---

† Sometimes also associated with the names of Cauchy and Buniakovsky.

b. In the space R_n the Schwarz inequality takes the form

$$\left|\sum_{j=1}^{n}\xi_j\eta_j\right| \leqslant \sqrt{\sum_{j=1}^{n}\xi_j^2}\sqrt{\sum_{j=1}^{n}\eta_j^2}, \tag{9}$$

and is valid for any pair of vectors $x = (\xi_1, \xi_2, \ldots, \xi_n)$ and $y = (\eta_1, \eta_2, \ldots, \eta_n)$, or equivalently, for any two sets of real numbers $\xi_1, \xi_2, \ldots, \xi_n$ and $\eta_1, \eta_2, \ldots, \eta_n$.

c. In the space $R_2(a, b)$, the Schwarz inequality takes the form

$$\left|\int_a^b x(t)y(t)\,dt\right| \leqslant \sqrt{\int_a^b x^2(t)\,dt}\sqrt{\int_a^b y^2(t)\,dt}. \tag{10}$$

8.35. Orthogonality. Two vectors x and y are said to be *orthogonal* if $(x, y) = 0$. Thus the notion of orthogonality of the vectors x and y is the same as the notion of x and y being conjugate (Sec. 7.41a) with respect to the bilinear form (x, y). If $x \neq 0$ and $y \neq 0$, then, by the general definition of the angle between two vectors, $(x, y) = 0$ means that x and y make an angle of 90° with each other. The zero vector is orthogonal to every vector $x \in \mathbf{R}$.

Examples

a. In the space R_n the orthogonality condition for the vectors $x = (\xi_1, \xi_2, \ldots, \xi_n)$ and $y = (\eta_1, \eta_2, \ldots, \eta_n)$ takes the form

$$\xi_1\eta_1 + \xi_2\eta_2 + \cdots + \xi_n\eta_n = 0.$$

For example, the vectors

$$\begin{aligned} e_1 &= (1, 0, \ldots, 0), \\ e_2 &= (0, 1, \ldots, 0), \\ &\cdots\cdots\cdots\cdots \\ e_n &= (0, 0, \ldots, 1) \end{aligned}$$

are orthogonal (in pairs).

b. In the space $R_2(a, b)$ the orthogonality condition for the vectors $x = x(t)$ and $y = y(t)$ takes the form

$$\int_a^b x(t)y(t)\,dt = 0.$$

The reader can easily verify, by calculating the appropriate integrals, that in the space $R_2(-\pi, \pi)$ any two vectors of the "trigonometric system"

$$1, \cos t, \sin t, \cos 2t, \sin 2t, \ldots, \cos nt, \sin nt, \ldots$$

are orthogonal.

8.36. We now derive some simple propositions associated with the concept of orthogonality.

a. LEMMA. *If the nonzero vectors* $x_1, x_2, \ldots, x_k$ *are orthogonal, then they are linearly independent.*

Proof. Suppose the vectors are linearly dependent. Then a relation of the form

$$\alpha_1 x_1 + \alpha_2 x_2 + \cdots + \alpha_k x_k = 0$$

holds, where $\alpha_1 \neq 0$, say. Taking the scalar product of this equation with x_1, we obtain $\alpha_1(x_1, x_1) = 0$, since by hypothesis the vectors $x_1, x_2, \ldots, x_k$ are orthogonal. It follows that $(x_1, x_1) = 0$ and hence that x_1 is the zero vector, contrary to hypothesis. ∎

The result of this lemma is often used in the following form: *If a sum of orthogonal vectors is zero, then each term in the sum is zero.*

b. LEMMA. *If the vectors* $y_1, y_2, \ldots, y_k$ *are orthogonal to the vector* x, *then any linear combination* $\alpha_1 y_1 + \alpha_2 y_2 + \cdots + \alpha_k y_k$ *is also orthogonal to* x.

Proof. We need merely note that

$$(\alpha_1 y_1 + \alpha_2 y_2 + \cdots + \alpha_k y_k, x)$$
$$= \alpha_1(y_1, x) + \alpha_2(y_2, x) + \cdots + \alpha_k(y_k, x) = 0. \quad ∎$$

The set of all linear combinations $\alpha_1 y_1 + \alpha_2 y_2 + \cdots + \alpha_k y_k$ forms a subspace $\mathbf{L} = \mathbf{L}(y_1, y_2, \ldots, y_k)$, namely the linear manifold spanned by the vectors $y_1, y_2, \ldots, y_k$ (Sec. 2.51). Therefore if x is orthogonal to the vectors $y_1, y_2, \ldots, y_k$, it is orthogonal to every vector of the subspace $\mathbf{L}$. In this case, we say that the vector x is *orthogonal to the subspace* $\mathbf{L}$. In general, if $\mathbf{F} \subset \mathbf{R}$ is any set of vectors in a Euclidean space $\mathbf{R}$, we say that the vector x is *orthogonal to the set* $\mathbf{F}$ if x is orthogonal to every vector in $\mathbf{F}$. According to Lemma 8.36b, the set $\mathbf{G}$ of all vectors x orthogonal to a set $\mathbf{F}$ is itself a subspace of the space $\mathbf{R}$. The most common situation is the case where $\mathbf{F}$ is a subspace. Then the subspace $\mathbf{G}$ is called the *orthogonal complement* of the subspace $\mathbf{F}$.

8.37. The Pythagorean theorem and its generalization. Let the vectors x and y be orthogonal. Then, by analogy with elementary geometry, we can call the vector $x + y$ the *hypotenuse* of the right triangle determined by the vectors x and y. Taking the scalar product of $x + y$ with itself, and using the orthogonality of the vectors x and y, we obtain

$$|x + y|^2 = (x + y, x + y) = (x, x) + 2(x, y) + (y, y)$$
$$= (x, x) + (y, y) = |x|^2 + |y|^2.$$

This proves the *Pythagorean theorem* in a general Euclidean space, i.e., *the square of the hypotenuse equals the sum of the squares of the sides.* It is easy to generalize this theorem to the case of any number of summands. In fact, let the vectors $x_1, x_2, \ldots, x_k$ be orthogonal and let

$$z = x_1 + x_2 + \cdots + x_k.$$

Then we have

$$|z|^2 = (x_1 + x_2 + \cdots + x_k, x_1 + x_2 + \cdots + x_k)$$
$$= |x_1|^2 + |x_2|^2 + \cdots + |x_k|^2. \quad (11)$$

8.38. The triangle inequalities. If x and y are arbitrary vectors, then by analogy with elementary geometry, it is natural to call $x + y$ the *third side of the triangle determined by the vectors x and y.* Using the Schwarz inequality, we get

$$|x + y|^2 = (x + y, x + y) = (x, x) + 2(x, y) + (y, y)$$
$$\begin{cases} \leqslant |x|^2 + 2\,|x|\,|y| + |y|^2 = (|x| + |y|)^2, \\ \geqslant |x|^2 - 2\,|x|\,|y| + |y|^2 = (|x| - |y|)^2, \end{cases}$$

or

$$|x + y| \leqslant |x| + |y|, \quad (12)$$

$$|x + y| \geqslant ||x| - |y||. \quad (13)$$

The inequalities (12) and (13) are called the *triangle inequalities.* Geometrically, they mean that *the length of any side of a triangle is no greater than the sum of the lengths of the two other sides and no less than the absolute value of the difference of the lengths of the two other sides.*

8.39. We could now successively carry over all the theorems of elementary geometry to any Euclidean space. But there is no need to do so. Instead we introduce the concept of a *Euclidean isomorphism* between two Euclidean spaces, i.e., two Euclidean spaces $\mathbf{R}'$ and $\mathbf{R}''$ are said to be *Euclidean-isomorphic* if they are isomorphic regarded as real linear spaces (see Sec. 2.71) and if in addition

$$(x', y') = (x'', y'')$$

whenever the vectors $x'', y'' \in \mathbf{R}''$ correspond to the vectors $x', y' \in \mathbf{R}'$. Then it is obvious that every geometric theorem (by which we mean any theorem based on the concepts of a linear space and a scalar product) proved for a space $\mathbf{R}'$ is also valid for any space $\mathbf{R}''$ which is Euclidean-isomorphic to $\mathbf{R}'$. According to Sec. 7.97, any two Euclidean spaces with the same dimension n are Euclidean-isomorphic. Hence any geometric theorem valid in an n-dimensional Euclidean space $\mathbf{R}_n'$ is also valid in any other n-dimensional Euclidean space $\mathbf{R}_n''$. In particular, the theorems of elementary geometry, i.e., the geometric theorems in the space $\mathbf{R}_3$, remain valid in any three-dimensional subspace of any Euclidean space. In this sense, the theorems of elementary geometry are all valid in any Euclidean space.

8.4. Orthogonal Bases

8.41. THEOREM. *In any n-dimensional Euclidean space $\mathbf{R}_n$ there exists a basis consisting of n nonzero orthogonal vectors.*

Proof. There exists a canonical basis $e_1, e_2, \ldots, e_n$ for the bilinear form (x, y), just as for any other symmetric bilinear form in an n-dimensional space (see Sec. 7.43). The condition

$$(e_i, e_k) = 0 \qquad (i \neq k)$$

satisfied by the vectors of the canonical basis is in this case just the condition for orthogonality of the vectors e_i and e_k. Thus the canonical basis $e_1, e_2, \ldots, e_n$ consists of n (pairwise) orthogonal vectors. ▮

In Sec. 8.6 we will consider a practical method for constructing such an *orthogonal basis.*

8.42. It is often convenient to normalize the vectors of an orthogonal basis by dividing each of them by its length. The resulting orthogonal basis in $\mathbf{R}_n$ is said to be *orthonormal.*

Let $e_1, e_2, \ldots, e_n$ be an arbitrary orthonormal basis in an n-dimensional Euclidean space $\mathbf{R}_n$. Then every vector $x \in \mathbf{R}_n$ can be represented in the form

$$x = \xi_1 e_1 + \xi_2 e_2 + \cdots + \xi_n e_n, \tag{14}$$

where $\xi_1, \xi_2, \ldots, \xi_n$ are the components of the vector x with respect to the basis $e_1, e_2, \ldots, e_n$. We will also call these components *Fourier coefficients* of the vector x with respect to the orthonormal system $e_1, e_2, \ldots, e_n$. Taking the scalar product of (14) with e_i, we find that

$$\xi_i = (x, e_i) \qquad (i = 1, 2, \ldots, n). \tag{15}$$

Let $y = \eta_1 e_1 + \eta_2 e_2 + \cdots + \eta_n e_n$ be any other vector of the space $\mathbf{R}_n$. Then it follows from (1) that

$$(x, y) = \xi_1 \eta_1 + \xi_2 \eta_2 + \cdots + \xi_n \eta_n. \tag{16}$$

Thus *in an orthonormal basis the scalar product of two vectors equals the sum of the products of the components (Fourier coefficients) of the vectors.* In particular, setting $y = x$, we obtain

$$|x|^2 = (x, x) = \xi_1^2 + \xi_2^2 + \cdots + \xi_n^2. \tag{17}$$

8.5. Perpendiculars

8.51. Let $\mathbf{R}'$ be a finite-dimensional subspace of a Euclidean space $\mathbf{R}$, and let f be a vector which is in general not an element of $\mathbf{R}'$. We now pose the problem of representing f in the form

$$f = g + h, \tag{18}$$

where the vector g belongs to the subspace $\mathbf{R}'$ and the vector h is orthogonal to $\mathbf{R}'$. The vector g appearing in the expansion (18) is called the *projection of f onto the subspace* $\mathbf{R}'$, and the vector h is called the *perpendicular dropped from the end of f onto the subspace* $\mathbf{R}'$. This terminology calls to mind certain familiar geometric associations, but it is not intended to do more than just suggest these associations.†

The solution of this problem has in effect already been given in Sec. 7.54 for any symmetric bilinear form which is nonsingular in the subspace $\mathbf{R}'$. Since the positive definite form (x, y) is nonsingular in every subspace $\mathbf{R}' \subset \mathbf{R}$ (Sec. 7.94), the existence and uniqueness of a solution of our problem follows from Sec. 7.54. Moreover, as shown in Sec. 7.55, the existence of the expansion (18) shows that the whole space $\mathbf{R}$ is the direct sum of the subspace $\mathbf{R}'$ and its orthogonal complement $\mathbf{R}''$. A direct sum whose terms are orthogonal is called an *orthogonal direct sum.* Thus we have expanded the space $\mathbf{R}$ as an orthogonal direct sum of the subspaces $\mathbf{R}'$ and $\mathbf{R}''$. If $\mathbf{R}$ and $\mathbf{R}'$ have dimensions n and k, respectively, then the dimension of $\mathbf{R}''$ equals $n - k$, since the dimension of the direct sum is the sum of the dimensions of its terms (Sec. 2.47).

We note that the problem is also solved in the case where f lies in the subspace $\mathbf{R}'$, since then

$$f = f + 0.$$

This solution is obviously unique. In fact, if

$$f = g + h \qquad (g \in \mathbf{R}', h \in \mathbf{R}''),$$

then $h = f - g \in \mathbf{R}'$ which implies $h = 0$, $g = f$.

8.52. Applying the Pythagorean theorem (Sec. 8.37) to the expansion (18), we obtain

$$|f|^2 = |g|^2 + |h|^2, \tag{19}$$

which implies the formula

$$0 \leqslant |h| \leqslant |f|, \tag{20}$$

---

† Since the concept of the "end of a vector" plays no role in our axiomatics, it is inappropriate to look for any logical content in this terminology.

expressing the geometric fact that the *length of a perpendicular does not exceed the length of the line segment from which it is dropped.* Consider the cases where one of the inequalities in (20) becomes an equality. The first equality sign holds if $|h| = 0$; this means that $f = g + 0$, i.e., f is an element of the subspace $\mathbf{R}'$. The second equality sign holds if $|h| = |f|$; according to (19), this means that $g = 0$ or $f = 0 + h$, i.e., f is orthogonal to the subspace $\mathbf{R}'$. Thus $|h| = 0$ *means that* f *belongs to* $\mathbf{R}'$, *while* $|h| = |f|$ *means that* f *is orthogonal to* $\mathbf{R}'$. In any other configuration of f, the (inherently positive) length of h is less than that of f.

Now let $e_1, e_2, \ldots, e_k$ be an orthonormal basis in the subspace $\mathbf{R}'$, and let

$$g = \sum_{j=1}^{k} a_j e_j.$$

Then, by Sec. 8.42,

$$|g|^2 = \sum_{j=1}^{k} a_j^2.$$

Substituting this value of $|g|^2$ into (19), we get

$$|f|^2 = |h|^2 + \sum_{j=1}^{k} a_j^2.$$

In particular, for any (finite) orthonormal system $e_1, e_2, \ldots, e_k$ and any vector f, we have the inequality

$$\sum_{j=1}^{k} a_j^2 \leqslant |f|^2,$$

known as *Bessel's inequality*. The geometric meaning of this inequality is clear: *The square of the length of the vector* f *is no less than the sum of the squares of its projections onto any* k *mutually orthogonal directions.*

8.53. In the applications, we sometimes need an explicit solution of the problem of dropping a perpendicular onto a subspace $\mathbf{R}'$, given some basis $\{b\} = \{b_1, b_2, \ldots, b_k\}$ in $\mathbf{R}'$ (in general, not an orthonormal basis). To solve this problem, we first expand the required vector g (the "foot of the perpendicular") with respect to the basis $\{b\}$, i.e., we write

$$g = \beta_1 b_1 + \beta_2 b_2 + \cdots + \beta_k b_k.$$

We then impose on the vector $h = f - g$ the condition that it be orthogonal to all the vectors $b_1, b_2, \ldots, b_k$, thereby obtaining the system of equations

$$\begin{aligned}
(h, b_1) &= (f - g, b_1) = (f, b_1) - \beta_1(b_1, b_1) - \beta_2(b_2, b_1) - \cdots - \beta_k(b_k, b_1) = 0,\\
(h, b_2) &= (f - g, b_2) = (f, b_2) - \beta_1(b_1, b_2) - \beta_2(b_2, b_2) - \cdots - \beta_k(b_k, b_2) = 0,\\
&\cdots\cdots\cdots\cdots\cdots\cdots\cdots\cdots\cdots\cdots\cdots\cdots\\
(h, b_k) &= (f - g, b_k) = (f, b_k) - \beta_1(b_1, b_k) - \beta_2(b_2, b_k) - \cdots - \beta_k(b_k, b_k) = 0,
\end{aligned}$$

with determinant

$$D = \begin{vmatrix} (b_1, b_1) & (b_2, b_1) & \cdots & (b_k, b_1) \\ (b_1, b_2) & (b_2, b_2) & \cdots & (b_k, b_2) \\ \cdot & \cdot & \cdots & \cdot \\ \cdot & \cdot & \cdots & \cdot \\ \cdot & \cdot & \cdots & \cdot \\ (b_1, b_k) & (b_2, b_k) & \cdots & (b_k, b_k) \end{vmatrix}.$$

But D is nonzero, being the determinant of the matrix of the positive definite form (x, y) in the basis $b_1, b_2, \ldots, b_k$ (see Sec. 7.96). Hence we can solve the system by Cramer's rule, obtaining the following expression for the coefficients β_j $(j = 1, 2, \ldots, k)$:

$$\beta_j = \frac{1}{D} \begin{vmatrix} (b_1, b_1) & (b_2, b_1) & \cdots & (b_{j-1}, b_1) & (f, b_1) & (b_{j+1}, b_1) & \cdots & (b_k, b_1) \\ (b_1, b_2) & (b_2, b_2) & \cdots & (b_{j-1}, b_2) & (f, b_2) & (b_{j+1}, b_2) & \cdots & (b_k, b_2) \\ \cdot & \cdot & \cdots & \cdot & \cdot & \cdot & \cdots & \cdot \\ \cdot & \cdot & \cdots & \cdot & \cdot & \cdot & \cdots & \cdot \\ \cdot & \cdot & \cdots & \cdot & \cdot & \cdot & \cdots & \cdot \\ (b_1, b_k) & (b_2, b_k) & \cdots & (b_{j-1}, b_k) & (f, b_k) & (b_{j+1}, b_k) & \cdots & (b_k, b_k) \end{vmatrix}.$$

8.54. The problem of dropping a perpendicular can be posed not only for a subspace, but also for a hyperplane, in which case the problem is formulated as follows: Suppose that in a Euclidean space $\mathbf{R}$, we are given a vector f and a hyperplane $\mathbf{R}''$, generated by parallel displacement of a subspace $\mathbf{R}'$. We wish to show that there exists a unique expansion

$$f = g + h, \tag{21}$$

where the vector g belongs to the hyperplane $\mathbf{R}''$ and the vector h is orthogonal to the subspace $\mathbf{R}'$.† The geometric meaning of the expansion (21) is illustrated in Figure 1(a). Note that the terms in the expansion (21) are in general no longer orthogonal.

The problem is now easily reduced to the problem of Sec. 8.51. In fact, if we fix any vector in the hyperplane $\mathbf{R}''$ and subtract it from both sides of (21), we obtain the problem of representing the vector $f - f_0$ as a sum of two vectors $g - f_0$ and h, of which the first belongs to the subspace $\mathbf{R}'$ and the second is orthogonal to $\mathbf{R}'$ (see Figure 1(b)). By the result of Sec. 8.51, such a representation exists. Therefore the representation (21) also exists. It

† Saying that g belongs to the hyperplane $\mathbf{R}''$ means geometrically that the end point of g lie in the hyperplane $\mathbf{R}''$, while its initial point is, as usual, at the origin of coordinates. One must not imagine that the whole vector g lies in the hyperplane $\mathbf{R}''$!

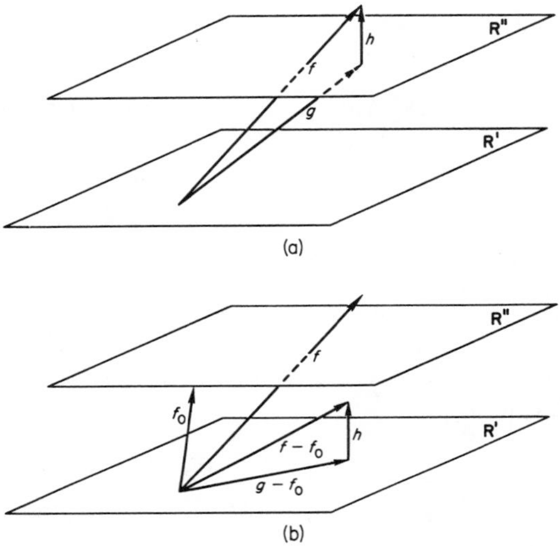

FIGURE 1

remains only to prove the uniqueness of the representation (21). If there were two such representations

$$f = g_1 + h_1 = g_2 + h_2,$$

then we would have

$$0 = (g_1 - g_2) + (h_1 - h_2),$$

where $g_1 - g_2$ belongs to the subspace $\mathbf{R}'$ and $h_1 - h_2$ is orthogonal to $\mathbf{R}'$. It follows that $g_1 - g_2 = h_1 - h_2 = 0$, as required.

8.6. The Orthogonalization Theorem

8.61. The following theorem is of fundamental importance in constructing orthogonal systems in a Euclidean space:

THEOREM ***(Orthogonalization theorem).*** *Let* $x_1, x_2, \ldots, x_k, \ldots$ *be a finite or infinite sequence of vectors in a Euclidean space* $\mathbf{R}$, *and let* $\mathbf{L}_k = \mathbf{L}(x_1, x_2, \ldots, x_k)$ *be the linear manifold spanned by the first k of these vectors. Then there exists a system of vectors* $y_1, y_2, \ldots, y_k, \ldots$ *such that*

1) *The linear manifold* $\mathbf{L}'_k = \mathbf{L}(y_1, y_2, \ldots, y_k)$ *spanned by the vectors* $y_1, y_2, \ldots, y_k$ *coincides with the linear manifold* $\mathbf{L}_k$ *for every positive integer k;*
2) *The vector* y_{k+1} *is orthogonal to* $\mathbf{L}_k$ *for every positive integer k.*

Proof. We will prove the theorem by induction, i.e., assuming that k vectors $y_1, y_2, \ldots, y_k$ have been constructed which satisfy the conditions of the theorem, we will construct a vector y_{k+1} such that the vectors $y_1, y_2, \ldots, y_k, y_{k+1}$ also satisfy the conditions of the theorem. First let $y_1 = x_1$. Then the condition $\mathbf{L}_1' = \mathbf{L}_1$ is obviously satisfied. The subspace $\mathbf{L}_k$ is finite-dimensional, and hence by Sec. 8.51 there exists an expansion

$$x_{k+1} = g_k + h_k, \tag{22}$$

where g_k is an element of $\mathbf{L}_k$ and h_k is orthogonal to $\mathbf{L}_k$. Setting $y_{k+1} = h_k$, we now verify that the conditions of the theorem are satisfied for this choice of y_{k+1}. By the induction hypothesis, the subspace $\mathbf{L}_k$ contains the vectors $y_1, y_2, \ldots, y_k$, and hence the larger subspace $\mathbf{L}_{k+1}$ also contains these vectors. Moreover, it follows from (22) that $\mathbf{L}_{k+1}$ contains the vector $h_k = y_{k+1}$. Therefore the subspace $\mathbf{L}_{k+1}$ contains all the vectors $y_1, y_2, \ldots, y_k, y_{k+1}$, and hence also contains the linear manifold $\mathbf{L}_{k+1}'$ spanned by these vectors. Conversely, the subspace $\mathbf{L}_{k+1}'$ contains the vectors $x_1, x_2, \ldots, x_k$, and moreover by (22), $\mathbf{L}_{k+1}'$ contains the vector x_{k+1} as well. It follows that $\mathbf{L}_{k+1}'$ contains the whole subspace $\mathbf{L}_{k+1}$. Therefore $\mathbf{L}_{k+1}' = \mathbf{L}_{k+1}$, and the first assertion of the theorem is proved. The second assertion is an obvious consequence of the construction of the vector $y_{k+1} = h_k$. This completes the induction, thereby proving the theorem. ▮

8.62. In the present case, the inequality (20) takes the form

$$0 \leqslant |y_{k+1}| \leqslant |x_{k+1}|. \tag{23}$$

As shown in Sec. 8.52, the equality $|y_{k+1}| = 0$ means that the vector x_{k+1} belongs to the subspace $\mathbf{L}_k$, and is therefore a linear combination of the vectors $x_1, x_2, \ldots, x_k$. The opposite equality $|y_{k+1}| = |x_{k+1}|$ means that the vector x_{k+1} is orthogonal to the subspace $\mathbf{L}_k$, and hence is orthogonal to each of the vectors $x_1, x_2, \ldots, x_k$.

8.63. *Remark.* *Every system of vectors $z_1, z_2, \ldots, z_k, \ldots$ satisfying the conditions of the orthogonalization theorem coincides to within numerical factors with the system $y_1, y_2, \ldots, y_k, \ldots$ constructed in the proof of the theorem.* In fact, the vector z_{k+1} must belong to the subspace $\mathbf{L}_{k+1}$, and at the same time z_{k+1} must be orthogonal to the subspace $\mathbf{L}_k$. The first of these conditions implies the existence of an expansion

$$z_{k+1} = c_1 y_1 + c_2 y_2 + \cdots + c_k y_k + c_{k+1} y_{k+1} = \tilde{y}_k + c_{k+1} y_{k+1},$$

where $\tilde{y}_k = c_1 y_1 + c_2 y_2 + \cdots + c_k y_k \in \mathbf{L}_k$ and $c_{k+1} y_{k+1}$ is orthogonal to $\mathbf{L}_k$. The second condition implies that $y_k = 0$ and hence that

$$z_{k+1} = c_{k+1} y_{k+1},$$

as required.

***8.64. Legendre polynomials.** Suppose we apply the orthogonalization theorem to the system of functions

$$x_0(t) = 1, x_1(t) = t, \ldots, x_k(t) = t^k, \ldots$$

in the Euclidean space $R_2(-1, 1)$. Then the subspace $L_k = L(1, t, \ldots, t^k)$ is obviously the set of all polynomials in t of degree $n \leqslant k$. The functions $x_0(t), x_1(t), \ldots, x_k(t)$ are linearly independent (see Sec. 2.22d), and hence the functions $y_0(t), y_1(t), \ldots$ obtained by the orthogonalization method are all nonzero, by Sec. 8.62. By its very construction, $y_k(t)$ must be a polynomial in t of degree k. In particular, direct calculation by the orthogonalization method gives

$$y_0(t) = 1, \quad y_1(t) = t, \quad y_2(t) = t^2 - \tfrac{1}{3}, \quad y_3(t) = t^3 - \tfrac{3}{5}t, \ldots.$$

These polynomials were introduced in 1785 by the French mathematician Legendre, in connection with certain problems of potential theory. The general formula for the Legendre polynomials was found by Rodrigues in 1814, who showed that the polynomial $y_n(t)$ is given by

$$p_n(t) = \frac{d^n}{dt^n}[(t^2 - 1)^n] \qquad (n = 0, 1, 2, \ldots) \tag{24}$$

to within a numerical factor. We now prove this formula, using the remark of Sec. 8.63, i.e., we will show that the polynomial $p_n(t)$ satisfies the conditions of the orthogonalization theorem, whence it will follows from the remark in question that $p_n(t)$ must equal $c_n y_n(t)$ for every n, as required.

a. *The linear manifold spanned by the vectors $p_0(t), p_1(t), \ldots, p_n(t)$ coincides with the set of all polynomials of degree no greater than n.* In fact, it is obvious from (24) that the polynomial $p_k(t)$ is clearly a polynomial in t of degree k. In particular,

$$\begin{aligned}
p_0(t) &= a_{00}, \\
p_1(t) &= a_{10} + a_{11}t, \\
p_2(t) &= a_{20} + a_{21}t + a_{22}t^2, \\
&\cdots\cdots\cdots\cdots \\
p_k(t) &= a_{k0} + a_{k1}t + \cdots + a_{kk}t^k, \\
&\cdots\cdots\cdots\cdots\cdots\cdots \\
p_n(t) &= a_{n0} + a_{n1}t + \cdots + a_{nk}t^k + \cdots + a_{nn}t^n,
\end{aligned} \tag{25}$$

where the leading coefficients $a_{00}, a_{11}, \ldots, a_{nn}$ are nonzero. Thus all the polynomials $p_0(t), p_1(t), \ldots, p_n(t)$ are elements of the linear manifold spanned by the functions $1, t, \ldots, t^n$, which is obviously just the set L_n of all polynomials in t of degree no greater than n. Conversely, the functions $1, t, \ldots, t^n$ can be expressed as linear combinations of $p_0(t), p_1(t), \ldots, p_n(t)$, since the matrix of the linear relations (25) has the nonvanishing determinant $a_{00}a_{11} \cdots a_{nn}$. Hence the linear manifold $L(p_0(t), p_1(t), \ldots, p_n(t))$ coincides with the linear manifold $L(1, t, \ldots, t^n)$ and therefore coincides with the set L_n, as required.

b. *The vector $p_n(t)$ is orthogonal to the subspace L_{n-1}.* It is sufficient to verify that the polynomial $p_n(t)$ is orthogonal in the sense of the space $R_2(-1, 1)$ to the functions $1, t, \ldots, t^{n-1}$. To show this, we use the formula for integration by parts, familiar from elementary calculus, which in the case of polynomials involves derivatives of the type considered in Sec. 6.73c from a purely algebraic point of view. In particular, the derivatives of the polynomial

$$(t^2 - 1)^n = (t - 1)^n(t + 1)^n$$

of orders $0, 1, \ldots, n - 1$ vanish for $t = \pm 1$.† Thus, calculating the scalar product of t^k and $p_n(t)$ for $k < n$ and integrating by parts, we obtain

$$(t^k, p_n(t)) = \int_{-1}^{+1} t^k[(t^2 - 1)^n]^{(n)}\, dt$$

$$= t^k[(t^2 - 1)^n]^{(n-1)}\Big|_{-1}^{+1} - k\int_{-1}^{+1} t^{k-1}[(t^2 - 1)^n]^{(n-1)}\, dt,$$

where the first term on the right vanishes. Integrating the second term by parts again, and continuing this process until the exponent of t becomes zero, we get

$$(t^k, p_n(t)) = -kt^{k-1}[(t^2 - 1)^n]^{(n-2)}\Big|_{-1}^{+1} + k(k-1)\int_{-1}^{+1} t^{k-2}[(t^2 - 1)^n]^{(n-2)}\, dt$$

$$= \cdots = \pm k!\int_{-1}^{+1} [(t^2 - 1)^n]^{(n-k)}\, dt$$

$$= \pm k![(t^2 - 1)^n]^{(n-k-1)}\Big|_{-1}^{+1} = 0,$$

i.e., $p_n(t)$ is orthogonal to L_{k-1}, as asserted.

Thus, finally, we have proved that for every n the polynomial $y_n(t)$ is the same as the polynomial $p_n(t) = [(t^2 - 1)^n]^{(n)}$, except for a numerical factor.

---

† Cf. formula (21), p. 163.

We now calculate $p_n(1)$, by applying the formula for n-fold differentiation of a product to the function

$$(t^2 - 1)^n = (t + 1)^n(t - 1)^n.$$

The result is

$$\begin{aligned} p_n(t) &= [(t + 1)^n(t - 1)^n]^{(n)} \\ &= (t + 1)^n[(t - 1)^n]^{(n)} + C_1^n[(t + 1)^n]'[(t - 1)^n]^{(n-1)} + \cdots \\ &= (t + 1)^n n! + C_1^n n(t + 1)^{n-1} n(n - 1) \cdots 2(t - 1) + \cdots, \end{aligned}$$

where $C_k^n = n!/k!(n - k)!$. The substitution $t = 1$ makes all the terms of this sum vanish from the second term on, and we get

$$p_n(1) = 2^n n!.$$

For numerical purposes, it is convenient to make the values of our orthogonal functions equal 1 for $t = 1$. To achieve this, we need only multiply $p_n(t)$ by the factor $1/2^n n!$. In fact, it is actually these normalized polynomials which are called the *Legendre polynomials*, i.e., the Legendre polynomial of degree n, denoted by $P_n(t)$, is given by the formula

$$P_n(t) = \frac{1}{2^n n!} [(t^2 - 1)^n]^{(n)}.$$

8.7. The Gram Determinant

8.71. By a *Gram determinant* is meant a determinant of the form

$$G(x_1, x_2, \ldots, x_k) = \begin{vmatrix} (x_1, x_1) & (x_1, x_2) & \cdots & (x_1, x_k) \\ (x_2, x_1) & (x_2, x_2) & \cdots & (x_2, x_k) \\ \cdot & \cdot & \cdots & \cdot \\ (x_k, x_1) & (x_k, x_2) & \cdots & (x_k, x_k) \end{vmatrix},$$

where $x_1, x_2, \ldots, x_k$ are arbitrary vectors of a Euclidean space **R**. In Sec. 7.96 we saw that this determinant is positive in the case of linearly independent vectors $x_1, x_2, \ldots, x_k$. To calculate the value of $G(x_1, x_2, \ldots, x_k)$, we apply the orthogonalization process to the vectors $x_1, x_2, \ldots, x_k$. Thus let $y_1 = x_1$ and suppose the vector

$$y_2 = \alpha_1 y_1 + x_2$$

is orthogonal to y_1. Replacing the vector x_1 by y_1 everywhere in the determinant $G(x_1, x_2, \ldots, x_k)$, we multiply the first column of $G(x_1, x_2, \ldots, x_k)$ by α_1 (associating α_1 with the second factors of the scalar products) and add it to the second column. Then we multiply the first row of the determinant by α_1 (associating α_1 with the first factors of the scalar products) and add it to the second row. As a result, the vector y_2 appears at every place in the determinant where x_2 appeared formerly.

Next let

$$y_3 = \beta_1 y_1 + \beta_2 y_2 + x_3$$

be orthogonal to y_1 and y_2. Multiply the first column by β_1 and the second column by β_2, and add them to the third column. Then carry out the same operations on the rows. As a result, x_3 is replaced by y_3 everywhere in $G(x_1, x_2, \ldots, x_k)$. We can continue this process until we arrive at the last column (and row). Since these operations do not change the value of the determinant, we finally obtain

$$G(x_1, x_2, \ldots, x_k) = \begin{vmatrix} (y_1, y_1) & 0 & \cdots & 0 \\ 0 & (y_2, y_2) & \cdots & 0 \\ \cdot & \cdot & \cdots & \cdot \\ \cdot & \cdot & \cdots & \cdot \\ \cdot & \cdot & \cdots & \cdot \\ 0 & 0 & \cdots & (y_k, y_k) \end{vmatrix} \tag{26}$$

$$= (y_1, y_1)(y_2, y_2) \cdots (y_k, y_k).$$

Moreover, by the result of Sec. 8.62, we have the inequality

$$0 \leqslant G(x_1, x_2, \ldots, x_k) \leqslant (x_1, x_1)(x_2, x_2) \cdots (x_k, x_k). \tag{27}$$

Next we examine the conditions under which the quantity $G(x_1, x_2, \ldots, x_k)$ can take the values 0 or $(x_1, x_1)(x_2, x_2) \cdots (x_k, x_k)$. It follows from the form (26) of the Gram determinant that it vanishes if and only if one of the vectors $y_1, y_2, \ldots, y_k$ vanishes. But according to Sec. 8.62, this implies that the vectors $x_1, x_2, \ldots, x_k$ are linearly dependent. Moreover, according to (26) and Sec. 8.62, the second equality sign holds in the inequality (27) only in the case where the vectors $x_1, x_2, \ldots, x_k$ are already orthogonal. Thus we have proved the following

THEOREM. *The Gram determinant of the vectors $x_1, x_2, \ldots, x_k$ vanishes if the vectors are linearly dependent and is positive if they are linearly independent. It equals the product of the squares of the lengths of the vectors $x_1, x_2, \ldots, x_k$ if they are orthogonal and is less than this quantity otherwise.*

8.72. The volume of a k-dimensional hyperparallelepiped. As is well known from elementary geometry, the area of a parallelogram equals the product of a base and the corresponding altitude. If the parallelogram is determined by two vectors x_1 and x_2, then for the base we can take the length of the vector x_1 and for the altitude we can take the length of the perpendicular dropped from the end of the vector x_2 onto the line containing the vector x_1. Similarly, the volume of the parallelepiped determined by the vectors x_1, x_2 and x_3 equals the product of the area of a base and the corresponding altitude; for the area of the base we choose the area of the parallelogram determined by the vectors x_1 and x_2, and for the altitude we take the length of the perpendicular dropped from the end of the vector x_3 onto the plane of the vectors x_1 and x_2.

These considerations make the following a very natural inductive definition of the volume of a k-dimensional hyperparallelepiped in a Euclidean space: Given a system of vectors $x_1, x_2, \ldots, x_k$ in a Euclidean space $\mathbf{R}$, let h_j denote the perpendicular dropped from the end of the vector x_{j+1} onto the subspace

$$\mathbf{L}(x_1, x_2, \ldots, x_j) \qquad (j = 1, 2, \ldots, k-1),$$

and introduce the following notation:

$V_1 = |x_1|$ (a one-dimensional volume, i.e., the length of the vector x_1),

$V_2 = V_1\,|h_1|$ (a two-dimensional volume, i.e., the area of the parallelogram determined by the vectors x_1, x_2),

$V_3 = V_2\,|h_2|$ (a three-dimensional volume, i.e., the volume of the parallelepiped determined by the vectors x_1, x_2, x_3),

. . . .

$V_k = V_{k-1}\,|h_{k-1}|$ (a k-dimensional volume, i.e., the volume of the hyperparalleliped determined by the vectors $x_1, x_2, \ldots, x_k$).

Obviously the volume V_k can be written in the form

$$V_k \equiv V[x_1, x_2, \ldots, x_k] = |x_1|\,|h_1| \cdots |h_{k-1}|.$$

Using equation (26), we can express the quantity V_k in terms of the vectors $x_1, x_2, \ldots, x_k$ as follows:

$$V_k^2 = \begin{vmatrix} (x_1, x_1) & (x_1, x_2) & \cdots & (x_1, x_k) \\ (x_2, x_1) & (x_2, x_2) & \cdots & (x_2, x_k) \\ \cdots & \cdots & \cdots & \cdots \\ (x_k, x_1) & (x_k, x_2) & \cdots & (x_k, x_k) \end{vmatrix}.$$

Thus *the Gram determinant of the k vectors* $x_1, x_2, \ldots, x_k$ *equals the square of the volume of the k-dimensional hyperparallelepiped determined by these vectors.*

8.73. Let

$$\xi_i^{(j)} \qquad (j = 1, 2, \ldots, k;\ i = 1, 2, \ldots, n)$$

be the components of the vector x_j with respect to an orthonormal basis $e_1, e_2, \ldots, e_n$. Expressing the scalar products in terms of the components of the vectors involved, we obtain the following formula for V_k^2:

$$V_k^2 = \begin{vmatrix} \xi_1^{(1)}\xi_1^{(1)} + \cdots + \xi_n^{(1)}\xi_n^{(1)} & \cdots & \xi_1^{(1)}\xi_1^{(k)} + \cdots + \xi_n^{(1)}\xi_n^{(k)} \\ \xi_1^{(2)}\xi_1^{(1)} + \cdots + \xi_n^{(2)}\xi_n^{(1)} & \cdots & \xi_1^{(2)}\xi_1^{(k)} + \cdots + \xi_n^{(2)}\xi_n^{(k)} \\ \cdot \quad \cdots \quad \cdot & \cdots & \cdot \quad \cdots \quad \cdot \\ \xi_1^{(k)}\xi_1^{(1)} + \cdots + \xi_n^{(k)}\xi_n^{(1)} & \cdots & \xi_1^{(k)}\xi_1^{(k)} + \cdots + \xi_n^{(k)}\xi_n^{(k)} \end{vmatrix}.$$

We now use an argument similar to that used in Sec. 4.54. Every column of the determinant just written is the sum of n "elementary columns" with elements of the form $\xi_i^{(j)}\xi_i^{(\alpha)}$, where the indices α and i are fixed in each elementary column, while j ranges from 1 to k. Therefore the whole determinant equals the sum of n^k "elementary determinants" consisting only of elementary columns. In each elementary column the factor $\xi_i^{(\alpha)}$ is constant and hence can be factored out of the elementary determinant. As a result, each elementary determinant takes the form

$$\xi_{i}^{(1)}\xi_{i_2}^{(2)} \cdots \xi_{i_k}^{(k)} \begin{vmatrix} \xi_{i_1}^{(1)} & \xi_{i_2}^{(1)} & \cdots & \xi_{i_k}^{(1)} \\ \xi_{i_1}^{(2)} & \xi_{i_2}^{(2)} & \cdots & \xi_{i_k}^{(2)} \\ \cdot & \cdot & \cdots & \cdot \\ \xi_{i_1}^{(k)} & \xi_{i_2}^{(k)} & \cdots & \xi_{i_k}^{(k)} \end{vmatrix}, \qquad (28)$$

where $i_1, i_2, \ldots, i_k$ are numbers from 1 to n. If some of these numbers are the same, then the corresponding elementary determinant obviously vanishes. Thus we need only consider the case where $i_1, i_2, \ldots, i_k$ are all different. In the entire sum we group together those terms of the form (28) which have the same indices $i_1, i_2, \ldots, i_k$ but arranged in different orders. Let

$$M^2[j_1, j_2, \ldots, j_k]$$

denote the sum of all such terms, where $j_1, j_2, \ldots, j_k$ are the indices $i_1, i_2, \ldots, i_k$ rearranged in increasing order. An argument similar to that used in Sec. 4.54 then leads to the following result: In the $n \times k$ matrix

$$\|\xi_i^{(j)}\| \qquad (i = 1, 2, \ldots, n; j = 1, 2, \ldots, k),$$

the quantity $M^2[j_1, j_2, \ldots, j_k]$ is the square of the minor of order k formed from the columns of this matrix with indices $j_1, j_2, \ldots, j_k$. The sum of all the terms (28) equals the sum of the squares of all the minors of order k of the matrix $\|\xi_i^{(j)}\|$. Thus the square of the volume of the k-dimensional hyperparallelepiped determined by the vectors $x_1, x_2, \ldots, x_k$ equals the sum of the squares of all the minors of order k in the matrix consisting of the components of the vectors $x_1, x_2, \ldots, x_k$ with respect to any orthonormal basis $e_1, e_2, \ldots, e_n$.

8.74. In the case $k = n$, the matrix $\|\xi_i^{(j)}\|$ has only one minor of order k, equal to the determinant of the matrix. Hence *the volume of the n-dimensional hyperparallelepiped determined by the vectors $x_1, x_2, \ldots, x_n$ equals the absolute value of the determinant formed from the components of the vectors $x_1, x_2, \ldots, x_n$ with respect to any orthonormal basis.*

8.75. Hadamard's inequality. Using the results of the preceding section, we can obtain an important estimate for the absolute value of an arbitrary determinant

$$D = \begin{vmatrix} \xi_{11} & \xi_{12} & \cdots & \xi_{1k} \\ \xi_{21} & \xi_{22} & \cdots & \xi_{2k} \\ \cdot & \cdot & \cdots & \cdot \\ \xi_{k1} & \xi_{k2} & \cdots & \xi_{kk} \end{vmatrix}$$

of order k. If we regard the numbers $\xi_{i1}, \xi_{i2}, \ldots, \xi_{ik}$ $(i = 1, 2, \ldots, k)$ as the components of a vector x_i with respect to an orthonormal basis in a k-dimensional Euclidean space, then the result of Sec. 8.74 allows us to interpret the absolute value of the determinant D as the volume of the k-dimensional hyperparallelepiped determined by the vectors $x_1, x_2, \ldots, x_k$. Then, using the expression for this volume in terms of the Gram determinant, we have

$$D^2 = G(x_1, x_2, \ldots, x_k).$$

Applying Theorem 8.71, we obtain

$$D^2 \leqslant (x_1, x_1)(x_2, x_2) \cdots (x_k, x_k) = \prod_{i=1}^{k} \sum_{j=1}^{k} \xi_{ij}^2,$$

an inequality known as *Hadamard's inequality*. Moreover, we note that according to Theorem 8.71, the equality holds if and only if the vectors $x_1, x_2, \ldots, x_k$ are pairwise orthogonal.

The geometric meaning of Hadamard's inequality is clear, i.e., *the volume of a hyperparallelelepiped does not exceed the product of the lengths of its sides, and it equals this product if and only if its sides are orthogonal.*

8.8. Incompatible Systems and the Method of Least Squares

8.81. Suppose we are given an incompatible system of linear equations

$$\begin{aligned} a_{11}x_1 + a_{12}x_2 + \cdots + a_{1m}x_m &= b_1, \\ a_{21}x_1 + a_{22}x_2 + \cdots + a_{2m}x_m &= b_2, \\ \cdots\cdots\cdots\cdots\cdots\cdots\cdots\cdots \\ a_{n1}x_1 + a_{n2}x_2 + \cdots + a_{nm}x_m &= b_n. \end{aligned} \tag{29}$$

Since the system is incompatible, it cannot be solved, i.e., we cannot find numbers $c_1, c_2, \ldots, c_m$ which satisfy all the equations of the system when substituted for the unknowns $x_1, x_2, \ldots, x_m$. Thus if we substitute the numbers $\xi_1, \xi_2, \ldots, \xi_m$ for the unknowns $x_1, x_2, \ldots, x_m$ in the left-hand side of the system (29), we obtain numbers $\gamma_1, \gamma_2, \ldots, \gamma_n$ which differ from the numbers $b_1, b_2, \ldots, b_n$. This suggests the following problem: *Given real numbers a_{jk} and b_k ($j = 1, \ldots, m$; $k = 1, \ldots, n$) find the numbers $\xi_1, \xi_2, \ldots, \xi_m$ which when substituted into* (29) *give the numbers $\gamma_1, \gamma_2, \ldots, \gamma_n$ with the smallest possible* ***mean square deviation***

$$\delta^2 = \sum_{j=1}^{n}(\gamma_j - b_j)^2 \tag{30}$$

from the numbers $b_1, b_2, \ldots, b_n$, and find the corresponding minimum value of δ^2.

An example of a situation where this problem arises in practice is the following: Suppose we want to determine the coefficients ξ_j in the linear relation

$$b = \xi_1 a_1 + \xi_2 a_2 + \cdots + \xi_m a_m$$

connecting the quantity b and the quantities $a_1, a_2, \ldots, a_m$, given the results of measurements of the a_j ($j = 1, 2, \ldots, m$) and the corresponding values of b. If the ith measurement gives the value a_{ij} for the quantity a_j and the value b_i for the quantity b, then clearly

$$\xi_1 a_{i1} + \xi_2 a_{i2} + \cdots + \xi_m a_{im} = b_i. \tag{31}$$

Thus n measurements lead to a system of n equations of the form (31), i.e., a system of the form (29). As a result of unavoidable measurement errors, this system will generally be incompatible, and then the problem of finding the coefficients $\xi_1, \xi_2, \ldots, \xi_m$ does not reduce to the problem of solving the system (29). This suggests determining the coefficients ξ_j in such a way that every equation is at least approximately valid and the total error is as small as possible. If we take as a measure of the error the mean square deviation of the quantities

$$\gamma_j = \sum_{i=1}^{m} a_{ij}\xi_i$$

from the known quantities b_j, i.e., if we take formula (30) as a measure of the error, then we arrive at the problem formulated at the beginning of this section. Moreover, in this case, it is also useful to know the quantity δ^2, since this helps to estimate the accuracy of the measurements.

8.82. We can immediately solve the problem just stated, if we interpret it geometrically in the real space R_n. Consider the m vectors $a_1, a_2, \ldots, a_m$

whose components form the columns of the system (29), i.e.,

$$\begin{aligned} a_1 &= (a_{11}, a_{21}, \ldots, a_{n1}), \\ a_2 &= (a_{12}, a_{22}, \ldots, a_{n2}), \\ &\ldots\ldots\ldots\ldots\ldots \\ a_m &= (a_{1m}, a_{2m}, \ldots, a_{nm}). \end{aligned}$$

Forming the linear combinations $\xi_1 a_1 + \xi_2 a_2 + \cdots + \xi_m a_m$, we obtain the vector $\gamma = (\gamma_1, \gamma_2, \ldots, \gamma_n)$. Our problem is to determine the numbers $\xi_1, \xi_2, \ldots, \xi_m$ in such a way that the vector γ has the smallest possible deviation in norm from the given vector $b = (b_1, b_2, \ldots, b_n)$. Now the set of *all* linear combinations of the vectors $a_1, a_2, \ldots, a_m$ forms a subspace $L = L(a_1, a_2, \ldots, a_m)$, and the projection of the vector b onto the subspace L is the vector in L which is the closest to b. Therefore the numbers $\xi_1, \xi_2, \ldots, \xi_m$ must be chosen in such a way that the linear combination

$$\xi_1 a_1 + \xi_2 a_2 + \cdots + \xi_m a_m$$

reduces to the projection of b onto L. But, as we know, the solution of this problem is given by the last equation in Sec. 8.53, i.e.,

$$\xi_j = \frac{1}{D} \begin{vmatrix} (a_1, a_1) & \cdots & (a_{j-1}, a_1) & (b, a_1) & (a_{j+1}, a_1) & \cdots & (a_m, a_1) \\ \cdot & \cdots & \cdot & \cdot & \cdot & \cdots & \cdot \\ \cdot & \cdots & \cdot & \cdot & \cdot & \cdots & \cdot \\ \cdot & \cdots & \cdot & \cdot & \cdot & \cdots & \cdot \\ (a_1, a_m) & \cdots & (a_{j-1}, a_m) & (b, a_m) & (a_{j+1}, a_m) & \cdots & (a_m, a_m) \end{vmatrix},$$

where D is the Gram determinant $G(a_1, a_2, \ldots, a_m)$.

8.83. The results of Sec. 8.72 also allow us to evaluate the deviation δ itself. In fact, δ is just the altitude of the $(m+1)$-dimensional hyperparallelepiped determined by the vectors $a_1, a_2, \ldots, a_m, b$, and hence is equal to the ratio of volumes

$$\frac{V[a_1, a_2, \ldots, a_m, b]}{V[a_1, a_2, \ldots, a_m]}.$$

Using the Gram determinant to write each of these volumes, we finally obtain

$$\delta^2 = \frac{G(a_1, a_2, \ldots, a_m, b)}{G(a_1, a_2, \ldots, a_m)}.$$

Thus the problem posed in Sec. 8.81 is now completely solved.

8.84. In numerical analysis the following problem is often encountered (*interpolation with the least mean square error*): *Given a function $f_0(t)$ defined in the interval $a \leqslant t \leqslant b$, find the polynomial $P(t)$ of degree k $(k < n)$ for which the mean square deviation from the function $f_0(t)$, defined by*

$$\delta^2(f_0, P) = \sum_{j=0}^{n} [f_0(t_j) - P(t_j)]^2$$

is the smallest. Here $t_0, t_1, \ldots, t_n$ are certain fixed points of the interval $a \leqslant t \leqslant b$. Using geometric considerations, M. A. Krasnosyelski has given the following simple solution of the problem: Introduce a Euclidean space R consisting of functions $f(t)$ considered *only* at the points $t_0, t_1, \ldots, t_n$, and define the scalar product by

$$(f, g) = \sum_{j=0}^{n} f(t_j)g(t_j).$$

Then the problem reduces to finding the projection of the vector $f_0(t)$ onto the subspace of all polynomials of degree not exceeding k. The coefficients of the desired polynomial

$$P(t) = \xi_0 + \xi_1 t + \cdots + \xi_k t^k$$

are given by the same formulas as in the problem analyzed previously, i.e.,

$$\xi_j = \frac{1}{D}\begin{vmatrix} (1, 1) & (t, 1) & \cdots & (t^{j-1}, 1) & (f_0, 1) & (t^{j+1}, 1) & \cdots & (t^k, 1) \\ (1, t) & (t, t) & \cdots & (t^{j-1}, t) & (f_0, t) & (t^{j+1}, t) & \cdots & (t^k, t) \\ \cdot & \cdot & \cdots & \cdot & \cdot & \cdot & \cdots & \cdot \\ \cdot & \cdot & \cdots & \cdot & \cdot & \cdot & \cdots & \cdot \\ \cdot & \cdot & \cdots & \cdot & \cdot & \cdot & \cdots & \cdot \\ (1, t^k) & (t, t^k) & \cdots & (t^{j-1}, t^k) & (f_0, t^k) & (t^{j+1}, t^k) & \cdots & (t^k, t^k) \end{vmatrix},$$

where D is the Gram determinant $G(1, t, \ldots, t^k)$. The least square deviation itself is given by the formula

$$\delta^2(f_0, P) = \frac{G(1, t, \ldots, t^k, P)}{G(1, t, \ldots, t^k)}.$$

8.9. Adjoint Operators and Isometry

8.91. Adjoint operators with respect to the form (x, y). We now apply the results of Sec. 7.6 on the connection between linear operators and bilinear forms to the case where the fixed form (x, y) is the scalar product of the vectors x and y. Let $\mathbf{A}$ and $\mathbf{B}$ be linear operators in a Euclidean space $\mathbf{R}_n$, and use the formulas

$$\mathbf{A}(x, y) = (\mathbf{A}x, y), \qquad \mathbf{B}(x, y) = (x, \mathbf{B}y) \tag{32}$$

to construct bilinear forms $\mathbf{A}(x, y)$ and $\mathbf{B}(x, y)$. Since any orthogonal basis is a canonical basis of the form (x, y), and since the canonical coefficients of (x, y) all equal 1 in any such basis, it follows from Sec. 7.61 that the matrix $\|a_{jk}\|$ of the form $\mathbf{A}(x, y)$ in any orthonormal basis coincides with the matrix $\|a_k^{(j)}\|$ of the operator $\mathbf{A}$, while the matrix $\|b_{jk}\|$ of the form $\mathbf{B}(x, y)$ is the *transpose* of the matrix $\|b_j^{(k)}\|$ of the operator $\mathbf{B}$. Conversely, given bilinear forms $\mathbf{A}(x, y)$ and $\mathbf{B}(x, y)$ in the space $\mathbf{R}_n$, there exist unique linear operators $\mathbf{A}$ and $\mathbf{B}$ such that the formulas (32) hold (see Sec. 7.62). Moreover, applying Theorem 7.63 to the form (x, y), we get the following

THEOREM. *Given any linear operator* $\mathbf{A}$ *acting in an n-dimensional Euclidean space* $\mathbf{R}_n$, *there exists a unique linear operator* $\mathbf{A}'$ *(the* ***adjoint*** *of* $\mathbf{A}$*) acting in* $\mathbf{R}_n$ *such that*

$$(\mathbf{A}x, y) = (x, \mathbf{A}'y)$$

for arbitrary $x, y \in \mathbf{R}_n$. *The matrix of the operator* $\mathbf{A}'$ *in any orthonormal basis of the space* $\mathbf{R}_n$ *is the transpose of the matrix of the operator* $\mathbf{A}$.

8.92. Using the operation of taking the adjoint in a Euclidean space, we now introduce the following classes of operators:

a. *Symmetric operators*, defined by the relation

$$\mathbf{A}' = \mathbf{A}.$$

A symmetric operator is characterized by the fact that transposition does not change its matrix in any orthonormal basis.

b. *Antisymmetric operators*, defined by the relation

$$\mathbf{A}' = -\mathbf{A}.$$

An antisymmetric operator is characterized by the fact that transposition changes the sign of its matrix in any orthonormal basis.

c. *Normal operators*, defined by the relation

$$\mathbf{A}'\mathbf{A} = \mathbf{A}\mathbf{A}'.$$

The class of normal operators obviously contains the class of symmetric operators and the class of antisymmetric operators. The study of these classes of operators will be pursued in Secs. 9.3–9.4.

8.93. We now formulate the results of Secs. 7.73–7.76 on invariant operators for the case of a Euclidean space $\mathbf{R}_n$. Consider a linear invertible mapping $y = \mathbf{Q}x$ of the space $\mathbf{R}_n$ into itself which does not change the scalar product:

$$(\mathbf{Q}x, \mathbf{Q}y) = (x, y).$$

A mapping of this kind, which in Sec. 7.73 was said to be invariant with respect to the form (x, y), will now be called *isometric*. Thus an isometric operator $\mathbf{Q}$ is characterized by the relation

$$\mathbf{Q}'\mathbf{Q} = \mathbf{E}$$

(cf. formula (33), p. 201), where $\mathbf{E}$ is the unit operator and $\mathbf{Q}'$ is the operator adjoint to $\mathbf{Q}$ with respect to the form (x, y), i.e., the operator adjoint to $\mathbf{Q}$ in the sense of Sec. 8.91. The inverse $\mathbf{Q}^{-1} = \mathbf{Q}'$ of an isometric operator is itself isometric, and so is the product of two isometric operators (see Sec. 7.74).

According to Sec. 7.75, an isometric operator $\mathbf{Q}$ is characterized by the fact that it carries every orthonormal basis $e_1, \ldots, e_n$ into another orthonormal basis $f_1 = \mathbf{Q}e_1, \ldots, f_n = \mathbf{Q}e_n$. The matrix $Q = \|q_i^{(j)}\|$ of an isometric operator $\mathbf{Q}$ in any orthonormal basis is called an *orthogonal matrix*. An orthonormal matrix is characterized by the conditions (35), p. 202, which in the present case take the form

$$\sum_{i=1}^{n} q_j^{(i)} q_k^{(i)} = \begin{cases} 1 & \text{if } j = k, \\ 0 & \text{if } j \neq k, \end{cases}$$

or by the conditions (35′), p. 202, which take the form

$$\sum_{k=1}^{n} q_k^{(j)} q_k^{(m)} = \begin{cases} 1 & \text{if } j = m, \\ 0 & \text{if } j \neq m, \end{cases}$$

i.e., the sum of the squares of the elements of any row (or column) equals 1, while the sum of the products of the corresponding elements of two different rows (or columns) equals 0.

8.94. It follows from the relation $Q^{-1} = Q'$ that the formulas

$$\begin{aligned} f_1 &= q_1^{(1)}e_1 + \cdots + q_n^{(1)}e_n, \\ &\cdots\cdots\cdots\cdots\cdots\cdots \\ f_n &= q_1^{(n)}e_1 + \cdots + q_n^{(n)}e_n \end{aligned} \tag{33}$$

for the transformation from one orthonormal basis $e_1, \ldots, e_n$ to another orthonormal basis $f_1, \ldots, f_n$ (such a transformation is called an *orthogonal transformation*) are "inverted" by the formulas

$$\begin{aligned} e_1 &= q_1^{(1)}f_1 + \cdots + q_1^{(n)}f_n, \\ &\cdots\cdots\cdots\cdots\cdots\cdots \\ e_n &= q_n^{(1)}f_1 + \cdots + q_n^{(n)}f_n \end{aligned} \tag{34}$$

By Sec. 5.31, the components η_k of a vector x with respect to the basis $f_1, \ldots, f_n$ are expressed in terms of the components ξ_j of the same vector with respect to the basis $e_1, \ldots, e_n$ by the formulas

$$\begin{aligned} \eta_1 &= q_1^{(1)}\xi_1 + \cdots + q_n^{(1)}\xi_n, \\ &\cdots\cdots\cdots\cdots\cdots\cdots \\ \eta_n &= q_1^{(n)}\xi_1 + \cdots + q_n^{(n)}\xi_n, \end{aligned} \tag{35}$$

with inverse formulas

$$\begin{aligned} \xi_1 &= q_1^{(1)}\eta_1 + \cdots + q_1^{(n)}\eta_n, \\ &\cdots\cdots\cdots\cdots\cdots\cdots \\ \xi_n &= q_n^{(1)}\eta_1 + \cdots + q_n^{(n)}\eta_n. \end{aligned} \tag{36}$$

8.95. Given $m < n$ rows of numbers $q_i^{(j)}$ $(i = 1, \ldots, n; j = 1, \ldots, m)$ satisfying the conditions

$$\sum_{i=1}^{n} q_i^{(j)} q_i^{(k)} = \begin{cases} 1 & \text{if } j = k, \\ 0 & \text{if } j \neq k, \end{cases}$$

consider the problem of finding $n - m$ more rows of numbers $q_i^{(j)}$ $(j = m + 1, \ldots, n)$ such that the $n \times n$ matrix $\|q_i^{(j)}\|$ $(i, j = 1, \ldots, n)$ is orthogonal. This problem is easily solved by using a geometrical argument. Suppose the given rows $q_i^{(j)}$ are interpreted as components of m vectors in a Euclidean space $\mathbf{R}_n$ with scalar product

$$((\xi_1, \ldots, \xi_n), (\eta_1, \ldots, \eta_n)) = \sum_{i=1}^{n} \xi_i \eta_i$$

(recall Example 8.22b). Then our problem consists of augmenting m given orthonormal vectors $q_1, \ldots, q_m$ with further vectors to make an orthonormal basis for the space $\mathbf{R}_n$. With this geometrical interpretation, the problem is obviously solvable. For example, we can augment $q_1, \ldots, q_m$ with any other vectors $q_{m+1}, \ldots, q_n$ such that the resulting system of n vectors is linearly independent, and then use Theorem 8.61 to make the whole system of n vectors orthonormal.

8.96. We now consider some further properties of symmetric operators.

a. *If the subspace* $\mathbf{R}' \subset \mathbf{R}$ *is invariant under the operator* $\mathbf{A}$, *then, by Sec.* 7.65, *the orthogonal complement of* $\mathbf{R}'$ *is invariant under the adjoint operator* $\mathbf{A}'$. Therefore, in the case of a symmetric operator $\mathbf{A}$, if the subspace $\mathbf{R}'$ is invariant under $\mathbf{A}$, then so is the orthogonal complement of $\mathbf{R}'$.

b. THEOREM. *Every symmetric operator in the plane* $(n = 2)$ *has an eigenvector.*

Proof. In this case, the equation determining the eigenvectors is just

$$\begin{vmatrix} a_{11} - \lambda & a_{12} \\ a_{21} & a_{22} - \lambda \end{vmatrix} = 0.$$

The discriminant of this quadratic equation is

$$(a_{11} + a_{22})^2 - 4(a_{11}a_{22} - a_{21}a_{12}) = (a_{11} - a_{22})^2 + 4a_{12}^2 \geqslant 0,$$

and hence has real roots. ∎

c. From these considerations and the fact that every operator in a real space has an invariant plane (see Sec. 6.66), it follows that *every symmetric operator in the space* $\mathbf{R}_n$ *has an orthogonal basis consisting of eigenvectors.* In Sec. 9.45 we will deduce this result in a more general way, without recourse to the real Jordan canonical form.

PROBLEMS

1. Suppose we define the scalar product of two vectors of the space V_3 as the product of the lengths of the vectors. Is the resulting space Euclidean?

2. Answer the same question if the scalar product is defined as the product of the lengths of the vectors and the cube of the cosine of the angle between them.

3. Answer the same question if the scalar product is defined as twice the usual scalar product.

4. Find the angle between opposite edges of a regular tetrahedron.

5. Find the angles of the "triangle" formed in the space $R_2(-1, 1)$ by the vectors $x_1(t) = 1$, $x_2(t) = t$, $x_3(t) = 1 - t$.

6. Write the triangle inequalities in the space $R_2(a, b)$.

7. Find the cosines of the angles between the line $\xi_1 = \xi_2 = \cdots = \xi_n$ and the coordinate axes in the space R_n.

8. In the space R_4 expand the vector f as the sum of two vectors, a vector g lying in the linear manifold spanned by the vectors b_i and a vector h orthogonal to this subspace:

a) $f = (5, 2, -2, 2)$, $b_1 = (2, 1, 1, -1)$, $b_2 = (1, 1, 3, 0)$;
b) $f = (-3, 5, 9, 3)$, $b_1 = (1, 1, 1, 1)$, $b_2 = (2, -1, 1, 1)$, $b_3 = (2, -7, -1, -1)$.

9. Prove that of all the vectors in the subspace $\mathbf{R}'$, the vector g of Sec. 8.51 (the projection of f onto $\mathbf{R}'$) makes the smallest angle with f.

10. Show that if the vector g_0 in the space $\mathbf{R}'$ is orthogonal to g (the projection of f onto $\mathbf{R}'$), then g_0 is orthogonal to f itself.

11. Show that the perpendicular dropped from the origin of coordinates onto a hyperplane **H** has the smallest length of all the vectors joining the origin with **H**.

12. Given the system of vectors $x_1 = \mathbf{i}$, $x_2 = 2\mathbf{i}$, $x_3 = 3\mathbf{i}$, $x_4 = 4\mathbf{i} - 2\mathbf{j}$, $x_5 = -\mathbf{i} + 10\mathbf{j}$, $x_6 = \mathbf{i} + \mathbf{j} + 5\mathbf{k}$ in the space V_3 with basis $\mathbf{i}, \mathbf{j}, \mathbf{k}$, construct the vectors $y_1, y_2, \ldots, y_6$ figuring in the orthogonalization theorem.

13. Using the method of the orthogonalization theorem, construct an orthogonal basis in the three-dimensional subspace of the space R_4 spanned by the vectors $(1, 2, 1, 3)$, $(4, 1, 1, 1)$, $(3, 1, 1, 0)$.

14. Given two subspaces $\mathbf{R}'$ and $\mathbf{R}''$ of a Euclidean space $\mathbf{R}$, let $m(\mathbf{R}', \mathbf{R}'')$ denote the maximum length of the perpendiculars dropped onto $\mathbf{R}''$ from the ends of the unit vectors $e' \in \mathbf{R}'$, and define the quantity $m(\mathbf{R}'', \mathbf{R}')$ similarly. Then the quantity

$$\theta = \max \{m(\mathbf{R}', \mathbf{R}''), m(\mathbf{R}'', \mathbf{R}')\}$$

is called the *spread* of the subspaces $\mathbf{R}'$ and $\mathbf{R}''$. Show that the subspaces $\mathbf{R}'$ and $\mathbf{R}''$ have the same dimension if $\theta < 1$. (M. A. Krasnosyelski and M. G. Krein)

15. Find the leading coefficient A_n of the Legendre polynomial $P_n(t)$.

16. Show that $P_n(t)$ is an even function for even n and an odd function for odd n. In particular, find $P_n(-1)$.

17. Show that if the polynomial $tP_{n-1}(t)$ is expanded in terms of the Legendre polynomials, so that

$$tP_{n-1}(t) = a_0P_0(t) + a_1P_1(t) + \cdots + a_nP_n(t),$$

then the coefficients $a_0, a_1, \ldots, a_{n-3}$ and a_{n-1} are zero.

18. Find the coefficients a_{n-2} and a_n of the expansion of the polynomial $tP_{n-1}(t)$ given in the preceding problem, thereby obtaining the recurrence formula

$$nP_n(t) = (2n - 1)tP_{n-1}(t) - (n - 1)P_{n-2}(t).$$

19. Find the polynomial

$$Q(t) = t^n + b_1t^{n-1} + \cdots + b_{n-1}t + b_n$$

for which the integral

$$\int_{-1}^{1} Q^2(t)\, dt$$

has the smallest value.

20. Find the norm of the Legendre polynomial $P_n(t)$.

21. Let $\mathbf{A}$ be any linear operator acting in an n-dimensional Euclidean space $\mathbf{R}_n$. Show that the ratio

$$k(\mathbf{A}) = \frac{V[\mathbf{A}x_1, \mathbf{A}x_2, \ldots, \mathbf{A}x_n]}{V[x_1, x_2, \ldots, x_n]}$$

is a constant (i.e., is independent of the choice of the vectors $x_1, x_2, \ldots, x_n$), and find the value of $k(\mathbf{A})$ (the "distortion coefficient").

22. Show that $k(\mathbf{AB}) = k(\mathbf{A})k(\mathbf{B})$ for any two linear operators **A** and **B**.

23. Let $x_1, x_2, \ldots, x_k, y, z$ be vectors in a Euclidean space **R**. Prove the inequality

$$\frac{V[x_1, x_2, \ldots, x_k, y, z]}{V[x_1, x_2, \ldots, x_k, y]} \leqslant \frac{V[x_1, x_2, \ldots, x_k, z]}{V[x_1, x_2, \ldots, x_k]}. \tag{37}$$

24. Let $x_1, x_2, \ldots, x_m$ be vectors in a Euclidean space **R**. Prove the inequality

$$V[x_1, x_2, \ldots, x_m] \leqslant \prod_{k=1}^{m} \{V[x_1, \ldots, x_{k-1}, x_{k+1}, \ldots, x_m]\}^{1/(m-1)}. \tag{38}$$

What is the geometric meaning of this inequality?

25 (*Continuation*). Prove the following inequalities, which sharpen Hadamard's inequality:

$$\begin{aligned}
&V[x_1, x_2, \ldots, x_m] \\
&\quad \leqslant \prod_{k=1}^{m} \{V[x_1, \ldots, x_{k-1}, x_{k+1}, \ldots, x_m]\}^{1/(m-1)} \\
&\quad \leqslant \prod_{1\leqslant k<l\leqslant m}^{m} \{V[x_1, \ldots, x_{k-1}, x_{k+1}, \ldots, x_{l-1}, x_{l+1}, \ldots, x_m]\}^{1\cdot 2(m-1)/(m-2)} \\
&\quad \leqslant \cdots \leqslant \prod_{1\leqslant s_1<s_2<\cdots<s_r\leqslant m} \{V[x_{s_1}, x_{s_2}, \ldots, x_{s_r}]\}^{1\cdot 2\cdots(n-r)/(m-1)(m-2)\cdots r} \\
&\quad \leqslant \cdots \leqslant \prod_{1\leqslant s_1<s_2\leqslant m} \{V[x_{s_1}, x_{s_2}]\}^{1/(m-1)} \leqslant \prod_{s=1}^{m} |x_s|.
\end{aligned}$$

(M. K. Faguet)

26. If $|a_{ik}| \leqslant M$, then

$$\det \|a_{ik}\| \leqslant M^n n^{n/2},$$

by Hadamard's inequality. Show that this estimate cannot be improved for $n = 2^m$.

27. Show that if $\mathbf{N}(\mathbf{A})$ and $\mathbf{T}(\mathbf{A})$ are the null space and range, respectively, of the operator **A**, then the orthogonal complements of these subspaces are the range and null space, respectively, of the adjoint operator $\mathbf{A}'$.

28. Let A be an orthogonal matrix. Show that $A_{ik} = a_{ik} \det A$ is the cofactor of the element a_{ik} of A.

29. Show that the sum of the squares of all the minors of order k appearing in k fixed rows of an orthogonal matrix equals 1. Show that the sum of the products of all the minors of order k appearing in one group of k rows with the corresponding minors in another group of k rows equals 0.

30. A linear operator **Q** preserves the length of every vector. Show that **Q** is isometric.

31. An operator **A** which preserves the orthogonality of any pair of vectors x and y, i.e., such that $(x, y) = 0$ implies $(\mathbf{A}x, \mathbf{A}y) = 0$, is called an *isogonal operator*. Isometric operators and similarity operators ($\mathbf{A}x = \lambda x$ for every x) are isogonal, and so is the product of any similarity operator and any isometric operator. Show that every isogonal operator is the product of a similarity operator and an isometric operator.

32. Let **Q** be a linear operator acting in an n-dimensional Euclidean space $\mathbf{R}_n$ ($n \geqslant 3$). Suppose **Q** does not change the area of any parallelogram, so that

$$V[x, y] = V[\mathbf{Q}x, \mathbf{Q}y].$$

Show that **Q** is an isometric operator.

33. Let **Q** be a linear operator acting in an n-dimensional Euclidean space $\mathbf{R}_n$, and suppose **Q** does not change the volume of any k-dimensional hyperparallelepiped ($k < n$). Show that **Q** is isometric. (M. A. Krasnosyelski)

Comment. For $k = n$ the assertion of Problem 33 fails to be valid, since then every operator **Q** with $\det Q = \pm 1$ will satisfy the condition of the problem.

34. Let $F = \{x_1, x_2, \ldots, x_k\}$ and $G = \{y_1, y_2, \ldots, y_k\}$ be two finite systems of vectors in a Euclidean space $\mathbf{R}_n$. Show that a necessary and sufficient condition for the existence of an isometric operator **Q** taking every vector x_i into the corresponding vector y_i ($i = 1, 2, \ldots, k$) is that the relations

$$(x_i, x_j) = (y_i, y_j) \qquad (i, j = 1, 2, \ldots, k)$$

hold.

35 (*The angles between two subspaces*). Let $\mathbf{R}'$ and $\mathbf{R}''$ be two subspaces of a Euclidean space **R**. Let the unit vector e' vary over the unit sphere of the subspace $\mathbf{R}'$, and let the unit vector e'' vary (independently of e') over the unit sphere of the subspace $\mathbf{R}''$. For some pair of vectors $e' = e_1'$, $e'' = e_1''$, the angle between e' and e'' achieves a minimum, which we denote by φ_1. Now let e' vary over its unit sphere while remaining orthogonal to e_1', and let e'' vary over its unit sphere while remaining orthogonal to e_1''. With these constraints, the angle between e' and e'' achieves a minimum $\varphi_2 \geqslant \varphi_1$ for some pair $e' = e_2'$, $e'' = e_2''$. Then let e' vary over its unit sphere while remaining orthogonal to e_1' and e_2', and let e'' vary over its unit sphere while remaining orthogonal to e_1'' and e_2''. In this way, we get a new minimum angle $\varphi_3 \geqslant \varphi_2$ and a new pair e_3' and e_3''. Continuing this process, we obtain a set of angles $\varphi_1, \varphi_2, \ldots, \varphi_k$, the number of which equals the smaller of the dimensions of $\mathbf{R}'$ and $\mathbf{R}''$. The angles $\varphi_1, \varphi_2, \ldots, \varphi_k$ are called the *angles between the subspaces* $\mathbf{R}'$ *and* $\mathbf{R}''$. Prove the following facts:

a) The angles $\varphi_1, \varphi_2, \ldots, \varphi_k$ are uniquely defined and do not depend on the choice of the vectors $e_1', e_1'', e_2', e_2'', \ldots$ if these vectors are not uniquely defined by the construction;
b) The angles $\varphi_1, \varphi_2, \ldots, \varphi_k$ determine the subspaces $\mathbf{R}'$ and $\mathbf{R}''$ to within their spatial orientation, i.e., if there are two pairs of subspaces $\mathbf{R}'$, $\mathbf{R}''$ and $\mathbf{S}'$, $\mathbf{S}''$ such that the angles between $\mathbf{R}'$ and $\mathbf{R}''$ are the same as those

between $\mathbf{S}'$ and $\mathbf{S}''$, then there exists an isometric operator which simultaneously carries $\mathbf{S}'$ into $\mathbf{R}'$ and $\mathbf{S}''$ into $\mathbf{R}''$;

c) Given any preassigned angles $\varphi_1 \leqslant \varphi_2 \leqslant \cdots \leqslant \varphi_k \leqslant \pi/2$, we can construct a pair of spaces $\mathbf{R}'$ and $\mathbf{R}''$ such that $\varphi_1, \varphi_2, \ldots, \varphi_k$ are the angles between $\mathbf{R}'$ and $\mathbf{R}''$.

36. Let $y_1, y_2, \ldots, y_m$ be the projections of the vectors $x_1, x_2; \ldots, x_m$ onto some subspace. Show that the volume of the hyperparallelepiped determined by the vectors $y_1, y_2, \ldots, y_m$ does not exceed the volume of the hyperparallelepiped determined by the vectors $x_1, x_2, \ldots, x_m$.

37 (*Continuation*). In Problem 36 suppose that both the vectors $x_1, x_2, \ldots, x_m$ and the vectors $y_1, y_2, \ldots, y_m$ are linearly independent. Show that the formula

$$V[y_1, y_2, \ldots, y_m] = V[x_1, x_2, \ldots, x_m] \cos \alpha_1 \cos \alpha_2 \cdots \cos \alpha_m$$

holds, where $\alpha_1, \alpha_2, \ldots, \alpha_m$ are the angles between the subspaces $\mathbf{L}_1 = \mathbf{L}(x_1, x_2, \ldots, x_m)$ and $\mathbf{L}_2 = \mathbf{L}(y_1, y_2, \ldots, y_m)$ (see Prob. 35).

38. A set of k vectors in a Euclidean space $\mathbf{R}$ will be called a *k-vector*, and we will say that two k-vectors $\{x_1, x_2, \ldots, x_k\}$ and $\{y_1, y_2, \ldots, y_k\}$ are *equal* if

1) The volume $V[x_1, x_2, \ldots, x_k]$ equals the volume $V[y_1, y_2, \ldots, y_k]$;
2) The linear manifold $\mathbf{L}(x_1, x_2, \ldots, x_k)$ coincides with the linear manifold $\mathbf{L}(y_1, y_2, \ldots, y_k)$;
3) The systems $x_1, x_2, \ldots, x_k$ and $y_1, y_2, \ldots, y_k$ have the same orientation, i.e., the operator in the space $\mathbf{L}(x_1, x_2, \ldots, x_k)$ carrying the system $x_1, x_2, \ldots, x_k$ into the system $y_1, y_2, \ldots, y_k$ has a positive determinant.

Show that a k-vector $\{x_1, x_2, \ldots, x_k\}$ in an n-dimensional space $\mathbf{R}_n$ is uniquely determined if we know the values of all the minors of order k of the $n \times k$ matrix

$$\|\xi_i^{(j)}\| \qquad (i = 1, 2, \ldots, n; j = 1, 2, \ldots, k)$$

formed from the components of the vectors $x_1, x_2, \ldots, x_k$ with respect to any orthonormal basis $e_1, e_2, \ldots, e_n$ of the space $\mathbf{R}_n$.

39. If the k-vector $\{x_1, x_2, \ldots, x_k\}$ equals the k-vector $\{y_1, y_2, \ldots, y_k\}$ (Prob. 38), show that the minors of order k of the matrix formed from the components of the vectors $x_1, x_2, \ldots, x_k$ equal the corresponding minors of the matrix formed from the components of the vectors $y_1, y_2, \ldots, y_k$.

40. By the *angles* between two k-vectors $\{x_1, x_2, \ldots, x_k\}$ and $\{y_1, y_2, \ldots, y_k\}$ we mean the angles between the subspaces $\mathbf{L}_1 = \mathbf{L}(x_1, x_2, \ldots, x_k)$ and $\mathbf{L}_2 = \mathbf{L}(y_1, y_2, \ldots, y_k)$ (see Prob. 35) subject, however, to the supplementary condition that the vectors $e_1, e_2, \ldots, e_k$ chosen in the subspace $\mathbf{L}_1$ (when constructing the angles) have the same orientation as the vectors $x_1, x_2, \ldots, x_k$ (this condition plays a role only in constructing the last vector e_k), and similarly for the subspace $\mathbf{L}_2$. Show that the angles $\beta_1, \beta_2, \ldots, \beta_k$ between the k-vectors and the angles $\alpha_1, \alpha_2, \ldots, \alpha_k$ between the corresponding subspaces are connected by the following relations:

$$\alpha_j = \beta_j \qquad (j < k),$$
$$\alpha_k = \beta_k \quad \text{or} \quad \alpha_k = \pi - \beta_k.$$

41. By the *scalar product* of two k-vectors $X = \{x_1, x_2, \ldots, x_k\}$ and $Y = \{y_1, y_2, \ldots, y_k\}$, specified by the matrices X and Y made up of the components of the vectors x_i and y_i with respect to some orthonormal basis of the space $\mathbf{R}_n$, we mean the sum of all the products of the minors of order k of the matrix X with the corresponding minors of the matrix Y. Show that this scalar product equals

$$V[x_1, x_2, \ldots, x_k]V[y_1, y_2, \ldots, y_k] \cos \beta_1 \cos \beta_2 \cdots \cos \beta_k,$$

where $\beta_1, \beta_2, \ldots, \beta_k$ are the angles between the k-vectors X and Y.

42. Show that the scalar product of the two k-vectors $X = \{x_1, x_2, \ldots, x_k\}$ and $Y = \{y_1, y_2, \ldots, y_k\}$ can be written in the form

$$\{X, Y\} = \begin{vmatrix} (x_1, y_1) & (x_1, y_2) & \ldots & (x_1, y_k) \\ (x_2, y_1) & (x_2, y_2) & \ldots & (x_2, y_k) \\ \cdot & \cdot & \cdots & \cdot \\ (x_k, y_1) & (x_k, y_2) & \ldots & (x_k, y_k) \end{vmatrix}.$$

43. Show that if the polynomial $[P(t)]^k$ is an annihilating polynomial of the isometric operator $\mathbf{A}$, then so is the polynomial $P(t)$.

chapter 9

UNITARY SPACES

9.1. Hermitian Forms

9.11. A numerical function $\mathbf{A}(x, y)$ of two arguments x and y in a complex space $\mathbf{C}$ is called a *Hermitian bilinear form* or simply a *Hermitian form* if it is a linear form of the first kind in x for every fixed value of y and a linear form of the second kind (Sec. 4.14) in y for every fixed value of x. In other words, $\mathbf{A}(x, y)$ is said to be a Hermitian form in x and y if the following conditions are satisfied for arbitrary x, y, z in $\mathbf{C}$ and arbitrary complex α:†

$$\begin{aligned} \mathbf{A}(x + z, y) &= \mathbf{A}(x, y) + \mathbf{A}(z, y), \\ \mathbf{A}(\alpha x, y) &= \alpha \mathbf{A}(x, y), \\ \mathbf{A}(x, y + z) &= \mathbf{A}(x, y) + \mathbf{A}(x, z), \\ \mathbf{A}(x, \alpha y) &= \bar{\alpha} \mathbf{A}(x, y). \end{aligned} \tag{1}$$

Using induction and (1), we easily obtain the general formula

$$\mathbf{A}\left(\sum_{i=1}^{k} \alpha_i x_i, \sum_{j=1}^{m} \beta_j y_j\right) = \sum_{i=1}^{k} \sum_{j=1}^{m} \alpha_i \bar{\beta}_j \mathbf{A}(x_i, y_j), \tag{2}$$

where $x_1, \ldots, x_k, y_1, \ldots, y_m$ are arbitrary vectors of the space $\mathbf{C}$ and $\alpha_1, \ldots, \alpha_k, \beta_1, \ldots, \beta_m$ are arbitrary complex numbers.

† As usual, the overbar denotes the complex conjugate.

9.12. *Examples*

a. If $L_1(x)$ is a linear form of the first kind and $L_2(x)$ is a linear form of the second kind (Sec. 4.14), then $\mathbf{A}(x, y) = L_1(x)L_2(y)$ is a Hermitian form.

b. An example of a Hermitian form in an n-dimensional space $\mathbf{C}_n$ with a fixed basis $e_1, e_2, \ldots, e_n$ is the function

$$\mathbf{A}(x, y) = \sum_{i=1}^{n}\sum_{k=1}^{n} a_{ik}\xi_i\bar{\eta}_k, \tag{3}$$

where

$$x = \sum_{i=1}^{n} \xi_i e_i, \qquad y = \sum_{k=1}^{n} \eta_k e_k$$

are arbitrary vectors and a_{ik} $(i, k = 1, 2, \ldots, n)$ are fixed complex numbers. In fact, (3) is the general representation of a Hermitian form in an n-dimensional complex space. This is proved in the same way as the analogous proposition for bilinear forms in a space $\mathbf{K}_n$ (see Sec. 7.13).

9.13. A Hermitian form $\mathbf{A}(x, y)$ is said to be *Hermitian–symmetric* (or simply *symmetric*) if

$$\mathbf{A}(y, x) = \overline{\mathbf{A}(x, y)} \tag{4}$$

for arbitrary vectors x and y. Given a symmetric Hermitian form $\mathbf{A}(x, y)$ in an n-dimensional complex space $\mathbf{C}_n$, suppose we use (3) to write $\mathbf{A}(x, y)$ in terms of the components of the vectors x and y with respect to the basis $e_1, \ldots, e_n$. Then

$$a_{ik} = \mathbf{A}(e_i, e_k) = \overline{\mathbf{A}(e_k, e_i)} = \bar{a}_{ki}, \tag{5}$$

i.e., the matrix $\|a_{ik}\|$ of the form $\mathbf{A}(x, y)$ in the basis $e_1, \ldots, e_n$ is carried into itself by transposing the matrix and replacing all its elements by their complex conjugates. Conversely, if the coefficients of a Hermitian form $\mathbf{A}(x, y)$ satisfy the condition (5), then $\mathbf{A}(x, y)$ is symmetric, since

$$\mathbf{A}(y, x) = \sum_{i,k=1}^{n} a_{ik}\eta_i\bar{\xi}_k = \sum_{i,k=1}^{n} \bar{a}_{ki}\bar{\xi}_k\eta_i = \overline{\sum_{i,k=1}^{n} a_{ki}\xi_k\bar{\eta}_i} = \overline{\mathbf{A}(x, y)}.$$

A matrix $\|a_{ik}\|$ such that $a_{ik} = \bar{a}_{ki}$ $(i, k = 1, \ldots, n)$ will henceforth be called *Hermitian–symmetric* (or simply *Hermitian*).

9.14. a. Suppose the Hermitian form $\mathbf{A}(x, y)$ has the matrix $A_{(e)} = \|a_{ik}\|$ in the basis $e_1, \ldots, e_n$ of the space $\mathbf{C}$ and the matrix $A_{(f)} = \|b_{ik}\|$ in the basis $f_1, \ldots, f_n$, where the relation between the two bases is given by

$$f_i = \sum_{j=1}^{n} p_j^{(i)} e_j \qquad (i = 1, \ldots, n).$$

Then, reasoning as in Sec. 7.15, we find that the relation between the matrices $A_{(e)}$ and $A_{(f)}$ is given by the formula

$$A_{(f)} = P^* A_{(e)} P, \tag{6}$$

where $P = \|p_j^{(i)}\|$ is the matrix of the transformation from the basis $e_1, \ldots, e_n$ to the basis $f_1, \ldots, f_n$, and P^* is the matrix obtained from P by transposing and then replacing elements by their complex conjugates. Writing $P^* = \|p_j^{*(i)}\|$, we have

$$p_j^{*(i)} = \overline{p_i^{(j)}} \qquad (i, j = 1, \ldots, n).$$

b. Just as in Sec. 7.23, it follows from (6) that the rank of the matrix $A_{(e)}$ of the Hermitian form $\mathbf{A}(x, y)$ is independent of the choice of the basis $\{e\}$. The form $\mathbf{A}(x, y)$ is said to be *nonsingular* if its *rank* (i.e., the rank of the matrix $A_{(e)}$ in any basis $\{e\}$) equals the dimension n of the space $\mathbf{C}_n$. If the form $\mathbf{A}(x, y)$ is nonsingular, then, given any vector $x_0 \neq 0$, there is a vector $y_0 \in \mathbf{C}_n$ such that $\mathbf{A}(x_0, y_0) \neq 0$ (cf. Sec. 7.15c).

9.15. a. By a *Hermitian quadratic form* in a complex space $\mathbf{C}$ we mean the function of one variable $x \in \mathbf{C}$ obtained by changing y to x in any Hermitian bilinear form $\mathbf{A}(x, y)$. It follows from Sec. 9.12b that in an n-dimensional complex space $\mathbf{C}_n$ with basis $e_1, \ldots, e_n$, a Hermitian quadratic form can be expanded in terms of the components $\xi_1, \ldots, \xi_n$ of the vector x by the formula

$$\mathbf{A}(x, x) = \sum_{i,k=1}^{n} a_{ik} \xi_i \bar{\xi}_k \tag{7}$$

with complex coefficients a_{ik}. Conversely, a function $\mathbf{A}(x, x)$ of the form (7) is the Hermitian quadratic form obtained by changing y to x in the Hermitian bilinear form

$$\mathbf{A}(x, y) = \sum_{i,k=1}^{n} a_{ik} \xi_i \bar{\eta}_k.$$

b. If a Hermitian bilinear form $\mathbf{A}(x, y)$ is symmetric, so that $a_{ik} = \bar{a}_{ki}$, then the corresponding Hermitian quadratic form $\mathbf{A}(x, x)$ is also said to be *symmetric*. A symmetric Hermitian quadratic form $\mathbf{A}(x, x)$ can only take real values, since it follows from (4) that

$$\overline{\mathbf{A}(x, x)} = \mathbf{A}(x, x).$$

Unlike the situation in Sec. 7.22, there is a unique Hermitian bilinear form $\mathbf{A}(x, y)$ corresponding to a given Hermitian quadratic form $\mathbf{A}(x, x)$. In fact,

$$\mathbf{A}(x + y, x + y) = \mathbf{A}(x, x) + \mathbf{A}(x, y) + \mathbf{A}(y, x) + \mathbf{A}(y, y),$$

$$\mathbf{A}(x + iy, x + iy) = \mathbf{A}(x, x) - i\mathbf{A}(x, y) + i\mathbf{A}(y, x) + \mathbf{A}(y, y).$$

Multiplying the first equation by i and then subtracting the second equation from the first, we easily find that

$$\mathbf{A}(x, y) = \frac{1}{2}\,[\mathbf{A}(x + y, x + y) + i\mathbf{A}(x + iy, x + iy)] - \frac{1+i}{2}\,[\mathbf{A}(x, x) + \mathbf{A}(y, y)],$$

so that $\mathbf{A}(x, y)$ is uniquely determined in terms of the values $\mathbf{A}(x, x)$, $\mathbf{A}(y, y)$, $\mathbf{A}(x + y, x + y)$ and $\mathbf{A}(x + iy, x + iy)$ of the given Hermitian quadratic form.

If the Hermitian quadratic form $\mathbf{A}(x, x)$ has the representation

$$\mathbf{A}(x, x) = \sum_{i,k=1}^{n} a_{ik}\xi_i\bar{\xi}_k$$

in some basis $e_1, \ldots, e_n$, then the Hermitian bilinear form

$$\mathbf{A}(x, y) = \sum_{i,k=1}^{n} a_{ik}\xi_i\bar{\eta}_k$$

obviously reduces to $\mathbf{A}(x, x)$ if we make the substitution $y = x$. Moreover, as just shown, this is the unique Hermitian bilinear form reducing to $\mathbf{A}(x, x)$ under this substitution.

9.16. a. *Given a symmetric Hermitian quadratic form* $\mathbf{A}(x, x)$ *in an* n*-dimensional complex space* $\mathbf{C}_n$*, there exists a basis in* $\mathbf{C}_n$ *in which* $\mathbf{A}(x, x)$ *can be written in the canonical form*

$$\mathbf{A}(x, x) = \sum_{k=1}^{n} \lambda_k \eta_k \eta_k = \sum_{k=1}^{n} \lambda_k\,|\eta_k|^2 \tag{8}$$

with real coefficients $\lambda_1, \lambda_2, \ldots, \lambda_n$.

The proof of this proposition is analogous to that of Theorem 7.31. Instead of equation (13), p. 186, we have

$$\begin{aligned} b_{1m}\xi_1\bar{\xi}_m &+ b_{2m}\xi_2\bar{\xi}_m + \cdots + b_{m-1,m}\xi_{m-1}\bar{\xi}_m + b_{mm}\xi_m\bar{\xi}_m \\ &+ \bar{b}_{1m}\bar{\xi}_1\xi_m + \cdots + \bar{b}_{m-1,m}\bar{\xi}_{m-1}\xi_m \\ &= b_{mm}\left|\frac{b_{1m}}{b_{mm}}\,\xi_1 + \frac{b_{2m}}{b_{mm}}\,\xi_2 + \cdots + \frac{b_{m-1,m}}{b_{mm}}\,\xi_{m-1} + \xi_m\right|^2 + \mathbf{A}_1(x, x) \end{aligned}$$

($b_{mm} \neq 0$), where $\mathbf{A}_1(x, x)$ is a symmetric Hermitian quadratic form in the variables $\xi_1, \xi_2, \ldots, \xi_{m-1}$. Instead of the transformation (14), p. 187, we now have the transformation

$$\begin{aligned}\xi_1 &= \xi_1' + \xi_2',\\ \xi_2 &= \xi_1' + i\xi_2',\\ \xi_3 &= \xi_3',\\ &\cdots\cdots\cdots\\ \xi_m &= \xi_m'\end{aligned}$$

which carries the sum $a_{12}\xi_1\bar{\xi}_2 + \bar{a}_{12}\bar{\xi}_1\xi_2$ ($a_{12} \neq 0$) into the expression

$$(a_{12} + \bar{a}_{12})\xi_1'\bar{\xi}_1' - i(a_{12} - \bar{a}_{12})\xi_2'\bar{\xi}_2' + \cdots,$$

where at least one of the two (real) coefficients $a_{12} + \bar{a}_{12}$ and $i(a_{12} - \bar{a}_{12})$ is nonzero.

b. The law of inertia (Theorem 7.91) continues to hold for a symmetric Hermitian quadratic form $\mathbf{A}(x, x)$ in a complex space, i.e., *the total number p of positive coefficients and the total number q of negative coefficients among the numbers $\lambda_1, \lambda_2, \ldots, \lambda_n$ do not depend on the choice of the canonical basis.* The proof of this proposition is the exact analogue of that of Theorem 7.91. As in the real case, the number p is called the *positive index of inertia* and the number q the *negative index of inertia* of the form $\mathbf{A}(x, x)$.

It should be noted that the law of inertia does not hold for quadratic (as opposed to Hermitian quadratic) forms in a complex space $\mathbf{C}_n$. For example, the quadratic form

$$\mathbf{A}(x, x) = \xi_1^2 + \xi_2^2$$

is transformed into

$$\mathbf{A}(x, x) = \eta_1^2 - \eta_2^2$$

by the coordinate transformation

$$\eta_1 = \xi_1, \qquad \eta_2 = i\xi_2.$$

c. Given a symmetric Hermitian quadratic form $\mathbf{A}(x, x)$ in a space $\mathbf{C}_n$, a canonical basis can always be found such that the corresponding canonical coefficients can only take the values ± 1. In fact, having reduced the form $\mathbf{A}(x, x)$ to the form

$$\mathbf{A}(x, x) = \lambda_1 |\eta_1|^2 + \cdots + \lambda_p |\eta_p|^2 - \mu_1 |\eta_{p+1}|^2 - \cdots - \mu_q |\eta_{p+q}|^2,$$

where the numbers $\lambda_1, \ldots, \lambda_p, \mu_1, \ldots, \mu_q$ are all positive, we make another coordinate transformation

$$\tau_1 = \sqrt{\lambda_1}\, \eta_1, \ldots, \tau_p = \sqrt{\lambda_p}\, \eta_p, \tau_{p+1} = \sqrt{\mu_1}\, \eta_{p+1}, \ldots, \tau_{p+q} = \sqrt{\mu_q}\, \eta_{p+q},$$

thereby reducing $\mathbf{A}(x, x)$ to the form

$$\mathbf{A}(x, x) = |\tau_1|^2 + \cdots + |\tau_p|^2 - |\tau_{p+1}|^2 - \cdots - |\tau_{p+q}|^2$$

(cf. Sec. 7.93).

9.17. a. The vector x_1 is said to be *conjugate to the vector y_1 with respect to the Hermitian bilinear form* $\mathbf{A}(x, y)$ if

$$\mathbf{A}(x_1, y_1) = 0.$$

If the vectors $x_1, x_2, \ldots, x_k$ are all conjugate to the vector y_1, then every vector of the linear manifold $\mathbf{L}(x_1, x_2, \ldots, x_k)$ spanned by $x_1, x_2, \ldots, x_k$ is also conjugate to y_1 (cf. Sec. 7.42c). In general, a vector y_1 conjugate to every vector of a subspace $\mathbf{C}' \subset \mathbf{C}$ is said to be *conjugate to the subspace* $\mathbf{C}'$. The set $\mathbf{C}''$ of all vectors $y_1 \in \mathbf{C}$ conjugate to the subspace $\mathbf{C}'$ is obviously a subspace of the space $\mathbf{C}$. This subspace $\mathbf{C}''$ is said to be *conjugate to* $\mathbf{C}'$.

A basis $e_1, e_2, \ldots, e_n$ of the space $\mathbf{C}_n$ is said to be a *canonical basis* of the form $\mathbf{A}(x, y)$ if

$$\mathbf{A}(e_i, e_k) = 0 \quad \text{for} \quad i \neq k.$$

Every symmetric Hermitian bilinear form $\mathbf{A}(x, y)$ *has a canonical basis.* In fact, let $e_1, e_2, \ldots, e_n$ be a basis in which the corresponding quadratic form $\mathbf{A}(x, x)$ can be written in the canonical form

$$\mathbf{A}(x, x) = \sum_{i=1}^{n} \lambda_i \xi_i \bar{\xi}_i,$$

where

$$x = \sum_{i=1}^{n} \xi_i e_i.$$

Then, by Sec. 9.15b, the bilinear form $\mathbf{A}(x, y)$ takes the canonical form

$$\mathbf{A}(x, y) = \sum_{i=1}^{n} \lambda_i \xi_i \bar{\eta}_i$$

in this basis, where

$$y = \sum_{i=1}^{n} \eta_i e_i,$$

and hence

$$\mathbf{A}(e_i, e_k) = \begin{cases} \lambda_i & \text{if} \quad i = k, \\ 0 & \text{if} \quad i \neq k. \end{cases}$$

b. Suppose the principal descending minors $\delta_1, \delta_2, \ldots, \delta_{n-1}$ of the matrix $\|a_{ik}\|$ of a symmetric Hermitian quadratic form $\mathbf{A}(x, x)$ are all nonvanishing. Then, just as in Sec. 7.52, we can use Jacobi's method to construct a canonical basis for $\mathbf{A}(x, x)$, and the canonical coefficients of $\mathbf{A}(x, x)$ are given by the same formulas

$$\lambda_1 = \delta_1, \qquad \lambda_2 = \frac{\delta_2}{\delta_1}, \ldots, \lambda_n = \frac{\delta_n}{\delta_{n-1}}$$

($\delta_n = \det \|a_{ik}\|$) as on p. 195.

c. A symmetric Hermitian bilinear form $\mathbf{A}(x, y)$ is said to be *positive definite* if $\mathbf{A}(x, x) > 0$ for every $x \neq 0$. Just as in the real case (Sec. 7.94), an equivalent condition is that all the canonical coefficients of $\mathbf{A}(x, x)$ be positive, or alternatively, that $p = n$, where p is the positive index of inertia of the form $\mathbf{A}(x, x)$.

Just as in Theorem 7.96, a necessary and sufficient condition for the form $\mathbf{A}(x, y)$ to be positive definite is that

$$\delta_1 > 0, \delta_2 > 0, \ldots, \delta_n > 0$$

(*Sylvester's conditions*). The proof given on p. 209 carries over without change to the complex case.

9.18. a. Given a nonsingular symmetric Hermitian bilinear form (x, y), we can introduce the concept of the adjoint of a linear operator (with respect to the form (x, y)), just as in Sec. 7.6. First we note that if $\mathbf{A}$ and $\mathbf{B}$ are linear operators in the space $\mathbf{C}_n$, then the forms

$$\mathbf{A}(x, y) = (\mathbf{A}x, y), \qquad \mathbf{B}(x, y) = (x, \mathbf{B}y)$$

are Hermitian bilinear forms, whose matrices are related to the matrices of the operators $\mathbf{A}$ and $\mathbf{B}$ (in any canonical basis of the form (x, y) with canonical coefficients ε_j) by the formulas

$$a_{jm} = \varepsilon_m a_m^{(j)}, \qquad b_{jm} = \varepsilon_j \overline{b_j^{(m)}}$$

(the notation is the same as in Sec. 7.61). Conversely, given two Hermitian bilinear forms $\mathbf{A}(x, y)$ and $\mathbf{B}(x, y)$, then, just as in Sec. 7.62, there exist unique linear operators $\mathbf{A}$ and $\mathbf{B}$ such that

$$\mathbf{A}(x, y) = (\mathbf{A}x, y), \qquad \mathbf{B}(x, y) = (x, \mathbf{B}y).$$

b. It follows, just as in Sec. 7.63, that given any linear operator $\mathbf{A}$ acting in the space $\mathbf{C}_n$, there exists a unique linear operator $\mathbf{A}^*$ acting in $\mathbf{C}_n$ such that

$$(\mathbf{A}x, y) = (x, \mathbf{A}^*y)$$

for arbitrary $x, y \in \mathbf{C}_n$. The matrices $\|a_m^{(j)}\|$ and $\|a_m^{*(j)}\|$ of the operators $\mathbf{A}$ and $\mathbf{A}^*$ in any canonical basis of the form (x, y) with canonical coefficients ε_j are related by the formula

$$a_j^{*(m)} = \frac{\varepsilon_m}{\varepsilon_j} \overline{a_m^{(j)}}.$$

The operator $\mathbf{A}^*$ is called the *adjoint (or Hermitian conjugate) of the operator* $\mathbf{A}$ *with respect to the form* (x, y).

c. The operation leading from an operator $\mathbf{A}$ to its adjoint $\mathbf{A}^*$ has the following properties (cf. Sec. 7.64):

1) $(\mathbf{A}^*)^* = \mathbf{A}$ for every operator $\mathbf{A}$;
2) $(\mathbf{A} + \mathbf{B})^* = \mathbf{A}^* + \mathbf{B}^*$ for every pair of operators $\mathbf{A}$ and $\mathbf{B}$;
3) $(\lambda\mathbf{A})^* = \bar{\lambda}\mathbf{A}^*$ for every operator $\mathbf{A}$ and every number $\lambda \in C$;
4) $(\mathbf{AB})^* = \mathbf{B}^*\mathbf{A}^*$ for every pair of operators $\mathbf{A}$ and $\mathbf{B}$.

9.19. a. As in Sec. 7.71, two complex spaces $\mathbf{C}'$ and $\mathbf{C}''$ equipped with nonsingular symmetric Hermitian bilinear forms $\mathbf{A}(x', y')$ and $\mathbf{A}(x'', y'')$, respectively, are said to be $\mathbf{A}$-*isomorphic* if the spaces $\mathbf{C}'$ and $\mathbf{C}''$ are isomorphic regarded as linear spaces over the field C (see Sec. 2.71) and if

$$\mathbf{A}(x', y') = \mathbf{A}(x'', y'')$$

for all corresponding pairs of elements $x', y' \in \mathbf{C}$ and $x'', y'' \in \mathbf{C}''$.

b. THEOREM. *Two finite-dimensional complex spaces* $\mathbf{C}'$ *and* $\mathbf{C}''$, *equipped with nonsingular symmetric Hermitian bilinear forms* $\mathbf{A}(x', y')$ *and* $\mathbf{A}(x'', y'')$, *respectively, are* $\mathbf{A}$-*isomorphic if and only if they have the same dimension and the indices of inertia* p', q' *of the form* $\mathbf{A}(x', y')$ *coincide with the corresponding indices of inertia* p'', q'' *of the form* $\mathbf{A}(x'', y'')$.

Proof. Precisely the same as that of the analogous proposition for real spaces (Theorem 7.93). ∎

c. In particular, two n-dimensional complex spaces $\mathbf{C}_n'$ and $\mathbf{C}_n''$, equipped with positive definite forms $\mathbf{A}(x', y')$ and $\mathbf{A}(x'', y'')$, respectively, are always $\mathbf{A}$-isomorphic (cf. Sec. 7.97).

9.2. The Scalar Product in a Complex Space

9.21. It will be recalled from Sec. 8.21 that the scalar product of two vectors x and y in a real space is taken to be any fixed symmetric positive definite bilinear form (x, y). The corresponding quadratic form (x, x) is then positive for every nonzero vector x, and can be used to define the length of x (see Sec. 8.31). In a complex space, any symmetric positive definite Hermitian bilinear form has the analogous property (see Sec. 9.17c). This leads to the following definition: A complex linear space $\mathbf{C}$ is said to be a *unitary space* if it is equipped with a symmetric positive definite Hermitian bilinear form (x, y), called the (*complex*) *scalar product* of the vectors x and

y, i.e., if there is a rule assigning to every pair of vectors $x, y \in \mathbf{C}$ a complex number (x, y) such that

a) $(y, x) = \overline{(x, y)}$ for every $x, y \in \mathbf{C}$;
b) $(x, y + z) = (x, y) + (x, z)$ for every $x, y, z \in \mathbf{C}$;
c) $(\lambda x, y) = \lambda(x, y)$ for every $x, y \in \mathbf{C}$ and every complex number λ;
d) $(x, x) > 0$ for every $x \neq 0$ and $(x, x) = 0$ for $x = 0$.

Axioms a)–c) imply the general formula

$$\left(\sum_{i=1}^{k} \alpha_i x_i, \sum_{j=1}^{m} \beta_j y_j\right) = \sum_{i=1}^{k} \sum_{j=1}^{m} \alpha_i \bar{\beta}_j (x_i, y_j),$$

where $x_1, \ldots, x_k, y_1, \ldots, y_m$ are arbitrary vectors of the space $\mathbf{C}$ and $\alpha_1, \ldots, \alpha_k, \beta_1, \ldots, \beta_m$ are arbitrary complex numbers.

9.22. *Examples*

a. In the n-dimensional space C_n (Sec. 2.15b) we define the scalar product of the vectors $x = (\xi_1, \xi_2, \ldots, \xi_n)$ and $y = (\eta_1, \eta_2, \ldots, \eta_n)$ by the formula

$$(x, y) = \xi_1 \bar{\eta}_1 + \xi_2 \bar{\eta}_2 + \cdots + \xi_n \bar{\eta}_n.$$

The reader can easily verify that axioms a)–d) are satisfied in this case.

b. In the space $C(a, b)$ of all continuous complex-valued functions on the interval $a \leqslant t \leqslant b$ (Sec. 2.15d) we define the scalar product of the functions $x = x(t)$ and $y = y(t)$ by the formula

$$(x, y) = \int_a^b x(t)\overline{y(t)}\, dt.$$

Axioms a)–d) are then immediate consequences of the basic properties of the integral.

9.23. Basic metric concepts. Next we introduce various metric concepts in a unitary space $\mathbf{C}$, just as was done in the case of a real Euclidean space (Sec. 8.3).

a. The length of a vector. As in the real case, by the *length* (or *norm*) of a vector x in a unitary space $\mathbf{C}$ we mean the quantity

$$|x| = +\sqrt{(x, x)}.$$

Every nonzero vector has a positive length, and the length of the zero vector equals 0. For any complex λ, we have the equality

$$|\lambda x| = \sqrt{(\lambda x, \lambda x)} = \sqrt{\lambda \bar{\lambda}(x, x)} = |\lambda| \sqrt{(x, x)} = |\lambda|\, |x|,$$

which shows that the length of a vector x multiplied by a numerical factor λ equals the absolute value of λ times the length of x.

A vector x of length 1 is said to be a *unit vector*. Every nonzero vector can be *normalized*, i.e., multiplied by a number λ such that the result is a unit vector. In fact, we need only choose λ such that

$$|\lambda| = \frac{1}{|x|},$$

just as on p. 217.

The set of all vectors $x \in \mathbf{C}$ such that $|x| \leqslant 1$ is called the *unit ball* in $\mathbf{C}$, while the set of all $x \in \mathbf{C}$ such that $|x| = 1$ is called the *unit sphere*.

b. The Schwarz inequality. The inequality

$$|(x, y)| \leqslant |x|\,|y| \tag{9}$$

holds for every pair of vectors x and y in $\mathbf{C}$. The idea of the proof is the same as in the real case (Sec. 8.33), except that we must now be careful about complex numbers. The inequality (9) is obvious if $(x, y) = 0$. Thus let $(x, y) \neq 0$. Clearly,

$$(\lambda x - y, \lambda x - y) \geqslant 0$$

for arbitrary complex λ. Expanding the left-hand side, we get

$$|\lambda|^2 (x, x) - \lambda(x, y) - \bar{\lambda}\,\overline{(x, y)} + (y, y) \geqslant 0. \tag{10}$$

Let γ be the line in the complex plane determined by the origin and the complex number (x, y), and let γ' be the line symmetric to γ with respect to the real axis. Suppose λ varies over the line γ', so that $\lambda = tz_0$, where t is real and

$$z_0 = \frac{\overline{(x, y)}}{|(x, y)|}$$

is the unit vector determining the direction of γ'. Then

$$\lambda(x, y) = t\,|(x, y)|$$

is real, and hence

$$\bar{\lambda}\,\overline{(x, y)} = \lambda(x, y),$$

so that the inequality (10) becomes

$$t^2(x, x) - 2t\,|(x, y)| + (y, y) \geqslant 0. \tag{11}$$

The same argument as in Sec. 8.33 now leads to the desired inequality (9).

If equality holds in (9), then the trinomial in the left-hand side of (11) has a unique real root t_0 (of multiplicity two). Replacing tz_0 by λ, we find that the trinomial in the left-hand side of (10) has the root $\lambda_0 = t_0 z_0$. Therefore

$$(\lambda_0 x - y, \lambda_0 x - y) = 0$$

and hence $y = \lambda_0 x$, so that the vectors x and y differ only by a (complex) numerical factor.

c. Orthogonality. Although the concept of the angle between two vectors is not introduced in a unitary space, we still consider the case where two vectors x and y are *orthogonal*, which means, just as in the real case, that

$$(x, y) = 0.$$

If x and y are orthogonal, then obviously

$$(y, x) = \overline{(x, y)} = 0.$$

It is easily verified that the analogues of Lemmas 8.36a–b and the Pythagorean theorem (Sec. 8.37) remain valid for orthogonal vectors in a unitary space. Moreover, the analogue of the expansion theorem of Sec. 8.51 also holds, i.e., given a finite-dimensional subspace $\mathbf{C}' \subset \mathbf{C}$ and a vector f which is in general not an element of $\mathbf{C}'$, there exists a unique representation

$$f = g + h,$$

where $g \in \mathbf{C}'$ and h is orthogonal to $\mathbf{C}'$. The set of all vectors h orthogonal to the subspace $\mathbf{C}'$ is itself a subspace, which we call the *orthogonal complement* of the subspace $\mathbf{C}'$ and denote by $\mathbf{C}''$. Just as in Sec. 8.51, we see that the original space $\mathbf{C}$ is the direct sum of the subspace $\mathbf{C}'$ and its orthogonal complement $\mathbf{C}''$.

d. The triangle inequalities. If x and y are two vectors in a unitary space $\mathbf{C}$, then, by Schwarz's inequality (9),

$$|x + y|^2 = (x + y, x + y) = (x, x) + (x, y) + \overline{(x, y)} + (y, y)$$

$$\begin{cases} \leqslant (x, x) + 2\,|(x, y)| + (y, y) \leqslant (|x| + |y|)^2, \\ \geqslant (x, x) - 2\,|(x, y)| + (y, y) \geqslant (|x| - |y|)^2, \end{cases}$$

or

$$|x + y| \leqslant |x| + |y|, \tag{12}$$

$$|x + y| \geqslant \big||x| - |y|\big|. \tag{13}$$

As in the real case, these inequalities are called the *triangle inequalities*.

9.24. Orthogonal bases in an n-dimensional unitary space $\mathbf{C}_n$. According to Sec. 9.16a, the symmetric Hermitian bilinear form (x, y) has a canonical basis $e_1, e_2, \ldots, e_n$ in the n-dimensional space $\mathbf{C}_n$, and in this case the condition

$$(e_i, e_k) = 0 \qquad (i \neq k)$$

for the basis to be canonical reduces to the orthogonality condition. Moreover, the orthogonal basis vectors $e_1, e_2, \ldots, e_n$ can be regarded as normalized, so that

$$|e_1| = |e_2| = \cdots = |e_n| = 1.$$

Let

$$x = \sum_{k=1}^{n} \xi_k e_k, \qquad y = \sum_{k=1}^{n} \eta_k e_k$$

be any two vectors in $\mathbf{C}_n$, with components ξ_k, η_k $(k = 1, \ldots, n)$ with respect to the basis $e_1, e_2, \ldots, e_n$. We then get the following formula for the scalar product (x, y) in terms of the components of x and y:

$$(x, y) = \sum_{k=1}^{n} \xi_k \bar{\eta}_k.$$

9.25. a. As shown in Sec. 9.18a, the formula

$$\mathbf{A}(x, y) = (\mathbf{A}x, y)$$

establishes a one-to-one correspondence between Hermitian bilinear forms $\mathbf{A}(x, y)$ and linear operators $\mathbf{A}$ acting in the space $\mathbf{C}_n$. In any orthonormal basis $e_1, e_2, \ldots, e_n$ of the space $\mathbf{C}_n$, the matrix $\|a_{jm}\|$ of the form $\mathbf{A}(x, y)$ and the matrix $\|a_j^{(m)}\|$ of the operator $\mathbf{A}$, where

$$a_{jm} = \mathbf{A}(e_j, e_m),$$

$$\mathbf{A}e_j = \sum_{k=1}^{n} a_k^{(j)} e_k,$$

are related by the formula

$$a_{jm} = a_m^{(j)}.$$

b. Let $\mathbf{A}$ be any linear operator acting in the space $\mathbf{C}_n$. Then, as shown in Sec. 9.18b, there is a unique operator $\mathbf{A}^*$, the adjoint of $\mathbf{A}$ with respect to the scalar product (x, y), such that

$$(\mathbf{A}x, y) = (x, \mathbf{A}^*y)$$

for arbitrary $x, y \in \mathbf{C}_n$. Since any orthonormal basis is a canonical basis for the form (x, y), with canonical coefficients $\varepsilon_j = 1$, the matrices $\|a_m^{(j)}\|$ and $\|a_m^{*(j)}\|$ of the operators $\mathbf{A}$ and $\mathbf{A}^*$ are related by the formula

$$a_j^{*(m)} = \overline{a_m^{(j)}}.$$

In other words, the matrix of the operator $\mathbf{A}^*$ is obtained from that of the operator $\mathbf{A}$ by "Hermitian transposition," i.e., by transposition followed by replacing all elements of the matrix by their complex conjugates. Correspondingly, we call the matrix of $\mathbf{A}^*$ the *Hermitian conjugate* (or *adjoint*) of that of $\mathbf{A}$.

c. As in Sec. 8.96a, *if the subspace* $\mathbf{C}' \subset \mathbf{C}$ *is invariant under the operator* $\mathbf{A}$, *then the orthogonal complement of* $\mathbf{C}'$ *is invariant under the adjoint operator* $\mathbf{A}^*$.

9.26. A coordinate transformation in an n-dimensional unitary space $\mathbf{C}_n$ leading from one orthonormal basis to another is called a *unitary transformation*. Unitary transformations are analogous to orthogonal transformations in a Euclidean space (see Sec. 8.94). If $e_1, \ldots, e_n$ and $f_1, \ldots, f_n$ are orthonormal bases in $\mathbf{C}_n$ and if $U = \|u_k^{(i)}\|$ is the matrix of the corresponding unitary transformation, so that

$$f_i = \sum_{k=1}^{n} u_k^{(i)} e_k,$$

then obviously

$$(f_i, f_j) = \sum_{k=1}^{n} u_k^{(i)} \overline{u_k^{(j)}} = \begin{cases} 1 & \text{if} \quad i = j, \\ 0 & \text{if} \quad i \neq j. \end{cases} \tag{14}$$

Conversely, if the numbers $u_k^{(i)}$ satisfy the conditions (14), then the matrix $\|u_k^{(i)}\|$ is a *unitary matrix*, i.e., the matrix of a unitary transformation.

The linear operator $\mathbf{U}$ corresponding to a unitary matrix is called a *unitary operator*. Just like an isometric operator in a real space, a unitary operator in a complex space does not "change the metric." In other words, if

$$x = \sum_{i=1}^{n} \xi_i e_i, \qquad y = \sum_{j=1}^{n} \eta_j e_j,$$

then

$$(\mathbf{U}x, \mathbf{U}y) = \sum_{i,j=1}^{n} \xi_i \bar{\eta}_j (\mathbf{U}e_i, \mathbf{U}e_j) = \sum_{i,j=1}^{n} \xi_i \bar{\eta}_j (f_i, f_j) = \sum_{i=1}^{n} \xi_i \bar{\eta}_i = (x, y).$$

The matrix V of the inverse transformation from the basis $f_1, \ldots, f_n$ to the basis $e_1, \ldots, e_n$ is also unitary. Moreover, if $V = \|v_k^{(i)}\|$, we have

$$u_k^{(i)} = (f_i, e_k), \qquad v_k^{(i)} = (e_i, f_k) = \overline{u_i^{(k)}}.$$

Thus the inverse of a unitary matrix is obtained by first transposing and then going over to complex conjugate elements. Therefore

$$\mathbf{U}^{-1} = \mathbf{U}^*$$

for a unitary operator $\mathbf{U}$, or equivalently,

$$\mathbf{U}^*\mathbf{U} = \mathbf{U}\mathbf{U}^* = \mathbf{E}.$$

9.3. Normal Operators

9.31. Definition. An operator $\mathbf{A}$ acting in an n-dimensional unitary space $\mathbf{C}_n$ is said to be *normal* if it commutes with its own adjoint, i.e., if

$$\mathbf{A}^*\mathbf{A} = \mathbf{A}\mathbf{A}^* \tag{15}$$

(cf. Sec. 8.92c). An example of a normal operator is given by any operator **A** whose eigenvectors $e_1, \ldots, e_n$, satisfying the relation

$$\mathbf{A}e_j = \lambda_j e_j \qquad (j = 1, \ldots, n),$$

form an orthogonal basis in $\mathbf{C}_n$. In fact, the matrix of the operator **A** in the basis $e_1, \ldots, e_n$ is then of the form

$$\begin{Vmatrix} \lambda_1 & 0 & \ldots & 0 \\ 0 & \lambda_2 & \ldots & 0 \\ \cdot & \cdot & \ldots & \cdot \\ 0 & 0 & \ldots & \lambda_n \end{Vmatrix}. \tag{16}$$

But, by Sec. 9.25, the matrix of the operator $\mathbf{A}^*$ in the same basis $e_1, \ldots, e_n$ is just

$$\begin{Vmatrix} \bar{\lambda}_1 & 0 & \ldots & 0 \\ 0 & \bar{\lambda}_2 & \ldots & 0 \\ \cdot & \cdot & \ldots & \cdot \\ 0 & 0 & \ldots & \bar{\lambda}_n \end{Vmatrix}, \tag{17}$$

from which is is obvious that the operators **A** and $\mathbf{A}^*$ commute.

9.32. THEOREM. *Every eigenvector x of a normal operator* **A** *with eigenvalue λ is an eigenvector of the operator* $\mathbf{A}^*$ *with eigenvalue* $\bar{\lambda}$.

Proof. Let $\mathbf{P} \subset \mathbf{C}_n$ be the subspace consisting of all eigenvectors of the operator **A** with eigenvalue λ. If $x \in \mathbf{P}$, then

$$\mathbf{A}\mathbf{A}^*x = \mathbf{A}^*\mathbf{A}x = \mathbf{A}^*(\lambda x) = \lambda\mathbf{A}^*x,$$

which implies $\mathbf{A}^*x \in P$. Hence **P** is invariant under the operator $\mathbf{A}^*$. Moreover,

$$(\mathbf{A}^*x, y) = (x, \mathbf{A}y) = (x, \lambda y) = (\bar{\lambda}x, y)$$

for arbitrary $x, y \in \mathbf{P}$, and hence

$$\mathbf{A}^*x = \bar{\lambda}x. \quad \blacksquare$$

9.33. a. THEOREM. *Given any normal operator* **A** *acting in a unitary space* $\mathbf{C}_n$, *there exists an orthonormal basis* $e_1, \ldots, e_n$ *in* $\mathbf{C}_n$ *consisting of eigenvectors of* **A**.

Proof. The normal operator **A**, like every linear operator in the space $\mathbf{C}_n$, has an eigenvector (see Sec. 4.95b). Let e_1 be an eigenvector of **A** with eigenvalue λ, and let $\mathbf{P} \subset \mathbf{C}_n$ be the subspace consisting of *all* eigenvectors of **A** with this eigenvalue λ. If **P** is the whole space $\mathbf{C}_n$, then we need only arbitrarily augment e_1 with vectors $e_2, \ldots, e_n$ to make an orthonormal basis for $\mathbf{C}_n$, thereby proving the theorem. Thus suppose $\mathbf{P} \neq \mathbf{C}_n$, and let $\mathbf{Q}$ be the orthogonal complement of $\mathbf{P}$ in $\mathbf{C}_n$. The subspace $\mathbf{P}$ is invariant under the operator $\mathbf{A}^*$, as in the proof of Theorem 9.32 (in fact, $\mathbf{A}^*$ carries every vector $x \in \mathbf{P}$ into the vector $\bar{\lambda}x$). It follows that $\mathbf{Q}$ is invariant under the operator $\mathbf{A}$ itself, because of Sec. 9.25c and the fact that $(\mathbf{A}^*)^* = \mathbf{A}$ (see Sec. 9.18c). We can now prove the theorem by induction. In fact, suppose the theorem is true for every space $\mathbf{C}_n$ of dimension $n \leqslant k$. Then it is also true for $\mathbf{C}_{k+1}$, since to get an orthonormal basis for $\mathbf{C}_{k+1}$ consisting of eigenvectors of $\mathbf{A}$, we need only choose such a basis in the subspace $\mathbf{Q}$ (such exists by the induction hypothesis, since the dimension of $\mathbf{Q}$ is $\leqslant k$) and then augment this basis by any orthonormal basis in $\mathbf{P}$. The proof is now complete, since the theorem is obviously true for the one-dimensional space $\mathbf{C}_1$. ∎

b. It follows from Theorem 9.33a that every normal operator $\mathbf{A}$ is diagonalizable (see Sec. 4.72f). In fact, $\mathbf{A}$ has the diagonal matrix

$$A = \left\| \begin{matrix} \lambda_1 & 0 & \cdots & 0 \\ 0 & \lambda_2 & \cdots & 0 \\ \cdot & \cdot & \cdots & \cdot \\ 0 & 0 & \cdots & \lambda_n \end{matrix} \right\|$$

in the orthonormal basis constructed in Theorem 9.33a, consisting of eigenvectors of $\mathbf{A}$. The eigenvalues of $\mathbf{A}$ lie on the principal diagonal of this matrix, each appearing a number of times equal to the dimension of the corresponding characteristic subspace (cf. p. 110). Hence the characteristic polynomial $\det \|A - \lambda E\|$ of the operator $\mathbf{A}$, which as we know is independent of the choice of basis (see Sec. 5.53), has the form

$$\det \|A - \lambda E\| = \prod_{k=1}^{m} (\lambda_k - \lambda)^{r_k}, \qquad \sum_{k=1}^{m} r_k = n, \tag{18}$$

where $\lambda_1, \ldots, \lambda_m$ are the distinct eigenvalues of the operator $\mathbf{A}$ and $r_1, \ldots, r_m$ are the dimensions of the corresponding characteristic subspaces.

c. On the other hand, suppose it is known that a normal operator $\mathbf{A}$ has a characteristic polynomial of the form

$$\det \|A - \lambda E\| = \prod_{k=1}^{s} (\mu_k - \lambda)^{p_k}, \qquad \sum_{k=1}^{s} p_k = n, \tag{19}$$

where $\mu_1, \ldots, \mu_s$ are distinct complex numbers and $p_1, \ldots, p_k$ are certain positive integers (multiplicities). Then it can be asserted that *the operator* $\mathbf{A}$ *has an orthonormal basis consisting of eigenvectors with eigenvalues* $\mu_1, \ldots, \mu_s$, *where the dimension of the characteristic subspace corresponding to the eigenvalue* μ_j *is just* p_j. In fact, the polynomials (18) and (19) must coincide, by the uniqueness of the characteristic polynomial. But then our assertion follows from the familiar theorem on the uniqueness of the factorization of a polynomial.

9.34. Self-adjoint operators. An operator **A** acting in a unitary space **C** is said to be *self-adjoint* if $\mathbf{A}^* = \mathbf{A}$, i.e., if

$$(\mathbf{A}x, y) = (x, \mathbf{A}y) \tag{20}$$

for arbitrary vectors $x, y \in \mathbf{C}$. Note that **A** is self-adjoint if and only if the bilinear form $(\mathbf{A}x, y)$ corresponding to **A** is Hermitian–symmetric.† According to Sec. 9.25, the matrix of a self-adjoint operator **A** in any orthonormal basis coincides with its own Hermitian conjugate, i.e., with the matrix obtained from that of **A** by transposition followed by taking complex conjugates of all elements. Conversely, every operator **A** with a Hermitian–symmetric matrix (i.e., a matrix equal to its own Hermitian conjugate) in some orthonormal basis is self-adjoint.

Since a self-adjoint operator **A** is obviously normal, it follows from Theorem 9.33a that there exists an orthonormal basis $e_1, \ldots, e_n$ in the space $\mathbf{C}_n$ in which the matrix of the operator **A** takes the form (16) and that of $\mathbf{A}^*$ takes the form (17). Hence $\bar{\lambda}_j = \lambda_j$ $(j = 1, \ldots, n)$, since $\mathbf{A}^* = \mathbf{A}$, i.e., the numbers λ_j are all real. This proves the following

THEOREM. *Given any self-adjoint operator* **A** *in a unitary space* $\mathbf{C}_n$, *there exists an orthonormal basis* $e_1, \ldots, e_n$ *consisting of eigenvectors of* **A** *with eigenvalues that are all real.*

Conversely, *every linear operator* **A** *in the space* $\mathbf{C}_n$ *with the indicated property is self-adjoint.* In fact, **A** is normal by Sec. 9.31, and comparing (16) and (17) we find that $\mathbf{A}^* = \mathbf{A}$, since the numbers λ_j are all real.

9.35. Antiself-adjoint operators. An operator **A** acting in a unitary space $\mathbf{C}_n$ is said to be *antiself-adjoint* if $\mathbf{A}^* = -\mathbf{A}$. The matrix of an antiself-adjoint operator **A** in any orthonormal basis $e_1, \ldots, e_n$ has the following characteristic property:

$$a_{ik} = (\mathbf{A}e_i, e_k) = (e_i, \mathbf{A}^*e_k) = (e_i, -\mathbf{A}e_k) = -\overline{(\mathbf{A}e_k, e_i)} = -\bar{a}_{ki}$$
$$(i, k = 1, \ldots, n).$$

An antiself-adjoint operator **A** is obviously normal. Applying Theorem 9.33a, we find that there exists an orthonormal basis $e_1, \ldots, e_n$ in the space $\mathbf{C}_n$ in which the matrix of the operator **A** takes the form (16) and that of $\mathbf{A}^*$ takes the form (17). Hence $\bar{\lambda}_j = -\lambda_j$ $(j = 1, \ldots, n)$, since $\mathbf{A}^* = -\mathbf{A}$,

† In fact, the condition $(\mathbf{A}y, x) = \overline{(\mathbf{A}x, y)}$ is equivalent to (20). For this reason, a self-adjoint operator might also be called *Hermitian–symmetric*.

i.e., the numbers λ_j are all purely imaginary. This proves the following

THEOREM. *Given any antiself-adjoint operator* **A** *in a unitary space* $\mathbf{C}_n$, *there exists an orthonormal basis* $e_1, \ldots, e_n$ *consisting of eigenvectors of* **A** *with eigenvalues that are all purely imaginary.*

Conversely, *every linear operator* **A** *in the space* $\mathbf{C}_n$ *with the indicated property is antiself-adjoint.*

9.36. As in Sec. 9.26, an operator **U** acting in a unitary space $\mathbf{C}_n$ is said to be *unitary* if $\mathbf{U}^*\mathbf{U} = \mathbf{U}\mathbf{U}^* = \mathbf{E}$. In particular, *every unitary operator is normal.* Applying Theorem 9.33a, we find that there exists an orthonormal basis $e_1, \ldots, e_n$ in the space $\mathbf{C}_n$ in which the matrix of the operator **U** takes the form (16) and that of $\mathbf{U}^*$ takes the form (17). Hence $\bar{\lambda}_j\lambda_j = 1$ $(j = 1, \ldots, n)$, since $\mathbf{U}^*\mathbf{U} = \mathbf{E}$, or equivalently,

$$|\lambda_j| = 1 \qquad (j = 1, \ldots, n).$$

This proves the following

THEOREM. *Given any unitary operator* **U** *in a unitary space* $\mathbf{C}_n$, *there exists an orthonormal basis* $e_1, \ldots, e_n$ *consisting of eigenvectors of the operator* **U** *with eigenvalues that are all of absolute value* 1.

Conversely, *every linear operator* **U** *in the space* $\mathbf{C}_n$ *with the indicated property is unitary.*

9.4. Applications to Operator Theory in Euclidean Space

9.41. Embedding of a Euclidean space in a unitary space. As in Sec. 8.21, let **R** be a (real) Euclidean space with scalar product (x, y). Consider the complex space **C** consisting of the formal sums $x + iy$ where $x, y \in \mathbf{R}$, with the following natural operations of addition and multiplication by arbitrary complex numbers:

$$(x_1 + iy_1) + (x_2 + iy_2) = (x_1 + x_2) + i(y_1 + y_2),$$
$$(\alpha + i\beta)(x + iy) = (\alpha x - \beta y) + i(\alpha y + \beta x).$$

Then it is easily verified that **C** has all the properties of a complex linear space.

We now identify the vectors $x + i0$ with the vectors $x \in \mathbf{R}$, calling them *real vectors* of the space **C**. The vectors $0 + iy$ will be denoted simply by iy and called *purely imaginary vectors*. By the *complex conjugate* of the vector $x + iy$, written $\overline{x + iy}$, we mean the vector $x - iy$.

Next we introduce a scalar product in the space **C**, defined by the formula

$$(x_1 + iy_1, x_2 + iy_2) = [(x_1, x_2) + (y_1, y_2)] + i[(y_1, x_2) - (x_1, y_2)].$$

It is easily verified that this scalar product satisfies axioms a)–d) of Sec. 9.21. In particular,

$$(x + iy, x + iy) = (x, x) + (y, y).$$

Thus the space **C** contains the space **R** as a subset, equipped with the same scalar product, and subject to the same operations of addition and multiplication by real numbers. Note that every orthonormal system (or basis) $e_1, \ldots, e_n$ in the space **R** is also an orthonormal system (or basis) in the space **C**.

9.42. Every linear operator **A** specified in the space **R** can be extended into the space **C** by the formula

$$\hat{\mathbf{A}}(x + iy) = \mathbf{A}x + i\mathbf{A}y, \tag{21}$$

where the operator $\hat{\mathbf{A}}$ is obviously a linear operator in the space **C**. The matrix of the operator $\hat{\mathbf{A}}$ in the space **C** relative to a basis $e_1, \ldots, e_n \in \mathbf{R}$ coincides with the matrix of the operator **A** in the space **R** relative to the same basis, since, according to (21),

$$\hat{\mathbf{A}}e_j = \mathbf{A}e_j \qquad (j = 1, \ldots, n).$$

This extension from **A** to $\hat{\mathbf{A}}$ preserves algebraic relations between linear operators, i.e., if $\mathbf{A} + \mathbf{B} = \mathbf{D}$ in the space **R**, then $\hat{\mathbf{A}} + \hat{\mathbf{B}} = \hat{\mathbf{D}}$ in the space **C**, while if $\mathbf{AB} = \mathbf{D}$ in the space **R**, then $\hat{\mathbf{A}}\hat{\mathbf{B}} = \hat{\mathbf{D}}$ in the space **C**. This follows for example from the fact that matrices are preserved under the extension from **A** to $\hat{\mathbf{A}}$.

9.43. Let $\mathbf{A}'$ be the adjoint of the operator **A** in the real space **R** (see Sec. 8.91). Then the extension $\hat{\mathbf{A}}'$ of the operator $\mathbf{A}'$ into the space **C** is just the operator $\mathbf{A}^*$ adjoint to the extension $\hat{\mathbf{A}}$ of **A**. In fact, given arbitrary vectors $z = x + iy$, $w = u + iv \in \mathbf{C}$, we have

$$\begin{aligned}(\mathbf{A}'(x + iy), u + iv) &= (\mathbf{A}'x, u) + i(\mathbf{A}'y, u) - i(\mathbf{A}'x, v) + (\mathbf{A}'y, v)\\ &= (x, \mathbf{A}u) + i(y, \mathbf{A}u) - i(x, \mathbf{A}v) + (y, \mathbf{A}v)\\ &= (x + iy, \hat{\mathbf{A}}(u + iv)),\end{aligned}$$

as required.

In particular, the extension of a symmetric operator ($\mathbf{A}' = \mathbf{A}$) is a self-adjoint operator ($\mathbf{A}^* = \mathbf{A}$), the extension of an antisymmetric operator ($\mathbf{A}' = -\mathbf{A}$) is an antiself-adjoint operator ($\mathbf{A}^* = -\mathbf{A}$), and the extension of an isometric operator ($\mathbf{U}' = \mathbf{U}^{-1}$) is a unitary operator ($\mathbf{U}^* = \mathbf{U}^{-1}$). Finally, the extension of a normal operator ($\mathbf{A}'\mathbf{A} = \mathbf{A}\mathbf{A}'$) is again a normal operator ($\mathbf{A}^*\mathbf{A} = \mathbf{A}\mathbf{A}^*$).

9.44. Structure of a real normal operator. Let σ and τ be real numbers. Then the easily verified matrix equality

$$\left\| \begin{matrix} \sigma & \tau \\ -\tau & \sigma \end{matrix} \right\| \left\| \begin{matrix} \sigma & -\tau \\ \tau & \sigma \end{matrix} \right\| = \left\| \begin{matrix} \sigma & -\tau \\ \tau & \sigma \end{matrix} \right\| \left\| \begin{matrix} \sigma & \tau \\ -\tau & \sigma \end{matrix} \right\| = \left\| \begin{matrix} \sigma^2 + \tau^2 & 0 \\ 0 & \sigma^2 + \tau^2 \end{matrix} \right\| \qquad (22)$$

shows that the matrix

$$\left\| \begin{matrix} \sigma & \tau \\ -\tau & \sigma \end{matrix} \right\|$$

commutes with its own transpose (and hence a fortiori with its own adjoint); more generally, the same is true of the quasi-diagonal (real) matrix

$$\left\| \begin{matrix} \begin{matrix} \sigma_1 & \tau_1 \\ -\tau_1 & \sigma_1 \end{matrix} & & & & & & & \\ & \begin{matrix} \sigma_2 & \tau_2 \\ -\tau_2 & \sigma_2 \end{matrix} & & & & & & \\ & & \ddots & & & & & \\ & & & \begin{matrix} \sigma_m & \tau_m \\ -\tau_m & \sigma_m \end{matrix} & & & & \\ & & & & \lambda_{m+1} & & & \\ & & & & & \lambda_{m+2} & & \\ & & & & & & \ddots & \\ & & & & & & & \lambda_r \end{matrix} \right\| \qquad (23)$$

of order $2m + r - m = m + r$.

THEOREM. *Given any normal operator* $\mathbf{A}$ *in a real Euclidean space* $\mathbf{R}_n$, *there exists an orthonormal basis* $f_1, \ldots, f_n \in \mathbf{R}_n$ *in which the matrix of* $\mathbf{A}$ *is of the form* (23), *with* $m + r = n$, *where the numbers* $\lambda_j = \sigma_j + i\tau_j$ $(j = 1, \ldots, m)$ *and* $\lambda_{m+1}, \ldots, \lambda_r$ *are uniquely determined by* $\mathbf{A}$. *In fact, these numbers are the roots of the characteristic equation*

$$\det \|A - \lambda E\| = 0, \qquad (24)$$

and each root of (24) *appears in the matrix* (23) *a number of times equal to its multiplicity.*

Proof. As in Sec. 9.41, we construct the unitary space $\mathbf{C}_n$ whose scalar product is the extension of the scalar product (x, y) defined in the space $\mathbf{R}_n$.

We then use (21) to extend the operators $\mathbf{A}$ and $\mathbf{A}'$ into the space $\mathbf{C}_n$. As shown in Sec. 9.43, the extensions of $\mathbf{A}$ and $\mathbf{A}'$ are the normal operator $\hat{\mathbf{A}}$ and its adjoint $\hat{\mathbf{A}}^*$. Let $\|a_{ik}\|$ denote the matrix of the operator $\mathbf{A}$ relative to any orthonormal basis $e_1, \ldots, e_n$ in the space $\mathbf{R}_n$ (the numbers a_{ik} are real).† Then the operator $\hat{\mathbf{A}}$ has the same matrix relative to the basis $e_1, \ldots, e_n$ in the whole space $\mathbf{C}_n$. Since the characteristic equation (24) has real coefficients, if λ_j is an imaginary root of (24), then so is the complex conjugate $\bar{\lambda}_j$. Bearing this in mind, we write the sequence of *distinct* roots of (24) in the form

$$\lambda_1, \bar{\lambda}_1, \ldots, \lambda_p, \bar{\lambda}_p, \lambda_{p+1}, \ldots, \lambda_q,$$

where the roots $\lambda_1, \ldots, \lambda_p$ are imaginary and the roots $\lambda_{p+1}, \ldots, \lambda_q$ are real. Then, by Sec. 9.33b, the space $\mathbf{C}_n$ can be represented as a direct sum of orthogonal subspaces

$$\Lambda_1, \bar{\Lambda}_1, \ldots, \Lambda_p, \bar{\Lambda}_p, \Lambda_{p+1}, \ldots, \Lambda_q,$$

where Λ_j consists of all eigenvectors of the operator $\mathbf{A}$ corresponding to the eigenvalue λ_j and $\bar{\Lambda}_j$ consists of all eigenvectors of $\hat{\mathbf{A}}$ corresponding to the eigenvalue $\bar{\lambda}_j$, while

$$\bar{\Lambda}_{p+1} = \Lambda_{p+1}, \ldots, \bar{\Lambda}_q = \Lambda_q.$$

If $z = x + iy \in \Lambda_j$, then the equation $\mathbf{A}z = \lambda_j z$ becomes

$$\sum_{k=1}^{n} a_{jk}\zeta_k = \lambda_j\zeta_j$$

in component form (with respect to the original basis $e_1, \ldots, e_n$), where

$$z = (\zeta_1, \ldots, \zeta_n) = (\xi_1 + i\eta_1, \ldots, \xi_n + i\eta_n).$$

Taking the complex conjugate and recalling that the numbers a_{jk} are real, we get

$$\sum_{k=1}^{n} a_{jk}\bar{\zeta}_k = \bar{\lambda}_j\bar{\zeta}_j$$

This means that the vector $\bar{z} = (\bar{\zeta}_1, \ldots, \bar{\zeta}_n)$ is also an eigenvector of the operator $\hat{\mathbf{A}}$ with eigenvalue $\bar{\lambda}_j$. It follows that the operation of taking the complex conjugate carries the space Λ_j into the space $\bar{\Lambda}_j$.

Now let $\lambda_1 = \sigma_1 + i\tau_1$, where $\tau_1 \neq 0$ since $\bar{\lambda}_1 \neq \lambda_1$, and let g_1 be any unit vector in Λ_1, so that $\bar{g}_1 \in \bar{\Lambda}_1$. Moreover, let

$$f_1 = \frac{1}{\sqrt{2}}(g_1 + \bar{g}_1), \qquad f_2 = \frac{1}{\sqrt{2}\,i}(g_1 - \bar{g}_1),$$

so that

$$g_1 = \frac{1}{\sqrt{2}}(f_1 + if_2), \qquad \bar{g}_1 = \frac{1}{\sqrt{2}}(f_1 - if_2),$$

† In the course of the proof, we will construct a new orthonormal basis $f_1, \ldots, f_n$ for $\mathbf{R}_n$ in which $\mathbf{A}$ has a matrix of the form (23).

where the vectors f_1 and f_2 are obviously real, and moreover orthonormal, since it follows from

$$(g_1, g_1) = (\bar{g}_1, \bar{g}_1) = 1, \qquad (g_1, \bar{g}_1) = 0$$

that

$$(f_1, f_1) = (f_2, f_2) = \frac{1}{2}[(g_1, g_1) + (\bar{g}_1, \bar{g}_1)] = 1,$$

$$(f_1, f_2) = -\frac{1}{2i}(g_1 + \bar{g}_1, g_1 - \bar{g}_1) = -\frac{1}{2i}[(g_1, g_1) - (\bar{g}_1, \bar{g}_1)] = 0.$$

Since

$$\begin{aligned} \mathbf{A}f_1 = \hat{\mathbf{A}}f_1 &= \frac{1}{\sqrt{2}}(\hat{\mathbf{A}}g_1 + \hat{\mathbf{A}}\bar{g}_1) = \frac{1}{\sqrt{2}}(\lambda_1 g_1 + \bar{\lambda}_1 \bar{g}_1) \\ &= \frac{1}{2}[(\sigma_1 + i\tau_1)(f_1 + if_2) + (\sigma_1 - i\tau_1)(f_1 - if_2)] = \sigma_1 f_1 - \tau_1 f_2, \\ \mathbf{A}f_2 = \hat{\mathbf{A}}f_2 &= \frac{1}{\sqrt{2}\,i}(\hat{\mathbf{A}}g_1 - \hat{\mathbf{A}}\bar{g}_1) = \frac{1}{\sqrt{2}\,i}(\lambda_1 g_1 - \bar{\lambda}_1 \bar{g}_1) = \tau_1 f_1 + \sigma_1 f_2, \end{aligned}$$

we see that the operator $\mathbf{A}$ transforms the plane of the vectors f_1, f_2 into itself and has the matrix

$$\left\| \begin{matrix} \sigma_1 & \tau_1 \\ -\tau_1 & \sigma_1 \end{matrix} \right\| \tag{25}$$

in the basis f_1, f_2. If the dimension of Λ_1 is greater than 1, we choose another unit vector $g_2 \in \Lambda_1$ orthogonal to g_1, with complex conjugate $\bar{g}_2 \in \bar{\Lambda}_1$ (the latter is automatically orthogonal to g_2). Repeating the above construction for g_2 and $\bar{g}_2$, we get a new pair of real vectors f_3, f_4 which are linear combinations of g_2, $\bar{g}_2$ and hence orthogonal to the vectors f_1, f_2 (themselves linear combinations of g_1, $\bar{g}_1$). Clearly $\mathbf{A}$ transforms the plane of the vectors f_3, f_4 into itself and has the same matrix (25). Continuing this construction, we eventually get $2m$ orthonormal real vectors $f_1, f_2, \ldots, f_{2m-1}, f_{2m}$, where m is the sum of the dimensions of the subspaces $\Lambda_1, \ldots, \Lambda_p$ and $\mathbf{A}$ transforms the plane of the vectors f_{2j-1}, f_{2j} into itself, with either the same matrix (25) or the analogous matrix obtained by replacing σ_1, τ_1 by σ_k, τ_k $(k = 2, \ldots, p)$.

Next consider the subspace Λ_{p+1} corresponding to the real root $\lambda_{p+1} = \bar{\lambda}_{p+1}$. The operation of taking the complex conjugate obviously carries the subspace Λ_{p+1} into itself. Let g be any vector in Λ_{p+1}, and let $\bar{g}$ be its complex conjugate. There are just two possibilities, namely, the vectors g and $\bar{g}$ are either linearly independent (in $\mathbf{C}_n$) or linearly dependent. If g and $\bar{g}$ are linearly independent, then so are the *real* vectors

$$f = \frac{1}{\sqrt{2}}(g + \bar{g}), \qquad f' = \frac{1}{\sqrt{2}\,i}(g - \bar{g}).$$

Like g and $\bar{g}$, these vectors belong to Λ_{p+1}, and hence are eigenvectors of the operator **A** with the same eigenvalue λ_{p+1}. On the other hand, if g and $\bar{g}$ are linearly dependent, then

$$\bar{g} = e^{2i\varphi}g \qquad (0 \leqslant \varphi < \pi),$$

since g and $\bar{g}$ have the same length. Therefore

$$e^{i\varphi}g = e^{-i\varphi}\bar{g} = \overline{e^{i\varphi}g},$$

so that the vector

$$f = e^{i\varphi}g$$

is real. Moreover, since f belongs to Λ_{p+1}, like g itself, f is an eigenvector of **A** with the same eigenvalue λ_{p+1}. Thus, in any event (continuing this construction if necessary), we can always find a basis in Λ_{p+1} consisting of real vectors. Applying the orthogonalization theorem (Theorem 8.61) to this basis, we finally get first an orthogonal and then an orthonormal basis in Λ_{p+1}. Clearly the operator **A** transforms Λ_{p+1} into itself and has the diagonal matrix

$$\left\|\begin{matrix} \lambda_{p+1} & 0 & \cdots & 0 \\ 0 & \lambda_{p+1} & \cdots & 0 \\ \cdot & \cdot & \cdots & \cdot \\ 0 & 0 & \cdots & \lambda_{p+1} \end{matrix}\right\| \tag{26}$$

in the orthonormal basis. Repeating this construction for the remaining subspaces $\Lambda_{p+1}, \ldots, \Lambda_q$, we eventually obtain a set of orthonormal vectors $f_{2m+1}, f_{2m+2}, \ldots, f_n$, which together with the previously constructed vectors $f_1, f_2, \ldots, f_{2m}$ form a full orthonormal basis for $\mathbf{R}_n$. To complete the proof, we need only take account of the special form of the typical blocks (25) and (26), compensating for the somewhat different indices in (23) which refer to roots which are not necessarily distinct. ∎

The geometric meaning of a normal operator can be deduced from this theorem. First we observe that the operator with matrix

$$\left\|\begin{matrix} \sigma & \tau \\ -\tau & \sigma \end{matrix}\right\|$$

in the basis f_1, f_2 can be interpreted as a rotation accompanied by an expansion in the plane of the vectors f_1, f_2. In fact, we need only note that

$$\left\|\begin{matrix} \sigma & \tau \\ -\tau & \sigma \end{matrix}\right\| = \sqrt{\sigma^2+\tau^2}\left\|\begin{matrix} \dfrac{\sigma}{\sqrt{\sigma^2+\tau^2}} & \dfrac{\tau}{\sqrt{\sigma^2+\tau^2}} \\ -\dfrac{\tau}{\sqrt{\sigma^2+\tau^2}} & \dfrac{\sigma}{\sqrt{\sigma^2+\tau^2}} \end{matrix}\right\| = M\left\|\begin{matrix} \cos\alpha & \sin\alpha \\ -\sin\alpha & \cos\alpha \end{matrix}\right\|,$$

$$M = \sqrt{\sigma^2+\tau^2}, \qquad \cos\alpha = \frac{\sigma}{\sqrt{\sigma^2+\tau^2}}, \qquad \sin\alpha = \frac{\tau}{\sqrt{\sigma^2+\tau^2}},$$

where the effect of the matrix

$$\begin{Vmatrix} \cos\alpha & \sin\alpha \\ -\sin\alpha & \cos\alpha \end{Vmatrix}$$

is to rotate every vector in the f_1, f_2 plane through the angle α, while M is clearly the expansion coefficient. Recalling (23), we now see that the total effect of the normal operator **A** is to produce rotations accompanied by expansions in m mutually orthogonal planes and expansions only (by factors of $\lambda_{m+1}, \ldots, \lambda_r$, respectively) in the $r - m$ directions orthogonal to these planes and to each other.†

9.45. The structure of a real symmetric operator. Let **A** be a symmetric operator acting in a real space $\mathbf{R}_n$, so that $\mathbf{A}' = \mathbf{A}$. Then the extension $\hat{\mathbf{A}}$ of the operator **A** into the unitary space $\mathbf{C}_n$ is self-adjoint, i.e., $\hat{\mathbf{A}}^* = \hat{\mathbf{A}}$. The eigenvalues $\lambda_1, \ldots, \lambda_n$ of a self-adjoint operator are all real (see Sec. 9.34). Hence there are no blocks of the form (25) in the representation (23), and all that remain are diagonal elements. This proves the following

THEOREM. *Given any symmetric operator* **A** *in a real Euclidean space* $\mathbf{R}_n$, *there exists an orthonormal basis in* $\mathbf{R}_n$ *consisting of eigenvectors of* **A**.

Geometrically, a symmetric operator produces expansions (by factors of $\lambda_1, \ldots, \lambda_n$, respectively) along each of n orthogonal directions. The numbers $\lambda_1, \ldots, \lambda_n$ are the roots of the characteristic equation (24). Hence the characteristic equation corresponding to a symmetric matrix $A = \|a_{ik}\|$ must have n (not necessarily distinct) real roots and no imaginary roots at all.

9.46. The structure of a real antisymmetric operator. If **A** is an antisymmetric operator acting in $\mathbf{R}_n$, so that $\mathbf{A}' = -\mathbf{A}$, then the extension $\hat{\mathbf{A}}$ of the operator **A** into the space $\mathbf{C}_n$ is antiself-adjoint, i.e., $\hat{\mathbf{A}}^* = -\hat{\mathbf{A}}$. The eigenvalues $\lambda_1, \ldots, \lambda_n$ of an antiself-adjoint operator are all purely imaginary (see Sec. 9.35). Hence the blocks (25) in the representation (23) take the special form

$$\begin{Vmatrix} 0 & \tau_j \\ -\tau_j & 0 \end{Vmatrix} \qquad (j = 1, 2, \ldots, m),$$

while the numbers $\lambda_{m+1}, \lambda_{m+2}, \ldots, \lambda_r$ must all be 0. This proves the following

† The expansion is actually a contraction if $0 < \sqrt{\sigma^2 + \tau^2} < 1$ or if $0 < \lambda_k < 1$. Moreover, expansion by a factor $\lambda_k < 0$ is actually an expansion accompanied by a reflection.

THEOREM. *Given any antisymmetric operator* **A** *in a real Euclidean space* $\mathbf{R}_n$, *there exists an orthonormal basis in* $\mathbf{R}_n$ *in which the matrix of* **A** *takes the quasi-diagonal form*

$$\left\| \begin{array}{ccccccccc} 0 & \tau_1 & & & & & & & \\ -\tau_1 & 0 & & & & & & & \\ & & 0 & \tau_2 & & & & & \\ & & -\tau_2 & 0 & & & & & \\ & & & & \ddots & & & & \\ & & & & & 0 & \tau_m & & \\ & & & & & -\tau_m & 0 & & \\ & & & & & & & 0 & \\ & & & & & & & & 0 \\ & & & & & & & & & \ddots \\ & & & & & & & & & & 0 \end{array} \right\|. \qquad (27)$$

Conversely, if the matrix of an operator **A** is of the form (27) in some orthonormal basis, then **A** is antisymmetric (Sec. 8.92b).

Geometrically, an antisymmetric operator produces rotations through 90° followed by expansions (by factors of $\tau_1, \ldots, \tau_m$, respectively) in m mutually orthogonal planes, while mapping into 0 all vectors orthogonal to these planes.

9.47. The structure of a real isometric operator. If **A** is an isometric operator acting in $\mathbf{R}_n$, so that $\mathbf{A}' = \mathbf{A}^{-1}$, then the extension $\hat{\mathbf{A}}$ of the operator **A** into the space $\mathbf{C}_n$ is unitary, i.e., $\hat{\mathbf{A}}^* = \hat{\mathbf{A}}^{-1}$. The eigenvalues $\lambda_1, \ldots, \lambda_n$ of a unitary operator are all of absolute value 1 (see Sec. 9.36). Hence the blocks (25) in the representation (23) take the special form

$$\left\| \begin{array}{cc} \cos \alpha_j & \sin \alpha_j \\ -\sin \alpha_j & \cos \alpha_j \end{array} \right\|,$$

and the numbers $\lambda_{m+1}, \ldots, \lambda_r$ must all be ± 1. This proves the following

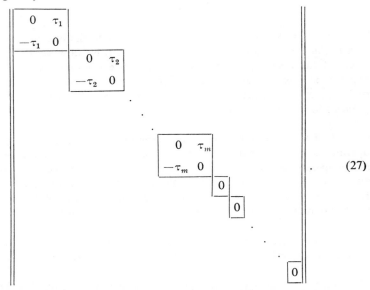

THEOREM. *Given any isometric operator* **A** *in a real Euclidean space* $\mathbf{R}_n$, *there exists an orthonormal basis in* $\mathbf{R}_n$ *in which the matrix of* **A** *takes the*

quasidiagonal form

$$\left\|\begin{array}{cccccccccc}
\cos\alpha_1 & \sin\alpha_1 & & & & & & & & \\
-\sin\alpha_1 & \cos\alpha_1 & & & & & & & & \\
 & & \cos\alpha_2 & \sin\alpha_2 & & & & & & \\
 & & -\sin\alpha_2 & \cos\alpha_2 & & & & & & \\
 & & & & \ddots & & & & & \\
 & & & & & \cos\alpha_m & \sin\alpha_m & & & \\
 & & & & & -\sin\alpha_m & \cos\alpha_m & & & \\
 & & & & & & & \pm 1 & & \\
 & & & & & & & & \pm 1 & \\
 & & & & & & & & & \ddots \\
 & & & & & & & & & \pm 1
\end{array}\right\|.$$

Geometrically, an isometric operator **A** produces a rotation through a certain angle (with no accompanying expansion) in each of m mutually orthogonal planes, and acts like the operator **E** or $-$**E** in each of the $r-m$ directions orthogonal to these planes and to each other. However, we can combine every pair of such directions with identical expansion coefficients (both $+1$ or both -1) into a plane in which the operator **A** also produces a rotation (through 0° or 180°). Making all such combinations, we find that if n is odd, then some last direction has the coefficient $+1$ or -1, while if n is even, there may be two ungrouped directions with coefficients $+1$ and -1. The presence of -1 among these remaining coefficients shows that besides the indicated rotations there is an additional reflection with respect to some coordinate plane, for example, the plane orthogonal to the basis vector e_n. We then have $\det A = -1$, whereas $\det A = +1$ if there is no such reflection.

PROBLEMS

1. A self-adjoint operator acting in a unitary space $\mathbf{C}_n$ is said to be *nonnegative* (or *positive*) if all its eigenvalues $\lambda_1, \ldots \lambda_n$ are nonnegative (or positive). Show that the square of every symmetric operator is nonnegative.

2. Show that given any self-adjoint nonnegative (or positive) operator **A**, we can find a unique nonnegative (or positive) operator **B**, the "square root of the operator **A**," such that $\mathbf{B}^2 = \mathbf{A}$.

3. Take the square root of the operator **A** specified by the matrix

$$A = \begin{Vmatrix} 13 & 14 & 4 \\ 14 & 24 & 18 \\ 4 & 18 & 29 \end{Vmatrix}$$

in an orthonormal basis e_1, e_2, e_3.

4. Let **A** be an arbitrary linear operator acting in a unitary space $\mathbf{C}_n$, and let $\mathbf{A}^*$ be its adjoint. Prove that $\mathbf{A}^*\mathbf{A}$ is a nonnegative operator. Prove that $\mathbf{A}'\mathbf{A}$ is a positive operator if **A** is nonsingular.

5. Given that a linear operator **A** is the product **SQ** of a self-adjoint operator **S** and a unitary operator **Q**, prove that $\mathbf{S}^2 = \mathbf{A}\mathbf{A}^*$.

6. Show that every nonsingular linear operator **A** can be represented as the product **SQ** of a self-adjoint operator and a unitary operator.

7. Prove that the representation of the operator **A** as a product **SQ** in Problem 6 is unique.

8. A linear operator **V** acting in $\mathbf{C}_n$ is said to be *nonexpanding* if $|\mathbf{V}x| \leqslant |x|$ for every x. Prove that every linear operator **A** can be represented as the product of a self-adjoint operator and a nonexpanding operator.

9. Show that two self-adjoint operators **A** and **B** commute if and only if they have a common system of n mutually orthogonal eigenvectors.

10. Given a linear operator **A** acting in the space $\mathbf{C}_n$, find an orthonormal basis in which the matrix of **A** has the triangular form

$$A = \begin{Vmatrix} a_1^{(1)} & a_2^{(1)} & \dots & a_n^{(1)} \\ 0 & a_2^{(2)} & \dots & a_n^{(2)} \\ \cdot & \cdot & \dots & \cdot \\ 0 & 0 & \dots & a_n^{(n)} \end{Vmatrix}.$$

chapter 10

QUADRATIC FORMS IN EUCLIDEAN AND UNITARY SPACES

10.1. Basic Theorem on Quadratic Forms in a Euclidean Space

10.11. We begin with the following theorem concerning symmetric bilinear forms in a Euclidean space:

THEOREM. *Every symmetric bilinear form* $\mathbf{A}(x, y)$ *in an n-dimensional Euclidean space* $\mathbf{R}_n$ *has a canonical basis consisting of orthogonal vectors.*

Proof. Consider the linear operator $\mathbf{A}$ corresponding to the given symmetric bilinear form (see Sec. 8.91). The operator $\mathbf{A}$ is also symmetric. According to the theorem on symmetric operators (Theorem 9.45), the space $\mathbf{R}_n$ has an orthonormal basis consisting of the eigenvectors of the operator $\mathbf{A}$, and the matrix of $\mathbf{A}$ is diagonal in this basis. Since this matrix is also the matrix of the bilinear form $\mathbf{A}(x, y)$, the orthonormal basis just found is a canonical basis of $\mathbf{A}(x, y)$. ▮

10.12. We now apply this result to the study of quadratic forms. Given a quadratic form

$$\mathbf{A}(x, x) = \sum_{i,k=1}^{n} a_{ik}\xi_i\xi_k \qquad (a_{ik} = a_{ki}), \tag{1}$$

we will regard the numbers $\xi_1, \xi_2, \ldots, \xi_n$ as the components of a vector x in an n-dimensional Euclidean space $\mathbf{R}_n$, with a scalar product defined by the formula

$$(x, y) = \sum_{i=1}^{n} \xi_i\eta_i,$$

where $y = (\eta_1, \eta_2, \ldots, \eta_n)$. The basis

$$\begin{aligned} e_1 &= (1, 0, \ldots, 0), \\ e_2 &= (0, 1, \ldots, 0), \\ &\ldots\ldots\ldots\ldots \\ e_n &= (0, 0, \ldots, 1) \end{aligned}$$

is an orthonormal basis in $\mathbf{R}_n$, and clearly

$$x = \sum_{i=1}^{n} \xi_i e_i, \qquad y = \sum_{i=1}^{n} \eta_i e_i.$$

Now consider the bilinear form

$$\mathbf{A}(x, y) = \sum_{i,k=1}^{n} a_{ik} \xi_i \eta_k$$

corresponding to the quadratic form (1). By Theorem 10.11, this form has an orthonormal basis $f_1, f_2, \ldots, f_n$. If the components of the vectors x and y are $\tau_1, \tau_2, \ldots, \tau_n$ and $\theta_1, \theta_2, \ldots, \theta_n$, respectively, in this basis, then we can write the bilinear form $\mathbf{A}(x, y)$ as

$$\mathbf{A}(x, y) = \sum_{i=1}^{n} \lambda_i \tau_i \theta_i$$

and the quadratic form $\mathbf{A}(x, x)$ as

$$\mathbf{A}(x, x) = \sum_{i=1}^{n} \lambda_i \tau_i^2. \tag{2}$$

The transformation from the basis $e_1, e_2, \ldots, e_n$ to the basis $f_1, f_2, \ldots, f_n$ is given by

$$f_j = \sum_{i=1}^{n} q_i^{(j)} e_i \qquad (j = 1, 2, \ldots, n),$$

where $Q = \|q_i^{(j)}\|$ is an orthogonal matrix (Sec. 8.93). According to the formulas (36), p. 240, the relation between the components $\tau_1, \tau_2, \ldots, \tau_n$ and $\xi_1, \xi_2, \ldots, \xi_n$ is given by the system of equations

$$\xi_j = \sum_{i=1}^{n} q_j^{(i)} \tau_i \qquad (j = 1, 2, \ldots, n), \tag{3}$$

involving the transposed matrix Q'. Thus we have proved the following important

THEOREM. *Every quadratic form* (1) *in an n-dimensional Euclidean space* $\mathbf{R}_n$ *can be reduced to the canonical form* (2) *by making an isometric coordinate transformation* (3).

10.13. The sequence of operations which must be performed in order to construct the coordinate transformation (3) and the canonical form (2) of the quadratic form (1) can be deduced from the results of Secs. 4.94 and 9.45. We now give this sequence of operations in final form:

a) *Use the quadratic form* (1) *to construct the symmetric matrix* $A = \|a_{ik}\|$.

b) *Form the characteristic polynomial* $\Delta(\lambda) = \det(A - \lambda E)$ *and find its roots.* By Sec. 9.45, this polynomial has n (not necessarily distinct) real roots.

c) *From a knowledge of the roots of the polynomial* $\Delta(\lambda)$, *we can already write the quadratic form* (1) *in canonical form* (2); *in particular, we can determine its positive and negative indices of inertia.*

d) *Substitute the root* λ_1 *into the system* (28), *p. 110.* For the given root λ_1, the system must have a number of linearly independent solutions equal to the multiplicity of the root λ_1. *Find these linearly independent solutions by using the rules for solving homogeneous systems of linear equations.*

e) *If the multiplicity of the root* λ_1 *is greater than unity, orthogonalize the resulting linearly independent solutions by using the method of Sec. 8.61.*

f) *Carrying out the indicated operations for every root, we finally obtain a system of n orthogonal vectors. We then normalize them by dividing each vector by its length. The resulting vectors*

$$f_1 = (q_1^{(1)}, q_2^{(1)}, \ldots, q_n^{(1)}),$$

$$f_2 = (q_1^{(2)}, q_2^{(2)}, \ldots, q_n^{(2)}),$$

$$\cdots\cdots\cdots\cdots\cdots\cdots$$

$$f_n = (q_1^{(n)}, q_2^{(n)}, \ldots, q_n^{(n)})$$

form an orthonormal system.

g) *Using the numbers* $q_i^{(j)}$, *we can write the coordinate transformation* (3).

h) *To express the new components* $\tau_1, \tau_2, \ldots, \tau_n$ *in terms of the old components* $\xi_1, \xi_2, \ldots, \xi_n$, *we write*

$$\tau_j = \sum_{i=1}^{n} q_i^{(j)} \xi_i \qquad (j = 1, 2, \ldots, n),$$

recalling that the inverse of the orthogonal matrix Q *is the transposed matrix* Q'.

10.14. In Sec. 7.33a we saw that neither the canonical form nor the canonical basis of a quadratic form is uniquely defined in an affine space; in general, any preassigned vector can be included in the canonical basis of the quadratic form. The situation is quite different in a Euclidean space, provided that only orthonormal bases are considered. The point is that the matrix of the quadratic form and the matrix of the corresponding symmetric linear operator transform in the same way, as already noted in Sec. 8.91.

Thus a canonical basis for the quadratic form is at the same time a basis consisting of the eigenvectors of the symmetric operator, and the coefficients of the quadratic form relative to the canonical basis (the "canonical coefficients") coincide with the eigenvalues of the operator. But the eigenvalues of the operator **A** are the roots of the equation $\det(A - \lambda E) = 0$, an equation which does not depend on the choice of a basis and is an invariant of the operator **A**. Hence *the set of canonical coefficients of the form* $(\mathbf{A}x, x)$ *is uniquely defined.* As for the canonical basis of the quadratic form $(\mathbf{A}x, x)$, it is defined with the same arbitrariness as in the definition of a complete orthonormal system of eigenvectors of the operator **A**, i.e., apart from permutations of the eigenvectors, we can multiply any of them by -1, or more generally, we can subject them to any isometric transformation in the characteristic subspace corresponding to a fixed eigenvalue λ.

10.2. Extremal Properties of a Quadratic Form

10.21. Next, given a quadratic form $\mathbf{A}(x, x)$ in a Euclidean space $\mathbf{R}_n$, we examine the values of $\mathbf{A}(x, x)$ on the unit sphere $(x, x) = 1$ of the space $\mathbf{R}_n$, and inquire at what points of the unit sphere the values of $\mathbf{A}(x, x)$ are stationary. It will be recalled that by definition a differentiable numerical function $f(x)$, defined at the points of a surface U, takes a *stationary value* at the point $x_0 \in U$ if the derivative of the function $f(x)$ along any direction on the surface U vanishes at the point x_0. In particular, the function $f(x)$ is stationary at the points where it has a maximum or a minimum.

The problem of determining the stationary values of a quadratic form on the unit sphere is a problem involving conditional extrema. One method of solving the problem is to use Lagrange's method,† as follows: We construct an orthonormal basis in the space $\mathbf{R}_n$ and denote the components of the vector x in this basis by $\xi_1, \xi_2, \ldots, \xi_n$. In this coordinate system, our quadratic form becomes

$$\mathbf{A}(x, x) = \sum_{i,k=1}^{n} a_{ik}\xi_i\xi_k,$$

and the condition $(x, x) = 1$ becomes

$$\sum_{i=1}^{n} \xi_i^2 = 1.$$

Using Lagrange's method, we construct the function

$$F(\xi_1, \xi_2, \ldots, \xi_n) = \sum_{i,k=1}^{n} a_{ik}\xi_i\xi_k - \lambda \sum_{i=1}^{n} \xi_i^2,$$

† See e.g., R. Courant, *Differential and Integral Calculus, Vol. II* (translated by E. J. McShane), Interscience Publishers, Inc., New York (1956), p. 190.

and equate to zero its partial derivatives with respect to ξ_i $(i = 1, 2, \ldots, n)$, recalling that $a_{ik} = a_{ki}$:

$$2\sum_{k=1}^{n} a_{ik}\xi_k - 2\lambda\xi_i = 0 \qquad (i = 1, 2, \ldots, n).$$

After dividing by 2, we obtain the familiar system

$$\begin{aligned}
(a_{11} - \lambda)\xi_1 + a_{12}\xi_2 + \cdots + a_{1n}\xi_n &= 0,\\
a_{21}\xi_1 + (a_{22} - \lambda)\xi_2 + \cdots + a_{2n}\xi_n &= 0,\\
\cdots\cdots\cdots\cdots\cdots\cdots\cdots\cdots\cdots\cdots\cdots\cdots\\
a_{n1}\xi_1 + a_{n2}\xi_2 + \cdots + (a_{nn} - \lambda)\xi_n &= 0
\end{aligned}$$

(cf. p. 110), which serves to define the eigenvectors of the symmetric operator corresponding to the quadratic form $\mathbf{A}(x, x)$. It follows that *the quadratic form* $\mathbf{A}(x, x)$ *takes stationary values at those vectors of the unit sphere which are eigenvectors of the symmetric operator* $\mathbf{A}$ *corresponding to the form* $\mathbf{A}(x, x)$.

10.22. We now calculate the values which the form takes at its stationary points. To do this, we introduce the corresponding symmetric operator $\mathbf{A}$ and write the quadratic form as

$$\mathbf{A}(x, x) = (\mathbf{A}x, x).$$

Suppose that $\mathbf{A}(x, x)$ takes a stationary value at the vector e_i. Since we have just shown that e_i is an eigenvector of the operator $\mathbf{A}$, i.e., $\mathbf{A}e_i = \lambda_i e_i$, we have

$$\mathbf{A}(e_i, e_i) = (\mathbf{A}e_i, e_i) = \lambda_i(e_i, e_i) = \lambda_i.$$

Hence *the stationary value of the form* $\mathbf{A}(x, x)$ *at* $x = e_i$ *equals the corresponding eigenvalue of the operator* $\mathbf{A}$. Since the eigenvalues of the operator $\mathbf{A}$ are the same as the canonical coefficients of the form $\mathbf{A}(x, x)$, we can conclude that the stationary values of the form $\mathbf{A}(x, x)$ coincide with its canonical coefficients. In particular, the maximum of the form $\mathbf{A}(x, x)$ on the unit sphere is equal to its largest canonical coefficient, and the minimum of $\mathbf{A}(x, x)$ on the unit sphere is equal to its smallest canonical coefficient.

10.23. Quadratic forms and bilinear forms can both be considered not only on the whole n-dimensional space $\mathbf{R}_n$, but also on a k-dimensional subspace $\mathbf{R}_k \subset \mathbf{R}_n$, and we can then look for an orthonormal canonical basis in $\mathbf{R}_k$. Let the quadratic form $\mathbf{A}(x, x)$ have the canonical form

$$\mathbf{A}(x, x) = \lambda_1\xi_1^2 + \lambda_2\xi_2^2 + \cdots + \lambda_n\xi_n^2 \tag{4}$$

in the whole space $\mathbf{R}_n$, and the canonical form

$$\mathbf{A}(x, x) = \mu_1\tau_1^2 + \mu_2\tau_2^2 + \cdots + \mu_k\tau_k^2$$

in the subspace $\mathbf{R}_k$. We now find the relation between the coefficients $\mu_1, \mu_2, \ldots, \mu_k$ and the coefficients $\lambda_1, \lambda_2, \ldots, \lambda_n$. For convenience, we assume that the canonical coefficients are arranged in decreasing order, i.e., that

$$\lambda_1 \geqslant \lambda_2 \geqslant \cdots \geqslant \lambda_n, \qquad \mu_1 \geqslant \mu_2 \geqslant \cdots \geqslant \mu_k.$$

As we know, the quantity λ_1 is the maximum value of the quadratic form $\mathbf{A}(x, x)$ on the unit sphere of the space $\mathbf{R}_n$; similarly, μ_1 is the maximum value of $\mathbf{A}(x, x)$ on the unit sphere of the subspace $\mathbf{R}_k$. This implies that $\mu_1 \leqslant \lambda_1$. Moreover, we also have $\mu_1 \geqslant \lambda_{n-k+1}$. To see this, let $e_1, e_2, \ldots, e_n$ be the canonical basis in which $\mathbf{A}(x, x)$ takes the form (4). Consider the $(n - k + 1)$-dimensional subspace $\mathbf{R}'$ spanned by the vectors $e_1, e_2, \ldots, e_{n-k+1}$. Since $k + (n - k + 1) > n$, then, by Corollary 2.47c, the subspaces $\mathbf{R}'$ and $\mathbf{R}_k$ have at least one nonzero vector in common. Let this vector be

$$x_0 = (\xi_1^{(0)}, \ldots, \xi_{n-k+1}^{(0)}, 0, \ldots, 0),$$

and assume that x_0 is normalized, i.e., that $|x_0| = 1$. According to (4), we have

$$\begin{aligned}\mathbf{A}(x_0, x_0) &= \lambda_1(\xi_1^{(0)})^2 + \cdots + \lambda_{n-k+1}(\xi_{n-k+1}^{(0)})^2 \\ &\geqslant \lambda_{n-k+1}(\xi_1^{(0)})^2 + \cdots + \lambda_{n-k+1}(\xi_{n-k+1}^{(0)})^2 = \lambda_{n-k+1}.\end{aligned}$$

This implies that μ_1, the *maximum* value of the quadratic form $\mathbf{A}(x, x)$ on the unit sphere of the subspace $\mathbf{R}_k$, cannot be less than λ_{n-k+1}, as asserted. Thus the quantity μ_1 satisfies the inequalities

$$\lambda_1 \geqslant \mu_1 \geqslant \lambda_{n-k+1}. \tag{5}$$

10.24. Naturally, the quantity μ_1 takes different values for different k-dimensional subspaces. We now show that *there exist k-dimensional subspaces for which the equality signs hold in* (5). Let $\mathbf{R}'$ be the subspace spanned by the first k vectors $e_1, e_2, \ldots, e_k$ of the canonical basis of the form $\mathbf{A}(x, x)$. Then $\mathbf{A}(x, x)$ is just

$$\mathbf{A}(x, x) = \lambda_1\xi_1^2 + \lambda_2\xi_2^2 + \cdots + \lambda_k\xi_k^2$$

in the basis $e_1, e_2, \ldots, e_k$ of $\mathbf{R}'$. In particular,

$$\mathbf{A}(e_1, e_1) = \lambda_1 = \max_{\substack{|x|=1 \\ x\in\mathbf{R}'}} \mathbf{A}(x, x).$$

Thus the quantity

$$\mu_1 = \mu_1(\mathbf{R}_k) = \max_{\substack{|x|=1 \\ x\in\mathbf{R}_k}} \mathbf{A}(x, x)$$

takes its maximum value λ_1 for $\mathbf{R}_k = \mathbf{R}'$.

Next let $\mathbf{R}''$ be the subspace spanned by the last k vectors e_{n-k+1}, $e_{n-k+2}, \ldots, e_n$ of the canonical basis of the form $\mathbf{A}(x, x)$. Then $\mathbf{A}(x, x)$ is just

$$\mathbf{A}(x, x) = \lambda_{n-k+1}\xi_{n-k+1}^2 + \cdots + \lambda_n\xi_n^2$$

in the basis $e_{n-k+1}, \ldots, e_n$ of $\mathbf{R}''$. In particular,

$$\mathbf{A}(e_{n-k+1}, e_{n-k+1}) = \lambda_{n-k+1} = \max_{\substack{|x|=1 \\ x \in \mathbf{R}''}} \mathbf{A}(x, x),$$

and, just as before, we conclude that μ_1 takes its minimum value λ_{n-k+1} for $\mathbf{R}_k = \mathbf{R}''$. Thus we obtain the following new definition of the coefficient λ_{n-k+1}: *The coefficient λ_{n-k+1} in the canonical representation of the quadratic form $\mathbf{A}(x, x)$ equals the smallest value of the maximum of $\mathbf{A}(x, x)$ on the unit spheres of all possible k-dimensional subspaces of the space $\mathbf{R}_n$.*

10.25. Using this result, we can estimate the other canonical coefficients of the quadratic form $\mathbf{A}(x, x)$ on the subspace $\mathbf{R}_k$. For example, if the subspace $\mathbf{R}_k$ is fixed, then μ_2 is the smallest value of the maximum of $\mathbf{A}(x, x)$ on the unit spheres of all the $(k - 1)$-dimensional subspaces of $\mathbf{R}_k$, while λ_{n-k+2} is the smallest value of the maximum of $\mathbf{A}(x, x)$ on the unit spheres of all the $(k - 1)$-dimensional subspaces of the *whole* space $\mathbf{R}_n$. Hence we have $\mu_2 \geqslant \lambda_{n-k+2}$, and similarly

$$\mu_3 \geqslant \lambda_{n-k+3},\ \mu_4 \geqslant \lambda_{n-k+4}, \ldots, \mu_k \geqslant \lambda_n.$$

On the other hand, λ_2 is the smallest value of the maximum of the quadratic form $\mathbf{A}(x, x)$ on the unit spheres of all the $(n - 1)$-dimensional subspaces of the whole space $\mathbf{R}_n$. But, according to Corollary 2.47c, the intersection of every $(n - 1)$-dimensional subspace with the subspace $\mathbf{R}_k$ is a subspace of no less than $(n - 1) + k - n = k - 1$ dimensions, so that λ_2 is no less than the smallest value of the maximum of $\mathbf{A}(x, x)$ on the unit spheres of all such subspaces; in particular, λ_2 is no less than μ_2, the smallest value of the maximum of $\mathbf{A}(x, x)$ on the unit spheres of all the $(k - 1)$-dimensional subspaces of $\mathbf{R}_k$. Therefore we have $\lambda_2 \geqslant \mu_2$, and similarly $\lambda_3 \geqslant \mu_3, \ldots$, $\lambda_k \geqslant \mu_k$. Thus the canonical coefficients $\mu_1, \mu_2, \ldots, \mu_k$ satisfy the inequalities

$$\begin{aligned} \lambda_1 \geqslant \mu_1 &\geqslant \lambda_{n-k+1}, \\ \lambda_2 \geqslant \mu_2 &\geqslant \lambda_{n-k+2}, \\ &\cdots\cdots \\ \lambda_k \geqslant \mu_k &\geqslant \lambda_n. \end{aligned} \tag{6}$$

For $k = n - 1$, the inequalities (6) become

$$\begin{aligned} \lambda_1 \geqslant \mu_1 &\geqslant \lambda_2, \\ \lambda_2 \geqslant \mu_2 &\geqslant \lambda_3, \\ &\cdots\cdots \\ \lambda_{n-1} \geqslant \mu_{n-1} &\geqslant \lambda_n. \end{aligned} \tag{7}$$

***10.26.** Consider the behavior of the quadratic form

$$\mathbf{A}(x, x) = \sum_{i=1}^{n} \lambda_i \xi_i^2$$

in the $(n - 1)$-dimensional subspace $\mathbf{R}_{n-1}$ specified by the equation

$$\alpha_1\xi_1 + \alpha_2\xi_2 + \cdots + \alpha_n\xi_n = 0 \qquad (\alpha_1^2 + \alpha_2^2 + \cdots + \alpha_n^2 = 1). \tag{8}$$

Assuming that all the coefficients $\lambda_1, \lambda_2, \ldots, \lambda_n$ are different, we can calculate the coefficients $\mu_1, \mu_2, \ldots, \mu_{n-1}$ by using a method due to M. G. Krein. At least one of the coefficients $\alpha_1, \alpha_2, \ldots, \alpha_n$ is nonzero. For example, suppose $\alpha_n \neq 0$. Then (8) implies

$$\xi_n = -\frac{1}{\alpha_n}\sum_{j=1}^{n-1} \alpha_j \xi_j.$$

Substituting this expression for ξ_n into $\mathbf{A}(x, x)$, we find that $\mathbf{A}(x, x)$ has the form

$$\mathbf{A}(x, x) = \lambda_1\xi_1^2 + \lambda_2\xi_2^2 + \cdots + \lambda_{n-1}\xi_{n-1} + \frac{\lambda_n}{\alpha_n^2}\left(\sum_{j=1}^{n-1} \alpha_j\xi_j\right)^2$$

in the subspace $\mathbf{R}_{n-1}$, in terms of the variables $\xi_1, \xi_2, \ldots, \xi_{n-1}$. The canonical coefficients of this quadratic form are the same as its stationary values on the unit sphere of the subspace $\mathbf{R}_{n-1}$ (Sec. 10.22). In the variables $\xi_1, \xi_2, \ldots \xi_{n-1}$ this sphere has the equation

$$\mathbf{B}(x, x) = \xi_1^2 + \xi_2^2 + \cdots + \xi_{n-1}^2 + \frac{1}{\alpha_n^2}\left(\sum_{j=1}^{n-1} \alpha_j\xi_j\right)^2 = 1.$$

Just as before, we determine these stationary values by using Lagrange's method. Thus we form the function

$$\mathbf{A}(x, x) - \lambda\mathbf{B}(x, x) = \sum_{i=1}^{n-1}(\lambda_i - \lambda)\xi_i^2 + \frac{\lambda_n - \lambda}{\alpha_n^2}\left(\sum_{j=1}^{n-1} \alpha_j\xi_j\right)^2,$$

and equate to zero its partial derivatives with respect to ξ_k $(k = 1, 2, \ldots, n - 1)$, obtaining

$$\xi_k(\lambda_k - \lambda) + \frac{\lambda_n - \lambda}{\alpha_n^2}\left(\sum_{j=1}^{n-1} \alpha_j\xi_j\right)\alpha_k = 0. \tag{9}$$

The required coefficients $\mu_1, \mu_2, \ldots, \mu_{n-1}$ are the roots of the equation obtained by equating to zero the determinant $D(\lambda)$ of the system of linear equations (9). The coefficient matrix of this system is clearly the sum of two matrices; the first matrix is diagonal with the numbers $\lambda_k - \lambda$ ($k = 1, 2, \ldots, n - 1$) along the diagonal, while the second matrix has the form

$$\frac{\lambda_n - \lambda}{\alpha_n^2}\left\|\begin{matrix} \alpha_1\alpha_1 & \alpha_2\alpha_1 & \cdots & \alpha_{n-1}\alpha_1 \\ \alpha_1\alpha_2 & \alpha_2\alpha_2 & \cdots & \alpha_{n-1}\alpha_2 \\ \cdot & \cdot & \cdots & \cdot \\ \alpha_1\alpha_{n-1} & \alpha_2\alpha_{n-1} & \cdots & \alpha_{n-1}\alpha_{n-1} \end{matrix}\right\|.$$

By the linear property of determinants (Sec. 1.44), the determinant $D(\lambda)$ is the sum of the determinant of the first matrix and all the determinants obtained by replacing one or more columns of the determinant of the first matrix by the corresponding columns of the second matrix and taking account of the factor $(\lambda_n - \lambda)/\alpha_n^2$. Since any two columns of the second matrix are proportional, we need only consider the case where *one* of the columns of the determinant of the first matrix is replaced by the corresponding column of the second matrix.

In particular, if the kth column of the first matrix is replaced by the kth column of the second matrix, the resulting determinant has the form

$$\frac{\lambda_n - \lambda}{\alpha_n^2}\left|\begin{matrix} \lambda_1 - \lambda & 0 & \cdots & 0 & \alpha_k\alpha_1 & 0 & \cdots & 0 \\ 0 & \lambda_2 - \lambda & \cdots & 0 & \alpha_k\alpha_2 & 0 & \cdots & 0 \\ \cdot & \cdot & \cdots & \cdot & \cdot & \cdot & \cdots & \cdot \\ 0 & 0 & \cdots & \lambda_{k-1} - \lambda & \alpha_k\alpha_{k-1} & 0 & \cdots & 0 \\ 0 & 0 & \cdots & 0 & \alpha_k\alpha_k & 0 & \cdots & 0 \\ 0 & 0 & \cdots & 0 & \alpha_k\alpha_{k+1} & \lambda_{k+1} - \lambda & \cdots & 0 \\ \cdot & \cdot & \cdots & \cdot & \cdot & \cdot & \cdots & \cdot \\ 0 & 0 & \cdots & 0 & \alpha_k\alpha_{n-1} & 0 & \cdots & \lambda_{n-1} - \lambda \end{matrix}\right|$$

$$= \frac{\alpha_k^2}{\alpha_n^2}\frac{\prod_{j=1}^{n}(\lambda_j - \lambda)}{\lambda_k - \lambda}.$$

Denote the determinant of the first matrix by

$$F(\lambda) = \prod_{k=1}^{n-1}(\lambda_k - \lambda),$$

and let

$$G(\lambda) = \prod_{k=1}^{n}(\lambda_k - \lambda).$$

Then the required determinant $D(\lambda)$ becomes

$$D(\lambda) = F(\lambda) + \frac{1}{\alpha_n^2} G(\lambda) \sum_{k=1}^{n-1} \frac{\alpha_k^2}{\lambda_k - \lambda}. \tag{10}$$

Solving the equation $D(\lambda) = 0$, we find the quantities $\mu_1, \mu_2, \ldots, \mu_{n-1}$ in which we are interested. Note that these quantities depend on the squares of the numbers α_k rather than on the numbers α_k themselves. Thus changing the sign of one or more coefficients in (8) does not change the canonical coefficients of the form $\mathbf{A}(x, x)$ in the subspace $\mathbf{R}_{n-1}$.

***10.27.** Equation (10) is of particular interest in that *it allows us to construct from given numbers* $\mu_1, \mu_2, \ldots, \mu_{n-1}$ *satisfying the inequalities* (7) *a subspace* $\mathbf{R}_{n-1}$ *in which the form* $\mathbf{A}(x, x)$ *has the canonical coefficients* $\mu_1, \mu_2, \ldots, \mu_{n-1}$. (Again it is assumed that the numbers $\lambda_1, \lambda_2, \ldots, \lambda_n$ are distinct.) We now show how this is done.

First we note that (10) can be written in the form

$$\alpha_n^2 \frac{D(\lambda)}{G(\lambda)} = \alpha_n^2 \frac{F(\lambda)}{G(\lambda)} + \sum_{k=1}^{n-1} \frac{\alpha_k^2}{\lambda_k - \lambda} = \sum_{k=1}^{n} \frac{\alpha_k^2}{\lambda_k - \lambda}. \tag{11}$$

Thus the numbers $\alpha_1^2, \alpha_2^2, \ldots, \alpha_n^2$ are proportional to the coefficients obtained when we expand the rational function $D(\lambda)/G(\lambda)$ in partial fractions. Now suppose we are given numbers $\mu_1, \mu_2, \ldots, \mu_{n-1}$ satisfying the inequalities

$$\begin{gathered} \lambda_1 > \mu_1 > \lambda_2, \\ \lambda_2 > \mu_2 > \lambda_3, \\ \cdots\cdots\cdot \\ \lambda_{n-1} > \mu_{n-1} > \lambda_n. \end{gathered} \tag{12}$$

Let

$$D_1(\lambda) = \prod_{k=1}^{n-1} (\mu_k - \lambda),$$

and expand the rational function $D_1(\lambda)/G(\lambda)$ in partial fractions

$$\frac{D_1(\lambda)}{G(\lambda)} = \frac{c_1}{\lambda_1 - \lambda} + \frac{c_2}{\lambda_2 - \lambda} + \cdots + \frac{c_n}{\lambda_n - \lambda}. \tag{13}$$

The coefficients $c_1, c_2, \ldots, c_n$ are given by the familiar formula†

$$c_k = \frac{D_1(\lambda_k)}{(\lambda_1 - \lambda_k) \cdots (\lambda_{k-1} - \lambda_k)(\lambda_{k+1} - \lambda_k) \cdots (\lambda_n - \lambda_k)} = -\frac{D_1(\lambda_k)}{G'(\lambda_k)},$$

† See e.g., R. A. Silverman, *Modern Calculus and Analytic Geometry*, The Macmillan Co., New York (1969), p. 861.

and all have the same sign. To see this, we note that the numbers $D_1(\lambda_1)$, $D_1(\lambda_2), \ldots, D_1(\lambda_n)$ alternate in sign, since, by hypothesis, the roots of the polynomial $D_1(\lambda)$ alternate with the roots of the polynomial $G(\lambda)$. Thus the numbers $D_1(\lambda_k)/G'(\lambda_k)$, and hence the coefficients c_k ($k = 1, \ldots, n$), all have the same sign. By supplying an extra factor, we can assume that the c_k are all positive and add up to 1. We can then define the numbers $\alpha_1, \alpha_2, \ldots, \alpha_n$ by the formulas

$$\alpha_1^2 = c_1, \qquad \alpha_2^2 = c_2, \ldots, \alpha_n^2 = c_n, \tag{14}$$

where each α_k can have either sign.

Finally we show that the subspace $\mathbf{R}_{n-1}$ defined by the equation

$$\alpha_1\xi_1 + \alpha_2\xi_2 + \cdots + \alpha_n\xi_n = 0$$

is the required subspace, in which the quadratic form $\mathbf{A}(x, x)$ has the canonical coefficients $\mu_1, \mu_2, \ldots, \mu_{n-1}$. In fact, as proved above, the polynomial $D(\lambda)$ whose roots are the canonical coefficients of $\mathbf{A}(x, x)$ in the subspace $\mathbf{R}_{n-1}$ is given by formula (10) or the equivalent formula (11). Comparing (11) with (13) and using (14), we find that the polynomial $D(\lambda)$ differs only by a numerical factor from the polynomial $D_1(\lambda)$ just constructed. But then the roots of $D(\lambda)$ coincide with the numbers $\mu_1, \mu_2, \ldots, \mu_{n-1}$, as required.

Remark. It can be shown that the numbers $\alpha_1, \ldots, \alpha_n$ depend continuously on the numbers $\lambda_1, \ldots, \lambda_n, \mu_1, \ldots, \mu_{n-1}$. Using this fact, we can verify that the problem can still be solved if the numbers $\mu_1, \ldots, \mu_{n-1}$ satisfy the inequalities (7) instead of (12) or if the numbers $\lambda_1, \ldots, \lambda_n$ are no longer distinct.

10.3. Simultaneous Reduction of Two Quadratic Forms

10.31. The following question plays an important role in certain problems of mathematics and physics: *Given two quadratic forms* $\mathbf{A}(x, x)$ *and* $\mathbf{B}(x, x)$ *defined in an n-dimensional affine space* $\mathbf{R}_n$, *how does one find a basis in which both* $\mathbf{A}(x, x)$ *and* $\mathbf{B}(x, x)$ *are reduced to canonical form* (i.e., to sums of squares of the components of x with certain coefficients)? The following example in the plane ($n = 2$) shows that this problem does not always have a solution:

Consider the two forms

$$\mathbf{A}(x, x) = \xi_1^2 - \xi_2^2,$$
$$\mathbf{B}(x, x) = \xi_1\xi_2.$$

Finding a common canonical basis for these two forms is the same as finding a common pair of conjugate vectors for the hyperbolas $\mathbf{A}(x, x) = 1$ and $\mathbf{B}(x, x) = 1$ (see Sec. 7.42). Since these are equilateral hyperbolas, we know from analytic geometry that the conjugate directions of the hyperbolas are symmetric with respect to their asymptotes. Therefore the polar angles φ_1 and φ_2 corresponding to the pair of conjugate directions satisfy the relation

$$\varphi_1 + \varphi_2 = \frac{\pi}{2}$$

for the first hyperbola and the relation

$$\varphi_1 + \varphi_2 = 0$$

for the second hyperbola (both relations hold only to within an integral multiple of π). Since the two relations are mutually exclusive, there does not exist a common pair of conjugate vectors in this case.

It turns out that the problem of simultaneous reduction of two quadratic forms does have a solution if we make the supplementary assumption that *one of the forms, say* $\mathbf{B}(x, x)$, *is positive definite*, i.e., that $\mathbf{B}(x, x) > 0$ for $x \neq 0$. In this case, the existence of a solution is easily proved as follows: Let $\mathbf{B}(x, y)$ be the symmetric bilinear form corresponding to the quadratic form $\mathbf{B}(x, x)$, and introduce a Euclidean metric in the affine space $\mathbf{R}_n$ by writing

$$(x, y) = \mathbf{B}(x, y).$$

The fact that $\mathbf{B}(x, y)$ is symmetric and positive definite guarantees that (x, y) satisfies the axioms for a scalar product. By Sec. 10.11 there exists an orthonormal basis (with respect to this metric) in which $\mathbf{A}(x, x)$ takes the canonical form

$$\mathbf{A}(x, x) = \lambda_1\xi_1^2 + \lambda_2\xi_2^2 + \cdots + \lambda_n\xi_n^2, \tag{15}$$

where $\xi_1, \xi_2, \ldots, \xi_n$ denote the components of the vector x in the basis just found. In the same basis, the second quadratic form $\mathbf{B}(x, x)$ becomes

$$\mathbf{B}(x, x) = (x, x) = \eta_1^2 + \eta_2^2 + \cdots + \eta_n^2,$$

by formula (17), p. 222. Hence, as asserted, there exists a basis in which both $\mathbf{A}(x, x)$ and $\mathbf{B}(x, x)$ have canonical form.

10.32. To construct the components of the vectors $e_1, \ldots, e_n$ of the basis which is simultaneously canonical for both quadratic forms, we use the extremal properties of quadratic forms. As shown in Sec. 10.21, the vectors $e_1, \ldots, e_n$ of the required basis are the vectors obeying the condition

$$(x, x) = \mathbf{B}(x, x) = 1$$

for which the form $\mathbf{A}(x, x)$ takes stationary values. Suppose $\mathbf{A}(x, x)$ and $\mathbf{B}(x, x)$ are given by

$$\mathbf{A}(x, x) = \sum_{i,k=1}^{n} a_{ik}\xi_i\xi_k,$$

$$\mathbf{B}(x, x) = \sum_{i,k=1}^{n} b_{ik}\xi_i\xi_k$$

in the original basis. Using Lagrange's method, we form the function

$$F(\xi_1, \xi_2, \ldots, \xi_n) = \sum_{i,k=1}^{n} a_{ik}\xi_i\xi_k - \mu \sum_{i,k=1}^{n} b_{ik}\xi_i\xi_k,$$

and then equate to zero its partial derivatives with respect to all the ξ_i:

$$\sum_{k=1}^{n} a_{ik}\xi_k - \mu \sum_{k=1}^{n} b_{ik}\xi_k = 0 \qquad (i = 1, 2, \ldots, n). \tag{16}$$

The resulting system of homogeneous equations

$$\begin{aligned} (a_{11} - \mu b_{11})\xi_1 + (a_{12} - \mu b_{12})\xi_2 + \cdots + (a_{1n} - \mu b_{1n})\xi_n &= 0, \\ (a_{21} - \mu b_{21})\xi_1 + (a_{22} - \mu b_{22})\xi_2 + \cdots + (a_{2n} - \mu b_{2n})\xi_n &= 0, \\ \cdots\cdots\cdots\cdots\cdots\cdots\cdots\cdots\cdots\cdots\cdots\cdots \\ (a_{n1} - \mu b_{n1})\xi_1 + (a_{n2} - \mu b_{n2})\xi_2 + \cdots + (a_{nn} - \mu b_{nn})\xi_n &= 0 \end{aligned} \tag{17}$$

has a nontrivial solution if and only if its determinant vanishes:

$$\begin{vmatrix} a_{11} - \mu b_{11} & a_{12} - \mu b_{12} & \cdots & a_{1n} - \mu b_{1n} \\ a_{21} - \mu b_{21} & a_{22} - \mu b_{22} & \cdots & a_{2n} - \mu b_{2n} \\ \cdots & \cdots & \cdots & \cdots \\ a_{n1} - \mu b_{n1} & a_{n2} - \mu b_{n2} & \cdots & a_{nn} - \mu b_{nn} \end{vmatrix} = 0. \tag{18}$$

Solving (18), we find n solutions $\mu = \mu_k$ $(k = 1, 2, \ldots, n)$. Then substituting μ_k into the system (17), we find the components $\xi_1^{(k)}, \xi_2^{(k)}, \ldots, \xi_n^{(k)}$ of the corresponding basis vector e_k. The results of Sec. 10.31 guarantee that (18) has n real roots and that every root of multiplicity r corresponds to r linearly independent solutions of the system (17).

10.33. Turning to the calculation of the canonical coefficients, we now show that the coefficients $\lambda_1, \lambda_2, \ldots, \lambda_n$ in the canonical representation (15) of the form $\mathbf{A}(x, x)$ coincide with the corresponding roots $\mu_1, \mu_2, \ldots, \mu_n$ of the determinant (18). We could use an argument like that given in Sec. 10.22, but we prefer to carry out a direct calculation. Given the root μ_m, we multiply the ith equation of the system (16) by $\xi_i^{(m)}$ (the ith component of the solution corresponding to μ_m) for $i = 1, 2, \ldots, n$ and then add all the resulting equations, obtaining

$$\mathbf{A}(e_m, e_m) = \sum_{i,k=1}^{n} a_{ik}\xi_i^{(m)}\xi_k^{(m)} = \mu_m \sum_{i,k=1}^{n} b_{ik}\xi_i^{(m)}\xi_k^{(m)} = \mu_m\mathbf{B}(e_m, e_m) = \mu_m, \tag{19}$$

since $\mathbf{B}(e_m, e_m) = 1$. On the other hand, if $\eta_1^{(m)}, \eta_2^{(m)}, \ldots, \eta_m^{(m)}$ are the canonical components of the vector e_m, then obviously $\eta_i^{(m)} = 0$ if $i \neq m$

while $\eta_m^{(m)} = 1$, and hence

$$\mathbf{A}(e_m, e_m) = \sum_{i=1}^{n} \lambda_i (\eta_i^{(m)})^2 = \lambda_m. \tag{20}$$

Comparing (19) and (20), we get $\mu_m = \lambda_m$, as asserted. This result allows us to write $\mathbf{A}(x, x)$ in canonical form, without calculating the canonical basis.

10.34. The problem posed in Sec. 10.31 of simultaneously reducing two quadratic forms $\mathbf{A}(x, x)$ and $\mathbf{B}(x, x)$ to canonical form, where one of the forms, say $\mathbf{B}(x, x)$, is positive definite, was solved in a rather strong form, i.e., we reduced $\mathbf{B}(x, x)$ to a sum of squares with coefficients equal to 1. In general, this is not required, and hence the coefficients of the canonical forms are not uniquely determined. Nevertheless, as we now show, *the ratios of the corresponding canonical coefficients are still independent of the means used to simultaneously reduce* $\mathbf{A}(x, x)$ *and* $\mathbf{B}(x, x)$ *to canonical form.*

Suppose that $\mathbf{A}(x, x)$ and $\mathbf{B}(x, x)$ have been simultaneously reduced to canonical form in two different ways, i.e., suppose that in the variables $\xi_1, \xi_2, \ldots, \xi_n$ we have

$$\mathbf{A}(x, x) = \sum_{i=1}^{n} \lambda_i \xi_i^2, \qquad \mathbf{B}(x, x) = \sum_{i=1}^{n} \nu_i \xi_i^2,$$

while in the variables $\eta_1, \eta_2, \ldots, \eta_n$ we have

$$\mathbf{A}(x, x) = \sum_{i=1}^{n} \rho_i \eta_i^2, \qquad \mathbf{B}(x, x) = \sum_{i=1}^{n} \tau_i \eta_i^2.$$

Since the form $\mathbf{B}(x, x)$ is positive definite, the numbers ν_i and τ_i ($i = 1, 2, \ldots, n$) are all positive. Consider the new coordinate transformation

$$\tilde{\xi}_i = \sqrt{\nu_i}\, \xi_i, \qquad \tilde{\eta}_i = \sqrt{\tau_i}\, \eta_i.$$

Then the forms $\mathbf{A}(x, x)$ and $\mathbf{B}(x, x)$ become

$$\mathbf{A}(x, x) = \sum_{i=1}^{n} \frac{\lambda_i}{\nu_i} \tilde{\xi}_i^2, \qquad \mathbf{B}(x, x) = \sum_{i=1}^{n} \tilde{\xi}_i^2$$

in the variables $\tilde{\xi}_i$ and

$$\mathbf{A}(x, x) = \sum_{i=1}^{n} \frac{\rho_i}{\tau_i} \tilde{\eta}_i^2, \qquad \mathbf{B}(x, x) = \sum_{i=1}^{n} \tilde{\eta}_i^2$$

in the variables $\tilde{\eta}_i$. Let $e_1, e_2, \ldots, e_n$ be the basis corresponding to the variables $\tilde{\xi}_i$, and let $f_1, f_2, \ldots, f_n$ be the basis corresponding to the variables $\tilde{\eta}_i$. Both these bases are orthonormal in the metric determined by the form $\mathbf{B}(x, x)$. Moreover, according to Sec. 10.14, the set of canonical coefficients of the quadratic form $\mathbf{A}(x, x)$ is uniquely determined. Hence the two sets of numbers $\lambda_1/\nu_1, \lambda_2/\nu_2, \ldots, \lambda_n/\nu_n$ and $\rho_1/\tau_1, \rho_2/\tau_2, \ldots, \rho_n/\tau_n$ must coincide, except possibly for order, and our assertion is proved.

10.4. Reduction of the General Equation of a Quadric Surface

10.41. In this and subsequent sections, we will call the elements of the n-dimensional linear space $\mathbf{R}_n$ *points* rather than vectors (cf. Sec. 2.17), which is more in keeping with the geometry of the situation. *By a* ***quadric*** *(or* ***second-degree****)* ***surface*** *in* $\mathbf{R}_n$ *is meant the locus of the points* $x = (\xi_1, \xi_2, \ldots, \xi_n) \in \mathbf{R}_n$ *which satisfy an equation of the form*

$$\sum_{i,k=1}^{n} a_{ik}\xi_i\xi_k + 2\sum_{i=1}^{n} b_i\xi_i + c = 0 \tag{21}$$

or

$$\mathbf{A}(x, x) + 2L(x) + c = 0,$$

where

$$\mathbf{A}(x, x) = \sum_{i,k=1}^{n} a_{ik}\xi_i\xi_k$$

is a quadratic form in the components of the radius vector of the point x,

$$L(x) = \sum_{i=1}^{n} b_i\xi_i$$

is a linear form, and c *is a constant.*†

We will assume that the space $\mathbf{R}_n$ is Euclidean and that the numbers $\xi_1, \xi_2, \ldots, \xi_n$ are the coordinates of the point x with respect to an orthonormal basis. The problem of this section is then to choose a new orthonormal basis in $\mathbf{R}_n$ such that our quadric surface is specified by a particularly simple equation, called the *canonical equation* of the surface. Subsequently, we will use the canonical equation to study the properties of the surface.

10.42. First of all, as in Sec. 10.12, we make an orthogonal coordinate transformation

$$\xi_i = \sum_{j=1}^{n} q_i^{(j)}\eta_j \qquad (i = 1, 2, \ldots, n) \tag{22}$$

in $\mathbf{R}_n$, reducing the quadratic form $\mathbf{A}(x, x)$ to the canonical form

$$\mathbf{A}(x, x) = \sum_{i=1}^{n} \lambda_i\eta_i^2.$$

---

† In the case $n = 2$, the geometric object defined by (21) is called a second-degree *curve*. However, we will henceforth always use the word "surface," despite the fact that, strictly speaking, it should be changed to "curve" whenever $n = 2$.

Substituting (22) into (21), we get

$$\sum_{i=1}^{n} \lambda_i \eta_i^2 + 2 \sum_{i=1}^{n} l_i \eta_i + c = 0, \tag{23}$$

where the l_i $(i = 1, 2, \ldots, n)$ are the new coefficients of the linear form $L(x)$.

If $\lambda_i \neq 0$ for some i in (23), we can eliminate the corresponding linear term by appropriately shifting the origin of coordinates. For example, if $\lambda_1 \neq 0$, we have

$$\lambda_1 \eta_1^2 + 2 l_1 \eta_1 = \lambda_1 \left(\eta_1 + \frac{l_1}{\lambda_1} \right)^2 - \frac{l_1^2}{\lambda_1}.$$

We then set

$$\eta_1' = \eta_1 + \frac{l_1}{\lambda_1},$$

which is equivalent to shifting the origin to the point

$$\left(-\frac{l_1}{\lambda_1}, 0, 0, \ldots, 0 \right).$$

As a result of this substitution, the pair of terms $\lambda_1 \eta_1^2 + 2 l_1 \eta_1$ is changed to

$$\lambda_1 \eta_1'^2 - \frac{l_1^2}{\lambda_1^2},$$

i.e., the quadratic term has the same coefficient as before, the linear term disappears, and l_1^2/λ_1^2 is subtracted from the constant term. After making all such transformations, the equation of the surface becomes

$$\lambda_1 \eta_1^2 + \lambda_2 \eta_2^2 + \cdots + \lambda_r \eta_r^2 + 2 l_{r+1} \eta_{r+1} + \cdots + 2 l_n \eta_n + c = 0.$$

Here, for simplicity, we have dropped the primes on the variables η_i', and we have renumbered the variables in such a way that the variables appearing in the quadratic form come first, i.e., $\lambda_1, \lambda_2, \ldots, \lambda_r$ are nonzero and $\lambda_k = 0$ for $k > r$. If $r = n$ or if the numbers $l_{r+1}, l_{r+2}, \ldots, l_n$ all turn out to be zero, we obtain the equation

$$\lambda_1 \eta_1^2 + \lambda_2 \eta_2^2 + \cdots + \lambda_r \eta_r^2 + c = 0, \tag{24}$$

called the *canonical equation of a central surface*. A quadric surface is said to be *nondegenerate* if all n variables appear in its canonical equation, and *degenerate* if less than n variables appear in its canonical equation. A nondegenerate central surface, with canonical equation

$$\lambda_1 \eta_1^2 + \lambda_2 \eta_2^2 + \cdots + \lambda_n \eta_n^2 + c = 0 \tag{25}$$

(i.e., such that $r = n$), is said to be a *proper central surface* if $c \neq 0$ and a *conical surface* if $c = 0$. The meaning of this terminology will be apparent later.

Now suppose at least one of the numbers $l_{r+1}, l_{r+2}, \ldots, l_n$ is nonzero, and carry out a new orthogonal coordinate transformation by using the formulas

$$\begin{aligned} \tau_1 &= \eta_1, \\ \tau_2 &= \eta_2, \\ &\cdots\cdots \\ \tau_r &= \eta_r, \\ \tau_{r+1} &= -\frac{1}{M}(l_{r+1}\eta_{r+1} + \cdots + l_n\eta_n), \\ &\cdots\cdots\cdots\cdots\cdots\cdots \end{aligned} \tag{26}$$

where M is a positive factor guaranteeing the orthogonality of the transformation matrix. Since the sum of the squares of the elements of every row of an orthogonal matrix must equal 1, we have

$$M^2 = l_{r+1}^2 + l_{r+2}^2 + \cdots + l_n^2.$$

The remaining rows (i.e., rows $r + 2, r + 3, \ldots, n$) can be arbitrary, provided only that the resulting matrix is orthogonal (see Sec. 8.95). As a result of the transformation (26), the equation of the surface takes the form

$$\lambda_1\tau_1^2 + \cdots + \lambda_r\tau_r^2 = 2M\tau_{r+1} - c.$$

If $c \neq 0$, another shift of the origin given by the formula

$$\tau_{r+1}' = \tau_{r+1} - \frac{c}{2M},$$

or

$$2M\tau_{r+1}' = 2M\tau_{r+1} - c,$$

allows us to eliminate the constant term. Then, dropping the prime on τ_{r+1}', we obtain the equation

$$\lambda_1\tau_1^2 + \cdots + \lambda_r\tau_r^2 = 2M\tau_{r+1}, \tag{27}$$

called the *canonical equation of a noncentral surface.*

10.5. Geometric Properties of a Quadric Surface

10.51. The center of a surface. By a *center* of a surface is meant a point

$$x_0 = (\xi_1^0, \xi_2^0, \ldots, \xi_n^0)$$

with the following property: If the point

$$(\xi_1^0 + \xi_1, \xi_2^0 + \xi_2, \ldots, \xi_n^0 + \xi_n)$$

lies on the surface, then the point

$$(\xi_1^0 - \xi_1, \xi_2^0 - \xi_2, \ldots, \xi_n^0 - \xi_n),$$

which is symmetric with respect to x_0, also lies on the surface. A surface with the canonical equation (24) has at least one center, since every point for which

$$\eta_1 = \eta_2 = \cdots = \eta_r = 0 \tag{28}$$

is obviously a center. This explains why such surfaces are called *central* surfaces.

We now show that a surface with the canonical equation (24) has no centers other than the point (28), a fact that will be used later. To see this, let $(\xi_1^0, \xi_2^0, \ldots, \xi_r^0)$ be a center of the surface. Then the relation

$$\lambda_1(\xi_1^0 + \xi_1)^2 + \lambda_2(\xi_2^0 + \xi_2)^2 + \cdots + \lambda_r(\xi_r^0 + \xi_r)^2 + c = 0$$

implies

$$\lambda_1(\xi_1^0 - \xi_1)^2 + \lambda_2(\xi_2^0 - \xi_2)^2 + \cdots + \lambda_r(\xi_r^0 - \xi_r)^2 + c = 0.$$

Subtracting the first equation from the first, we obtain the equation

$$\lambda_1\xi_1^0\xi_1 + \lambda_2\xi_2^0\xi_2 + \cdots + \lambda_r\xi_r^0\xi_r = 0, \tag{29}$$

satisfied for arbitrary $\xi_1, \xi_2, \ldots, \xi_n$ corresponding to points on the surface (24). If the point $(\xi_1^0 + \xi_1, \xi_2^0 + \xi_2, \ldots, \xi_n^0 + \xi_n)$ lies on the surface (24), then so does the point $(-\xi_1^0 - \xi_1, \xi_2^0 + \xi_2, \ldots, \xi_n^0 + \xi_n)$. But

$$-\xi_1^0 - \xi_1 = \xi_1^0 + (-2\xi_1^0 - \xi_1),$$

and hence we have

$$\lambda_1\xi_1^0(-2\xi_1^0 - \xi_1) + \lambda_2\xi_2^0\xi_2 + \cdots + \lambda_r\xi_r^0\xi_r = 0, \tag{29'}$$

as well as (29). Subtracting (29′) from (29), we get

$$2\lambda_1\xi_1^0(\xi_1 + \xi_1^0) = 0,$$

which implies $\xi_1 = -\xi_1^0$ if $\xi_1^0 \neq 0$. But since ξ_1 can be replaced by $-\xi_1$, we also have $-\xi_1 = -\xi_1^0$. This, together with $\xi_1 = -\xi_1^0$, contradicts the assumption that $\xi_1^0 \neq 0$, thereby proving that $\xi_1^0 = 0$. Similarly, we find that

$$\xi_2^0 = \cdots = \xi_r^0 = 0,$$

as required.

10.52. Proper central surfaces. Consider a proper central surface, i.e., a surface with canonical equation (25), where $c \neq 0$. Dividing by c, we transform (25) into the form

$$\pm\frac{\eta_1^2}{a_1^2} \pm \frac{\eta_2^2}{a_2^2} \pm \cdots \pm \frac{\eta_n^2}{a_n^2} = 1,$$

where the numbers a_i are defined by

$$a_i = +\sqrt{\frac{c}{\lambda_i}} \qquad (i = 1, 2, \ldots, n),$$

and are called the *semiaxes* of the surface. Renumbering the coordinates in such a way that the positive terms appear first, we get

$$\frac{\eta_1^2}{a_1^2} + \frac{\eta_2^2}{a_2^2} + \cdots + \frac{\eta_k^2}{a_k^2} - \frac{\eta_{k+1}^2}{a_{k+1}^2} - \cdots - \frac{\eta_n^2}{a_n^2} = 1. \tag{30}$$

It is natural to exclude the case $k = 0$ from consideration, since there are no real values $\eta_1, \eta_2, \ldots, \eta_n$ satisfying (30) if $k = 0$. (In this case, one sometimes says that (30) defines an "imaginary" surface.) This leaves n different types of proper central surfaces, corresponding to the values $k = 1, 2, \ldots, n$.

a. In the two-dimensional case ($n = 2$), we have $k = 1$, $k = 2$, and equation (30) leads to the two curves

$$(k = 1) \qquad \frac{\eta_1^2}{a_1^2} - \frac{\eta_2^2}{a_2^2} = 1 \qquad \text{(a hyperbola)},$$

$$(k = 2) \qquad \frac{\eta_1^2}{a_1^2} + \frac{\eta_2^2}{a_2^2} = 1 \qquad \text{(an ellipse)},$$

familiar from analytic geometry.

b. For $n = 3$ we have $k = 1, k = 2, k = 3$, and the corresponding proper central surfaces in three-dimensional space are given by the equations

$$(k = 1) \qquad \frac{\eta_1^2}{a_1^2} - \frac{\eta_2^2}{a_2^2} - \frac{\eta_3^2}{a_3^2} = 1,$$

$$(k = 2) \qquad \frac{\eta_1^2}{a_1^2} + \frac{\eta_2^2}{a_2^2} - \frac{\eta_3^2}{a_3^2} = 1,$$

$$(k = 3) \qquad \frac{\eta_1^2}{a_1^2} + \frac{\eta_2^2}{a_2^2} + \frac{\eta_3^2}{a_3^2} = 1.$$

We now remind the reader of the construction of each of these three surfaces.

Consider the sections of each of the surfaces made by the horizontal planes $\eta_3 = Ca_3$ $(-\infty < C < \infty)$. These sections are respectively hyperbolas

$$(k = 1) \qquad \frac{\eta_1^2}{a_1^2} - \frac{\eta_2^2}{a_2^2} = 1 + C^2$$

with the η_1-axis as transverse axis, ellipses

$$(k = 2) \qquad \frac{\eta_1^2}{a_1^2} + \frac{\eta_2^2}{a_2^2} = 1 + C^2$$

defined for all values of C, and ellipses

$$(k = 3) \qquad \frac{\eta_1^2}{a_1^2} + \frac{\eta_2^2}{a_2^2} = 1 - C^2$$

defined only for $|C| \leqslant 1$. To locate the vertices of these sections, we construct the sections of each surface made by the coordinate planes $\eta_1 = 0$, $\eta_2 = 0$. In the case $k = 1$, only the coordinate plane $\eta_2 = 0$ gives a real section, i.e., the hyperbola

$$\frac{\eta_1^2}{a_1^2} - \frac{\eta_3^2}{a_3^2} = 1.$$

The vertices of the hyperbola formed by the horizontal sections lie on this curve, and as a result of the construction we obtain the surface shown in Figure 2, called a *hyperboloid of two sheets.*

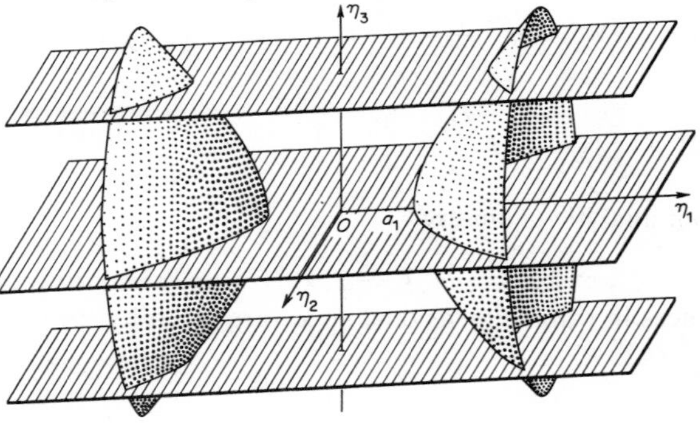

FIGURE 2

In the case $k = 2$, the sections made by both planes $\eta_1 = 0$ and $\eta_2 = 0$ are hyperbolas

$$\frac{\eta_2^2}{a_2^2} - \frac{\eta_3^2}{a_3^2} = 1, \qquad \frac{\eta_1^2}{a_1^2} - \frac{\eta_3^2}{a_3^2} = 1$$

with the η_3-axis as transverse axis. The set of ellipses formed by the horizontal sections have vertices lying on these hyperbolas, and form the surface shown in Figure 3, called a *hyperboloid of one sheet*. Finally, in the case $k = 3$, the sections made by the coordinate planes $\eta_1 = 0$, $\eta_2 = 0$ are ellipses. Drawing the ellipses made by the horizontal sections, we obtain an *ellipsoid* (see Figure 4).

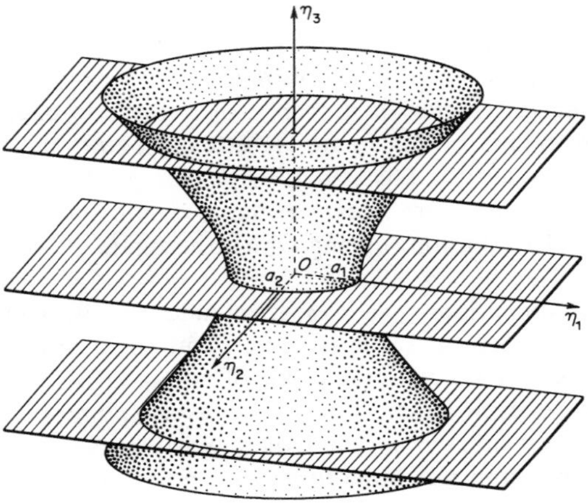

FIGURE 3

c. Quadric surfaces in spaces of more than three dimensions are not easily visualized. Nevertheless, even in the multidimensional case, we can show essential differences between the types of proper central surfaces corresponding to the different values $k = 1, 2, \ldots, n$. We begin by pointing out differences which are geometrically obvious in three dimensions. On the hyperboloid of two sheets ($k = 1$), there exists a pair of points which cannot be made to coincide by a continuous displacement of the points along the surface; to obtain such a pair of points, we need only take the first point on one sheet and the second point on the other sheet. On the hyperboloid of one sheet ($k = 2$), any two points can be made to coincide by means of a continuous displacement along the surface; however, there exists a closed curve, e.g., a curve going around the "throat" of the hyperboloid, which cannot be continuously deformed into a point. On the ellipsoid, ($k = 3$), any closed curve can be deformed into a point. These facts can

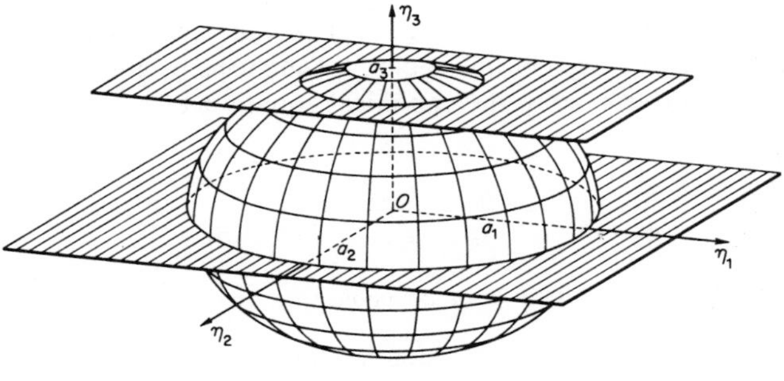

FIGURE 4

serve as the starting point for classifying the geometric differences between proper central surfaces in an n-dimensional space, as we now show.

We introduce the following definitions: A geometric figure A is said to be *homeomorphic* to a figure B if there exists a one-to-one, bicontinuous† mapping of the points of the figure A into the points of the figure B. A figure A lying on a surface S is said to be *homotopic* to a figure B lying on the same surface if the figure A can be mapped into the figure B by means of a continuous deformation, during which the figure A always remains on the surface S.

Using these definitions, we can formulate the geometric differences between the proper central surfaces corresponding to different values of k as follows: For $k = 1$ we can find a pair of points on the surface which are not homotopic to each other. For $k = 2$ every point on the surface is homotopic to every other point, but there exists a curve which is homeomorphic to a circle and not homotopic to a point. For $k = 3$ every curve which is homeomorphic to a circle is homotopic to a point, but there exists a part of the surface which is homeomorphic to a sphere (in three-dimensional space) and not homotopic to a point. Continuing in this way, we can formulate the following distinguishing property of the proper central surface corresponding to a given value of k: Every part of the surface which is homeomorphic to a sphere in $(k - 1)$-dimensional space is homotopic to a point, but there exists a part of the surface which is homeomorphic to a sphere in k-dimensional space and not homotopic to a point. In particular, this implies that the proper central surfaces in n-dimensional space (which are obviously homeomorphic to each other for equal values of k) are not homeomorphic to each other for distinct values of k. The proof of these facts will not be given here, and can be found in a course on elementary topology.

10.53. Conical surfaces. Next we consider a conical surface, i.e., a surface with canonical equation (25), where $c = 0$. In this case, equation (25) becomes homogeneous, i.e., if the point $(\eta_1, \eta_2, \ldots, \eta_n)$ satisfies (25), then so does the point $(t\eta_1, t\eta_2, \ldots, t\eta_n)$ for any t. This means that the surface is made up of straight lines going through the origin of coordinates.‡ Just as before, we can write the canonical equation of a conical surface in the form

$$\frac{\eta_1^2}{a_1^2} + \cdots + \frac{\eta_k^2}{a_k^2} - \frac{\eta_{k+1}^2}{a_{k+1}^2} - \cdots - \frac{\eta_n^2}{a_n^2} = 0. \tag{31}$$

† Equivalently, continuous in both directions, i.e., continuous with a continuous inverse.

‡ Except when all the terms in (25) have the same sign, in which case (25) defines a single point, namely the origin.

We now find the number of different types of conical surfaces corresponding to a given value of n. If the number of negative terms $m = n - k$ in the canonical equation (31) is greater than $n/2$, then, multiplying the equation by -1, we obtain an equation describing the same surface but which now has a number of negative terms less than $n/2$. Therefore it is sufficient to consider the cases corresponding to the values $m \leqslant n/2$. If m is even, then, excluding the case of a point ($m = 0$), we obtain $n/2$ different types of conical surfaces, corresponding to the values $m = 1, 2, \ldots, n/2$. If n is odd, there are $(n - 1)/2$ different types of conical surfaces, i.e., those corresponding to the values $m = 1, 2, \ldots, (n - 1)/2$.

a. In the plane ($n = 2$), besides a point, there is only one other type of conical surface ($m = 1$), with the canonical equation

$$\frac{\eta_1^2}{a_1^2} - \frac{\eta_2^2}{a_2^2} = 0.$$

The corresponding geometric figure is a pair of intersecting straight lines with the equations

$$\frac{\eta_1}{a_1} = \pm \frac{\eta_2}{a_2}.$$

In three-dimensional space ($n = 3$), besides a point, there is also only one other type of conical surface, corresponding to

$$m = \frac{n-1}{2} = \frac{3-1}{2} = 1,$$

with canonical equation

$$\frac{\eta_1^2}{a_1^2} + \frac{\eta_2^2}{a_2^2} - \frac{\eta_3^2}{a_3^2} = 0.$$

The corresponding geometric object is a cone. In the particular case where $a_1 = a_2$, this is a right circular cone (see Figure 5).

b. To visualize the form of a conical surface in the general case, we consider its intersection with the hyperplane

$$\eta_n = Ca_n \qquad (-\infty < C < \infty). \tag{32}$$

Substituting (32) into (31), we get

$$\frac{\eta_1^2}{a_1^2} + \cdots + \frac{\eta_k^2}{a_k^2} - \frac{\eta_{k+1}^2}{a_{k+1}^2} - \cdots - \frac{\eta_{n-1}^2}{a_{n-1}^2} = C^2.$$

This is the equation of a proper central surface in an $(n - 1)$-dimensional space. The surfaces corresponding to different values of C are all similar to

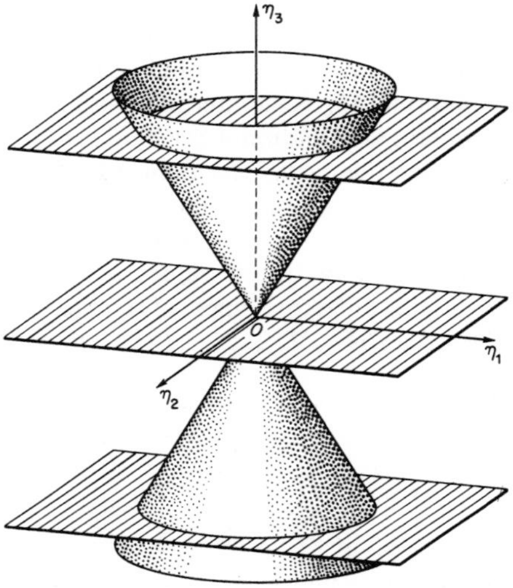

FIGURE 5

each other, with semiaxes proportional to the value of C. Thus *every conical surface in the n-dimensional space* $\mathbf{R}_n$ *can be obtained from a central surface in the* $(n-1)$*-dimensional space* $\mathbf{R}_{n-1}$ *by displacing the central surface along a perpendicular to* $\mathbf{R}_{n-1}$ *and at the same time proportionately stretching the surface in all directions.* Moreover, to obtain all possible types of conical surfaces in this way, we need only use the central surfaces in $\mathbf{R}_{n-1}$ for which the number of negative terms in the canonical equation does not exceed $(n-1)/2$.

10.54. Nondegenerate noncentral surfaces (paraboloids). Just as in Sec. 10.52, we can reduce the canonical equation of a nondegenerate noncentral surface to the form

$$\frac{\eta_1^2}{a_1^2} + \cdots + \frac{\eta_k^2}{a_k^2} - \frac{\eta_{k+1}^2}{a_{k+1}^2} - \cdots - \frac{\eta_{n-1}^2}{a_{n-1}^2} = 2\eta_n. \tag{33}$$

We now find the number of different types of nondegenerate noncentral surfaces. If the number of negative terms in the left-hand side of (33) is greater than $(n-1)/2$, then, multiplying (33) by -1, we obtain the equation of the same surface, but with a number of negative terms in the left-hand side which is less than $(n-1)/2$ and with a change of sign of the right hand side. The sign of the right-hand side is restored by the mirror reflection $\eta_n' = -\eta_n$. Thus, if we do not count surfaces obtained from each other by mirror reflections as being of different types, the number of different types of nondegenerate noncentral surfaces is equal to the number of integers satisfying the inequality $0 \leqslant m \leqslant (n-1)/2$. This number equals $n/2$ if n is even and $(n+1)/2$ if n is odd.

a. In the plane ($n = 2$) there is only one nondegenerate noncentral curve, i.e., the parabola with canonical equation

$$\eta_1^2 = 2a_1^2\eta_2 \qquad (m = 0).\dagger$$

b. In three dimensions there are two nondegenerate noncentral surfaces

$$\left(n = 3, \frac{n+1}{2} = \frac{3+1}{2} = 2\right)$$

$$\frac{\eta_1^2}{a_1^2} + \frac{\eta_2^2}{a_2^2} = 2\eta_3 \qquad (m = 0),$$

$$\frac{\eta_1^2}{a_1^2} - \frac{\eta_2^2}{a_2^2} = 2\eta_3 \qquad (m = 1).$$

In the first case ($m = 0$), the sections of the surface made by the plane $\eta_3 = C > 0$ is an ellipse. To find the position of the vertices of this ellipse, we construct the sections of the surface made by the coordinate planes $\eta_1 = 0$ and $\eta_2 = 0$. Each of these sections is a parabola, and the intersections of these parabolas with the plane $\eta_3 = C$ locate the vertices of the ellipse. The resulting surface, shown in Figure 6, is called an *elliptic paraboloid* (a *circular paraboloid* in the special case where $a_1 = a_2$).

In the second case ($m = 1$), the section of the surface made by the plane $\eta_3 = C > 0$ is a hyperbola with the η_1-axis as its transverse axis. To find

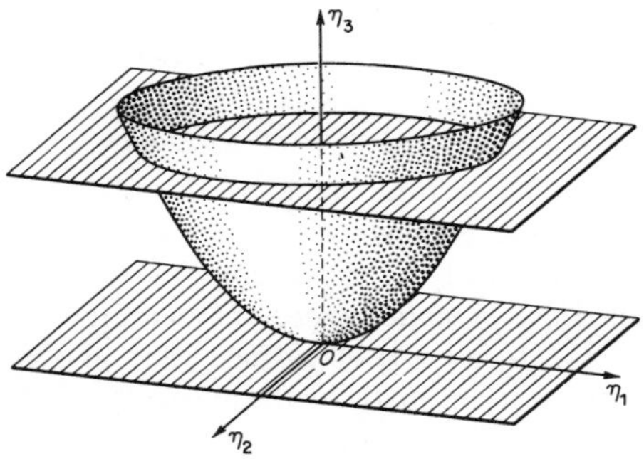

FIGURE 6

† Note that now $m = n - 1 - k$.

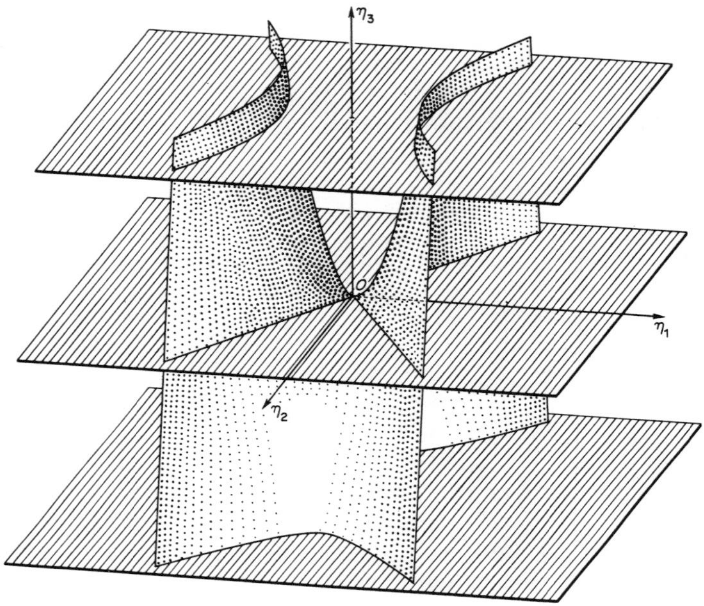

FIGURE 7

the position of the vertices, we note that the section of the surface made by the coordinate plane $\eta_2 = 0$ is the parabola

$$\eta_1^2 = 2a_1^2\eta_3,$$

whose intersection with the plane $\eta_3 = C$ gives the position of the vertices of the hyperbola. The section made by the plane $\eta_3 = C < 0$ is a hyperbola with the η_2-axis as its transverse axis. The vertices of this hyperbola lie on the parabola

$$\eta_2^2 = -2a_2^2\eta_3$$

in the plane $\eta_1 = 0$. The section made by the plane $\eta_3 = 0$ is a pair of straight lines, which serve as asymptotes for the projections on the plane $\eta_3 = 0$ of all the hyperbolas lying in horizontal sections of the surface. The surface itself is called a *hyperbolic paraboloid* (see Figure 7).

c. To visualize the form of the surface (33) in the general case, we investigate the way the sections made by the hyperplanes $\eta_n = C$ change when C varies from 0 to $+\infty$. Every such section is a central surface in $n - 1$ dimensions. All these surfaces are similar to each other, and their semiaxes (unlike the case of conical surfaces) vary according to a parabolic law, i.e., are proportional to the square root of C. For $C = 0$ the central surface becomes conical. For $C < 0$ the central surface goes into the *conjugate surface*, i.e., the positive and negative terms in the canonical equation exchange their roles. In the special case where the terms of (33) have the same sign, which, to be explicit, we take to be positive, the surface exists only in the half-space $\eta_n \geqslant 0$.

d. The reason for calling this class of nondegenerate surfaces *noncentral* is that such surfaces *actually have no centers*. For $n = 3$ this is obvious from Figures 6 and 7. To prove the assertion in the general case, assume the contrary, i.e., suppose that the surface (33) has a center $(\eta_1^0, \eta_2^0, \ldots, \eta_n^0)$. Since, in particular, this center must be a center of symmetry for the section $\eta_n = \eta_n^0$, which represents a nondegenerate central surface in $n - 1$ dimensions, we must have

$$\eta_1^0 = \eta_2^0 = \cdots = \eta_{n-1}^0 = 0$$

(cf. Sec. 10.51). Thus the center must lie on the η_n-axis. Now if we go from an arbitrary point $(\eta_1, \ldots, \eta_{n-1}, \eta_n^0 + \delta)$ lying on the surface to the symmetric point $(-\eta_1, \ldots, -\eta_{n-1}, \eta_n^0 - \delta)$, equation (33) must still be satisfied. But the left-hand side of (33) remains the same when we make this transition, and hence its right-hand side cannot change. It follows that $\delta = 0$, and hence that there are no points on the surface for which $\eta_n \neq \eta_n^0$. But (33) obviously has solutions $\eta_1, \eta_2, \ldots, \eta_n$ with $\eta_n \neq \eta_n^0$. This contradiction shows that our surface cannot have a center.

10.55. Degenerate surfaces. As in Sec. 10.42, by a degenerate surface we mean a surface whose canonical equation contains less than n coordinates. For example, suppose that the coordinate η_n is absent in the canonical equation. Then all the sections of the surface made by the $(n - 1)$-dimensional hyperplanes $\eta_n = C$ $(-\infty < C < \infty)$ give the same surface in $n - 1$ dimensions. Therefore *every degenerate surface in the n-dimensional space* $\mathbf{R}_n$ *is generated by translating a quadric surface in the* $(n - 1)$*-dimensional space* $\mathbf{R}_{n-1}$ *along a perpendicular to* $\mathbf{R}_{n-1}$.

a. We now find the appropriate curves in the plane $(n = 2)$. In this case, the canonical equation contains only one coordinate and hence is just

$$\frac{\eta_1^2}{a_1^2} = C.$$

For $C > 0$ we obtain a pair of parallel lines, for $C = 0$ a pair of coincident lines, and for $C < 0$ an "imaginary curve."

b. To construct degenerate surfaces in three-dimensional space $(n = 3)$, we must translate all the second-degree curves in the $\eta_1\eta_2$-plane along the η_3-axis. When this is done, ellipses, hyperbolas and parabolas give elliptic, hyperbolic and parabolic cylinders, respectively (see Figure 8), while pairs

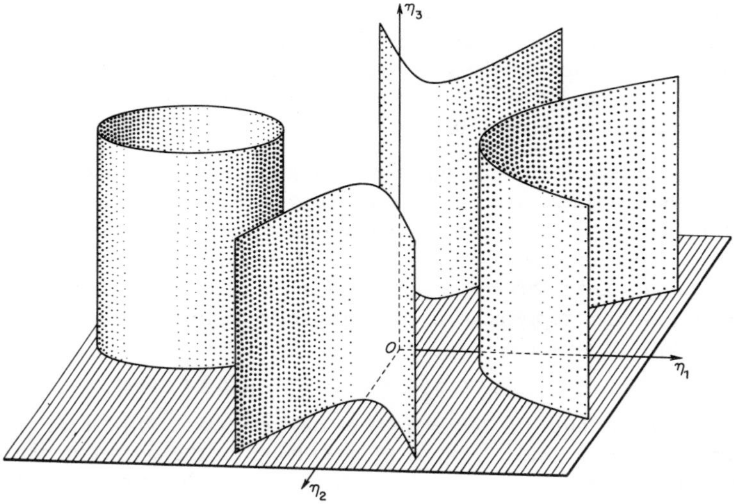

FIGURE 8

of intersecting, parallel and coincident lines lead to intersecting, parallel and coincident planes (see Figure 9).

*10.6. Analysis of a Quadric Surface from Its General Equation

10.61. We have just described all possible types of quadric surfaces in an n-dimensional Euclidean space, where the type of the surface was determined from its canonical equation. However, the surface is often specified by its general equation (21) rather than by its canonical equation, and it is sometimes important to determine the type of the surface, i.e., construct its

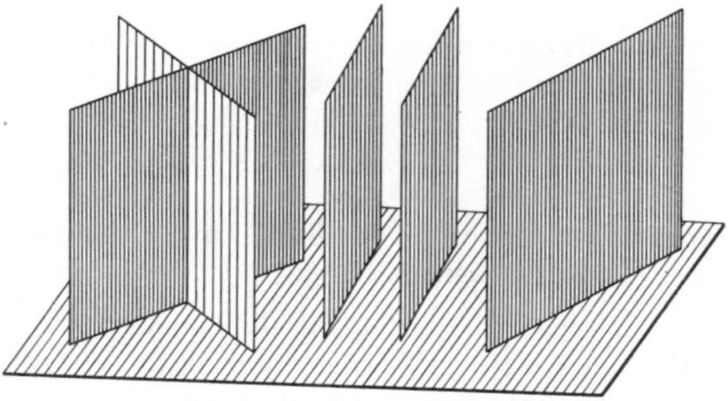

FIGURE 9

canonical equation, without carrying out all the transformations described in Sec. 10.42. It turns out that to write down the canonical equation of the surface specified by equation (21), we need only know the following two quantities:

a) The roots of the polynomial

$$\Delta(\lambda) = \begin{vmatrix} a_{11} - \lambda & a_{12} & \cdots & a_{1n} \\ a_{21} & a_{22} - \lambda & \cdots & a_{2n} \\ \cdot & \cdot & \cdots & \cdot \\ a_{n1} & a_{n2} & \cdots & a_{nn} - \lambda \end{vmatrix}$$

of degree n;

b) The coefficients of the polynomial

$$\Delta_1(\lambda) = \begin{vmatrix} a_{11} - \lambda & a_{12} & \cdots & a_{1n} & b_1 \\ a_{21} & a_{22} - \lambda & \cdots & a_{2n} & b_2 \\ \cdot & \cdot & \cdots & \cdot & \cdot \\ a_{n1} & a_{n2} & \cdots & a_{nn} - \lambda & b_n \\ b_1 & b_2 & \cdots & b_n & c \end{vmatrix}$$

of degree n.

To obtain explicit expressions for the coefficients of $\Delta_1(\lambda)$, we use the linear property of determinants (Sec. 1.44). Every column of the determinant $\Delta_1(\lambda)$, except the last one, can be written as a sum of two columns, the first consisting of the numbers a_{ij} ($i = 1, 2, \ldots, n; j$ fixed) and the number b_j, the second consisting of n zeros and the number $-\lambda$. As a result, the determinant $\Delta_1(\lambda)$ can be written as a sum of determinants, each of which is obtained by replacing certain columns (except the last one) in the matrix

$$A_1 = \begin{vmatrix} a_{11} & a_{12} & \cdots & a_{1n} & b_1 \\ a_{21} & a_{22} & \cdots & a_{2n} & b_2 \\ \cdot & \cdot & \cdots & \cdot & \cdot \\ a_{n1} & a_{n2} & \cdots & a_{nn} & b_n \\ b_1 & b_2 & \cdots & b_n & c \end{vmatrix} \tag{34}$$

by columns consisting of n zeros and the single element $-\lambda$, with the number $-\lambda$ appearing on the principal diagonal of the matrix. After expansion with respect to the columns containing the number $-\lambda$, each of these determinants becomes

$$(-\lambda)^k M_{n+1-k},$$

where k is the number of columns containing the element $-\lambda$, and M_{n+1-k} is a minor of order $n + 1 - k$ of the matrix A_1. This minor is characterized by the fact that if it uses the ith row ($i = 1, 2, \ldots, n$) of A_1, it also uses the ith column, and moreover, it must use the last row and column of A_1. Minors with this property will be called *bordered minors*. It is obvious that *every* bordered minor of the matrix A_1 appears in the expansion of the determinant $\Delta_1(\lambda)$. From this we immediately conclude that *the coefficient of* $(-\lambda)^k$ *in the expansion of the determinant* $\Delta_1(\lambda)$ *in powers of* $-\lambda$ *equals the sum of all the bordered minors of order* $n + 1 - k$. It is convenient to write the expansion of $\Delta_1(\lambda)$ in the form

$$\Delta_1(\lambda) = \alpha_{n+1} - \alpha_n\lambda + \alpha_{n-1}\lambda^2 - \cdots + \alpha_1(-\lambda)^n,$$

where the coefficient α_k is the sum of all the kth-order bordered minors of the matrix A_1.

10.62. As we already know, the roots of the characteristic polynomial $\Delta(\lambda)$ give us the coefficients of the squared variables in the canonical equation. To find the remaining term, which is of degree 0 if the canonical equation has the form (24) and of degree 1 if it has the form (27), we must examine the behavior of the polynomial $\Delta_1(\lambda)$ under coordinate transformations.

Thus consider the quadratic form

$$\mathbf{A}_1(x, x) = \sum_{i,k=1}^{n} a_{ik}\xi_i\xi_k + 2\sum_{i=1}^{n} b_i\xi_i\xi_{n+1} + c\xi_{n+1}^2 \qquad (35)$$

in the $(n + 1)$-dimensional Euclidean space $\mathbf{R}_{n+1}$, where $\xi_1, \xi_2, \ldots, \xi_n, \xi_{n+1}$ are the components of the vector $x \in \mathbf{R}_{n+1}$ with respect to some orthonormal basis $e_1, e_2, \ldots, e_n, e_{n+1}$. The operator corresponding to (35) is the symmetric operator $\mathbf{A}_1$ which has the matrix (34) in the basis $e_1, e_2, \ldots, e_n, e_{n+1}$; we will also denote this matrix by $A_{(e)}$. Besides this operator, consider the operator $\mathbf{E}_1$ defined by the relations

$$\mathbf{E}_1 e_k = e_k \qquad (k \leqslant n),$$

$$\mathbf{E}_1 e_{n+1} = 0.$$

This operator has the matrix

$$E_1 = \left\| \begin{matrix} 1 & 0 & 0 & \cdots & 0 & 0 \\ 0 & 1 & 0 & \cdots & 0 & 0 \\ 0 & 0 & 1 & \cdots & 0 & 0 \\ \cdot & \cdot & \cdot & \cdots & \cdot & \cdot \\ 0 & 0 & 0 & \cdots & 1 & 0 \\ 0 & 0 & 0 & \cdots & 0 & 0 \end{matrix} \right\| \qquad (36)$$

in the same basis $e_1, e_2, \ldots, e_n, e_{n+1}$. Let $\mathbf{R}_n$ denote the subspace with the vectors $e_1, e_2, \ldots, e_n$ as a basis. Then the operator $\mathbf{E}_1$ is obviously the identity operator in this subspace.

Now suppose we are given an isometric operator $\mathbf{Q}$ in the space $\mathbf{R}_n$ Then $\mathbf{Q}$ carries the orthonormal basis $e_1, e_2, \ldots, e_n$ into another orthonormal basis $f_1, f_2, \ldots, f_n$. We construct a new isometric operator $\mathbf{Q}_1$ in the space $\mathbf{R}_{n+1}$ by setting

$$\mathbf{Q}_1 e_k = f_k \qquad (k \leqslant n),$$

$$\mathbf{Q}_1 e_{n+1} = e_{n+1} = f_{n+1}.$$

If the matrix of the operator $\mathbf{Q}$ has the form

$$Q = \begin{Vmatrix} q_{11} & q_{12} & \cdots & q_{1n} \\ q_{21} & q_{22} & \cdots & q_{2n} \\ \cdot & \cdot & \cdots & \cdot \\ q_{n1} & q_{n2} & \cdots & q_{nn} \end{Vmatrix}$$

in the space $\mathbf{R}_n$, then the matrix of the operator $\mathbf{Q}_1$ just constructed has form

$$Q_1 = \begin{Vmatrix} q_{11} & q_{12} & \cdots & q_{1n} & 0 \\ q_{21} & q_{22} & \cdots & q_{2n} & 0 \\ \cdot & \cdot & \cdots & \cdot & \cdot \\ q_{n1} & q_{n2} & \cdots & q_{nn} & 0 \\ 0 & 0 & \cdots & 0 & 1 \end{Vmatrix}$$

in the space $\mathbf{R}_{n+1}$. This matrix corresponds to the following coordinate transformation (see Sec. 8.94):

$$\begin{aligned} \xi_1 &= q_{11}\eta_1 + q_{21}\eta_2 + \cdots + q_{n1}\eta_n, \\ \xi_2 &= q_{12}\eta_1 + q_{22}\eta_2 + \cdots + q_{n2}\eta_n, \\ &\cdots\cdots\cdots\cdots\cdots\cdots\cdots \\ \xi_n &= q_{1n}\eta_1 + q_{2n}\eta_2 + \cdots + q_{nn}\eta_n, \\ \xi_{n+1} &= \eta_{n+1}. \end{aligned} \tag{37}$$

In the new basis $f_1, f_2, \ldots, f_n, f_{n+1}$ the operator $\mathbf{A}$ has the matrix

$$A_{(f)} = Q^{-1}A_{(e)}Q$$

(see Sec. 5.51), while the operator $\mathbf{E}_1$ has the same matrix (36) as before. Moreover, according to Sec. 5.52,

$$\det (A_{(f)} - \lambda E_1) = \det (A_{(e)} - \lambda E_1).$$

We now assume that (37) is the transformation (see Sec. 10.42) which reduces the quadratic form

$$\mathbf{A}(x, x) = \sum_{i,k=1}^{n} a_{ik}\xi_i\xi_k$$

to the canonical form

$$\mathbf{A}(x, x) = \sum_{i=1}^{n} \lambda_i\eta_i^2.$$

It follows from (37) that $\mathbf{Q}_1$ transforms the quadratic form (35) in $n+1$ variables into

$$\sum_{i=1}^{n} \lambda_i\eta_i^2 + 2\sum_{i=1}^{n} l_i\eta_i\eta_{n+1} + c\eta_{n+1}^2. \tag{38}$$

After this transformation, the matrix of the operator $\mathbf{A}_1$, which, as we know, transforms in the same way as the matrix of the corresponding quadratic form, becomes

$$A_{(f)} = \left\| \begin{matrix} \lambda_1 & 0 & \cdots & 0 & 0 & \cdots & 0 & l_1 \\ 0 & \lambda_2 & \cdots & 0 & 0 & \cdots & 0 & l_2 \\ \cdot & \cdot & \cdots & \cdot & \cdot & \cdots & \cdot & \cdot \\ 0 & 0 & \cdots & \lambda_r & 0 & \cdots & 0 & l_r \\ 0 & 0 & \cdots & 0 & 0 & \cdots & 0 & l_{r+1} \\ \cdot & \cdot & \cdots & \cdot & \cdot & \cdots & \cdot & \cdot \\ 0 & 0 & \cdots & 0 & 0 & \cdots & 0 & l_n \\ l_1 & l_2 & \cdots & l_r & l_{r+1} & \cdots & l_n & c \end{matrix} \right\|,$$

and the polynomial $\Delta_1(\lambda) = \det (A_{(f)} - \lambda E_1)$ equals the determinant

$$\left| \begin{matrix} \lambda_1 - \lambda & 0 & \cdots & 0 & 0 & \cdots & 0 & l_1 \\ 0 & \lambda_2 - \lambda & \cdots & 0 & 0 & \cdots & 0 & l_2 \\ \cdot & \cdot & \cdots & \cdot & \cdot & \cdots & \cdot & \cdot \\ 0 & 0 & \cdots & \lambda_r - \lambda & 0 & \cdots & 0 & l_r \\ 0 & 0 & \cdots & 0 & -\lambda & \cdots & 0 & l_{r+1} \\ \cdot & \cdot & \cdots & \cdot & \cdot & \cdots & \cdot & \cdot \\ 0 & 0 & \cdots & 0 & 0 & \cdots & -\lambda & l_n \\ l_1 & l_2 & \cdots & l_r & l_{r+1} & \cdots & l_n & c \end{matrix} \right|.$$

The coefficients of this polynomial can be calculated by using the bordered minors of the matrix $A_{(f)}$, just as they were calculated before by using the bordered minors of the matrix $A_{(e)} = A_1$.

We note that for $r < n$ all the bordered minors of the matrix $A_{(f)}$ which are of order higher than $r + 2$ must vanish, since they contain two proportional columns. Thus for $r < n$ the coefficients $\alpha_{r+3}, \alpha_{r+4}, \ldots, \alpha_{n+1}$ vanish. Moreover, for $r < n$ the nonvanishing minors of order $r + 2$ must use the first r rows and first r columns of the matrix $A_{(f)}$. In general, the bordered minors of order $r + 1$ need not use these r rows and columns. However, we note the following two cases where a bordered minor of order $r + 1$ must in fact use the first r rows and columns:

1) $r = n$, in which case it is obvious that the matrix $A_{(f)}$ has only one minor of order $r + 1$ (i.e., of order $n + 1$), namely its determinant, made up of *all* the rows and columns of $A_{(f)}$;
2) $r < n$, $l_{r+1} = l_{r+2} = \cdots = l_n = 0$, in which case there is only one nonvanishing bordered minor of order $r + 1$, made up of elements from the rows and columns with numbers $1, 2, \ldots, r, n + 1$.

10.63. Next we show how the next step in the transformation of equation (38), made with the aim of eliminating the quantities $l_1, l_2, \ldots, l_r$, affects the matrix of the operator $\mathbf{A}_1$. First consider the transformation

$$\eta_1' = \eta_1 + \frac{l_1}{\lambda_1}\eta_{n+1}',$$

$$\eta_k' = \eta_k \qquad (k = 2, 3, \ldots, n + 1),$$

carrying the matrix $A_{(f)}$ into the matrix

$$A_{(f)}^{(1)} = \left\| \begin{matrix} \lambda_1 & 0 & \cdots & 0 & 0 & \cdots & 0 & 0 \\ 0 & \lambda_2 & \cdots & 0 & 0 & \cdots & 0 & l_2 \\ \cdot & \cdot & \cdots & \cdot & \cdot & \cdots & \cdot & \cdot \\ 0 & 0 & \cdots & \lambda_r & 0 & \cdots & 0 & l_r \\ 0 & 0 & \cdots & 0 & 0 & \cdots & 0 & l_{r+1} \\ \cdot & \cdot & \cdots & \cdot & \cdot & \cdots & \cdot & \cdot \\ 0 & 0 & \cdots & 0 & 0 & \cdots & 0 & l_n \\ 0 & l_2 & \cdots & l_r & l_{r+1} & \cdots & l_n & c - \dfrac{l_1^2}{\lambda_1} \end{matrix} \right\|.$$

This operation on $A_{(f)}$ can be described as follows: The first column is multiplied by l_1/λ_1 and subtracted from the last column, and then the first row is also multiplied by l_1/λ_1 and subtracted from the last row. The subsequent transformations required to eliminate the quantities $l_2, l_3, \ldots, l_r$

can be described similarly. As a result of all these transformations, the matrix $A_{(f)}$ goes into the matrix

$$A_{(f)}^{(r)} = \left\| \begin{matrix} \lambda_1 & 0 & \cdots & 0 & 0 & \cdots & 0 & 0 \\ 0 & \lambda_2 & \cdots & 0 & 0 & \cdots & 0 & 0 \\ \cdot & \cdot & \cdots & \cdot & \cdot & \cdots & \cdot & \cdot \\ 0 & 0 & \cdots & \lambda_r & 0 & \cdots & 0 & 0 \\ 0 & 0 & \cdots & 0 & 0 & \cdots & 0 & l_{r+1} \\ \cdot & \cdot & \cdots & \cdot & \cdot & \cdots & \cdot & \cdot \\ 0 & 0 & \cdots & 0 & 0 & \cdots & 0 & l_n \\ 0 & 0 & \cdots & 0 & l_{r+1} & \cdots & l_n & c' \end{matrix} \right\|.$$

Moreover, these transformations do not change the values of the bordered minors of the matrix $A_{(f)}$ which use the first r rows and columns of $A_{(f)}$.

Next consider the polynomial

$$\det (A_{(f)}^{(r)} - \lambda E_1) = \Delta_1^{(r)}(\lambda)$$

$$= \left| \begin{matrix} \lambda_1 - \lambda & 0 & \cdots & 0 & 0 & \cdots & 0 & 0 \\ 0 & \lambda_2 - \lambda & \cdots & 0 & 0 & \cdots & 0 & 0 \\ \cdot & \cdot & \cdots & \cdot & \cdot & \cdots & \cdot & \cdot \\ 0 & 0 & \cdots & \lambda_r - \lambda & 0 & \cdots & 0 & 0 \\ 0 & 0 & \cdots & 0 & -\lambda & \cdots & 0 & l_{r+1} \\ \cdot & \cdot & \cdots & \cdot & \cdot & \cdots & \cdot & \cdot \\ 0 & 0 & \cdots & 0 & 0 & \cdots & -\lambda & l_n \\ 0 & 0 & \cdots & 0 & l_{r+1} & \cdots & l_n & c \end{matrix} \right|$$

$$= \alpha'_{n+1} - \alpha'_n\lambda + \alpha'_{n-1}\lambda^2 - \cdots + \alpha'_1(-\lambda)^n,$$

where we have dropped the prime on c'. The coefficients of this polynomial are calculated by using the bordered minors of the matrix $A_{(f)}^{(r)}$ in just the same way as the coefficients of the polynomial $\Delta_1(\lambda)$ are calculated by using the bordered minors of the matrix $A_{(f)}$. Since the bordered minors of order $r + 2$ (where $r < n$) are invariant under the transformation leading from $A_{(f)}$ to $A_{(f)}^{(r)}$, as shown above, we find that $\alpha'_{r+2} = \alpha_{r+2}$. In the same way, we have $\alpha'_{r+1} = \alpha_{r+1}$ in the two special cases noted above.

10.64. First we consider the special case $r = n$. Here the coefficient α'_{n+1} of the polynomial $\Delta_1^{(r)}(\lambda)$ is obviously equal to the product $\lambda_1\lambda_2 \cdots \lambda_n c$, so that the quantity c in the canonical equation (25), p. 288 is just

$$c = \frac{\alpha'_{n+1}}{\lambda_1\lambda_2 \cdots \lambda_n} = \frac{\alpha_{n+1}}{\lambda_1\lambda_2 \cdots \lambda_n}.$$

10.65. Next suppose that $r < n$. Then we must determine the coefficient α_{r+2} of the polynomial $\Delta_1^{(r)}(\lambda)$, which will be needed in a moment.† The nonvanishing bordered minors of $A_{(f)}^{(r)}$ of order $r + 2$ have the form

$$\begin{vmatrix} \lambda_1 & 0 & \cdots & 0 & 0 & 0 \\ 0 & \lambda_2 & \cdots & 0 & 0 & 0 \\ \cdot & \cdot & \cdots & \cdot & \cdot & \cdot \\ 0 & 0 & \cdots & \lambda_r & 0 & 0 \\ 0 & 0 & \cdots & 0 & 0 & l_m \\ 0 & 0 & \cdots & 0 & l_m & c \end{vmatrix} = -\lambda_1\lambda_2 \cdots \lambda_r l_m^2 \qquad (m = r+1, \ldots, n),$$

and their sum, which equals the coefficient $\alpha'_{r+2} = \alpha_{r+2}$, is given by

$$-\lambda_1\lambda_2 \cdots \lambda_r(l_{r+1}^2 + l_{r+2}^2 + \cdots + l_n^2).$$

We recall that the condition for reducing equation (21) to the canonical form (27) is that at least one of the coefficients $l_{r+1}, l_{r+2}, \ldots, l_n$ be nonvanishing. We can now formulate this condition equivalently in the form of the inequality

$$\alpha_{r+2} \neq 0,$$

and at the same time give the following formula for calculating the coefficient M of the canonical form (27):

$$M^2 = l_{r+1}^2 + l_{r+2}^2 + \cdots + l_n^2 = -\frac{\alpha_{r+2}}{\lambda_1\lambda_2 \cdots \lambda_r}.$$

However, if $\alpha_{r+2} = 0$, then $l_{r+1} = l_{r+2} = \cdots = l_n = 0$, and (21) reduces to the canonical form (24). Thus we have arrived at another special case. In this case, the coefficient $\alpha'_{r+1} = \alpha_{r+1}$ is obviously equal to the product $\lambda_1\lambda_2 \cdots \lambda_r c$, so that the coefficient c of the canonical form (24) is just

$$\frac{\alpha'_{r+1}}{\lambda_1\lambda_2 \cdots \lambda_r} = \frac{\alpha_{r+1}}{\lambda_1\lambda_2 \cdots \lambda_r}.$$

† It is easily verified that in this case all the coefficients α_m of the polynomial $\Delta_1^{(r)}(\lambda)$ with $m > r + 2$ vanish.

10.66. We now summarize these results in the form of a table. As before, we agree to arrange the roots $\lambda_1, \lambda_2, \ldots, \lambda_n$ of the characteristic polynomial $\Delta(\lambda)$ in such a way that the nonzero roots $\lambda_1, \lambda_2, \ldots, \lambda_r$ come first, denoting the product $\lambda_1\lambda_2 \cdots \lambda_r$ by Λ_r.

Data		*Canonical Equation*
$\lambda_n \neq 0$		$\lambda_1\eta_1^2 + \lambda_2\eta_2^2 + \cdots + \lambda_n\eta_n^2 + \dfrac{\alpha_{n+1}}{\Lambda_n} = 0$
$\lambda_n = 0$, $\lambda_{n-1} \neq 0$	$\alpha_{n+1} \neq 0$	$\lambda_1\eta_1^2 + \lambda_2\eta_2^2 + \cdots + \lambda_{n-1}\eta_{n-1}^2 + 2\sqrt{-\dfrac{\alpha_{n+1}}{\Lambda_{n-1}}}\,\eta_n = 0$
	$\alpha_{n+1} = 0$	$\lambda_1\eta_1^2 + \lambda_2\eta_2^2 + \cdots + \lambda_{n-1}\eta_{n-1}^2 + \dfrac{\alpha_n}{\Lambda_{n-1}} = 0$
$\lambda_{n-1} = 0$, $\lambda_{n-2} \neq 0$	$\alpha_n \neq 0$	$\lambda_1\eta_1^2 + \lambda_2\eta_2^2 + \cdots + \lambda_{n-2}\eta_{n-2}^2 + 2\sqrt{-\dfrac{\alpha_n}{\Lambda_{n-2}}}\,\eta_{n-1} = 0$
	$\alpha_n = 0$	$\lambda_1\eta_1^2 + \lambda_2\eta_2^2 + \cdots + \lambda_{n-2}\eta_{n-2}^2 + \dfrac{\alpha_{n-1}}{\Lambda_{n-2}} = 0$
......		....
......		....
$\lambda_2 = 0$, $\lambda_1 \neq 0$	$\alpha_3 \neq 0$	$\lambda_1\eta_1^2 + 2\sqrt{-\dfrac{\alpha_3}{\lambda_1}}\,\eta_2 = 0$
	$\alpha_3 = 0$	$\lambda_1\eta_1^2 + \dfrac{\alpha_2}{\lambda_1} = 0$

10.7. Hermitian Quadratic Forms

10.71. Many of the theorems of the preceding sections carry over to the case of quadratic forms in a complex space. We begin with the following basic

THEOREM. *Every symmetric Hermitian bilinear form* $\mathbf{A}(x, y)$ *in an n-dimensional unitary space* $\mathbf{C}_n$ *has a canonical basis consisting of n orthogonal vectors.*

Proof. According to Sec. 9.34, the linear operator $\mathbf{A}$ associated with the form $\mathbf{A}(x, y)$ by the formula $\mathbf{A}(x, y) \equiv (\mathbf{A}x, y)$ is self-adjoint. Hence by Theorem 9.34, there is an orthonormal basis $e_1, \ldots, e_n$ in the space $\mathbf{C}_n$ consisting of eigenvectors of the operator $\mathbf{A}$. The matrix of the operator $\mathbf{A}$ is diagonal in this basis, and hence so is the matrix of the form $\mathbf{A}(x, y)$, since the operator and the form have the same matrix in any orthonormal basis of the space $\mathbf{C}_n$. Therefore $e_1, \ldots, e_n$ is a canonical basis of the form $\mathbf{A}(x, y)$. ∎

10.72. It follows from this theorem that every symmetric Hermitian quadratic form $\mathbf{A}(x, x)$ can be reduced to the canonical form

$$\mathbf{A}(x, x) = \sum_{j=1}^{n} \lambda_j \, |\xi_j|^2$$

by a unitary transformation. The sequence of operations leading to determination of the coefficients λ_j and the components of the vectors of the canonical basis is the same as in the real case (see Sec. 10.13).

10.73. Next we look for the stationary values of a symmetric Hermitian quadratic form $\mathbf{A}(x, x)$ on the unit sphere

$$\sum_{j=1}^{n} |\xi_j|^2 = 1$$

in $\mathbf{C}_n$, recalling from Sec. 9.15b that $\mathbf{A}(x, x)$ takes only real values. Let $e_1, \ldots, e_n$ be an orthonormal basis of the form $\mathbf{A}(x, x)$. Then in this basis we have

$$\mathbf{A}(x, x) = \sum_{j=1}^{n} \lambda_j \, |\xi_j|^2 = \sum_{j=1}^{n} \lambda_j(\sigma_j^2 + \tau_j^2),$$

$$(x, x) = \sum_{j=1}^{n} |\xi_j|^2 = \sum_{j=1}^{n} (\sigma_j^2 + \tau_j^2)$$

($\xi_j = \sigma_j + i\tau_j$). Using Lagrange's method, we equate to zero the partial derivatives of the function $\mathbf{A}(x, x) - \lambda(x, x)$ with respect to each of the $2n$ real variables σ_j, τ_j ($j = 1, \ldots, n$). This gives

$$2\lambda_j\sigma_j - 2\lambda\sigma_j = 0, \qquad 2\lambda_j\tau_j - 2\lambda\tau_j = 0 \qquad (j = 1, \ldots, n).$$

These equations are satisfied for a vector x with $|x| = 1$ if and only if λ coincides with one of the numbers $\lambda_1, \ldots, \lambda_n$. Suppose $\lambda = \lambda_k$. Then a solution of the equations is given by the vector x with components $\xi_j = \sigma_j + i\tau_j = 0$ for $j \neq k$ and $|\xi_k| = 1$. Hence, just as in the real case (Sec. 10.21), the Hermitian quadratic form $\mathbf{A}(x, x)$ takes stationary values at those vectors of the unit sphere which belong to its canonical basis $e_1, \ldots, e_n$, in other words at the eigenvectors of the corresponding self-adjoint operator $\mathbf{A}$. The values of the form at these points coincide with the corresponding canonical coefficients. In particular, the maximum of the form $\mathbf{A}(x, x)$ is the largest of the coefficients λ_j, and the minimum of $\mathbf{A}(x, x)$ is the smallest of these coefficients.

10.74. Next consider the problem of the simultaneous reduction to canonical form of two symmetric Hermitian quadratic forms $\mathbf{A}(x, x)$ and

$\mathbf{B}(x, x)$, one of which, say $\mathbf{B}(x, x)$, is positive definite. To solve this problem, we choose the Hermitian bilinear form $\mathbf{B}(x, y)$ as the scalar product. Then, by Sec. 10.72, there exists an orthonormal canonical basis for the form $\mathbf{A}(x, x)$, in the sense of the given scalar product. In this basis we have

$$\mathbf{A}(x, x) = \sum_{j=1}^{n} \lambda_j |\xi_j|^2, \qquad \mathbf{B}(x, x) = \sum_{j=1}^{n} |\xi_j|^2,$$

as required.

The calculation of the coefficients λ_j and the components of the vectors of the canonical basis (with respect to an arbitrary original basis) is carried out in the same way as in the real case (Sec. 10.32), after first writing the forms $\mathbf{A}(x, x)$ and $\mathbf{B}(x, x)$ as real functions of the real variables σ_j, τ_j $(j = 1, \ldots, n)$, where $\xi_j = \sigma_j + i\tau_j$. We leave the details as an exercise for the reader.

PROBLEMS

1. Use an orthogonal coordinate transformation to transform each of the following quadratic forms to canonical form:

a) $2\xi_1^2 + \xi_2^2 - 4\xi_1\xi_2 - 4\xi_2\xi_3$;
b) $2\xi_1^2 + 5\xi_2^2 + 5\xi_3^2 + 4\xi_1\xi_2 - 4\xi_1\xi_3 - 8\xi_2\xi_3$;
c) $2\xi_1^2 + 2\xi_2^2 + 2\xi_3^2 - 4\xi_1\xi_2 + 2\xi_1\xi_4 + 2\xi_2\xi_3 - 4\xi_3\xi_4$;
d) $2\xi_1\xi_2 + 2\xi_1\xi_3 - 2\xi_1\xi_4 - 2\xi_2\xi_3 + 2\xi_2\xi_4 + 2\xi_3\xi_4$.

2. What are the stationary values of the quadratic form

$$\mathbf{A}(x, x) = x_1^2 + \tfrac{1}{2}x_2^2 + \tfrac{1}{3}x_3^2$$

on the sphere $|x| = 1$, where $x = (x_1, x_2, x_3)$, and of what type are they (minimum, maximum, etc.)?

3. Show that each of the quantities $\mu_1, \mu_2, \ldots, \mu_k$ can actually attain the upper and lower bounds indicated in formula (6), p. 279.

4. Two quadratic forms $\mathbf{A}(x, x)$ and $\mathbf{B}(x, x)$ in $\mathbf{R}_n$ are said to be *comparable* if the inequality $\mathbf{A}(x, x) \leqslant \mathbf{B}(x, x)$ holds for every $x \in \mathbf{R}_n$. Let $\lambda_1 \geqslant \lambda_2 \geqslant \cdots \geqslant \lambda_n$ be the canonical coefficients of the form $\mathbf{A}(x, x)$, and let $\mu_1 \geqslant \mu_2 \geqslant \cdots \geqslant \mu_n$ be those of the form $\mathbf{B}(x, x)$. Show that the inequality

$$\lambda_k \leqslant \mu_k$$

holds for every $k = 1, 2, \ldots, n$. (This is obvious in the case where $\mathbf{A}(x, x)$ and $\mathbf{B}(x, x)$ have a common canonical basis.)

5. Find a common pair of conjugate directions for the curves

$$\frac{x^2}{4} + \frac{y^2}{1} = 1, \qquad 2xy = 1.$$

6. Construct the linear transformation which reduces both quadratic forms

$$\mathbf{A}(x, x) = \xi_1^2 + 2\xi_1\xi_2 + 2\xi_2^2 - 2\xi_1\xi_3 + 3\xi_3^2,$$
$$\mathbf{B}(x, x) = \xi_1^2 + 2\xi_1\xi_2 + 3\xi_2\xi_3 - 2\xi_1\xi_3 + 6\xi_3^2$$

to canonical form. What are the corresponding canonical forms?

7. Show that the basis in which the quadratic forms $\mathbf{A}(x, x)$ and $\mathbf{B}(x, x)$ both take canonical form, with canonical coefficients $\lambda_1, \lambda_2, \ldots, \lambda_n$ and $\nu_1, \nu_2, \ldots, \nu_n$, respectively, is uniquely determined to within numerical factors, provided that the ratios

$$\frac{\lambda_1}{\nu_1}, \frac{\lambda_2}{\nu_2}, \ldots, \frac{\lambda_n}{\nu_n}$$

are distinct.

8. Prove that the midpoints of the chords of a quadric surface parallel to the vector $y = (\eta_1, \eta_2, \ldots, \eta_n)$ lie on an $(n-1)$-dimensional hyperplane (the diametral plane conjugate to the vector y).

9. What quadric surfaces in three-dimensional space (with coordinates x, y, z) are represented by the following equations:

a) $\dfrac{x^2}{4} - \dfrac{y^2}{9} + \dfrac{z^2}{1} = 1$; b) $\dfrac{x^2}{4} - \dfrac{y^2}{9} - \dfrac{z^2}{1} = -1$; c) $x = y^2 + z^2$;

d) $y = x^2 + z^2 + 1$; e) $y = xz$?

10. Simplify the following equations of quadric surfaces in three-dimensional space, and give the corresponding coordinate transformations:

a) $5x^2 + 6y^2 + 7z^2 - 4xy + 4yz - 10x + 8y + 14z - 6 = 0$;
b) $x^2 + 2y^2 - z^2 + 12xy - 4xz - 8yz + 14x + 16y - 12z - 3 = 0$;
c) $4x^2 + y^2 + 4z^2 - 4xy + 8xz - 4yz - 12x - 12y + 6z = 0$.

11. Show that the intersection of an ellipsoid with semiaxes $a_1 \geqslant a_2 \geqslant \cdots \geqslant a_n$ with a k-dimensional hyperplane going through the center of the ellipsoid is another ellipsoid with semiaxes $b_1 \geqslant b_2 \geqslant \cdots \geqslant b_k$, where

$$a_1 \geqslant b_1 \geqslant a_{n-k+1},$$
$$a_2 \geqslant b_2 \geqslant a_{n-k+2},$$
$$\cdots\cdots\cdots\cdots$$
$$a_k \geqslant b_k \geqslant a_n.$$

*chapter 11

FINITE-DIMENSIONAL ALGEBRAS AND THEIR REPRESENTATIONS

11.1. More on Algebras

11.11. The concept of an algebra was introduced in Sec. 6.21, this being the name given to a linear space (over a field K) equipped with a (commutative or noncommutative) operation of multiplication of elements, obeying axioms 1)–3), p. 136. The algebras considered in Chapter 6 were for the most part commutative, but, in passing, we mentioned an important example of a noncommutative finite-dimensional algebra, namely, the algebra $\mathbf{B}(\mathbf{K}_n)$ of all linear operators acting in an n-dimensional space $\mathbf{K}_n$. This chapter is devoted to the study of $\mathbf{B}(\mathbf{K}_n)$ and its subalgebras. But first we will find it convenient to consider abstract finite-dimensional algebras.

11.12. Not every algebra has a unit, as shown by the example of the trivial algebra, i.e., any algebra such that $xy = 0$ for all elements x and y (Example 6.22a). Nevertheless, every algebra can be extended to an algebra with a unit in the following standard way:

Given any algebra $\mathbf{A}$, let $\mathbf{A}^+$ be the set of all formal sums $a + \lambda$, where $a \in \mathbf{A}$ and λ is a number from the field K. Then $\mathbf{A}^+$ is obviously a linear space with operations

$$(a + \lambda) + (b + \mu) = (a + b) + (\lambda + \mu)$$

and

$$\mu(a + \lambda) = \mu a + \lambda\mu$$

($a, b \in \mathbf{A}$; $\lambda, \mu \in K$). Moreover, $\mathbf{A}^+$ is an algebra with respect to the multiplication operation

$$(a + \lambda)(b + \mu) = (ab + \lambda b + \mu a) + \lambda\mu.$$

The algebra $\mathbf{A}^+$ certainly has a unit, i.e., the formal sum of the zero element of $\mathbf{A}$ and the number 1. We now need only note that the original algebra $\mathbf{A}$ can be regarded as a subset of $\mathbf{A}^+$ by simply identifying each element $a \in \mathbf{A}$ with the formal sum $a + 0 \in \mathbf{A}^+$.

11.2. Representations of Abstract Algebras

11.21. Let $\mathbf{A}$ be an abstract algebra over a field K, and let $\mathbf{B}(\mathbf{K})$ be the algebra of all linear operators acting in a linear space $\mathbf{K}$ over the same field K. We now consider morphisms of the algebra $\mathbf{A}$ into the algebra $\mathbf{B}(\mathbf{K})$, henceforth indicated by notation of the form $\mathbf{T}:\mathbf{A} \to \mathbf{B}(\mathbf{K})$.

a. Definition. A morphism $\mathbf{T}:\mathbf{A} \to \mathbf{B}(\mathbf{K})$ is called a *representation of the algebra* $\mathbf{A}$ *in the space* $\mathbf{K}$. A representation is called *trivial* if $\mathbf{T}a = 0$ for every $a \in \mathbf{A}$ and *exact* (or *faithful*) if $\mathbf{T}$ is a monomorphism, i.e., if the operators $\mathbf{T}_a$ and $\mathbf{T}_b$ corresponding to distinct elements a and b of the algebra $\mathbf{A}$ are themselves distinct elements of the algebra $\mathbf{B}(\mathbf{K})$.

The set of all elements $a \in \mathbf{A}$ which are carried into the zero operator by the representation $\mathbf{T}$ is called the *kernel* of the representation $\mathbf{T}$. The kernel of the trivial representation is the whole algebra $\mathbf{A}$, while the kernel of an exact representation consists of a single element, namely the zero element of the algebra. In the general case, the kernel of any representation is a two-sided ideal of the algebra $\mathbf{A}$ (see Example 6.25d).

b. Definition. Two representations $\mathbf{T}':\mathbf{A} \to \mathbf{B}(\mathbf{K}')$ and $\mathbf{T}'':\mathbf{A} \to \mathbf{B}(\mathbf{K}'')$ of an algebra $\mathbf{A}$ are said to be *equivalent* if there is an isomorphism $\mathbf{U}:\mathbf{K}' \to \mathbf{K}''$ between the linear spaces $\mathbf{K}'$ and $\mathbf{K}''$ such that

$$\mathbf{U}\mathbf{T}'_a = \mathbf{T}''_a\mathbf{U}$$

for every $a \in \mathbf{A}$. Obviously, in the case of finite-dimensional spaces $\mathbf{K}'$ and $\mathbf{K}''$, equivalence of the representations $\mathbf{T}'$ and $\mathbf{T}''$ means that the operators $\mathbf{T}'_a$ and $\mathbf{T}''_a$ ($a \in \mathbf{A}$) have identical matrices in suitable bases of the spaces $\mathbf{K}'$ and $\mathbf{K}''$.

c. Let $\mathbf{T}:\mathbf{A} \to \mathbf{B}(\mathbf{K})$ be a representation of the algebra $\mathbf{A}$. A subspace $\mathbf{K}' \subset \mathbf{K}$ is called an *invariant subspace of the representation* $\mathbf{T}$ if it is invariant with respect to all operators $\mathbf{T}_a$, $a \in \mathbf{A}$. By considering the operators $\mathbf{T}_a$ only on the space $\mathbf{K}'$, we obviously get a new representation $\mathbf{T}^{\mathbf{K}'}:\mathbf{A} \to \mathbf{B}(\mathbf{K}')$, called the *restriction* of the representation $\mathbf{T}$ onto $\mathbf{K}'$.

d. Finally let $\mathbf{T}:\mathbf{A} \to \mathbf{B}(\mathbf{K})$ be a representation of the algebra $\mathbf{A}$ such that $\mathbf{K}$ is the direct sum of subspaces $\mathbf{K}_k$ $(1 \leqslant k \leqslant n)$ invariant with respect to the representation $\mathbf{T}$, and let $\mathbf{T}^k$ denote the restriction of the representation $\mathbf{T}$ onto $\mathbf{K}_k$ $(1 \leqslant k \leqslant n)$. Then we say that *the representation* $\mathbf{T}$ *is the direct sum of the representations* $\mathbf{T}^k$ $(1 \leqslant k \leqslant n)$.

11.22. To every algebra $\mathbf{A}$ we can assign in a natural way a representation $\mathbf{T}:\mathbf{A} \to \mathbf{B}(\mathbf{A})$ in the linear space $\mathbf{A}$ itself which associates with each element $a \in \mathbf{A}$ the operator of left multiplication by a, i.e., the operator $\tilde{\mathbf{T}}_a$ defined by the formula $\tilde{\mathbf{T}}_a b = ab$ for every $b \in \mathbf{A}$. This representation is called the *left regular representation* of the algebra $\mathbf{A}$. The invariant subspaces of the left regular representation are obviously left ideals in $\mathbf{A}$ (Sec. 6.23a). Using this concept, we can establish the following important

THEOREM. *Every algebra is isomorphic to a subalgebra of the algebra* $\mathbf{B}(\mathbf{K})$, *for a suitable choice of* $\mathbf{K}$.

Proof. It is easy to see that the theorem is equivalent to the assertion that every algebra has an exact representation. Let $\mathbf{A}$ be the given algebra. As shown in Sec. 11.12, there exists an algebra $\mathbf{A}^+$ with a unit e which has $\mathbf{A}$ as a subalgebra. Let $\tilde{\mathbf{T}}:\mathbf{A}^+ \to \mathbf{B}(\mathbf{A}^+)$ be the left regular representation of this algebra. Then $\tilde{\mathbf{T}}$ is exact, since $\tilde{\mathbf{T}}_a e = ae = a \neq 0$ for every $a \in \mathbf{A}^+$, $a \neq 0$. Hence the restriction of the morphism $\tilde{\mathbf{T}}$ onto the subalgebra $\mathbf{A} \subset \mathbf{A}^+$ is an exact representation of the algebra $\mathbf{A}$ in the space $\mathbf{K} = \mathbf{A}^+$. ▮

11.3. Irreducible Representations and Schur's Lemma

11.31. Among all representations of a given algebra we now distinguish those with the simplest structure in a certain sense. Every representation $\mathbf{T}:\mathbf{A} \to \mathbf{B}(\mathbf{K})$ of an algebra $\mathbf{A}$ has at least two invariant subspaces, $\mathbf{K}$ itself and the subspace $\{0\}$ consisting of the zero element alone. Any other invariant subspace is said to be *proper*. Proper invariant subspaces which contain no other such subspaces are called *minimal invariant subspaces* of the representation $\mathbf{T}$.

Definition. A nontrivial representation $\mathbf{T}:\mathbf{A} \to \mathbf{B}(\mathbf{K})$ is said to be *irreducible* if it has no proper invariant subspaces.

11.32. Given any vector $z \in \mathbf{K}$, it is easy to see that the set $\mathbf{K}_z = \{\mathbf{T}_a z \in \mathbf{K} : a \in \mathbf{A}\}$ is an invariant subspace of the representation $\mathbf{T}$. A vector $z \in \mathbf{K}$ is said to be *cyclic* (with respect to the representation $\mathbf{T}$) if $\mathbf{K}_z = \mathbf{K}$. This definition, together with the definition of irreducibility, immediately implies the following

THEOREM. *A representation acting in the space* $\mathbf{K}$ *is irreducible if and only if every nonzero vector* $z \in \mathbf{K}$ *is cyclic.*

Despite its simplicity, this result will subsequently be found very useful.

11.33. The irreducible representations of algebras over the field C of complex numbers have the following important property:

THEOREM (***Schur's lemma***). *Let* $\mathbf{T}: \mathbf{A} \to \mathbf{B}(\mathbf{C})$ *be an irreducible representation of the algebra* $\mathbf{A}$ *over the field* C. *Then every operator in* $\mathbf{C}$ *which commutes with all the operators* $\mathbf{T}_a$, $a \in \mathbf{A}$, *is a multiple of the identity operator* $\mathbf{E}$.

Proof. Let $\mathbf{S}$ be an operator which commutes with all $\mathbf{T}_a$, $a \in \mathbf{A}$, and let x be an eigenvector of $\mathbf{S}$ (Sec. 4.9). Then $\mathbf{S}x = \lambda x$ for some complex λ, and hence $\mathbf{S}\mathbf{T}_a x = \mathbf{T}_a \mathbf{S}x = \lambda \mathbf{T}_a x$ for every $a \in \mathbf{A}$. But the representation $\mathbf{T}$ is irreducible, and hence, by Theorem 11.32, every vector $y \in \mathbf{C}$ can be represented in the form $y = \mathbf{T}_a x$, $a \in \mathbf{A}$. It follows that $\mathbf{S} = \lambda \mathbf{E}$. ▌

It should be noted that the proof makes essential use of the fact that every linear operator in a (finite-dimensional) complex linear space has an eigenvector (see Sec. 4.95b). In view of the decisive role of Schur's lemma, we will henceforth confine ourselves to a consideration of linear spaces and algebras over the field of complex numbers.

11.4. Basic Types of Finite-Dimensional Algebras

Beginning with this section, unless the contrary is explicitly stated, we will consider only finite-dimensional algebras (i.e., algebras which are finite-dimensional regarded as linear spaces) over the field C of complex numbers.

What is the structure of finite-dimensional algebras and their representations? Most of this chapter will be devoted to results along just these lines. In particular, we will distinguish some classes of algebras whose structure can be studied completely, i.e., we will succeed in describing all such algebras (to within an isomorphism) and all their representations. We refer to the classes of simple and semisimple algebras.

The various classes of algebras arise when we consider specific properties of their ideals and representations.

11.41. Definition. A nontrivial algebra is called *simple* if it contains no proper two-sided ideals (Sec. 6.23a). An example of a simple algebra is the algebra $\mathbf{B}(\mathbf{C}_n)$ of all linear operators in a finite-dimensional space. In fact, let $\mathbf{J}$ be a two-sided ideal in the algebra $\mathbf{B}(\mathbf{C}_n)$, and let $A = \|a_{jk}\| \in \mathbf{J}$ be a nonzero matrix such that $a_{rs} \neq 0$, say. Then, as shown in Sec. 4.44, by multiplying the matrix A from the right and from the left by certain matrices, i.e., by performing operations that do not leave the ideal $\mathbf{J}$, we can get a matrix E_{rs} whose only nonzero element 1 appears in the rth row and sth column. Moreover, by further multiplying E_{rs} from the right and from the left by certain matrices, we can get any matrix E_{jk} $(j, k = 1, \ldots, n)$ without leaving the ideal **J**. But linear combinations of the matrices E_{jk} give the matrix of any operator in $\mathbf{B}(\mathbf{C}_n)$, and hence $\mathbf{J} = \mathbf{B}(\mathbf{C}_n)$. As we will see later (Sec. 11.64), this example is unique in the class of all finite-dimensional algebras over the complex number field.

THEOREM. *Every simple algebra has an exact irreducible representation.*

Proof. Let **A** be a simple algebra, and consider its left regular representation $\tilde{\mathbf{T}}: \mathbf{A} \to \mathbf{B}(\mathbf{A})$. It follows at once from the fact that **A** is finite-dimensional that among the invariant subspaces of the representation $\tilde{\mathbf{T}}$ there is a minimal subspace $\mathbf{A}'$. The restriction **T** of the representation $\tilde{\mathbf{T}}$ onto $\mathbf{A}'$ is nontrivial. To show this, we need only prove that for every $b \in \mathbf{A}'$, the set $\mathbf{A}b = \{ab : a \in \mathbf{A}\} \neq \{0\}$, resorting to the following simple proof (due to A. S. Nemirovski): Suppose, to the contrary, that $\mathbf{A}b = \{0\}$. Then, as is easily seen, the set $b\mathbf{A} = \{b : a \in \mathbf{A}\}$ is a two-sided ideal in **A**, and hence, since **A** is simple, either $b\mathbf{A} = \mathbf{A}$ or $b\mathbf{A} = \{0\}$. But if $b\mathbf{A} = \mathbf{A}$, then $\mathbf{A}b = \{0\}$ implies that every product in **A** equals zero, while if $b\mathbf{A} = \{0\}$, the set $\{\lambda b : \lambda \in C\}$ is a two-sided ideal in **A** since $\mathbf{A}b = \{0\}$, and hence must coincide with the whole algebra since **A** is simple. Thus, in both cases, the algebra **A** turns out to be trivial, and hence cannot be simple.

Thus the representation $\mathbf{T}: \mathbf{A} \to \mathbf{B}(\mathbf{A}')$ is nontrivial. But then, on the one hand, it is irreducible, by the minimality of $\mathbf{A}'$, while on the other hand, its kernel, being a two-sided ideal distinct from the whole simple algebra **A**, consists of the zero element alone. Therefore **T** (like any irreducible representation of **A**) is at the same time exact. ∎

It turns out that the converse theorem is also true, i.e., every finite-dimensional algebra with an exact irreducible representation is simple. This will be shown at the end of Sec. 11.64.

11.42. An arbitrary algebra may not have exact irreducible representations. But it is natural to single out those algebras whose properties can be described in terms of their irreducible representations. This leads to the following wider class of algebras:

Definition. An algebra **A** is called *semisimple* if, given any nonzero element $a \in \mathbf{A}$, there exists an irreducible representation mapping a into a nonzero operator. In other words, the intersection of the kernels of all the irreducible representations of a semisimple algebra consists of the zero element alone.

It follows from Theorem 11.41 that every simple algebra is also semisimple. On the other hand, consider the n-dimensional $(n > 1)$ algebra C_n, consisting of the elements $a = (\alpha_1, \ldots, \alpha_n)$ where $\alpha_j \in C$, with multiplication component by component.† This algebra is obviously commutative. Moreover, the set of all $a = (\alpha_1, \ldots, \alpha_n)$ such that $\alpha_1 = 0$, say, is a two-sided ideal in C_n, so that the algebra C_n is not simple. Suppose that with every element $a = (\alpha_1, \ldots, \alpha_n)$ we associate the complex number α_k $(1 \leqslant k \leqslant n)$, or equivalently the operator of multiplication by α_k in the one-dimensional space C_1. Then we get an irreducible representation of the algebra C_n which maps every element of C_n with $\alpha_k \neq 0$ into an operator distinct from zero. Since every nonzero element $a \in C_n$ has at least one nonzero component, there exists an irreducible representation mapping a into a nonzero operator. Therefore the algebra C_n is semisimple.

In this example, C_n is a direct sum of simple (one-dimensional) algebras. The example can easily be generalized by considering a direct sum of simple noncommutative algebras. Then, as will be shown in Sec. 11.77, we get the general form of a finite-dimensional semisimple algebra over the field of complex numbers.

11.43. Next we introduce algebras whose properties are, in a certain sense, the opposite of those of a semisimple algebra:

Definition. An algebra **A** is called a *radical* algebra if every nontrivial representation of **A** has a proper invariant subspace. In other words, a radical algebra has no irreducible representations at all.

As an example, consider the algebra **A** of polynomials $P(z) = c_1 z + \cdots + c_n z^n$ with the usual operations but subject to the condition $z^{n+1} = 0$. Then every element of the algebra **A** vanishes when raised to the $(n + 1)$th power, so that no element of **A** has an inverse. The algebra **A** has no nontrivial one-dimensional representations, since every nonzero linear operator in a one-dimensional space is invertible. Let **T** be a nontrivial (and hence multidimensional) representation of the algebra **A**, and let **Z** be the operator corresponding to the element z. Since **Z** (like z itself) is noninvertible, there exists a vector $e \neq 0$ such that $\mathbf{Z}e = 0$. But then $P(\mathbf{Z})e = 0$ for every $P(z) \in \mathbf{A}$. Thus we have found a nontrivial invariant subspace (the straight line determined by the vector e) of the representation **T**. It follows that **A** is a radical algebra.

11.44. Definition. By the *radical* of an algebra **A** is meant the intersection of the kernels of all irreducible representations of **A** if such representations exist, or the whole algebra **A** if no such representations exist.

† I.e., if $a = (\alpha_1, \ldots, \alpha_n)$, $b = (\beta_1, \ldots, \beta_n)$, then $ab = (\alpha_1\beta_1, \ldots, \alpha_n\beta_n)$.

Since the kernel of every representation is a two-sided ideal of the algebra **A** (see Sec. 11.21a), the radical of **A**, being an intersection of two-sided ideals of **A**, is itself a two-sided ideal of **A**.

The study of algebras with nontrivial radicals (in particular, radical algebras) involves substantial difficulties, with results that, as a rule, are not in definitive form (some of these results will be found at the end of this chapter). On the other hand, semisimple algebras and their representations can be studied in complete detail. In fact, as we will see below, the study of semisimple algebras reduces to that of simple algebras.

We now turn to the detailed study of simple algebras and their representations.

11.5. The Left Regular Representation of a Simple Algebra

11.51. Thus let **A** be a simple algebra, and let $\mathbf{T}:\mathbf{A} \to \mathbf{B}(\mathbf{X})$ be a fixed exact irreducible representation of **A** (the existence of **T** follows from Theorem 11.41). This representation will henceforth be called *standard.*

THEOREM. *Let* $\tilde{\mathbf{T}}:\mathbf{A} \to \mathbf{B}(\mathbf{A})$ *be the left regular representation of a simple algebra* **A**, *and let* **I** *be a minimal invariant subspace of* $\tilde{\mathbf{T}}$. *Then*

a) *The restriction* $\tilde{\mathbf{T}}^{\mathbf{I}}$ *of the representation* $\tilde{\mathbf{T}}$ *onto* **I** *is equivalent to* **T**;

b) *The subspace* **I**, *regarded as a subalgebra of* **A**, *has a right unit.*

Proof. First we fix an element $a \in \mathbf{I}$, $a \neq 0$. Since the representation **T** is exact, $\mathbf{T}_a x \neq 0$ for some $x \in \mathbf{X}$. Consider the linear operator $\mathbf{U}:\mathbf{I} \to \mathbf{X}$ defined by the formula $\mathbf{U}b = \mathbf{T}_b x$ for every $b \in \mathbf{I}$. It is easy to see that the kernel of the operator **U** is a left ideal in **A** (or equivalently an invariant subspace of the representation $\tilde{\mathbf{T}}$) contained in **I** but not coinciding with **I**. Hence the kernel of **U** consists of the zero element alone. On the other hand, the image of **U** is obviously a nonzero invariant subspace of the irreducible representation **T**, and hence coincides with the whole space **X**. Thus **U** is an isomorphism of **I** onto **X**. Moreover, for arbitrary $b \in \mathbf{I}$ and $c \in \mathbf{A}$,

$$\mathbf{U}\tilde{\mathbf{T}}_c^{\mathbf{I}} b = \mathbf{U}(cb) = \mathbf{T}_{cb} x = \mathbf{T}_c(\mathbf{T}_b x) = \mathbf{T}_c \mathbf{U} b,$$

and hence

$$\mathbf{U}\tilde{\mathbf{T}}_c^{\mathbf{I}} = \mathbf{T}_c \mathbf{U},$$

which shows that the representations $\tilde{\mathbf{T}}^{\mathbf{I}}$ and **T** are equivalent (see Sec. 11.21b). Furthermore, since **U** maps **I** onto all of **X**, there exists an element $e \in \mathbf{I}$ such that $\mathbf{U}e = \mathbf{T}_e x = x$. It follows that

$$\mathbf{U}(be) = \mathbf{T}_{be} x = \mathbf{T}_b(\mathbf{T}_e x) = \mathbf{T}_b x = \mathbf{U}b$$

for every $b \in \mathbf{I}$. But $\mathbf{U}$ is a one-to-one mapping, and hence $be = b$. Thus e is a right unit in the algebra. ▮

It should be noted that any exact irreducible representation of a simple algebra can be chosen as the standard representation. Therefore an automatic consequence of this theorem is the fact that all exact irreducible representations of a simple algebra are equivalent.

11.52. LEMMA. *Given an arbitrary algebra* $\mathbf{A}$, *let* $\mathbf{I}_1$ *and* $\mathbf{I}_2$ *be left ideals of* $\mathbf{A}$ *with right units* e_1 *and* e_2', *respectively, where* $ae_1 = 0$ *for every* $a \in \mathbf{I}_2$ *Then there exists a right unit* e_2 *in* $\mathbf{I}_2$ *such that* $be_2 = 0$ *for every* $b \in \mathbf{I}_1$.

Proof. Let $e_2 = e_2' - e_1 e_2'$. Then for every $a \in \mathbf{I}_2$ we have

$$ae_2 = ae_2' - ae_1 e_2' = a,$$

since $ae_2' = a$ and $ae_1 = 0$. Moreover,

$$be_2 = be_2' - be_1 e_2' = be_2' - be_2' = 0$$

for every $b \in \mathbf{I}_1$. ▮

11.53. THEOREM. *The left regular representation of a simple algebra* $\mathbf{A}$ *is the direct sum of its irreducible representations.*

Proof. We will construct the desired set of minimal invariant subspaces of the representation $\tilde{\mathbf{T}}:\mathbf{A} \to \mathbf{B}(\mathbf{A})$ by induction, proving at each step that, as an algebra, the direct sum of the subspaces already found has a right unit. For the first subspace we take any minimal invariant subspace $\mathbf{I}_1$ of the representation $\tilde{\mathbf{T}}$. According to Theorem 11.51, $\mathbf{I}_1$ has a right unit e_1. Suppose we have already found minimal invariant subspaces $\mathbf{I}_1, \ldots, \mathbf{I}_k$ such that the left ideal $\mathbf{J}_k' = \mathbf{I}_1 + \cdots + \mathbf{I}_k$ has a right unit e_k. If $\mathbf{J}_k' = \mathbf{A}$, we have succeeded in constructing the desired invariant subspaces. Otherwise, let

$$\mathbf{J}_k'' = \{a \in \mathbf{A} : ae_k = 0\}.$$

Then it is easy to see that $\mathbf{J}_k''$ is an invariant subspace of the representation $\tilde{\mathbf{T}}$, whose intersection with $\mathbf{J}_k'$ is empty. Moreover, since every element $a \in \mathbf{A}$ can be represented in the form $a = ae_k + (a - ae_k)$, where $ae_k \in \mathbf{J}_k'$ and $a - ae_k \in \mathbf{J}_k''$, the algebra $\mathbf{A}$ is the direct sum of $\mathbf{J}_k'$ and $\mathbf{J}_k''$.

The finite-dimensional invariant subspace $\mathbf{J}_k''$ contains a minimal invariant subspace, which we denote by $\mathbf{I}_{k+1}$. According to Theorem 11.51, $\mathbf{I}_{k+1}$ contains a right unit e_{k+1}', where $ae_k = 0$ for every $a \in \mathbf{I}_{k+1}$ since $\mathbf{I}_{k+1} \subset \mathbf{J}_k''$. It follows from Lemma 11.52 that $\mathbf{I}_{k+1}$ contains a right unit e_k'' such that $be_k'' = 0$ for every $b \in \mathbf{J}_k'$. Let $e_{k+1} = e_k + e_k''$. Then, as is easily verified, e_{k+1} is a right unit in the ideal

$$\mathbf{J}_{k+1}' = \mathbf{I}_1 + \cdots + \mathbf{I}_k + \mathbf{I}_{k+1}.$$

This proves the legitimacy of making the induction from k to $k + 1$. The algebra $\mathbf{A}$ is finite-dimensional, and hence at some stage we get the set of minimal invariant subspaces $\mathbf{I}_1, \ldots, \mathbf{I}_m$ of the representation $\tilde{\mathbf{T}}$ whose direct sum is the whole algebra $\mathbf{A}$. Hence the left regular representation of $\mathbf{A}$ is the direct sum of its irreducible representations. ▮

11.54. We note that it was shown in the course of the proof that every simple algebra has a right unit. Actually, we have the following stronger

THEOREM. *Every simple algebra has a unit.*

Proof. Let $\mathbf{A}$ be a simple algebra, and let e be a right unit of $\mathbf{A}$. Consider the operator $\mathbf{T}_e$ in the standard representation $\mathbf{T}:\mathbf{A} \to \mathbf{B}(\mathbf{X})$. Then

$$\mathbf{T}_a(\mathbf{T}_e x - x) = \mathbf{T}_{ae} x - \mathbf{T}_a x = 0$$

for every $x \in \mathbf{X}$ and $a \in \mathbf{A}$. Since $\mathbf{T}$ is irreducible, every nonzero vector must be cyclic (Theorem 11.32). It follows that $\mathbf{T}_e x - x = 0$. In other words, $\mathbf{T}_e$ is the identity operator in the space $\mathbf{X}$. But then

$$\mathbf{T}_a \mathbf{T}_e = \mathbf{T}_e \mathbf{T}_a = \mathbf{T}_a$$

for every $a \in \mathbf{A}$, and hence $ae = ea = a$ by the exactness of the representation $\mathbf{T}$. Therefore e is a unit in $\mathbf{A}$. ▮

11.6. Structure of Simple Algebras

At the end of this section we will solve the problem of the structure of simple algebras. In so doing, we will find the following concept very useful:

11.61. Let $\mathbf{X}$ be a linear space, and let $\mathbf{A}_0$ be a subalgebra of $\mathbf{B}(\mathbf{X})$. The subset of $\mathbf{B}(\mathbf{X})$ consisting of the operators which commute with all operators in $\mathbf{A}_0$ will be called the *commutator* of the algebra $\mathbf{A}_0$, denoted by $\overline{\mathbf{A}}_0$.

It is easy to see that $\overline{\mathbf{A}}_0$ is itself a subalgebra of $\mathbf{B}(\mathbf{X})$. The commutator of this new subalgebra, denoted by $\overline{\overline{\mathbf{A}}}_0$, will be called the *second commutator* of the algebra $\mathbf{A}_0$. Obviously we have $\mathbf{A}_0 \subset \overline{\overline{\mathbf{A}}}_0$.

11.62. Given any algebra $\mathbf{A}$, every element $a \in \mathbf{A}$ defines two operators in $\mathbf{B}(\mathbf{A})$, the operator of left multiplication $\mathbf{T}_a$, specified by the formula $\mathbf{T}_a b = ab$, and the operator of right multiplication $\mathbf{R}_a$, specified by the formula $\mathbf{R}_a b = ba$. It is easy to see that the set of all operators of left multiplication and the set of all operators of right multiplication form subalgebras in $\mathbf{B}(\mathbf{A})$, which we denote by $\mathbf{A}_0^l$ and $\mathbf{A}_0^r$, respectively.

LEMMA. *If the algebra* $\mathbf{A}$ *has a unit, then* $\overline{\mathbf{A}_0^l} = \mathbf{A}_0^r$ *and* $\overline{\mathbf{A}_0^r} = \mathbf{A}_0^l$.

Proof. If $\mathbf{S} \in \overline{\mathbf{A}_0^l}$, then

$$\mathbf{S}(ab) = \mathbf{ST}_a b = \mathbf{T}_a\mathbf{S}b = a\mathbf{S}b.$$

Setting $b = e$, where e is the unit in $\mathbf{A}$, we get $\mathbf{S}a = a\mathbf{S}e$. Therefore $\mathbf{S}$ is the operator of right multiplication by the element $\mathbf{S}e \in \mathbf{A}$, i.e., $\mathbf{S} \in \mathbf{A}_0^r$. It follows that $\overline{\mathbf{A}_0^l} \subset \mathbf{A}_0^r$, and hence that $\overline{\mathbf{A}_0^l} = \mathbf{A}_0^r$, since obviously $\mathbf{A}_0^r \subset \overline{\mathbf{A}_0^l}$. The formula $\overline{\mathbf{A}_0^r} = \mathbf{A}_0^l$ is proved in just the same way. ∎

11.63. THEOREM. *Given a simple algebra* $\mathbf{A}$ *with standard representation* $\mathbf{T}:\mathbf{A} \to \mathbf{B}(\mathbf{X})$, *let* $\mathbf{A}_0$ *be the algebra of operators of* $\mathbf{T}$. *Then* $\overline{\overline{\mathbf{A}}}_0 = \mathbf{A}_0$.

Proof. The algebra $\mathbf{A}_0$ defined above can obviously be regarded as the algebra of operators of the left regular representation $\tilde{\mathbf{T}}:\mathbf{A} \to \mathbf{B}(\mathbf{A})$ of the algebra $\mathbf{A}$. According to Theorem 11.53, this representation is the direct sum of certain irreducible representations $\tilde{\mathbf{T}}^{\mathbf{I}_i}:\mathbf{A} \to \mathbf{B}(\mathbf{I}_i)$ $(1 \leqslant i \leqslant m)$, where, by Theorem 11.51, each representation is equivalent to the standard representation. This means the following: We can find a basis $x_1, \ldots, x_n$ in the space $\mathbf{X}$ and a basis $f_1^{(i)}, \ldots, f_n^{(i)}$ in each of the subspaces $\mathbf{I}_i$ $(1 \leqslant i \leqslant m)$ such that for every $a \in \mathbf{A}$, the matrix of the operator $\mathbf{T}_a$ in the basis $f_1^{(1)}$, $f_2^{(1)}, \ldots, f_n^{(m)}$ of the whole space $\mathbf{A}$ has the quasi-diagonal form

$$\tilde{T}_a = \left\| \begin{array}{cccc} \boxed{T_a} & & & \\ & \boxed{T_a} & & \\ & & \ddots & \\ & & & \boxed{T_a} \end{array} \right\|, \tag{1}$$

where each block along the principal diagonal is the matrix of the operator $\mathbf{T}_a$ in the basis $x_1, \ldots, x_n$ and the "off-diagonal" blocks consist entirely of zeros. It follows from the rule for multiplication of block matrices (Sec. 4.51) that every matrix commuting with all matrices of the form (1) is a matrix of the form

$$\left\| \begin{array}{ccc} S_{11} & \cdots & S_{1m} \\ \cdot & \cdots & \cdot \\ S_{m1} & \cdots & S_{mm} \end{array} \right\|, \tag{2}$$

where each block S_{ij} is an $n \times n$ matrix commuting with all the matrices T_a, $a \in \mathbf{A}$.

Now let $\mathbf{P}$ be an operator in $\overline{\overline{\mathbf{A}}}_0$, and let P be its matrix in the basis $x_1, \ldots, x_n$. Then the quasi-diagonal matrix

$$\tilde{P} = \left\| \begin{matrix} \boxed{P} & & & & \\ & \boxed{P} & & & \\ & & \cdot & & \\ & & & \cdot & \\ & & & & \cdot \; \boxed{P} \end{matrix} \right\|$$

obviously commutes with all matrices of the form (2), and hence determines in the basis $f_1^{(1)}, f_2^{(1)}, \ldots, f_n^{(m)}$ of the space $\mathbf{A}$ an operator belonging to the second commutator of the algebra $\mathbf{A}_0$. By Theorem 11.54, every simple algebra has a unit, and hence, by Lemma 11.62,

$$\overline{\overline{\mathbf{A}_0^l}} = \overline{\mathbf{A}_0^r} = \mathbf{A}_0^l,$$

which means that the matrix $\tilde{P}$ determines in the basis $f_1^{(1)}, f_2^{(1)}, \ldots, f_n^{(m)}$ an operator $\tilde{\mathbf{P}}$, equal to $\mathbf{T}_b$ for some $b \in \mathbf{A}$. But then $\mathbf{P} = \mathbf{T}_b$ for the same b, and hence $\mathbf{P}$ belongs to the algebra $\mathbf{A}_0$. The proof is now complete, since $\mathbf{P}$ is an arbitrary element of $\overline{\overline{\mathbf{A}}}_0$. ▌

11.64. We are now in a position to prove the basic theorem on simple algebras:

THEOREM (***First structure theorem***). *Every simple algebra is isomorphic to the algebra of all linear operators acting in some finite-dimensional space* $\mathbf{X}$.

Proof. Let $\mathbf{A}$ be a simple algebra, and let $\mathbf{T}: \mathbf{A} \to \mathbf{B}(\mathbf{X})$ be the standard representation of $\mathbf{A}$. It is sufficient to prove that the algebra $\mathbf{A}_0$ of operators of the representation $\mathbf{T}$ coincides with $\mathbf{B}(\mathbf{X})$. Since the representation $\mathbf{T}$ is irreducible, it follows at once from Schur's lemma (Theorem 11.33) that the commutator $\overline{\mathbf{A}}_0$ of the algebra $\mathbf{A}_0$ consists of just those operators which are multiples of the identity operator. But then the second commutator $\overline{\overline{\mathbf{A}}}_0$ coincides with the whole algebra $\mathbf{B}(\mathbf{X})$. At the same time $\overline{\overline{\mathbf{A}}}_0 = \mathbf{A}_0$, by Theorem 11.63, and hence $\mathbf{A}_0 = \mathbf{B}(\mathbf{X})$. ▌

It should be noted that behind all the considerations leading to the first structure theorem lies the fact that every simple algebra has an exact irreducible representation. Hence we have incidentally proved that every algebra with an exact irreducible representation is isomorphic to the algebra $\mathbf{B}(\mathbf{X})$. It follows at once that the converse of Theorem 11.41 holds: *Every algebra with an exact irreducible representation is simple.*

11.7. Structure of Semisimple Algebras

11.71. In this section we will show that the problem of the structure of a semisimple algebra reduces completely to the problem of the structure of a simple algebra (already studied above). To this end, we will find it useful to introduce some new concepts.

Definition. By a *normal series* of an algebra $\mathbf{A}$ is meant a chain of algebras†

$$\mathbf{A} = \mathbf{I}_0 \supseteq \mathbf{I}_1 \supseteq \cdots \supseteq \mathbf{I}_n \supseteq \mathbf{I}_{n+1} = \{0\}$$

in which each algebra is a two-sided ideal of the preceding algebra. By a *composition series* of an algebra $\mathbf{A}$ is meant a normal series of $\mathbf{A}$ in which each ideal is maximal (i.e., is not contained in any larger two-sided ideal) and $\mathbf{I}_n$ contains no proper two-sided ideals.

It is easy to see that every finite-dimensional algebra has a composition series. In fact, among the (proper) two-sided ideals of a finite-dimensional algebra $\mathbf{A}$ there is a maximal ideal $\mathbf{I}_1$, say. Similarly, the algebra $\mathbf{I}_1$ contains a maximal two-sided ideal $\mathbf{I}_2$, $\mathbf{I}_2$ contains a maximal two-sided ideal $\mathbf{I}_3$, and so on. Since the original algebra $\mathbf{A}$ is finite-dimensional, after a finite number of steps we finally arrive at an algebra $\mathbf{I}_n$ which contains no further proper ideals. The chain of algebras

$$\mathbf{A} = \mathbf{I}_0 \supset \mathbf{I}_1 \supset \cdots \supset \mathbf{I}_n \supset \mathbf{I}_{n+1} = \{0\}$$

so obtained is obviously a composition series of the algebra $\mathbf{A}$.

11.72. Before turning to the special properties of normal and composition series of semisimple algebras, we prove the following

LEMMA. *Given any element a of a semisimple algebra $\mathbf{A}$, there exists an element $b \in \mathbf{A}$ such that every power of the element ba is nonzero.*

Proof. By the definition of a semisimple algebra, there exists an irreducible representation $\mathbf{T}:\mathbf{A} \to \mathbf{B}(\mathbf{X})$ such that $\mathbf{T}_a \neq \mathbf{0}$. Then for some $x \in \mathbf{X}$, $x \neq 0$, the vector $y = \mathbf{T}_a x$ is nonzero and therefore, by Theorem 11.32, is a cyclic vector of the irreducible representation $\mathbf{T}$. Hence there is an element $b \in \mathbf{A}$ such that $\mathbf{T}_b y = x$, i.e., such that

$$\mathbf{T}_{ba} x = \mathbf{T}_b(\mathbf{T}_a x) = \mathbf{T}_b y = x.$$

It follows that every power of the operator $\mathbf{T}_{ba}$, and hence every power of the element $ba \in \mathbf{A}$, is nonzero. ∎

† Here and in the rest of this section (only) we write $A \subseteq B$ (equivalently, $B \supseteq A$) to mean that A is a subset of B, reserving the notation $A \subset B$ (equivalently, $B \supset A$) to mean that A is a *proper* subset of B (i.e., $A \subseteq B$ but $A \neq B$).

11.73. THEOREM. *A normal series of a semisimple algebra cannot contain nonzero trivial algebras.*

Proof. Let $\mathbf{A}$ be a semisimple algebra, and let

$$\mathbf{A} = \mathbf{I}_0 \supseteq \mathbf{I}_1 \supseteq \cdots \supseteq \mathbf{I}_n \supseteq \mathbf{I}_{n+1} = \{0\}$$

be a normal series of $\mathbf{A}$. It can be assumed without loss of generality that the algebra $\mathbf{I}_n$ contains an element a distinct from zero. Obviously, to prove the theorem, we need only find an element $c \in \mathbf{I}_n$ such that $ca \neq 0$.

By Lemma 11.72, there exists an element $b \in \mathbf{A}$ such that every power of ba is nonzero.

$$c_k = (ba)^{2^{k+1}-1}b \qquad (k = 0, 1, \ldots, n-1).$$

Then induction on k shows that $c_k \in \mathbf{I}_{k+1}$. In fact, for $k = 0$ we have $c_0 = bab \in \mathbf{I}_1$ since $a \in \mathbf{I}_1$, and the possibility of carrying out the induction follows at once from the obvious relation $c_{k+1} = c_k a c_k$ and the fact that $a \in \mathbf{I}_{k+2}$. Thus we see that the element $c = c_{n-1}$ belongs to the algebra $\mathbf{I}_n$, and moreover

$$ca = (ba)^{2^n-1}ba = (ba)^{2^n} \neq 0,$$

as required. ▮

11.74. Next we prove three simple propositions:

LEMMA. *Let* $\mathbf{A} \supseteq \mathbf{I}_1 \supseteq \mathbf{I}_2 \supset \{0\}$ *be a normal series of an algebra* $\mathbf{A}$, *where the algebra* $\mathbf{I}_2$ *is simple. Then* $\mathbf{I}_2$ *is a two-sided ideal in* $\mathbf{A}$.

Proof. By Theorem 11.54, the algebra $\mathbf{I}_2$ has a unit e. Since $e \in \mathbf{I}_1$, the elements ae and ea belong to $\mathbf{I}_1$ for every $a \in \mathbf{A}$. But then

$$ab = a(eb) = (ae)b \in \mathbf{I}_2,$$

$$ba = (be)a = b(ea) \in \mathbf{I}_2$$

for every $b \in \mathbf{I}_2$. ▮

11.75. LEMMA. *Let* $\mathbf{A}$ *be an arbitrary algebra, and let* $\mathbf{I}$ *be a two-sided ideal of* $\mathbf{A}$ *with a unit. Then* $\mathbf{A}$ *has a two-sided ideal* $\mathbf{J}$ *such that* $\mathbf{A}$ *is the direct sum of* $\mathbf{I}$ *and* $\mathbf{J}$.

Proof. Let $\mathbf{J} = \{a \in \mathbf{A} : ae = 0\}$, where e is the unit of the algebra $\mathbf{I}$. Then obviously $\mathbf{J}$ is a left ideal in $\mathbf{A}$. Moreover, $\mathbf{A}$ is the direct sum of $\mathbf{I}$ and $\mathbf{J}$, since $b = be + (b - be)$ and $b - be \in \mathbf{J}$.

We must still prove that $\mathbf{J}$ is a right ideal in $\mathbf{A}$. Clearly

$$ab = abe + a(b - be)$$

for arbitrary $a \in \mathbf{J}$ and $b \in \mathbf{A}$. But $be = ebe$ since $be \in \mathbf{I}$, and hence

$$abe = (ae)be = 0$$

since $ae = 0$. Therefore $ab = a(b - be)$, so that ab is the product of two elements of $\mathbf{J}$. It follows that $ab \in \mathbf{J}$. ∎

11.76. LEMMA. *Let* **I** *and* **J** *be two-sided ideals of an algebra* **A**, *and suppose* **A** *is the direct sum of* **I** *and* **J**, *with* **I** *the maximal two-sided ideal in* **A**. *Then the algebra* **J** *contains no proper two-sided ideals.*

Proof. Let $\mathbf{J}'$ be a two-sided ideal of $\mathbf{J}$ which does not coincide with $\mathbf{J}$. Then the algebra $\mathbf{J}'' = \mathbf{I} + \mathbf{J}'$ is a two-sided ideal in $\mathbf{A}$. But $\mathbf{I}$ is maximal, and hence $\mathbf{J}'' = \mathbf{I}$. It follows that $\mathbf{J}' = \{0\}$. ∎

11.77. We are now at last in a position to prove the basic theorem on the structure of semisimple algebras:

THEOREM (***Second structure theorem***). *Every semisimple algebra* **A** *is a direct sum of two-sided ideals of* **A**, *each of which is a simple algebra.*

Proof. As shown in Sec. 11.71, we can construct a composition series

$$\mathbf{A} = \mathbf{I}_0 \supset \mathbf{I}_1 \supset \cdots \supset \mathbf{I}_n \supset \mathbf{I}_{n+1} = \{0\}$$

for **A**. Our theorem is then obviously a special case of the following

Assertion. *For every* k $(0 \leqslant k \leqslant n)$ *the algebra* $\mathbf{I}_{n-k}$ *is a direct sum of two-sided ideals of* $\mathbf{I}_{n-k}$, *each a simple algebra, and moreover* $\mathbf{I}_{n-k}$ *has a unit.*

We now prove this assertion by induction on k. The algebra $\mathbf{I}_n$ has no proper two-sided ideals, and moreover is nontrivial, by Theorem 11.73. Hence the algebra $\mathbf{I}_n$ is simple and, in particular, has a unit (by Theorem 11.54). This proves the assertion for $k = 0$.

Suppose now that the assertion is true for some k $(0 \leqslant k \leqslant n - 1)$. This means, in particular, that the algebra $\mathbf{I}_{n-k}$ has a unit, and hence, by Lemma 11.75, $\mathbf{I}_{n-k-1}$ is a direct sum $\mathbf{I}_{n-k} + \mathbf{J}$ where $\mathbf{J}$ is a two-sided ideal in $\mathbf{I}_{n-k-1}$. Since $\mathbf{I}_{n-k}$ is a maximal two-sided ideal in $\mathbf{I}_{n-k-1}$, it follows from Lemma 11.76 that the algebra $\mathbf{J}$ contains no proper two-sided ideals. At the same time, applying Theorem 11.73 to the normal series

$$\mathbf{A} = \mathbf{I}_0 \supset \mathbf{I}_1 \supset \cdots \supset \mathbf{I}_{n-k-1} \supset \mathbf{J} \supset \{0\},$$

we find that $\mathbf{J}$ is nontrivial and hence simple. By the induction hypothesis, the algebra $\mathbf{I}_{n-k}$ is a direct sum of two-sided ideals of $\mathbf{I}_{n-k}$, each a simple algebra. Being simple, each of these subalgebras is also a two-sided ideal in $\mathbf{I}_{n-k-1}$, by Lemma 11.74. It follows at once from this fact and the relation $\mathbf{I}_{n-k-1} = \mathbf{I}_{n-k} + \mathbf{J}$ that $\mathbf{I}_{n-k-1}$ is also a direct sum of two-sided ideals of $\mathbf{I}_{n-k-1}$, each a simple algebra.

We must still show that the algebra $\mathbf{I}_{n-k-1}$ has a unit. Let e_1 be the unit of the algebra $\mathbf{I}_{n-k}$ (which exists by the induction hypothesis), and let e_2 be the unit of the simple algebra **J**. Then, since $ab = ba = 0$ for arbitrary $a \in \mathbf{I}_{n-k}$, $b \in \mathbf{J}$, it is easy to see that the element $e = e_1 + e_2$ is a unit in the whole algebra $\mathbf{I}_{n-k-1}$.

Thus we have justified the induction on k, thereby proving the italicized assertion. But, as already noted, our theorem is a special case of this assertion (corresponding to $k = n$). ▌

It should be noted that we have incidentally proved that *every semisimple algebra has a unit.*

The two-sided ideals found in the theorem, whose direct sum is the given semisimple algebra **A**, will henceforth be called the *simple components of the algebra* **A**.

11.78. It was shown in Sec. 11.64 that every simple algebra is isomorphic to the algebra $\mathbf{B}(\mathbf{X})$ for some finite-dimensional space **X** or, equivalently, to the algebra of all square matrices of a certain order. Now let $\mathbf{X}_1, \ldots, \mathbf{X}_n$ be a set of finite-dimensional spaces, and let $\mathbf{B}(\mathbf{X}_1, \ldots, \mathbf{X}_n)$ be the set of all rows of the form

$$a = (a_1, \ldots, a_n),$$

where a_k is an operator from the algebra $\mathbf{B}(\mathbf{X}_k)$ (or, if convenient, a matrix of the appropriate order). Obviously $\mathbf{B}(\mathbf{X}_1, \ldots, \mathbf{X}_n)$ is an algebra with respect to the "component-by-component" operations defined by the formulas

$$a + b = (a_1 + b_1, \ldots, a_n + b_n),$$

$$\lambda a = (\lambda a_1, \ldots, \lambda a_n),$$

$$ab = (a_1 b_1, \ldots, a_n b_n),$$

where $a, b \in \mathbf{B}(\mathbf{X}_1, \ldots, \mathbf{X}_n)$, $a = (a_1, \ldots, a_n)$, $b = (b_1, \ldots, b_n)$, and λ is a complex number. It follows from these considerations that Theorem 11.77 has the following equivalent form:

Every semisimple algebra is isomorphic to the algebra $\mathbf{B}(\mathbf{X}_1, \ldots, \mathbf{X}_n)$ *for some set of spaces* $\mathbf{X}_1, \ldots, \mathbf{X}_n$.

We note further that the simple components of the algebra $\mathbf{B}(\mathbf{X}_1, \ldots, \mathbf{X}_n)$ obviously consist of rows of the form $(0, \ldots, 0, a_k, 0, \ldots, 0)$, where the kth entry ranges over the whole algebra $\mathbf{B}(\mathbf{X}_k)$ and the remaining entries are all zero. We will identify each such component with the corresponding algebra $\mathbf{B}(\mathbf{X}_k)$.

11.79. We conclude this section by finding all two-sided ideals of a semisimple algebra:

THEOREM. *Every two-sided ideal of a semisimple algebra* **A** *is the direct* *sum of a certain number of simple components of* $\mathbf{A}$.

Proof. According to Sec. 11.78, the semisimple algebra $\mathbf{A}$ is isomorphic to some algebra of the form $\mathbf{B}(\mathbf{X}_1, \ldots, \mathbf{X}_n)$ with simple components $\mathbf{B}(\mathbf{X}_k)$, $1 \leqslant k \leqslant n$. Let $\mathbf{I}$ be a two-sided ideal in $\mathbf{B}(\mathbf{X}_1, \ldots, \mathbf{X}_n)$, and let $\mathbf{I}_k$ be the intersection of $\mathbf{I}$ with $\mathbf{B}(\mathbf{X}_k)$. If $\mathbf{I}$ contains the element

$$a = (a_1, \ldots, a_{k-1}, a_k, a_{k+1}, \ldots, a_n),$$

then $\mathbf{I}$ also contains the element

$$ae_k = (0, \ldots, 0, a_k, 0, \ldots, 0),$$

where e_k is the unit in $\mathbf{B}(\mathbf{X}_k)$. It follows that $\mathbf{I}$ can be written as the direct sum

$$\mathbf{I} = \mathbf{I}_1 + \cdots + \mathbf{I}_n.$$

But it is easily seen that $\mathbf{I}_k$ is a two-sided ideal in the simple algebra $\mathbf{B}(\mathbf{X}_k)$ for every k $(1 \leqslant k \leqslant n)$. Hence either $\mathbf{I}_k = \{\mathbf{0}\}$ or $\mathbf{I}_k$ coincides with the whole algebra $\mathbf{B}(\mathbf{X}_k)$. ∎

11.8. Representations of Simple and Semisimple Algebras

From a knowledge of the structure of simple and semisimple algebras, we can without particular difficulty find all their representations to within an equivalence.

11.81. Let $\mathbf{A}$ be a semisimple algebra. Then, by Sec. 11.78, we can identify $\mathbf{A}$ with the algebra $\mathbf{B}(\mathbf{X}_1, \ldots, \mathbf{X}_n)$ for some set of spaces $\mathbf{X}_k$ $(1 \leqslant k \leqslant n)$. Therefore, besides the given algebra $\mathbf{A}$, we are led in a natural way to consider n representations $\mathbf{T}^k: \mathbf{A} \to \mathbf{B}(\mathbf{X}_k)$, $1 \leqslant k \leqslant n$ of $\mathbf{A}$, defined by the formula

$$\mathbf{T}_a^k = a_k \in \mathbf{B}(\mathbf{X}_k)$$

for every $a = (a_1, \ldots, a_k, \ldots, a_n) \in \mathbf{A}$. Since the image of the representation $\mathbf{T}^k$ is the whole algebra $\mathbf{B}(\mathbf{X}_k)$, these representations are all irreducible.

THEOREM. *Every irreducible representation of a semisimple algebra* $\mathbf{A}$ *is equivalent to one of the representations* $\mathbf{T}^k$ $(1 \leqslant k \leqslant n)$.

Proof. Let $\mathbf{A} = \mathbf{B}(\mathbf{X}_1, \ldots, \mathbf{X}_n)$ be a semisimple algebra, with an irreducible representation $\mathbf{T}: \mathbf{A} \to \mathbf{B}(\mathbf{X})$, and let $\mathbf{Z}(\mathbf{T})$ be the kernel of the representation $\mathbf{T}$. Since $\mathbf{Z}(\mathbf{T})$ is a two-sided ideal in $\mathbf{A}$ (Sec. 11.21a), it follows from Theorem 11.79 that $\mathbf{Z}(\mathbf{T})$ is the direct sum of certain simple components of $\mathbf{A}$. Let $\mathbf{A}_1$ denote the direct sum of the remaining simple components of $\mathbf{A}$ which do not figure in $\mathbf{Z}(\mathbf{T})$, and let $\mathbf{T}^{(1)}: \mathbf{A}_1 - \mathbf{B}(\mathbf{X})$ be the restriction onto $\mathbf{A}_1$ of the original representation $\mathbf{T}$. The new representation $\mathbf{T}^{(1)}$ is now exact, and moreover irreducible since the images of the representations $\mathbf{T}^{(1)}$ and $\mathbf{T}$ obviously coincide. The algebra $\mathbf{A}_1$, having an exact irreducible representation, must be simple (see Sec. 11.64). Hence $\mathbf{A}_1$ reduces to a single simple component, i.e., $\mathbf{A}_1$ coincides with $\mathbf{B}(\mathbf{X}_k)$ for some k $(1 \leqslant k \leqslant n)$. But then, as is easily seen,

$$\mathbf{T}_a = \mathbf{T}^{(1)}_{a_k}, \qquad a_k \in \mathbf{B}(\mathbf{X}_k)$$

for every $a = (a_1, \ldots, a_k, \ldots, a_n) \in \mathbf{A}$.

Now, according to Sec. 11.51, all exact irreducible representations of a simple algebra are equivalent. In particular, the representation $\mathbf{T}^{(1)}:\mathbf{B}(\mathbf{X}_k) \to \mathbf{B}(\mathbf{X})$ and the identity representation $\mathbf{T}^{(2)}:\mathbf{B}(\mathbf{X}_k) \to \mathbf{B}(\mathbf{X}_k)$ are equivalent. This means that there exists an isomorphism $\mathbf{U}:\mathbf{X} \to \mathbf{X}_k$ such that $\mathbf{U}\mathbf{T}^{(1)}_{a_k} = \mathbf{T}^{(2)}_{a_k}\mathbf{U}$ for every $a_k \in \mathbf{B}(\mathbf{X}_k)$. But $\mathbf{T}_a = \mathbf{T}^{(1)}_{a_k}$ for every $a \in \mathbf{A}$, as just shown, while on the other hand it follows from the definition of the representation $\mathbf{T}^k$ that $\mathbf{T}^k_a = \mathbf{T}^{(2)}_{a_k}$. Therefore $\mathbf{U}\mathbf{T}_a = \mathbf{T}^k_a\mathbf{U}$ for every $a \in \mathbf{A}$, which proves the equivalence of the representations $\mathbf{T}$ and $\mathbf{T}^k$. ∎

11.82. Next we consider arbitrary representations of simple and semisimple algebras. In this regard, the following general proposition will be found useful:

LEMMA. *Given an arbitrary algebra* $\mathbf{A}$, *let* $\mathbf{T}:\mathbf{A} \to \mathbf{B}(\mathbf{X})$ *be any representation of* $\mathbf{A}$, *and let* $\mathbf{X}_1, \ldots, \mathbf{X}_n$ *be minimal invariant subspaces of* $\mathbf{T}$ *spanning a linear manifold which coincides with* $\mathbf{X}$.† *Then* $\mathbf{X}$ *is the direct sum of certain of the subspaces* $\mathbf{X}_1, \ldots, \mathbf{X}_n$.

Proof. An intersection of invariant subspaces of a representation is itself an invariant subspace. Therefore it follows from the minimality of the given subspaces that for any k, the intersection of the subspace $\mathbf{X}_{k+1}$ with the linear manifold spanned by the subspaces $\mathbf{X}_1, \ldots, \mathbf{X}_k$ is either empty or $\mathbf{X}_{k+1}$ itself. Hence by consecutively choosing those of the subspaces $\mathbf{X}_1, \ldots, \mathbf{X}_n$ which are not contained in the linear manifold spanned by the preceding subspaces, we get the subspaces whose direct sum is the whole linear manifold spanned by $\mathbf{X}_1, \ldots, \mathbf{X}_n$, namely the whole space $\mathbf{X}$. ∎

11.83. According to the second structure theorem, every semisimple algebra $\mathbf{A}$ is isomorphic to an algebra of the form $\mathbf{B}(\mathbf{X}_1, \ldots, \mathbf{X}_n)$. In what follows, we will find it convenient to consider the realization of $\mathbf{B}(\mathbf{X}_1, \ldots, \mathbf{X}_n)$ in the form of an algebra of rows, each made up of n matrices of the appropriate orders. The number appearing in the "ijth" place in the kth matrix of the row corresponding to the element $a \in \mathbf{A}$ will be denoted by $\lambda^{(k)}_{ij}(a)$. Moreover, we will use $e^{(k)}_{ij}$ to denote the element of the algebra $\mathbf{A}$ such that

† By the *linear manifold spanned by the spaces* $\mathbf{X}_1, \ldots, \mathbf{X}_n$ we mean the set of all linear combinations of the form $\alpha_1 x_1 + \cdots + \alpha_n x_n$ where $x_k \in \mathbf{X}_k$ (cf. Sec. 2.51).

$\lambda_{ij}^{(k)}(e_{ij}^{(k)}) = 1$ while all other elements in the matrices of the corresponding row equal zero. It should be noted that

$$\sum_{i,k} e_{ii}^{(k)} = e, \tag{3}$$

where e is the unit of the algebra $\mathbf{A}$.

LEMMA. *Let $\mathbf{T}:\mathbf{A} \to \mathbf{B}(\mathbf{X})$ be a representation of a semisimple algebra $\mathbf{A}$ and suppose the vector $y = \mathbf{T}_{e_{ii}^{(k)}}x$ is nonzero for some $x \in \mathbf{X}$ and certain indices i and k. Then y belongs to some minimal invariant subspace of the representation $\mathbf{T}$.*

Proof. Let $\mathbf{Y} = \{\mathbf{T}_a y : a \in \mathbf{A}\}$. Then, since $y = \mathbf{T}_{e_{ii}^{(k)}}x$, it follows from the rule for matrix multiplication that every element $z_1 \in \mathbf{Y}$ is of the form $z_1 = \mathbf{T}_b x$, where b is some linear combination of the elements $e_{ij}^{(k)}$ (with i and k fixed). It is sufficient to show that if $z_1 \neq 0$, then z_1 is a cyclic vector with respect to the restriction of the representation $\mathbf{T}$ onto $\mathbf{Y}$.

Now let $z_2 \in \mathbf{Y}$, so that $z_2 = \mathbf{T}_c x$, where c is another linear combination of the same elements $e_{ij}^{(k)}$. Using the realization of the algebra $\mathbf{A}$ as an algebra of matrix rows, we find an element $a \in \mathbf{A}$ such that $c = ab$. But then $z_2 = \mathbf{T}_c x = \mathbf{T}_a(\mathbf{T}_b x) = \mathbf{T}_a z_1$. Hence the vector z_1 is cyclic, as asserted. ∎

11.84. THEOREM. *Every representation of a semisimple algebra $\mathbf{A}$ is a direct sum of irreducible representations and the trivial representation.*

Proof. Given any representation $\mathbf{T}^0:\mathbf{A} \to \mathbf{B}(\mathbf{X}^0)$, consider the operator $\mathbf{T}_e^0$ where e is the unit in $\mathbf{A}$. Then the formula

$$x = \mathbf{T}_e^0 x + (x - \mathbf{T}_e^0 x)$$

obviously defines an expansion of $\mathbf{X}^0$ as a direct sum of subspaces $\mathbf{X}$ and $\mathbf{X}_0$ invariant with respect to $\mathbf{T}^0$, where the restriction of $\mathbf{T}^0$ onto $\mathbf{X}_0$ is the trivial representation. We must still show that the representation $\mathbf{T}:\mathbf{A} \to \mathbf{B}(\mathbf{X})$, the restriction of $\mathbf{T}^0$ onto $\mathbf{X}$, is a direct sum of irreducible representations.

Let $x_1, \ldots, x_m$ be a basis in $\mathbf{X}$. Then $\mathbf{T}_e$ is the identity operator in $\mathbf{X}$, and hence, because of (3), the linear manifold spanned by the vectors of the type $\mathbf{T}_{e_{ii}^{(k)}}x$ for all possible indices i, j and k coincides with the whole space $\mathbf{X}$. By Lemma 11.83, every nonzero vector of this type lies in some minimal irreducible subspace of the representation $\mathbf{T}$. Thus the conditions of Lemma 11.82 are in force. But then the space $\mathbf{X}$ is the direct sum of certain minimal invariant subspaces of the representation $\mathbf{T}$, so that $\mathbf{T}$ is a direct sum of irreducible representations. ∎

11.85. Theorems 11.81 and 11.84 together describe to within an equivalence all representations of semisimple (including simple) algebras. In particular, we see that the operators of a given representation of a simple algebra (singling out this case for greater clarity) are described in some basis by quasi-diagonal matrices of the form

$$\begin{Vmatrix} \boxed{M} & & & & \\ & \boxed{M} & & & \\ & & \ddots & & \\ & & & \boxed{M} & \\ & & & & \boxed{0} \end{Vmatrix}, \tag{4}$$

where M ranges over the whole set of matrices of the appropriate order and 0 denotes the zero matrix. In the more general case of a semisimple algebra, the corresponding matrices are quasi-diagonal matrices of the form

$$\begin{Vmatrix} \boxed{M_1} & & & & & & & \\ & \ddots & & & & & & \\ & & \boxed{M_1} & & & & & \\ & & & \ddots & & & & \\ & & & & \boxed{M_k} & & & \\ & & & & & \ddots & & \\ & & & & & & \boxed{M_k} & \\ & & & & & & & \boxed{0} \end{Vmatrix}, \tag{5}$$

where each of the matrices $M_1, \ldots, M_k$ appearing in the indicated larger blocks ranges independently over the whole set of matrices of the appropriate order (in general different for different matrices).

11.86. Incidentally *we have described all simple and semisimple matrix algebras* (i.e., algebras which themselves consist of matrices). In fact, by merely assigning each matrix of such an algebra its operator (in any basis), we get an exact representation of the algebra. This and the preceding considerations immediately imply the following assertion:

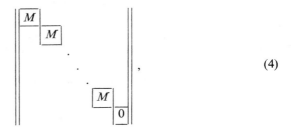

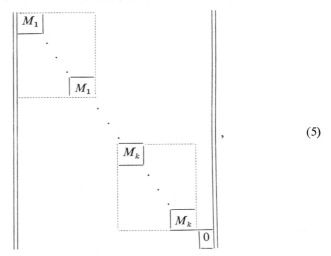

Every simple (or semisimple) matrix algebra consists of all matrices of the form $P^{-1}LP$, where P is a fixed nonsingular matrix and L ranges over the set of all matrices of the form (4) (*or of the form* (5)).

For algebras containing the unit matrix, we get a somewhat different result:

Every simple matrix algebra containing the unit matrix consists of all matrices of the form $P^{-1}LP$, where P is a fixed nonsingular matrix, L ranges over the set of all quasi-diagonal matrices of the form

$$\left\|\begin{array}{cccc} \boxed{M} & & & \\ & \boxed{M} & & \\ & & \ddots & \\ & & & \boxed{M} \end{array}\right\|, \tag{6}$$

and M ranges over the set of all matrices of the appropriate order. Every semisimple algebra containing the unit matrix consists of all matrices of the form $P^{-1}LP$, where P is a fixed nonsingular matrix, L ranges over the set of all quasi-diagonal matrices of the form

$$\left\|\begin{array}{ccccccc} \boxed{M_1} & & & & & & \\ & \ddots & & & & & \\ & & \boxed{M_1} & & & & \\ & & & \ddots & & & \\ & & & & \boxed{M_k} & & \\ & & & & & \ddots & \\ & & & & & & \boxed{M_k} \end{array}\right\|,$$

and each of the matrices $M_1, \ldots, M_k$ ranges independently over the whole set of matrices of the appropriate order.

11.9. Some Further Results

Thus we have completed the description of simple and semisimple finite-dimensional algebras, as well as their representations. Further investiga tion of finite-dimensional algebras lies beyond the scope of this chapter.

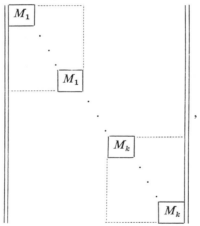

Nevertheless, to give perspective, we now cite some well-known results along these lines.

11.91. Wedderburn's theorem. *Every finite-dimensional algebra is the direct sum (regarded as a linear space) of its radical and some semisimple algebra.*†

11.92. The radical of a finite-dimensional algebra consists only of nilpotent elements. Moreover, for every such algebra there exists a positive integer n such that the product of any n elements of its radical equals zero.‡

11.93. Every representation of a radical algebra is described in some basis by matrices with zeros on and below the principal diagonal.§

PROBLEMS

1. Prove that every left ideal of the algebra $\mathbf{B}(\mathbf{K}_n)$ is the set of all operators whose null spaces contain some subspace $\mathbf{K}' \subset \mathbf{K}_n$.

2. Prove that every right ideal of the algebra $\mathbf{B}(\mathbf{K}_n)$ is the set of all operators whose ranges are contained in some subspace $\mathbf{K}' \subset \mathbf{K}_n$.

3. Find all maximal left and right ideals of the algebra $\mathbf{B}(\mathbf{K}_n)$.

4. Given any semisimple algebra $\mathbf{B}$ of linear operators over a space $\mathbf{C}_n$, introduce a scalar product (x, y) in $\mathbf{C}_n$ such that $\mathbf{A} \in \mathbf{B}$ implies $\mathbf{A}^* \in \mathbf{B}$.

5 (Converse of Problem 4). Given any algebra $\mathbf{B}$ of linear operators over a space $\mathbf{C}_n$, prove that if there exists a scalar product (x, y) in $\mathbf{C}_n$ such that $\mathbf{A} \in \mathbf{B}$ implies $\mathbf{A}^* \in \mathbf{B}$, then the algebra $\mathbf{B}$ is semisimple.

6. Suppose the conditions of Problem 5 are satisfied. Prove that $\mathbf{B}$ is a simple algebra if the intersection of the commutator $\bar{\mathbf{B}}$ (Sec. 11.61) and the algebra $\mathbf{B}$ itself consists only of operators which are multiples of the identity operator.

7. Let $\mathbf{B}$ be the simple algebra consisting of all matrices of the form (6) made up of m^2 blocks:

$$\left\|\begin{matrix} M & 0 & \cdots & 0 \\ 0 & M & \cdots & 0 \\ \cdot & \cdot & \cdots & \cdot \\ 0 & 0 & \cdots & M \end{matrix}\right\|.$$

† See e.g., N. Jacobson, *The Theory of Rings*, American Mathematical Society, New York (1943), p. 116.

‡ See e.g., N. G. Chebotarev, *Introduction to the Theory of Algebras* (in Russian), Gostekhizdat, Moscow (1949), Sec. 8.

§ Here, of course, it is not asserted that the matrices of the operators of the representation range over the whole set of matrices of this type. See e.g., A. Y. Khelemeski, *On algebras of nilpotent operators and related categories* (in Russian), Vestnik MGU, Ser. Mat. Mekh., no. 4 (1963), pp. 49–55.

Show that the commutator of **B** can be represented (in the same basis) by all matrices of the form

$$\left\| \begin{matrix} \lambda_{11}E & \lambda_{12}E & \cdots & \lambda_{1m}E \\ \lambda_{21}E & \lambda_{22}E & \cdots & \lambda_{2m}E \\ \cdot & \cdot & \cdots & \cdot \\ \lambda_{m1}E & \lambda_{m2}E & \cdots & \lambda_{mm}E \end{matrix} \right\|,$$

where the λ_{jk} ($j, k = 1, \ldots, m$) are arbitrary complex numbers. In particular, show that the intersection of **B** and $\bar{\mathbf{B}}$ consists only of matrices which are multiples of the unit matrix.

8. For what semisimple matrix algebra **B** does the commutator $\bar{\mathbf{B}}$ coincide with **B** itself?

9. Describe every semisimple commutative algebra **B** ($\mathbf{B} \subset \bar{\mathbf{B}}$).

10. Describe every semisimple matrix algebra **B** for which $\bar{\mathbf{B}} \subset \mathbf{B}$.

11. Prove that $\bar{\bar{\mathbf{B}}} = \mathbf{B}$ for every semisimple algebra **B**.

12. Let **B** be the algebra consisting of all polynomials in a single operator **A** (hence **B** is commutative, so that $\bar{\mathbf{B}} \supset \mathbf{B}$). Under what conditions does $\bar{\bar{\mathbf{B}}} = \mathbf{B}$?

13.* Show that if the algebra $\mathbf{B} \not\equiv \{\mathbf{0}\}$ consists only of nilpotent elements (i.e., if $\mathbf{A}^k = \mathbf{0}$ for some $k = k(\mathbf{A})$ for every $\mathbf{A} \in \mathbf{B}$), then the equality $\mathbf{CB} = \mathbf{B}$ cannot hold for any $\mathbf{C} \in \mathbf{B}$.

14. An algebra **B** is said to be *nilpotent* if there exists a number p such that the product of any p elements of **B** equals zero. Show that an algebra **B** equal to the direct sum $\mathbf{B}_1 + \cdots + \mathbf{B}_m$ of its right ideals is nilpotent if each ideal $\mathbf{B}_j$ ($j = 1, \ldots, m$) is nilpotent.

15. Prove that if a finite-dimensional algebra **B** consists only of nilpotent elements, then **B** itself is nilpotent.

16. Given a nilpotent algebra **B** of operators in the space $\mathbf{K}_n$, let $\mathbf{M}_1 \subset \mathbf{K}_n$ be the intersection of all null spaces of all the operators $\mathbf{A} \in \mathbf{B}$, let $\mathbf{M}_2 \subset \mathbf{K}_n$ be the intersection of all subspaces carried into $\mathbf{M}_1$ by the operators $\mathbf{A} \in \mathbf{B}$, let $\mathbf{M}_3 \subset \mathbf{K}_n$ be the intersection of all subspaces carried into $\mathbf{M}_2$ by the operators $\mathbf{A} \in \mathbf{B}$, and so on. Show that

$$\{0\} \subset \mathbf{M}_1 \subset \mathbf{M}_2 \subset \cdots \subset \mathbf{M}_p = \mathbf{K}_n,$$

where each set is a proper subset of the next and p is the *index of nilpotency* of **B**, i.e., the smallest number p such that the product of any p operators in **B** equals zero.

17. Prove that for every nilpotent algebra **B** of operators in a space $\mathbf{K}_n$, there exists a basis in which every operator $\mathbf{A} \in \mathbf{B}$ is specified by a matrix of the form

$$A = \begin{Vmatrix} 0 & A_{12} & A_{13} & \cdots & A_{1,p-1} \\ 0 & 0 & A_{23} & \cdots & A_{2,p-1} \\ 0 & 0 & 0 & \cdots & A_{3,p-1} \\ \cdot & \cdot & \cdot & \cdots & \cdot \\ 0 & 0 & 0 & \cdots & 0 \end{Vmatrix},$$

where p is the index of nilpotency of **B**. (A. Y. Khelemski) *appendix

CATEGORIES OF FINITE-DIMENSIONAL SPACES

A.1. Introduction

A.11. Recently the concept of a category and certain related ideas have begun to play an important role in various branches of mathematics.† An example of a category is a collection of sets together with mappings of the sets into one another. A collection of linear spaces or algebras together with their morphisms is another example of a category.

The exact definition of a category is as follows: Let $\mathscr{A}$ be a set of indices α, and let $\mathscr{K}$ be a set of elements $\mathbf{X}_\alpha$ ($\alpha \in \mathscr{A}$) called *objects of the category* $\mathscr{K}$. Suppose that for every pair of objects $\mathbf{X}_\beta$ and $\mathbf{X}_\alpha$ there is a set $\mathscr{B}_{\beta\alpha}$ of other elements $\mathbf{A}_{\beta\alpha}$ called *mappings* of the object $\mathbf{X}_\alpha$ into the object $\mathbf{X}_\beta$ such that the product of the mappings $\mathbf{A}_{\gamma\beta}$ and $\mathbf{A}_{\beta\alpha}$ is defined for arbitrary α, β, γ and belongs to $\mathscr{B}_{\gamma\alpha}$, where multiplication is associative, i.e.,

$$\mathbf{A}_{\delta\gamma}(\mathbf{A}_{\gamma\beta}\mathbf{A}_{\beta\alpha}) = (\mathbf{A}_{\delta\gamma}\mathbf{A}_{\gamma\beta})\mathbf{A}_{\beta\alpha}$$

for arbitrary α, β, γ, δ. In particular, the set $\mathscr{B}_{\alpha\alpha}$ of mappings of the objects $\mathbf{X}_\alpha$ into themselves is defined, and (associative) multiplication of mappings is defined in $\mathscr{B}_{\alpha\alpha}$. Finally, it is required that the set $\mathscr{B}_{\alpha\alpha}$ contain the unit

† See e.g., H. Cartan and S. Eilenberg, *Homological Algebra*, Princeton University Press, Princeton, N.J. (1956); Séminaire A. Grothendieck, *Algèbre Homologique*, Secrétariat Mathématique, Paris (1958); A. G. Kurosh et al., *Elements of the theory of categories* (in Russian), Uspekhi Mat. Nauk, vol. 15, no. 6 (1960), pp. 3–52.

element $\mathbf{1}_\alpha$, which has the property that

$$\mathbf{1}_\alpha \mathbf{A}_{\alpha\beta} = \mathbf{A}_{\alpha\beta}, \qquad \mathbf{A}_{\gamma\alpha}\mathbf{1}_\alpha = \mathbf{A}_{\gamma\alpha}$$

for arbitrary α, β and γ. Instead of $\mathscr{B}_{\alpha\alpha}$ we will usually write simply $\mathscr{B}_\alpha$.

A set $\mathscr{K}$ of objects $\mathbf{X}_\alpha$ and mappings $\mathbf{A}_{\beta\alpha}$ with the properties just enumerated is called a *category*. A category $\mathscr{K}$ is called *linear* if in the set $\mathscr{B}_{\beta\alpha}$ of mappings $\mathbf{A}_{\beta\alpha}$ (with arbitrary fixed α and β) there are defined operations of addition of mappings and multiplication of mappings by numbers (from the field K). This makes the set $\mathscr{B}_{\beta\alpha}$ into a linear space over the field K. Thus in a linear category the set $\mathscr{B}_\alpha$ becomes an algebra with a unit (over the field K).

A.12. In this appendix we will consider linear categories whose elements are finite-dimensional linear spaces (of dimension $\geqslant 1$) over the field C of complex numbers, while the mappings are linear mappings (morphisms) of one such space into another.

Thus we start with the following definition: Let $\mathbf{X}_\alpha$ ($\alpha \in \mathscr{A}$) be a set of finite-dimensional complex linear spaces, and for every α let $\mathscr{B}_\alpha$ be an algebra of linear operators carrying $\mathbf{X}_\alpha$ into itself. Moreover, suppose that for every pair of indices α and β there is a set $\mathscr{B}_{\beta\alpha}$ of linear operators $\mathbf{A}_{\beta\alpha}$ carrying $\mathbf{X}_\alpha$ into $\mathbf{X}_\beta$ such that 1) if $\mathscr{B}_{\beta\alpha}$ contains the operators $\mathbf{A}_{\beta\alpha}$ and $\mathbf{B}_{\beta\alpha}$, then $\mathscr{B}_{\beta\alpha}$ contains the operator sum $\mathbf{A}_{\beta\alpha} + \mathbf{B}_{\beta\alpha}$, and 2) if $\mathscr{B}_{\beta\alpha}$ contains the operator $\mathbf{A}_{\beta\alpha}$, then $\mathscr{B}_{\beta\alpha}$ contains the product $\lambda\mathbf{A}_{\beta\alpha}$ where λ is an arbitrary complex number. A family of linear operators with these two properties will be called a *linear family*. In particular, the linear family $\mathscr{B}_{\alpha\alpha}$ coincides with the algebra $\mathscr{B}_\alpha$. It is also assumed that

$$\mathscr{B}_{\gamma\beta}\mathscr{B}_{\beta\alpha} \subset \mathscr{B}_{\gamma\alpha} \tag{1}$$

for arbitrary α, β and γ, i.e., that every product

$$\mathbf{A}_{\gamma\beta}\mathbf{A}_{\beta\alpha} \qquad (\mathbf{A}_{\gamma\beta} \in \mathscr{B}_{\gamma\beta}, \mathbf{A}_{\beta\alpha} \in \mathscr{B}_{\beta\alpha})$$

belongs to $\mathscr{B}_{\gamma\alpha}$. Such a set of spaces $\mathbf{X}_\alpha$ together with algebras $\mathscr{B}_\alpha$ and linear families $\mathscr{B}_{\beta\alpha}$ will be called a *category of finite-dimensional spaces* or simply a *category*, and will be denoted by $\mathscr{K}$.

If we choose a basis in every space $\mathbf{X}_\alpha$, then the algebras $\mathscr{B}_\alpha$ and linear families $\mathscr{B}_{\beta\alpha}$ can be identified with the algebras and linear families of the corresponding matrices, a fact which will henceforth be exploited systematically.

In what follows, we will find the categories of linear spaces corresponding to given algebras $\mathscr{B}_\alpha$, confining ourselves to the case where the $\mathscr{B}_\alpha$ are semisimple algebras containing the unit matrix. According to Sec. 11.86, for such an algebra the space $\mathbf{X}_\alpha$ can be decomposed into a direct sum of subspaces $\mathbf{X}_{\alpha j}$ invariant under all the operators $\mathbf{A}_{\alpha\alpha}$, where in each subspace $\mathbf{X}_{\alpha j}$ the algebra $\mathscr{B}_\alpha$ is a simple algebra containing the unit matrix, i.e., is described in some basis by the set of all quasi-diagonal matrices of the form

$$\left\| \begin{matrix} \boxed{C} & & & \\ & \boxed{C} & & \\ & & \ddots & \\ & & & \boxed{C} \end{matrix} \right\|,$$

where C ranges over the set of all matrices of the appropriate order.

We begin with an analysis of some special cases for which general results can afterwards be stated. Thus in Sec. A.2 we consider the case where every algebra $\mathscr{B}_\alpha$ is *complete*, i.e., is the algebra of *all* linear operators acting in the space $\mathbf{X}_\alpha$. The opposite case where each $\mathscr{B}_\alpha$ is an algebra of operators of the form $\lambda\mathbf{E}$ (multiples of the identity operator $\mathbf{E}$) is considered in Sec. A.3. The results of Sec. A.4 pertain to the case of simple algebras $\mathscr{B}_\alpha$, this being a natural generalization of the case of the algebras $\{\lambda\mathbf{E}\}$. In Sec. A.5 we consider the case where each algebra $\mathscr{B}_\alpha$ is an algebra of all diagonal matrices of a given order, while in Sec. A.6 the general category with semi-simple algebras $\mathscr{B}_\alpha$ is reduced to the categories considered in the preceding sections.

A.13. We now recall the notation and rules of operation governing matrices of linear operators mapping a linear space $\mathbf{X}$ into a linear space $\mathbf{Y}$ (see Secs. 4.41–4.43). Let $\mathbf{X}$ be an n-dimensional space with basis $e_1, \ldots, e_n$, and let $\mathbf{Y}$ be an m-dimensional space with basis $f_1, \ldots, f_m$. Then with every linear operator $\mathbf{A}$ mapping $\mathbf{X}$ into $\mathbf{Y}$ we associate an $m \times n$ matrix

$$A = \left\| \begin{matrix} a_{11} & a_{12} & \cdots & a_{1n} \\ a_{21} & a_{22} & \cdots & a_{2n} \\ \cdot & \cdot & \cdots & \cdot \\ a_{m1} & a_{m2} & \cdots & a_{mn} \end{matrix} \right\|$$

(with m rows and n columns), where the numbers $a_{1j}, a_{2j}, \ldots, a_{mj}$ in the jth column are the coefficients of the expansion of the vector $\mathbf{A}e_j \in \mathbf{Y}$ with respect to the basis $f_1, \ldots, f_m$. Moreover, let $\mathbf{Z}$ be a k-dimensional space with basis $g_1, \ldots, g_k$. Then with every operator $\mathbf{B}$ mapping the space $\mathbf{Y}$ into the space $\mathbf{Z}$ we associate a $k \times m$ matrix

$$B = \left\| \begin{matrix} b_{11} & b_{12} & \cdots & b_{1m} \\ b_{21} & b_{22} & \cdots & b_{2m} \\ \cdot & \cdot & \cdots & \cdot \\ b_{k1} & b_{k2} & \cdots & b_{km} \end{matrix} \right\|.$$

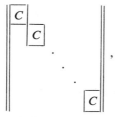

The operator $\mathbf{C} = \mathbf{BA}$ maps $\mathbf{X}$ into $\mathbf{Y}$ and has the $k \times n$ matrix

$$C = \begin{Vmatrix} c_{11} & c_{12} & \cdots & c_{1n} \\ c_{21} & c_{22} & \cdots & c_{2n} \\ \cdot & \cdot & \cdots & \cdot \\ c_{k1} & c_{k2} & \cdots & c_{kn} \end{Vmatrix},$$

obtained by multiplying the matrices B and A in accordance with the formula

$$c_{pq} = \sum_{j=1}^{m} b_{pj} a_{jq} \qquad (p = 1, \ldots, k;\ q = 1, \ldots, n).$$

A.14. The following fact, slightly generalizing Examples 4.44a–b (and proved in the same way), will often be found useful:

LEMMA. *Given an $m \times n$ matrix $A = \|a_{pq}\|$, suppose A is multiplied **from the left** by a $k \times m$ matrix $B = \|b_{rs}\|$ with all its elements equal to zero except the single element $b_{r_0 s_0} = 1$. Then the result is a $k \times n$ matrix BA whose r_0th row consists of the elements of the s_0th row of the matrix A while all other elements of BA vanish. On the other hand, if the matrix A is multiplied **from the right** by an $n \times l$ matrix $C = \|c_{rs}\|$ with all its elements equal to zero except the single element $c_{r_1 s_1}$, the result is an $m \times l$ matrix AC whose s_1th column consists of the elements of the r_1th column of the matrix A while all other elements of AC vanish.*

A.15. It follows from the lemma that if an $m \times n$ matrix A is multiplied from the left by a $k \times m$ matrix B and from the right by an $n \times l$ matrix C, where B and C have the indicated properties, then the result is a $k \times l$ matrix BAC all of whose elements vanish with the (possible) exception of the single element, equal to $a_{s_0 r_1}$, appearing in the r_0th row and s_1th column (cf. Example 4.44c).

A.2. The Case of Complete Algebras

A.21. Suppose the category $\mathscr{K}$ consists of finite-dimensional linear spaces $\mathbf{X}_\alpha$, where for every α the algebra $\mathscr{B}_\alpha$ of operators acting in $\mathbf{X}_\alpha$ is *complete*, i.e., is the algebra of *all* linear operators in $\mathbf{X}_\alpha$. Fixing arbitrary bases $e_1, \ldots, e_n$ in the space $\mathbf{X}_1$ and $f_1, \ldots, f_m$ in the space $\mathbf{X}_2$, we can identify the operators in the sets $\mathscr{B}_{11}$, $\mathscr{B}_{12}$, $\mathscr{B}_{21}$, $\mathscr{B}_{22}$ with the corresponding matrices.

Let n be the dimension of the space $\mathbf{X}_1$ and m the dimension of the space $\mathbf{X}_2$. Suppose the family $\mathscr{B}_{21}$ contains a nonzero operator $\mathbf{A}$, so that the corresponding $m \times n$ matrix $A = \|a_{pq}\|$ has at least one nonzero element, say $a_{p_0 q_0}$. We can assume without loss of generality that $a_{p_0 q_0} = 1$. It follows from the condition (1) and the assumption that $\mathscr{B}_1$ and $\mathscr{B}_2$ are complete matrix algebras that the product of A from the left by an $m \times m$ matrix and from the right by an $n \times n$ matrix is itself a matrix in the family $\mathscr{B}_{21}$. But, according to Sec. A.15, there is always an operation of this kind leading to an $m \times n$ matrix with a unique nonzero element equal to 1 in any preassigned position. Hence, since any $m \times n$ matrix is a linear combination of such matrices, we see that $\mathscr{B}_{21}$ contains all $m \times n$ matrices, i.e., *$\mathscr{B}_{21}$ is a complete family of operators mapping $\mathbf{X}_1$ into $\mathbf{X}_2$.*

A.22. As we will see below, the category $\mathscr{K}$ just described can be related to a certain *partially ordered set*.

Definition. A set S is said to be *partially ordered* if for every pair of elements $A, B \in S$ there is a relation, denoted by the symbol $\leqslant$ (and read "less than or equal") satisfying the following axioms:

a) If $A \leqslant B$ and $B \leqslant A$, then $A = B$;
b) If $A \leqslant B$ and $B \leqslant C$, then $A \leqslant C$;
c) $A \leqslant A$ for every A.

A somewhat more general concept is that of a *prepartially ordered set*, by which we mean a set S with a relation $\leqslant$ satisfying only axioms b) and c). In this case, if $A \leqslant B$ and $B \leqslant A$, we call A and B *equivalent* and write $A \sim B$. Then $A \sim B$ and $B \sim C$ together imply $A \sim C$. In fact, by axiom b), it follows from $A \leqslant B$, $B \leqslant C$ that $A \leqslant C$ and from $C \leqslant B$, $B \leqslant A$ that $C \leqslant A$. But $A \leqslant C$ and $C \leqslant A$ together imply $A \sim C$. Therefore the relation $\leqslant$ allows us to partition the whole set S into (*equivalence*) *classes* $\mathscr{A}, \mathscr{B}, \ldots$, where each class $\mathscr{A}$ contains all elements equivalent to A as well as a given element A, while elements A and B belonging to distinct classes are nonequivalent.

Next we introduce the relation $\leqslant$ for the classes $\mathscr{A}$ and $\mathscr{B}$ themselves, writing $\mathscr{A} \leqslant \mathscr{B}$ if there exist elements $A \in \mathscr{A}$, $B \in \mathscr{B}$ such that $A \leqslant B$. This definition is independent of the choice of the elements $A \in \mathscr{A}$, $B \in \mathscr{B}$. In fact, suppose $A_1 \in \mathscr{A}$, $B_1 \in \mathscr{B}$, so that $A \sim A_1$, $B \sim B_1$. Then $A_1 \leqslant A \leqslant B \leqslant B_1$ and hence $A_1 \leqslant B_1$ as required. The fact that axioms b) and c) for a partially ordered set hold for the classes $\mathscr{A}, \mathscr{B}, \ldots$ now follows from the fact that they hold for the elements $A, B, \ldots$. To show that axiom a) also holds for the classes $\mathscr{A}, \mathscr{B}, \ldots$, let $\mathscr{A} \leqslant \mathscr{B}$, $\mathscr{B} \leqslant \mathscr{A}$ and choose arbitrary elements $A \in \mathscr{A}$, $B \in \mathscr{B}$. Then $A \leqslant B$ and $B \leqslant A$, so that A and B are equivalent. But then $\mathscr{A}$ and $\mathscr{B}$ coincide, ie., $\mathscr{A} = \mathscr{B}$, as required.

Thus by introducing an equivalence relation in a prepartially ordered set S, in the way indicated, we arrive at a *partially ordered set of classes of equivalent elements*.

A.23. We now resume our study of the category $\mathscr{K}$. It follows from Sec. A.21 that given any pair of spaces $\mathbf{X}_1$ and $\mathbf{X}_2$, there are just four possibilities:

a) $\mathscr{B}_{12}$ and $\mathscr{B}_{21}$ are both complete sets of operators;
b) $\mathscr{B}_{12}$ is a complete set and $\mathscr{B}_{21}$ consists of the zero element alone;
c) $\mathscr{B}_{21}$ is a complete set and $\mathscr{B}_{12}$ consists of the zero element alone;
d) $\mathscr{B}_{12}$ and $\mathscr{B}_{21}$ both consist of the zero element alone.

If $\mathscr{B}_{12}$ is a complete set and no assumptions at all are made about $\mathscr{B}_{21}$, we write $\mathbf{X}_1 \leqslant \mathbf{X}_2$ (the relation $\mathbf{X}_2 \leqslant \mathbf{X}_1$ has the analogous meaning).

As we now show, the relation $\leqslant$ makes the category $\mathscr{K}$ into a prepartially ordered set. In fact, $\mathscr{B}_{11}$ is a complete set of operators for the given space $\mathbf{X}_1$, by hypothesis, and hence $\mathbf{X}_1 \leqslant \mathbf{X}_1$. Moreover, if $\mathbf{X}_1 \leqslant \mathbf{X}_2$ and $\mathbf{X}_2 \leqslant \mathbf{X}_3$, then $\mathscr{B}_{12}$ and $\mathscr{B}_{23}$ are complete sets of linear operators mapping $\mathbf{X}_1$ into $\mathbf{X}_2$ and $\mathbf{X}_2$ into $\mathbf{X}_3$, respectively. Since all our spaces have dimension $\geqslant 1$, there is obviously a nonzero operator in the set $\mathscr{B}_{13}$. In fact, let $e_1 \in \mathbf{X}_1$, $e_2 \in \mathbf{X}_2$, $e_3 \in \mathbf{X}_3$ be fixed nonzero vectors. Then such an operator can be obtained as the product $\mathbf{AB}$, where the operator $\mathbf{A} \in \mathscr{B}_{12}$ carries e_2 into e_1 and the operator $\mathbf{B} \in \mathscr{B}_{23}$ carries e_3 into e_2. By Sec. A.21, $\mathscr{B}_{13}$ is a complete set of operators carrying $\mathbf{X}_3$ into $\mathbf{X}_1$, so that $\mathbf{X}_1 \leqslant \mathbf{X}_3$. Thus axioms b) and c) are satisfied, and the category $\mathscr{K}$ has been made into a prepartially ordered set.

A.24. In accordance with Sec. A.22, we now introduce an equivalence relation in $\mathscr{K}$, writing $\mathbf{X}_1 \sim \mathbf{X}_2$ if $\mathbf{X}_2 \leqslant \mathbf{X}_1$ and $\mathbf{X}_1 \leqslant \mathbf{X}_2$, i.e., if both $\mathscr{B}_{12}$ and $\mathscr{B}_{21}$ are complete sets of the corresponding linear operators. Then the set of spaces $\mathbf{X}_\alpha$ decomposes into classes of equivalent spaces, and the set of all such classes becomes a partially ordered set when equipped with a relation as in Sec. A.22.

Conversely, *every partially ordered set of classes $\mathscr{X}_\alpha$ of finite-dimensional spaces defines a category of the type under consideration.* In fact, for spaces $\mathbf{X}_1$ and $\mathbf{X}_2$ belonging to the same class we specify $\mathscr{B}_{12}$ and $\mathscr{B}_{21}$ as complete sets of operators, while for spaces $\mathbf{X}_1$ and $\mathbf{X}_3$ belonging to classes $\mathscr{X}_1$ and $\mathscr{X}_3$ such that $\mathscr{X}_1 < \mathscr{X}_3$ (i.e., such that $\mathscr{X}_1 \leqslant \mathscr{X}_3$ but $\mathscr{X}_1 \neq \mathscr{X}_3$), we specify $\mathscr{B}_{13}$ as a complete set and $\mathscr{B}_{31}$ as the set consisting of the zero element alone. Moreover, if $\mathbf{X}_1$ and $\mathbf{X}_4$ belong to noncomparable classes $\mathscr{X}_1$ and $\mathscr{X}_4$, we specify that $\mathscr{B}_{14}$ and $\mathscr{B}_{41}$ both consist of the zero element alone.

The description of categories of the indicated type is now complete.

A.3. The Case of One-Dimensional Algebras

A.31. Turning to the case where the given algebras $\mathscr{B}_\alpha$ are all one-dimensional, we consider two simple examples:

a. Let the category $\mathscr{K}_1$ consist of two spaces $\mathbf{X}_1$ and $\mathbf{X}_2$ of the same dimension, and let the set $\mathscr{B}_{21}$ consist of an operator $\mathbf{A}$ mapping $\mathbf{X}_1$ onto $\mathbf{X}_2$

in a one-to-one fashion together with all its multiples $\lambda\mathbf{A}$, $\lambda \in C$, while the set $\mathscr{B}_{12}$ consists of the operator $\mathbf{B}$ which is the inverse of $\mathbf{A}$ together with all its multiples $\mu\mathbf{B}$, $\mu \in C$. Then obviously

$$\mathscr{B}_{12}\mathscr{B}_{21} = \{\lambda\mathbf{E}\}, \qquad \mathscr{B}_{21}\mathscr{B}_{12} = \{\lambda\mathbf{E}\}.$$

b. Let the category $\mathscr{K}_2$ consist of two arbitrary spaces $\mathbf{X}_1$ and $\mathbf{X}_2$ with fixed subspaces $\mathbf{X}_1' \subset \mathbf{X}_1$ and $\mathbf{X}_2' \subset \mathbf{X}_2$, and let the set $\mathscr{B}_{21}$ consist of all operators carrying $\mathbf{X}_1$ into $\mathbf{X}_2'$ with $\mathbf{X}_1'$ going into $\{0\}$, while the set $\mathscr{B}_{12}$ consists of all operators carrying $\mathbf{X}_2$ into $\mathbf{X}_1'$ with $\mathbf{X}_2'$ going into $\{0\}$. Then obviously

$$\mathscr{B}_{12}\mathscr{B}_{21} = \{0\}, \qquad \mathscr{B}_{21}\mathscr{B}_{12} = \{0\}.$$

It will now be shown that the categories $\mathscr{K}_1$ and $\mathscr{K}_2$ essentially exhaust all categories consisting of two spaces with $\mathscr{B}_j = \{\lambda\mathbf{E}\}$ $(j = 1, 2)$, i.e., that the following alternative holds for any such category $\mathscr{K}$: *Either* $\mathscr{B}_{12}\mathscr{B}_{21} = \{0\}$, *in which case* $\mathscr{B}_{21}\mathscr{B}_{12} = \{0\}$ *also and the category* $\mathscr{K}$ *is contained in a category of the type* $\mathscr{K}_2$, *or the spaces* $\mathbf{X}_1$ *and* $\mathbf{X}_2$ *have the same dimension and* $\mathscr{K}$ *is a category of the type* $\mathscr{K}_1$.

A.32. Thus let $\mathscr{K}$ be a category consisting of two spaces $\mathbf{X}_1$ and $\mathbf{X}_2$ subject to the condition $\mathscr{B}_1 = \{\lambda\mathbf{E}\}$, $\mathscr{B}_2 = \{\lambda\mathbf{E}\}$. Let $\mathbf{N}_1 \subset \mathbf{X}_1$ be the intersection of the null spaces (Sec. 4.62) of all operators $\mathbf{A}_{21} \in \mathscr{B}_{21}$, and let $\mathbf{N}_2 \subset \mathbf{X}_2$ be the intersection of the null spaces of all operators $\mathbf{A}_{12} \in \mathscr{B}_{12}$. If $\mathscr{B}_{21}\mathbf{X}_1 \subset \mathbf{N}_2$ and $\mathscr{B}_{12}\mathbf{X}_2 \subset \mathbf{N}_1$, we are dealing with a subcategory of a category of the type $\mathscr{K}_2$ in which $\mathbf{X}_1' = \mathbf{N}_1$, $\mathbf{X}_2' = \mathbf{N}_2$. Therefore we assume that $\mathscr{B}_{21}\mathbf{X}_1$ is not contained in $\mathbf{N}_2$, say, and hence that there is a vector $x_1 \in \mathbf{X}_1$ and an operator $\mathbf{A}_{21} \in \mathscr{B}_{21}$ such that $\mathbf{A}_{21}x_1 = x_2$ does not belong to $\mathbf{N}_2$.

Every operator $\mathbf{B}_{21} \in \mathscr{B}_{21}$ carries x_1 into a vector collinear with x_2, and every operator $\mathbf{C}_{12} \in \mathscr{B}_{12}$ carries x_2 into a vector collinear with x_1. In fact, let $\mathbf{A}_{21}x_1 = x_2$, $\mathbf{B}_{21}x_1 = y_2$, and consider an operator $\mathbf{C}_{12}^0 \in \mathscr{B}_{12}$ such that $\mathbf{C}_{12}^0x_2 \neq 0$. Then, by the basic condition, $\mathbf{C}_{12}^0x_2 = \mathbf{C}_{12}^0\mathbf{A}_{21}x_1 = \lambda x_1$, where $\lambda \neq 0$. Replacing $\mathbf{C}_{12}^0$ by a multiple of $\mathbf{C}_{12}^0$, we can assume that $\lambda = 1$. Moreover $\mathbf{B}_{12}\mathbf{C}_{12}^0x_2 = \mathbf{B}_{21}x_1 = y_2$, while at the same time $\mathbf{B}_{21}\mathbf{C}_{12}^0x_2 = \mu x_2$, and hence $y_2 = \mu x_2$. Since, conversely, $x_1 = \mathbf{C}_{12}^0x_2$ and $x_1 \notin \mathbf{N}_1$ by the definition of x_1, we have analogously $\mathbf{C}_{12}x_2 = \mu x_1$ for every $\mathbf{C}_{12} \in \mathscr{B}_{12}$.

Moreover, in the given case, $\mathbf{N}_1$ and $\mathbf{N}_2$ reduce to the set $\{0\}$ consisting of the zero vector alone. In fact, if $z_1 \in \mathbf{N}_1$, then $\mathbf{A}_{21}(x_1 + z_1) = \mathbf{A}_{21}x_1 = x_2$, i.e., the vector x_1 in the above construction can be replaced by $x_1 + z_1$. But then $\mathbf{C}_{12}^0x_2$ is a multiple of both x_1 and $x_1 + z_1$, so that x_1 and z_1 are collinear. Therefore $z_1 = 0$, since $x_1 \in \mathbf{N}_1$. It follows that $\mathbf{N}_1 = \{0\}$. Similarly, starting with x_2, we find that $\mathbf{N}_2 = \{0\}$.

We now see that x_1 can be chosen to be any nonzero vector of the space $\mathbf{X}_1$, since there is always an operator $\mathbf{A}_{21} \in \mathscr{B}_{21}$ carrying x_1 into a nonzero vector. Hence the operators of the set $\mathscr{B}_{21}$ establish a one-to-one correspondence between all the straight lines of the space $\mathbf{X}_1$ and some set of straight lines of the space $\mathbf{X}_2$, in fact the set of all straight lines of the space $\mathbf{X}_2$ by the symmetry of our construction.

Next we prove that the whole set $\mathscr{B}_{21}$ reduces to the set of multiples of a single operator. Let $x_1 \neq 0$ be an arbitrary vector of the space $\mathbf{X}_1$, and let x_2 be a nonzero vector determining the straight line in the space $\mathbf{X}_2$ corresponding to x_1. As we know, there is an operator $\mathbf{A}_{21}^0 \in \mathscr{B}_{21}$ carrying x_1 into precisely x_2. Every other operator $\mathbf{A}_{21} \in \mathscr{B}_{21}$ carries x_1 into λx_2 for some λ. First suppose $\mathbf{A}_{21}x_1 = \lambda x_2$, where $\lambda \neq 0$. Then the operator

$$\mathbf{B}_{21} = \frac{1}{\lambda}\mathbf{A}_{21}$$

carries x_1 into precisely x_2. Moreover, $\mathbf{B}_{21}$ coincides with $\mathbf{A}_{21}^0$ everywhere. In fact, suppose to the contrary that $\mathbf{A}_{21}^0 y_1 = y_2$, $\mathbf{B}_{21}y_1 = z_2 \neq y_2$. This can happen only if $y_2 \neq 0$, $z_2 = \mu y_2$, $\mu \neq 1$ or if $y_2 = 0$, $z_2 \neq 0$. Let $z_1 = \alpha x_1 + \beta y_1$ be a nonzero vector with $\alpha \neq 0$, $\beta \neq 0$. Then the vectors $\mathbf{A}_{21}^0 z_1$ and $\mathbf{B}_{21}z_1$ are collinear, as proved above. But this is impossible in our case, since

$$\mathbf{A}_{21}^0(\alpha x_1 + \beta y_1) = \alpha x_2 + \beta y_2, \qquad \mathbf{B}_{21}(\alpha x_1 + \beta y_1) = \alpha x_2 + \beta\mu y_2$$

if $y_2 \neq 0$, while

$$\mathbf{A}_{21}^0(\alpha x_1 + \beta y_1) = \alpha x_2, \qquad \mathbf{B}_{21}(\alpha x_1 + \beta y_1) = \alpha x_2 + \beta z_2$$

if $y_2 = 0$. This contradiction shows that if $\mathbf{A}_{21}x_1 = \lambda x_2$, $\lambda \neq 0$, then $\mathbf{A}_{21} = \lambda\mathbf{A}_{21}^0$. Now suppose $\mathbf{A}_{21}x_1 = 0$. Then, as just proved, $\mathbf{A}_{21}^0 + \mathbf{A}_{21} = \mathbf{A}_{21}^0$ and hence $\mathbf{A}_{21} = 0$. Thus $\mathscr{B}_{21}$ reduces to the set of multiples of a fixed operator $\mathbf{A}_{21}^0$, and similarly $\mathscr{B}_{12}$ reduces to the set of multiples of a fixed operator $\mathbf{B}_{12}^0$. The products $\mathbf{A}_{21}^0\mathbf{B}_{12}^0$ and $\mathbf{B}_{12}^0\mathbf{A}_{21}^0$ are nonzero and, by the basic assumption, give operators which are multiples of the identity operator. Hence the operators $\mathbf{A}_{21}^0$ and $\mathbf{B}_{12}^0$ are inverses of each other (apart from a numerical factor). But this is possible only if the spaces $\mathbf{X}_1$ and $\mathbf{X}_2$ have the same dimension. Thus, finally, we have proved that *every category $\mathscr{K}$ of the indicated type which is not a subcategory of a category of the type $\mathscr{K}_2$, is a category of the type $\mathscr{K}_1$.*

A.33. The categories of the types $\mathscr{K}_1$ and $\mathscr{K}_2$ are not the only possible categories with two spaces $\mathbf{X}_1$, $\mathbf{X}_2$ and algebras $\mathscr{B}_1 = \{\lambda\mathbf{E}\}$, $\mathscr{B}_2 = \{\lambda\mathbf{E}\}$. In fact, suppose that in the set $\mathscr{B}_{21}$ of a category $\mathscr{K}$ of the type $\mathscr{K}_2$ we choose a linear subset without increasing $\mathbf{N}_1$ or decreasing $\mathbf{N}_2$ (for example, by imposing a suitable extra linear homogeneous condition on the elements of the matrices of the operators $\mathbf{A}_{21}$). Then we get a category $\mathscr{K}'$ which satisfies the given conditions but does not coincide with $\mathscr{K}$. In the set of all categories with $\mathscr{B}_j = \{\lambda\mathbf{E}\}$ $(j = 1, 2)$, partially ordered with respect to set inclusion, the categories of the type $\mathscr{K}_2$ are characterized by the fact that they are maximal, in the sense that no category of the type $\mathscr{K}_2$, except for singular cases where $\mathbf{X}_1' = \{0\}$ or $\mathbf{X}_2' = \{0\}$, can be enlarged while preserving the properties of a category and the conditions $\mathscr{B}_j = \{\lambda\mathbf{E}\}$. In fact, suppose that a category $\mathscr{K}$ of the type $\mathscr{K}_2$ can be enlarged by including an operator $\mathbf{A}_{21}^0$ taking a value $y_2 \notin \mathbf{X}_2'$ for some $x_1 \in \mathbf{X}_1 - \mathbf{X}_1'$, where $\mathbf{X}_1' = \{0\}$. Let $\mathbf{B}_{12}^0 \in \mathscr{B}_{12}$ be an operator carrying y_2 into a nonzero vector $x_1' \in \mathbf{X}_1'$. Then $\mathbf{B}_{12}^0\mathbf{A}_{21}^0 x_1 = x_1'$, contrary to hypothesis.

Moreover, suppose that in the category $\mathscr{K}$ we include an operator $\mathbf{A}_{21}$ carrying a vector $x_1 \in \mathbf{X}_1'$ into a nonzero vector $y_2' \in \mathbf{X}_2'$. Then clearly $\mathbf{X}_2' \neq \mathbf{X}_2$, since otherwise $\mathbf{X}_1' = \mathscr{B}_{12}\mathbf{X}_2 = \mathscr{B}_{12}\mathbf{X}_2' = \{0\}$, and there cannot exist a vector x_1 mapped into a nonzero vector. Hence there is an operator $\mathbf{B}_{12} \in \mathscr{B}_{12}$ carrying a vector $y_2 \in \mathbf{X}_2 - \mathbf{X}_2'$ into x_1. But then $\mathbf{A}_{12}\mathbf{B}_{21}y_2 = y_2'$, contrary to hypothesis.

Similarly, assuming that $\mathbf{X}_2' \neq \{0\}$, we find that it is impossible to include a single extra operator in the set $\mathscr{B}_{12}$. Thus our category $\mathscr{K}$ of the type $\mathscr{K}_2$ is indeed maximal, under the assumption that $\mathbf{X}_1' \neq \{0\}$, $\mathbf{X}_2' \neq \{0\}$.

A.34. The singular cases must be considered separately. For example, suppose $\mathbf{X}_1' = \{0\}$, so that $\mathscr{B}_{12}$ consists of the zero operator alone. Then, if $\mathbf{X}_2' \neq \mathbf{X}_2$, the category is nonmaximal, and we can enlarge the set $\mathscr{B}_{21}$ to include all operators mapping $\mathbf{X}_1$ into $\mathbf{X}_2$ without dropping the conditions $\mathscr{B}_j = \{\lambda\mathbf{E}\}$ $(j = 1, 2)$. This gives a "trivial" maximal category, where $\mathscr{B}_{12} = \{0\}$ and $\mathscr{B}_{21}$ is a complete set of operators mapping $\mathbf{X}_1$ into $\mathbf{X}_2$. There is an analogous maximal category with $\mathscr{B}_{21} = \{0\}$ and $\mathscr{B}_{12}$ a complete set. Thus, finally, we find that the general category of the type $\mathscr{K}_2$ is maximal under the following conditions: 1) $\mathbf{X}_1' \neq \{0\}, \mathbf{X}_2' \neq \{0\}$; 2) $\mathbf{X}_1' = \{0\}, \mathbf{X}_2' = \mathbf{X}_2$; 3) $\mathbf{X}_1' = \mathbf{X}_1, \mathbf{X}_2' = \{0\}$.

A.35. We now turn to the general case of a category with an arbitrary number $N \leqslant \infty$ of spaces $\mathbf{X}_\alpha$, $\alpha \in \mathscr{A}$. Here we have the following analogue of the alternative proved in Sec. A.31:

THEOREM. *If* $\mathscr{B}_1 = \cdots = \mathscr{B}_k = \{\lambda\mathbf{E}\}$, *then either the product* $\mathscr{B}_{1k}\mathscr{B}_{k,k-1}\cdots\mathscr{B}_{32}\mathscr{B}_{21}$ *vanishes, or the spaces* $\mathbf{X}_1, \ldots, \mathbf{X}_k$ *all have the same dimension and* $\mathscr{B}_{ij} = \{\lambda\mathbf{A}_{ij}^0\}$ *where the* $\mathbf{A}_{ij}$ *are fixed invertible operators such that*

$$\mathbf{A}_{1k}^0\mathbf{A}_{k,k-1}^0\cdots\mathbf{A}_{32}^0\mathbf{A}_{21}^0 = \mathbf{E}.$$

Proof. Suppose the product $\mathscr{B}_{1k}\mathscr{B}_{k,k-1}\cdots\mathscr{B}_{32}\mathscr{B}_{21}$ contains a nonzero operator, which is therefore equal to $\lambda\mathbf{E}$ with $\lambda \neq 0$, and let r_j be the dimension of the space $\mathbf{X}_j$ $(j = 1, \ldots, k)$. Consider the category $\mathscr{K}_0$ made up of the two spaces $\mathbf{X}_1$, $\mathbf{X}_2$ and the following sets of operators $\mathscr{B}_{12}^0$, $\mathscr{B}_{21}^0$:

$$\mathscr{B}_{12}^0 = \mathscr{B}_{1k}\mathscr{B}_{k,k-1}\cdots\mathscr{B}_{32}, \qquad \mathscr{B}_{21}^0 = \mathscr{B}_{21}$$

($\mathscr{B}^0_{12}$ is the linear manifold spanned by the corresponding operator products $\mathbf{A}_{1k}\mathbf{A}_{k,k-1}\cdots\mathbf{A}_{32}$, each mapping $\mathbf{X}_2$ into $\mathbf{X}_1$). Since clearly $\mathscr{B}^0_{12}\mathscr{B}^0_{21} \neq \{0\}$, it follows from Secs. A.31–A.32 that $\mathbf{X}_1$ and $\mathbf{X}_2$ have the same dimension $r_1 = r_2$, while $\mathscr{B}^0_{21} = \mathscr{B}_{21} = \{\lambda\mathbf{A}^0_{21}\}$ where $\mathbf{A}^0_{21}$ is an invertible operator with inverse $(\mathbf{A}^0_{21})^{-1}$ and $\mathscr{B}^0_{12} = \{\mu(\mathbf{A}^0_{21})^{-1}\}$. Similarly, applying the same argument to the category $\mathscr{K}'_0$ made up of the two spaces $\mathbf{X}_2$, $\mathbf{X}_3$ and the linear manifolds $\mathscr{B}^0_{23}$, $\mathscr{B}^0_{32}$ spanned by the operators of the form

$$(\mathbf{A}^0_{21})^{-1}\mathbf{A}_{1k}\mathbf{A}_{k,k-1}\cdots\mathbf{A}_{43}$$

and $\mathbf{A}_{32}$, respectively, we find that $r_2 = r_3$ and $\mathscr{B}_{32} = \{\lambda\mathbf{A}^0_{32}\}$ where $\mathbf{A}^0_{32}$ is an invertible matrix. Continuing in this way, we arrive at the desired conclusion after k steps. ▌

A.36. In this section and the next, when considering a category made up of N spaces, we will assume that all the cyclic products $\mathscr{B}_{\alpha\beta}\mathscr{B}_{\beta\gamma}\cdots\mathscr{B}_{\delta\alpha}$ vanish. Otherwise, we would simply identify the corresponding spaces which are all of the same dimension.

First consider the following concrete category, which we denote by $\mathscr{K}_2^N$: Let $\mathbf{X}_{12}, \ldots, \mathbf{X}_{1N}$ be $N - 1$ arbitrary subspaces of the space $\mathbf{X}_1$, and for distinct $j, k, l, \ldots$ let

$$\mathbf{X}_{1jk} = \mathbf{X}_{1j} \cap \mathbf{X}_{1k}, \qquad \mathbf{X}_{1jkl} = \mathbf{X}_{1j} \cap \mathbf{X}_{1k} \cap \mathbf{X}_{1l}, \ldots,$$

where we successively form intersections of the spaces $\mathbf{X}_{1j}$ two at a time, three at a time, and so on. If N is finite, the last intersection will be $\mathbf{X}_{12\ldots N}$, the intersection of all $N - 1$ of the selected subspaces, while if N is infinite there will be no last intersection. Let the same construction be carried out in all the remaining spaces $\mathbf{X}_2, \mathbf{X}_3, \ldots$, where the index of the whole space is always the first of the indices appearing in the symbol used to denote any of its subspaces. Thus to any set of distinct indices $j, k, \ldots$ (in that order) there corresponds a unique subspace of $\mathbf{X}_j$. As for the sets $\mathscr{B}_{\beta\alpha}$, we define $\mathscr{B}_{21}$ as the set of all operators mapping $\mathbf{X}_1$ into $\mathbf{X}_2$ such that every subspace $\mathbf{X}_{1j\ldots k}$ goes into $\mathbf{X}_{21j\ldots k}$ if the sequence $j, \ldots, k$ does not contain the index 2 and into the set $\{0\}$ otherwise, with the other sets $\mathscr{B}_{\beta\alpha}$ being defined similarly.

We now prove that $\mathscr{K}_2^N$ is in fact a category. Given operators $\mathbf{A}_{21} \in \mathscr{B}_{21}$ and $\mathbf{B}_{32} \in \mathscr{B}_{32}$, consider the operator $\mathbf{B}_{32}\mathbf{A}_{21}$ carrying the space $\mathbf{X}_1$ into the space $\mathbf{X}_3$. The operator $\mathbf{A}_{21}$ carries the subspace $\mathbf{X}_{1j\ldots k}$ into $\mathbf{X}_{21j\ldots k}$ and then $\mathbf{B}_{32}$ carries $\mathbf{X}_{21j\ldots k}$ into $\mathbf{X}_{321j\ldots k} \subset \mathbf{X}_{31j\ldots k}$. Hence $\mathbf{B}_{32}\mathbf{A}_{21} \in \mathscr{B}_{31}$, as required. Moreover, if there is a sequence of operators $\mathbf{A}_{1j}, \ldots, \mathbf{A}_{k1}$ mapping the space $\mathbf{X}_1$ into itself, then the resulting operator carries $\mathbf{X}_1$ into $\mathbf{X}_{1j\ldots k1} = \{0\}$, in keeping with the requirement that $\mathscr{B}_{1j}\cdots\mathscr{B}_{k1} = \{0\}$.

A.37. Next we show that *every category*† *made up of* $N \leqslant \infty$ *spaces* $\mathbf{X}_1, \mathbf{X}_2, \ldots$ *with* $\mathscr{B}_j = \{\lambda\mathbf{E}\}$ *is contained in a category of the type* $\mathscr{K}_2^N$. Let $\mathbf{X}_{jk}$ be the total image in the space $\mathbf{X}_j$ of the space $\mathbf{X}_k$ under the action of all operators in $\mathscr{B}_{jk}$, and let $\mathbf{X}_{jkl\ldots sm}$ be the total image in the space $\mathbf{X}_j$ of the space $\mathbf{X}_m$ under the action of all operators of the form $\mathbf{A}_{jk}\mathbf{A}_{kl} \cdots \mathbf{A}_{sm}$ (in that order). Then $\mathbf{X}_{jkl\ldots sm}$ is contained in the intersection of $\mathbf{X}_{jk}, \mathbf{X}_{jl}, \ldots, \mathbf{X}_{jm}$. In fact, if $z \in \mathbf{X}_{jkl\ldots sm}$, then

$$z \in \sum_\alpha \mathbf{A}_{jk}^\alpha \mathbf{A}_{kl}^\alpha \cdots \mathbf{A}_{pq}^\alpha \mathbf{A}_{qr}^\alpha \cdots \mathbf{A}_{sm}^\alpha z_m^\alpha,$$

where $z_m^\alpha \in \mathbf{X}_m$, or equivalently,

$$z \in \sum_\alpha \mathbf{A}_{jk}^\alpha \mathbf{A}_{kl}^\alpha \cdots \mathbf{A}_{pq}^\alpha y_q^\alpha,$$

where

$$y_q^\alpha = \sum_\alpha \mathbf{A}_{qr}^\alpha \cdots \mathbf{A}_{sm}^\alpha z_m^\alpha \in \mathbf{X}_q.$$

But $\mathbf{A}_{jk}^\alpha \mathbf{A}_{kl}^\alpha \cdots \mathbf{A}_{pq}^\alpha \in \mathscr{B}_{jq}$, and hence $z \in \mathbf{X}_{jq}$, as required. Note also that $\mathbf{A}_{ij}$ carries $\mathbf{X}_{j\ldots m}$ into $\mathbf{X}_{ij\ldots m}$.

It is now clear that our category is contained in a category of the type $\mathscr{K}_2^N$ with defining subspaces $\mathbf{X}_{jk}$. In particular, all the maximal categories must be of the type $\mathscr{K}_2^N$. However, it is not clear what conditions on the subspaces $\mathbf{X}_{jk}$ make a category of the type $\mathscr{K}_2^N$ maximal. (Recall that in the case of two spaces $\mathbf{X}_1$ and $\mathbf{X}_2$, a necessary and sufficient condition for maximality of a category of the type $\mathscr{K}_2$ is that the spaces $\mathbf{X}_{12}$ and $\mathbf{X}_{21}$ either be both different from $\{0\}$, or else that one of the spaces be the whole space while the other is the space $\{0\}$.)

A.4. The Case of Simple Algebras

If the given algebras $\mathscr{B}_\alpha$ are all simple, then, by Sec. 11.86, each $\mathscr{B}_\alpha$ consists of the set of all quasi-diagonal matrices of the form

$$\left\| \begin{array}{cccc} \boxed{C} & & & \\ & \boxed{C} & & \\ & & \ddots & \\ & & & \boxed{C} \end{array} \right\| \qquad (2)$$

in some basis, where C ranges over the set of all square matrices of order r_α.

† Of the special type under consideration.

A.41. First we consider a category $\mathscr{K}$ with just two spaces, a space $\mathbf{X}_1$ of dimension n_1 with k_1 blocks of size m_1 (so that $n_1 = k_1 m_1$) and a space $\mathbf{X}_2$ of dimension n_2 with k_2 blocks of size m_2 (so that $n_2 = k_2 m_2$). Then every matrix $\mathfrak{A}_{21} \in \mathscr{B}_{21}$ can be partitioned into blocks as follows:

$$\mathfrak{A}_{21} = \underbrace{\left\| \begin{array}{cccc} \overbrace{A_{11}}^{m_1} & A_{12} & \cdots & A_{1k_1} \\ A_{21} & \cdot & \cdots & \cdot \\ \cdot & \cdot & \cdots & \cdot \\ A_{k_2 1} & \cdot & \cdots & A_{k_2 k_1} \end{array} \right\|}_{k_1} \Bigg\} k_2 \qquad (\text{block height } m_2)$$

Similarly, every matrix $\mathfrak{B}_{12} \in \mathscr{B}_{12}$ can be written in the form

$$\mathfrak{B}_{12} = \underbrace{\left\| \begin{array}{cccc} \overbrace{B_{11}}^{m_2} & B_{12} & \cdots & B_{1k_2} \\ B_{21} & \cdot & \cdots & \cdot \\ \cdot & \cdot & \cdots & \cdot \\ B_{k_1 1} & \cdot & \cdots & B_{k_1 k_2} \end{array} \right\|}_{k_2} \Bigg\} k_1 \qquad (\text{block height } m_1)$$

THEOREM. *Either* $\mathfrak{A}_{21}\mathfrak{B}_{12} = 0$ (*for arbitrary* $\mathfrak{A}_{21} \in \mathscr{A}_{21}$, $\mathfrak{B}_{12} \in \mathscr{B}_{12}$,), *or* $k_1 = k_2$ *and the matrices* A_{jk} *are all multiples of an (arbitrary) fixed matrix* Λ *and the matrices* B_{jk} *are all multiples of an (arbitrary) fixed matrix* M, *with the constants of proportionality making up a pair of mutually inverse matrices* $\tilde{A}$ *and* $\tilde{B}$ *of order* $k_1 = k_2$.†

Proof. If the matrices $\mathfrak{A}_{21}$ and $\mathfrak{B}_{12}$ belong to the category $\mathscr{K}$, then so does their product (from the appropriate sides) by matrices C_1 and C_2 of the form (2). Therefore, along with the equality

$$\mathfrak{A}_{21}\mathfrak{B}_{12} = C_2,$$

we also have

$$\mathfrak{A}_{21}C_1\mathfrak{B}_{12} = C_2$$

† A category of the second type will be denoted by $\mathscr{K}_3$.

for an arbitrary matrix C_1 of the form (2). Recalling the rule for multiplication of block matrices (Sec 4.51), we have

$$\begin{aligned} A_{11}CB_{11} + A_{12}CB_{21} + \cdots + A_{1k_1}CB_{k_11} \\ = A_{21}CB_{12} + A_{22}CB_{22} + \cdots + A_{2k_1}CB_{k_12} \\ = \cdots = A_{k_21}CB_{1k_2} + A_{k_22}CB_{2k_2} + \cdots + A_{k_2k_1}CB_{k_1k_2}, \end{aligned} \qquad (3)$$

$$A_{11}CB_{12} + A_{12}CB_{22} + \cdots + A_{1k_1}CB_{k_12} = 0,$$

$$\cdots\cdots\cdots\cdots\cdots\cdots$$

Let C be the matrix with a single nonzero element, equal to 1, appearing in the rth row and sth column ($r \leqslant m_1$, $s \leqslant m_1$). In general, if A is any $m_2 \times m_1$ matrix and B any $m_1 \times m_2$ matrix, then ACB is a $m_1 \times m_2$ matrix of rank 1, with the element $a_{pr}b_{sq}$ appearing in the pth row and qth column. With this choice of C, the formulas (3) become

$$\begin{aligned} a_{pr}^{11}b_{sq}^{11} + a_{pr}^{12}b_{sq}^{21} + \cdots + a_{pr}^{1k_1}b_{sq}^{k_11} \\ = a_{pr}^{21}b_{sq}^{12} + a_{pr}^{22}b_{sq}^{22} + \cdots + a_{pr}^{2k_1}b_{sq}^{k_12} \\ = \cdots = a_{pr}^{k_21}b_{sq}^{1k_2} + a_{pr}^{k_22}b_{sq}^{2k_2} + \cdots + a_{pr}^{k_2k_1}b_{sq}^{k_1k_2}, \end{aligned} \qquad (4)$$

$$a_{pr}^{11}b_{sq}^{12} + a_{pr}^{12}b_{sq}^{22} + \cdots + a_{pr}^{1k_1}b_{sq}^{k_12} = 0,$$

$$\cdots\cdots\cdots\cdots\cdots\cdots\cdots\cdots,$$

where the superscripts denote the indices of the corresponding matrices.

We can regard (4) as a single matrix equation

$$\tilde{A}_{pr}\tilde{B}_{sq} = \left\| \begin{matrix} a_{pr}^{11} & a_{pr}^{12} & \cdots & a_{pr}^{1k_1} \\ a_{pr}^{21} & a_{pr}^{22} & \cdots & a_{pr}^{2k_1} \\ \cdot & \cdot & \cdots & \cdot \\ a_{pr}^{k_21} & a_{pr}^{k_22} & \cdots & a_{pr}^{k_2k_1} \end{matrix} \right\| \left\| \begin{matrix} b_{sq}^{11} & b_{sq}^{12} & \cdots & b_{sq}^{1k_2} \\ b_{sq}^{21} & b_{sq}^{22} & \cdots & b_{sq}^{2k_2} \\ \cdot & \cdot & \cdots & \cdot \\ b_{sq}^{k_11} & b_{sq}^{k_12} & \cdots & b_{sq}^{k_1k_2} \end{matrix} \right\|$$

$$= \left. \underbrace{\left\| \begin{matrix} \lambda & 0 & \cdots & 0 \\ 0 & \lambda & \cdots & 0 \\ \cdot & \cdot & \cdots & \cdot \\ 0 & 0 & \cdots & \lambda \end{matrix} \right\|}_{k_2} \right\} k_2.$$

Similarly, we have

$$\tilde{B}_{sq}\tilde{A}_{pr} = \left\| \begin{matrix} b_{sq}^{11} & b_{sq}^{12} & \cdots & b_{sq}^{1k_2} \\ b_{sq}^{21} & b_{sq}^{22} & \cdots & b_{sq}^{2k_2} \\ \cdot & \cdot & \cdots & \cdot \\ b_{sq}^{k_1 1} & b_{sq}^{k_1 2} & \cdots & b_{sq}^{k_1 k_2} \end{matrix} \right\| \left\| \begin{matrix} a_{pr}^{11} & a_{pr}^{12} & \cdots & a_{pr}^{1k_1} \\ a_{pr}^{21} & a_{pr}^{22} & \cdots & a_{pr}^{2k_1} \\ \cdot & \cdot & \cdots & \cdot \\ a_{pr}^{k_2 1} & a_{pr}^{k_2 2} & \cdots & a_{pr}^{k_2 k_1} \end{matrix} \right\|$$

$$= \underbrace{\left\| \begin{matrix} \mu & 0 & \cdots & 0 \\ 0 & \mu & \cdots & 0 \\ \cdot & \cdot & \cdots & \cdot \\ 0 & 0 & \cdots & \mu \end{matrix} \right\|}_{k_1} \left.\vphantom{\begin{matrix} \mu \\ 0 \\ \cdot \\ 0 \end{matrix}}\right\} k_1.$$

Thus we see that the matrices $\tilde{A}_{pr}$ and $\tilde{B}_{sq}$ (with parameters p, r, s, q) form a category connecting the space $\mathbf{X}_1$ of dimension k_1 with the space $\mathbf{X}_2$ of dimension k_2, subject to the conditions

$$\mathscr{B}_1 = \{\lambda\mathbf{E}\}, \qquad \mathscr{B}_2 = \{\mu\mathbf{E}\}.$$

We can now apply the alternative proved in Sec. 12.32. Namely, if $k_1 \neq k_2$, then in fact $\lambda = 0$, $\mu = 0$, while if $\lambda \neq 0$ (or if $\mu \neq 0$) for at least one set of indices p, q, r, s, then $k_1 = k_2$ and the matrices $\tilde{A}_{pr}$ are all multiples of a single invertible matrix $\tilde{A}$, while the matrices $\tilde{B}_{sq}$ are all multiples of the inverse matrix $\tilde{B} = \tilde{A}^{-1}$:

$$\tilde{A}_{pr} = \lambda_{pr}\tilde{A}, \qquad \tilde{B}_{sq} = \mu_{sq}\tilde{B}.$$

The matrix $\tilde{A}_{pr}$ consists of the elements of the matrices A_{jk} appearing in the pth row and rth column. Hence

$$\tilde{a}_{pr}^{jk} = \lambda_{pr}\tilde{a}^{jk},$$

where the $\tilde{a}^{jk}$ are the elements of the matrix $\tilde{A}$ of order $k_1 = k_2$. It follows that the matrices $\tilde{A}_{jk}$ are all multiples with coefficients $\tilde{a}^{ij}$ of a fixed matrix $\Lambda = \|\lambda_{pr}\|$, and similarly for the matrices $\tilde{B}_{jk}$. Moreover, the matrices $\tilde{A}$ and $\tilde{B}$ are inverses of each other, as already noted. ∎

A.42. Thus if $\mathfrak{A}_{21}\mathfrak{B}_{12} \neq 0$, then $k_1 = k_2$ and the category $\mathscr{K}$ is of the form

$$\mathfrak{A}_{21} = \left\| \begin{matrix} \boxed{\tilde{a}^{11}\Lambda} & \boxed{\tilde{a}^{12}\Lambda} & \cdots & \boxed{\tilde{a}^{1k_1}\Lambda} \\ \boxed{\tilde{a}^{21}\Lambda} & \cdot & \cdots & \cdot \\ \cdot & \cdot & \cdots & \cdot \\ \boxed{\tilde{a}^{k_1 1}\Lambda} & \cdot & \cdots & \boxed{\tilde{a}^{k_1 k_1}\Lambda} \end{matrix} \right\|, \tag{5}$$

$$\mathfrak{B}_{12} = \left\| \begin{matrix} \boxed{\tilde{b}^{11}\mathrm{M}} & \boxed{\tilde{b}^{12}\mathrm{M}} & \cdots & \boxed{\tilde{b}^{1k_1}\mathrm{M}} \\ \boxed{\tilde{b}^{21}\mathrm{M}} & \cdot & \cdots & \cdot \\ \cdot & \cdot & \cdots & \cdot \\ \boxed{\tilde{b}^{k_1 1}\mathrm{M}} & \cdot & \cdots & \boxed{\tilde{b}^{k_1 k_1}\mathrm{M}} \end{matrix} \right\|,$$

where $\Lambda = \|\lambda_{pr}\|$ is an $m_2 \times m_1$ matrix and $M = \|\mu_{sq}\|$ is an $m_1 \times m_2$ matrix. Among the matrices Λ figuring in the given category there must be a nonzero matrix Λ_0 (since $\mathscr{B}_{21}\mathscr{B}_{12} \neq \{0\}$), and hence any $m_2 \times m_1$ matrix must be a matrix Λ since we can get any nonzero matrix by multiplying Λ_0 from the right by C_2 and from the left by C_1. Hence if $\mathscr{B}_{21}\mathscr{B}_{12} \neq \{0\}$, the set $\mathscr{B}_{21}$ consists of all matrices of the form (5), where $\tilde{A} = \|\tilde{a}^{jk}\|$ is a fixed invertible matrix and Λ ranges over the set of all $m_2 \times m_1$ matrices. The situation is similar if $\mathscr{B}_{12}\mathscr{B}_{21} \neq \{0\}$. It is now clear that the inequalities $\mathscr{B}_{12}\mathscr{B}_{21} \neq \{0\}$ and $\mathscr{B}_{21}\mathscr{B}_{12} \neq \{0\}$ either both hold or both fail to hold.

A.43. The above results can be formulated in terms of tensor products, an approach which allows us to explain some further facts as well. Thus we begin with the following definition:

Given a k-dimensional linear space $\mathbf{X}$ with a basis $e_1, \ldots, e_k$ and an m-dimensional linear space $\mathbf{Y}$ with a basis $f_1, \ldots, f_m$, by the *tensor product* $\mathbf{X} \times \mathbf{Y} = \mathbf{Z}$ of the spaces $\mathbf{X}$ and $\mathbf{Y}$ we mean the set of all finite formal sums

$$\sum_{\nu=1}^{p} x_\nu \times y_\nu,$$

where $x_\nu \in \mathbf{X}$, $y_\nu \in \mathbf{Y}$. Here it is assumed that

$$[x_1 \times y] + [x_2 \times y] = [(x_1 + x_2) \times y],$$

$$[x \times y_1] + [x \times y_2] = [x \times (y_1 + y_2)],$$

$$\sum_{\nu=1}^{p} \lambda_\nu x_\nu \times y_\nu = \sum_{\nu=1}^{p} x_\nu \times \lambda_\nu y_\nu = \sum_{\nu=1}^{p} \lambda_\nu [x_\nu \times y_\nu].$$

It follows that $\mathbf{Z}$ is a linear space of dimension $\leqslant km$, where all the vectors of $\mathbf{Z}$ can be expressed in terms of vectors of the form $e_i \times f_j$ ($i = 1, \ldots, k$; $j = 1, \ldots, m$). It is further assumed that the vectors $e_i \times f_j$ are linearly independent and hence form a basis for the space $\mathbf{Z}$, so that the coefficients c_{ij} in the expansion

$$g = \sum_{i=1}^{k} \sum_{j=1}^{m} c_{ij} [e_i \times f_j] \tag{6}$$

can be uniquely determined. We can write (6) somewhat differently by summing over the index i. This gives

$$g = \sum_{j=1}^{m} \left(\sum_{i=1}^{k} c_{ij} e_i \right) \times f_j = \sum_{j=1}^{m} x_j \times f_j,$$

where the

$$x_j = \sum_{i=1}^{k} c_{ij} e_i$$

are arbitrary vectors of the space $\mathbf{X}$ (no longer necessarily basis vectors).

A.44. Let $\mathbf{A}$ be an operator mapping a space $\mathbf{X}_1$ of dimension k_1 into a space $\mathbf{X}_2$ of dimension k_2, and let $\mathbf{B}$ be an operator mapping a space $\mathbf{Y}_1$ of dimension m_1 into a space $\mathbf{Y}_2$ of dimension m_2. Then by the *tensor product* $\mathbf{C} = \mathbf{A} \times \mathbf{B}$ of the operators $\mathbf{A}$ and $\mathbf{B}$ we mean the operator mapping the space $\mathbf{Z}_1 = \mathbf{X}_1 \times \mathbf{Y}_1$ into the space $\mathbf{Z}_2 = \mathbf{X}_2 \times \mathbf{Y}_2$ in accordance with the formula

$$\mathbf{C}[e_i^1 \times f_j^1] = \mathbf{A}e_i^1 \times \mathbf{B}e_j^1 \tag{7}$$

(the superscript is the index of the space). If

$$\mathbf{A}e_i^1 = \sum_{\lambda=1}^{k_2} a_{i\lambda} e_\lambda^2, \qquad \mathbf{B}f_j^1 = \sum_{\mu=1}^{m_2} b_{j\mu} f_\mu^2,$$

then (7) takes the form

$$\mathbf{C}[e_i^1 \times f_j^1] = \sum_{\lambda=1}^{k_2} \sum_{\mu=1}^{m_2} a_{i\lambda} b_{j\mu} [e_\lambda^2 \times f_\mu^2].$$

Next we find the structure of the matrix C of the operator $\mathbf{C}$ with respect to the bases $e_i^1 \times f_j^1$ and $e_k^2 \times f_l^2$ arranged in $\mathbf{X}_1 \times \mathbf{Y}_1$ in the order

$$e_1^1 \times f_1^1, e_2^1 \times f_1^1, \ldots, e_{k_1}^1 \times f_1^1, e_1^1 \times f_2^1, e_2^1 \times f_2^1, \ldots, e_{k_1}^1 \times f_2^1, \ldots,$$
$$e_1^1 \times f_{m_1}^1, e_2^1 \times f_{m_1}^1, \ldots, e_{k_1}^1 \times f_{m_1}^1,$$

and similarly in the space $\mathbf{X}_2 \times \mathbf{Y}_2$. According to Sec. A.13, the matrix C has the form

$$\left\| \begin{array}{ccccccc}
a_{11}b_{11} & a_{12}b_{11} & \cdots & a_{1k_2}b_{11} & \cdots & a_{11}b_{1m_2} & a_{12}b_{1m_2} & \cdots & a_{1k_2}b_{1m_2} \\
a_{21}b_{11} & a_{22}b_{11} & \cdots & a_{2k_2}b_{11} & \cdots & a_{21}b_{1m_2} & a_{22}b_{1m_2} & \cdots & a_{2k_2}b_{1m_2} \\
\cdot & \cdot & \cdots & \cdot & \cdots & \cdot & \cdot & \cdots & \cdot \\
a_{k_1 1}b_{11} & a_{k_1 2}b_{11} & \cdots & a_{k_1 k_2}b_{11} & \cdots & a_{k_1 1}b_{1m_2} & a_{k_1 2}b_{1m_2} & \cdots & a_{k_1 k_2}b_{1m_2} \\
\cdot & \cdot & \cdots & \cdot & \cdots & \cdot & \cdot & \cdots & \cdot \\
a_{11}b_{m_1 1} & a_{12}b_{m_1 1} & \cdots & a_{1k_2}b_{m_1 1} & \cdots & a_{11}b_{m_1 m_2} & a_{12}b_{m_1 m_2} & \cdots & a_{1k_2}b_{m_1 m_2} \\
\cdot & \cdot & \cdots & \cdot & \cdots & \cdot & \cdot & \cdots & \cdot \\
a_{k_1 1}b_{m_1 1} & a_{k_1 2}b_{m_1 1} & \cdots & a_{k_1 k_2}b_{m_1 1} & \cdots & a_{k_1 1}b_{m_1 m_2} & a_{k_1 2}b_{m_1 m_2} & \cdots & a_{k_1 k_2}b_{m_1 m_2}
\end{array} \right\|$$

or

$$\left\| \begin{array}{ccc}
\boxed{Ab_{11}} & \cdots & \boxed{Ab_{1m_2}} \\
\cdot & \cdots & \cdot \\
\boxed{Ab_{m_1 1}} & \cdots & \boxed{Ab_{m_1 m_2}}
\end{array} \right\|$$

when written as a block matrix.

A.45. Applying Secs. A.41 and A.44, we see that *the operators of the algebra $\mathscr{B}_1$ considered above are the operators in the tensor product of an m_1-dimensional space $\mathbf{X}_1$ and a k_1-dimensional space $\mathbf{Y}_1$ which are tensor products of an arbitrary operator $\mathbf{C} \in \mathscr{B}(\mathbf{X}_1)$ and the unit operator $\mathbf{E} \in \mathscr{B}(\mathbf{Y}_1)$. Moreover, the operators of the set $\mathscr{B}_{21}$ are the tensor products of an arbitrary operator $\Lambda \in \mathscr{B}(\mathbf{X}_2, \mathbf{X}_1)$ and a fixed invertible operator $\tilde{\mathbf{A}} \in \mathscr{B}(\mathbf{Y}_2, \mathbf{Y}_1)$, while the operators of the set $\mathscr{B}_{12}$ are the tensor products of an arbitrary operator $\mathrm{M} \in \mathscr{B}(\mathbf{X}_1, \mathbf{X}_2)$ and the inverse operator $\tilde{\mathbf{A}}^{-1}$.*

A.46. The following formula obviously holds for products of tensor products of operators:

$$(\mathbf{A} \times \mathbf{B})(\mathbf{C} \times \mathbf{D}) = (\mathbf{AC}) \times (\mathbf{BD}).$$

Hence, multiplying the operators $\mathbf{A}_{21} \in \mathscr{B}_{21}$ and $\mathbf{A}_{12} \in \mathscr{B}_{12}$, we find that

$$(\Lambda \times \tilde{\mathbf{A}})(\mathrm{M} \times \tilde{\mathbf{A}}^{-1}) = (\Lambda\mathrm{M}) \times (\tilde{\mathbf{A}}\tilde{\mathbf{A}}^{-1}) = (\Lambda\mathrm{M}) \times \mathbf{E} \in \mathscr{B}_2,$$

as must be the case for a category $\mathscr{K}$.

A.47. Next we find the invariant subspaces of the algebra $\mathscr{B} = \{\mathbf{C} \times \mathbf{E}\}$ of operators acting in the space $\mathbf{Z} = \mathbf{X} \times \mathbf{Y}$. These subspaces are tensor products of the form $\mathbf{X} \times \mathbf{Y}_0$, where $\mathbf{Y}_0$ is an arbitrary subspace of $\mathbf{Y}$, since

$$(\mathbf{C} \times \mathbf{E})(\mathbf{X} \times \mathbf{Y}_0) = \mathbf{CX} \times \mathbf{EY}_0 \in \mathbf{X} \times \mathbf{Y}_0.$$

To see that $\mathbf{Z}$ *has no other invariant subspaces*, let

$$z = \sum x_i \times y_j$$

be any vector in $\mathbf{Z}$ (it can be assumed that the vectors x_i are linearly independent), and suppose $\mathbf{C}$ carries the vectors x_i into given vectors $\tilde{x}_i \in \mathbf{X}$. Then

$$(\mathbf{C} \times \mathbf{E}) \sum x_i \times y_j = \sum \tilde{x}_i \times y_j,$$

and hence any subspace invariant under all the operators $\mathbf{C} \times \mathbf{E}$ which contains the vector $\sum x_i \times y_j$ also contains every vector $\sum x_i \times y_j$. This proves the italicized assertion.

If we apply every operator $\Lambda \times \tilde{\mathbf{A}}$ of the category $\mathscr{K}$ to an invariant subspace $\mathbf{X}_1 \times \mathbf{Y}_{10} \subset \mathbf{X}_1 \times \mathbf{Y}_1$, then, since the matrix Λ is arbitrary, the resulting image in the space $\mathbf{Z}_2 = \mathbf{X}_2 \times \mathbf{Y}_2$ is the subspace $\Lambda\mathbf{X}_1 \times \tilde{\mathbf{A}}\mathbf{Y}_{10} = \mathbf{X}_2 \times \mathbf{Y}_{20}$. Hence *the operators of the category $\mathscr{K}$ establish a one-to-one correspondence between the invariant subspaces of the spaces $\mathbf{Z}_1$ and $\mathbf{Z}_2$, at the same time establishing a one-to-one correspondence between the ordinary subspaces of the spaces $\mathbf{Y}_1$ and $\mathbf{Y}_2$.*

A.48. Everything said above is valid under the condition $\mathscr{B}_{21}\mathscr{B}_{12} \neq \{0\}$ (or equivalently $\mathscr{B}_{12}\mathscr{B}_{21} \neq \{0\}$). If $\mathscr{B}_{12}\mathscr{B}_{21} = \mathscr{B}_{21}\mathscr{B}_{12} = \{0\}$, the above scheme does not work, and the matrices of the category $\mathscr{K}$ do not in general consist of blocks which are multiples of a fixed matrix Λ. The situation is then the same as in Sec. A.32, and we can apply the result proved there, i.e., our category $\mathscr{K}$ is contained in some category of the type $\mathscr{K}_2$ (just which one to be explained below).

A.49. We now turn to the case of a category made up of an arbitrary number of spaces $\mathbf{Z}_\alpha$ ($\alpha \in \mathscr{A}$)† and simple algebras $\mathscr{B}_\alpha$. Two spaces $\mathbf{Z}_1$ and $\mathbf{Z}_2$ will be called *cognate* if $\mathscr{B}_{12}\mathscr{B}_{21} \neq \{0\}$, so that the matrices of $\mathscr{B}_{21}$ are of the form (5). It is clear that the relation of being cognate is transitive. In fact, if $\mathbf{Z}_1$ is cognate to $\mathbf{Z}_2$ and $\mathbf{Z}_2$ is cognate to $\mathbf{Z}_3$, then $\mathbf{Z}_1$ is cognate to $\mathbf{Z}_3$, since, by the arbitrariness of the matrices Λ, there are nonzero matrices in the product $\mathscr{B}_{32}\mathscr{B}_{21}$. Hence we can partition the whole set of spaces $\mathbf{Z}_\alpha$ into nonintersecting classes of cognate spaces. If $\mathbf{Z}_1$ and $\mathbf{Z}_2$ belong to distinct classes, then $\mathscr{B}_{12}\mathscr{B}_{21} = \mathscr{B}_{21}\mathscr{B}_{12} = \{0\}$.

We can now repeat the scheme of Sec. A.36 with certain modifications. Suppose our set of spaces $\mathbf{Z}_\alpha$ is partitioned into various classes $G_1, \ldots, G_r, \ldots$ of cognate spaces, where the spaces belonging to the class G_r are of the form $\mathbf{X}_{rj} \times \mathbf{Y}_r$, and $\mathbf{Y}_r$ denotes essentially one space in which invertible operators act. We first consider the spaces $\mathbf{Y}_r$ by themselves, and construct for them a category $\mathscr{K}_2^N$ just as in Sec. A.36 (satisfying the condition $\mathscr{B}_{is}\mathscr{B}_{si} = \{0\}$) by choosing arbitrary subspaces $\mathbf{Y}_{r\mu}$ and then forming their intersections $\mathbf{Y}_{r\mu\nu}, \mathbf{Y}_{r\mu\nu\tau}, \ldots$. This category consists of the operators A_{is} mapping $\mathbf{Y}_s$ into $\mathbf{Y}_i$ and at the same time carrying the subspaces $\mathbf{Y}_{s\mu\nu}, \ldots \subset \mathbf{Y}_s$ into the subspaces $\mathbf{Y}_{is\mu\nu}, \ldots \subset \mathbf{Y}_i$. Then for the spaces $\mathbf{Z}_\alpha$ we construct the following category, denoted by $\mathscr{K}_4^N$: *If $\mathbf{Z}_i$ and $\mathbf{Z}_k$ are cognate, the operators $\mathbf{A}_{jk} \in \mathscr{B}_{jk}$ are those previously constructed, while if $\mathbf{Z}_j = \mathbf{X}_j \times \mathbf{Y}_j$ and $\mathbf{Z}_k = \mathbf{X}_k \times \mathbf{Y}_k$ belong to distinct classes, the operators $\mathbf{A}_{jk}$ are arbitrary operators mapping $\mathbf{Z}_k$ into $\mathbf{Z}_j$ and at the same time carrying every invariant subspace $\mathbf{X}_k \times \mathbf{Y}_{k\mu\nu\ldots}$ into an invariant subspace $\mathbf{X}_j \times \mathbf{Y}_{jk\mu\nu\ldots}$.*

We now verify that every category $\mathscr{K}$ with simple algebras $\mathscr{B}_j$ is contained in a category of the type $\mathscr{K}_4^N$. Suppose $\mathbf{Z}_j = \mathbf{X}_j \times \mathbf{Y}_j$ and $\mathbf{Z}_k = \mathbf{X}_k \times \mathbf{Y}_k$ belong to distinct classes of cognate spaces. Let $\mathbf{Z}_{jk}$ be the total image in the space $\mathbf{Z}_j$ of the space $\mathbf{Z}_k$ under the action of all operators in $\mathscr{B}_{jk}$. Then $\mathbf{Z}_{jk}$ is obviously an invariant subspace of $\mathbf{Z}_j$, and hence is of the form $\mathbf{X}_j \times \mathbf{Y}_{jk}$ where $\mathbf{Y}_{jk}$ is some subspace of $\mathbf{Y}_j$. Similarly, let $\mathbf{Z}_{jkl\ldots sm}$ be the total image in the space $\mathbf{Z}_j$ of the space $\mathbf{Z}_m$ under the action of all operators of the form $A_{jk}A_{kl} \cdots A_{sm}$. Then $\mathbf{Z}_{jkl\ldots sm}$ is also an invariant subspace, which is easily seen to be contained in the intersection of $\mathbf{Z}_{jk}, \mathbf{Z}_{jl}, \ldots, \mathbf{Z}_{jm}$, by an argument like that given in Sec. A.37. It follows that our category $\mathscr{K}$ is contained in a category of the type $\mathscr{K}_4^N$, as asserted.

† We temporarily denote each space by $\mathbf{Z}_\alpha$ instead of $\mathbf{X}_\alpha$, reserving $\mathbf{X}_\alpha$ for the first factor in the tensor product $\mathbf{Z}_\alpha = \mathbf{X}_\alpha \times \mathbf{Y}_\alpha$ (cf. Sec. A.45).

A.5. The Case of Complete Algebras of Diagonal Matrices

Suppose the given algebras $\mathscr{B}_\alpha$ are all complete algebras of diagonal matrices. Then in each space $\mathbf{X}_\alpha$ there is a fixed basis in which the matrices of the operators $\mathbf{A}_\alpha \in \mathscr{B}_\alpha$ are all diagonal. Relative to these bases, the operators $\mathbf{A}_{\beta\alpha} \in \mathscr{B}_{\beta\alpha}$ are also specified by certain (rectangular) matrices, so that our problem can be stated as a problem in matrix theory.

A.51. First consider a category $\mathscr{K}$ with two spaces $\mathbf{X}_1$ and $\mathbf{X}_2$, and let A_{12} be the matrix of any operator in $\mathscr{B}_{12}$. Then, by the definition of a category, the product

$$B_{12} = A_1 A_{12} B_2, \tag{8}$$

where A_1 and B_2 are suitable diagonal matrices, is also the matrix of an operator in $\mathscr{B}_{12}$. Suppose A_1 is the (diagonal) matrix whose only nonzero element, equal to 1, appears in the jth row and jth column, while B_2 is the matrix whose only nonzero element, again equal to 1, appears in the kth row and kth column. Then, by Lemma A.14, all the elements of the matrix B_{12} vanish with the (possible) exception of the single element appearing in the jth row and kth column, and this element is just the element a_{jk} of the matrix A_{12}. Thus the operation (8) replaces every element of the matrix A_{12} by zero, except the element a_{jk} which it leaves unchanged.

This leads to the following conclusion about the structure of the family $\mathscr{B}_{12}$: *The family $\mathscr{B}_{12}$ consists of all matrices with arbitrary elements at a fixed set of positions and zeros everywhere else.*

A.52. Let S_{12} denote the fixed set of positions in the matrices of the family $\mathscr{B}_{12}$ at which arbitrary elements are allowed. We now explain the connection between the sets S_{12} and S_{21}. Let $A_{12} \in \mathscr{B}_{12}$ be a matrix whose only nonzero element, equal to 1, appears in the j_1th row and k_1th column,† so that $(j_1, k_1) \in S_{12}$, and let $B_{12} \in \mathscr{B}_{21}$ be any matrix with arbitrary nonzero elements at the positions of S_{21}. Then the products $C_1 = A_{12}B_{21}$ and $D_2 = B_{21}A_{12}$ are diagonal matrices, by hypothesis. On the other hand, by Lemma A.14, the j_1th row of C_1 consists of the elements of the k_1th row of B_{21} while all the other elements of C_1 vanish. Since C_1 must be diagonal, we see that all the elements of the k_1th row of B_{21} vanish with the (possible) exception of the element in the j_1th column. Similarly, the k_1th column of D_2 consists of the elements of the j_1th column of the matrix B_{21} while all other elements vanish, and since D_2 must be diagonal, all the elements of the j_1th column of B_{21} vanish with the (possible) exception of the element in

† For simplicity, if $\mathbf{A}_{12}$ is an operator in $\mathscr{B}_{12}$ and A_{12} is its matrix, we write $A_{12} \in \mathscr{B}_{12}$ as well as $\mathbf{A}_{12} \in \mathscr{B}_{12}$, and similarly for $\mathscr{B}_{21}$, etc.

the k_1th row. Thus, *if* $(j_1, k_1) \in S_{12}$, *all the elements of the* j_1*th column and* k_1*th row of an arbitrary matrix of* $\mathscr{B}_{21}$ *vanish with the (possible) exception of the element at the intersection of this row and column.*

We are now able to determine the structure of the set S_{21} from a knowledge of the set S_{12}. By suitably interchanging rows and columns of the matrices of $\mathscr{B}_{12}$ (which is equivalent to interchanging elements in the bases of the spaces $\mathbf{X}_1$ and $\mathbf{X}_2$), we can see to it that the rows and columns appearing first in the matrices of $\mathscr{B}_{12}$ contain no positions in the set S_{12}, while the rows and columns with only one position each in S_{12} come next and the rows and columns with at least two positions each in S_{12} come last. Thus a matrix $A_{12} \in \mathscr{B}_{12}$ has the form

$$A_{12} = \begin{array}{c} \\ 1 \\ \\ \alpha \\ \\ \\ \beta \\ \\ \\ m \end{array} \begin{array}{cc|ccc|ccc} 1 & \gamma & & & \delta & & & n \\ \hline 0 & \cdots & 0 & \cdots & \cdot & 0 & \cdot & \cdots \\ 0 & \cdots & \cdot & \cdots & \cdot & \cdot & \cdot & \cdots \\ \cdot & \cdots & \cdot & \cdots & \cdot & \cdot & \cdot & \cdots \\ \hline 0 & \cdots & 1 & \cdots & \cdot & 0 & \cdot & \cdots \\ \cdot & \cdots & \cdot & \cdots & \cdot & \cdot & \cdot & \cdots \\ 0 & \cdots & \cdot & \cdots & 1 & 0 & \cdot & \cdots \\ \hline 0 & \cdots & \cdot & \cdots & \cdot & 1 & 1 & \cdots \\ \cdot & \cdots & \cdot & \cdots & \cdot & 1 & \cdot & \cdots \\ 0 & \cdots & 0 & \cdots & \cdot & 1 & \cdot & \cdots \end{array}, \qquad (9)$$

where the positions corresponding to the set S_{12} are occupied by ones and all other positions are occupied by zeros.

Next we construct the general matrix $B_{21} \in \mathscr{B}_{21}$, with n rows and m columns:

$$B_{21} = \begin{array}{c} \\ 1 \\ \\ \gamma \\ \\ \\ \delta \\ \\ \\ n \end{array} \begin{array}{cc|ccc|cc} 1 & \alpha & & & \beta & & m \\ \hline \cdot & \cdots & 0 & \cdots & \cdot & 0 & \cdots \\ \cdot & \cdots & \cdot & \cdots & \cdot & \cdot & \cdots \\ \cdot & \cdots & \cdot & \cdots & \cdot & \cdot & \cdots \\ \hline 0 & \cdots & 1 & \cdots & \cdot & 0 & \cdots \\ \cdot & \cdots & \cdot & \cdots & \cdot & \cdot & \cdots \\ 0 & \cdots & \cdot & \cdots & 1 & 0 & \cdots \\ \hline 0 & \cdots & \cdot & \cdots & \cdot & 0 & \cdots \\ \cdot & \cdots & \cdot & \cdots & \cdot & \cdot & \cdots \\ 0 & \cdots & 0 & \cdots & \cdot & 0 & \cdots \end{array}. \qquad (10)$$

Since the matrix A_{12} has a one in row $\alpha + 1$ and column $\gamma + 1$, the matrix B_{21} can have a one in column $\alpha + 1$ and row $\gamma + 1$ but, in any event, the remaining elements of this row and column must vanish. The same is true of all the rows from $\gamma + 1$ to δ and columns from $\alpha + 1$ to β. If the matrix A_{12} has two ones in column $\delta + 1$, then all the elements of the corresponding row of the matrix B_{21} vanish, and the same is true of all columns from $\delta + 1$ to n (which contain at least two ones). However, if a column of the matrix A_{12} contains only a single one, then there are two ones in some suitable row with index $>\beta$, and this causes the column of the matrix B_{21} with the same index to vanish. As a result, the whole lower right-hand corner of the matrix B_{21} is occupied by zeros. In fact, let (j, k) be any position in this corner, and consider the corresponding position (k, j) in the matrix A_{12}. Then the kth row or jth column of A_{12} has at least two ones, since otherwise we would have put this row or column in an "earlier" position. This means that the kth column or jth row of B_{21} consists entirely of zeros, so that in any event there must be a zero at the position (j, k). The lower left-hand corner of the matrix B_{21} also consists entirely of zeros. In fact, if a one appeared anywhere in the lower left-hand corner of B_{21}, say at the position (j, k), then, by the symmetry of the construction, all the elements in the jth column of A_{12} except possibly the element in the kth row (i.e., in the upper right-hand corner of A_{12}) would have to vanish, which is impossible since this column must have a one in the lower right-hand corner. A similar argument shows that the upper right-hand corner of B_{21} also consists entirely of zeros. As for the elements in the upper left-hand corner of B_{21}, they can be arbitrary.

Thus it is clear that our category can be enlarged by including *all* elements of the lower right-hand corner of the matrix (9) in the set S_{12} (provided S_{12} does not already contain all these elements) and including *all* elements of the upper left-hand corner of the matrix (10) in the set S_{21}. The category then becomes maximal, since it is no longer possible to enlarge S_{12} without making S_{21} smaller. In geometric language, the maximal category made up of two spaces $\mathbf{X}_1$ and $\mathbf{X}_2$ is constructed as follows: *The space* $\mathbf{X}_1$ *is the direct sum of three subspaces* $\mathbf{X}_1^0$, $\mathbf{X}_1^1$, $\mathbf{X}_1^2$ *and the space* $\mathbf{X}_2$ *is the direct sum of three subspaces* $\mathbf{X}_2^0$, $\mathbf{X}_2^1$, $\mathbf{X}_2^2$, *where* $\mathbf{X}_1^1$ *and* $\mathbf{X}_2^1$ *have the same dimension. The effect of an operator* $\mathbf{A}_{12}$ *is such that* $\mathbf{X}_1^0$ *is mapped into* $\{0\}$, $\mathbf{X}_1^1$ *is mapped into* $\mathbf{X}_2^1$ *by a diagonal matrix and* $\mathbf{X}_1^2$ *is mapped into* $\mathbf{X}_2^2$ *in an arbitrary way, while the effect of an operator* $\mathbf{B}_{21}$ *is such that* $\mathbf{X}_2^0$ *is mapped into* $\mathbf{X}_1^0$ *in an arbitrary way,* $\mathbf{X}_2^1$ *is mapped into* $\mathbf{X}_1^1$ *by a diagonal matrix and* $\mathbf{X}_2^2$ *is mapped into* $\{0\}$. An arbitrary (nonmaximal) category differs from the maximal category in that the operators mapping $\mathbf{X}_1^2$ into $\mathbf{X}_2^2$ are not arbitrary, but rather correspond to matrices with zeros in certain fixed positions, while the same is true of the operators mapping $\mathbf{X}_2^0$ into $\mathbf{X}_1^0$ (there is no connection whatsoever between the positions occupied by these zeros).

A.53. Next we consider a category $\mathscr{K}$ involving arbitrarily many spaces $\mathbf{X}_\alpha$ ($\alpha \in \mathscr{A}$). First of all, it is clear that every subcategory of the category $\mathscr{K}$ made up of a pair of spaces $\mathbf{X}_\alpha$, $\mathbf{X}_\beta$ and corresponding families $\mathscr{B}_{\beta\alpha}$, $\mathscr{B}_{\alpha\beta}$ is constructed in the way just described, i.e., $\mathscr{B}_{\beta\alpha}$ is the family of all matrices with arbitrary elements at some prescribed set of positions $S_{\beta\alpha}$, while $\mathscr{B}_{\alpha\beta}$ is the family of all matrices with arbitrary elements at some other prescribed set of positions $S_{\alpha\beta}$. In this regard, we introduce the following notation: If S is any set of positions in an $m \times n$ matrix, then $\mathscr{B}_{mn}(S)$ is the set of all $m \times n$ matrices with arbitrary elements at the positions S and zeros everywhere else.

Now let S_1 be a set of positions in an $m \times n$ matrix and S_2 a set of positions in an $n \times p$ matrix. Suppose S is the *product* S_1S_2, defined as the set of all positions in an $m \times p$ matrix at which one can get nonzero elements in the product $\mathscr{B}_{mn}(S_1)\mathscr{B}_{np}(S_2)$. In other words, a position (i, k) belongs to the set S_1S_2 if and only if there exists an index j such that (i, j) belongs to S_1 and (j, k) belongs to S_2. Let $S_{11}, \ldots, S_{1p}$ be a collection of such sets of positions for an $m \times n$ matrix, and let $S_{21}, \ldots, S_{2q}$ be an analogous collection for an $n \times p$ matrix. Then the general formula

$$\bigcup_{i=1}^{p} S_{1i} \bigcup_{j=1}^{q} S_{2j} = \bigcup_{i=1}^{p} \bigcup_{j=1}^{q} S_{1i}S_{2j} \tag{11}$$

is an easy consequence of the definition of a product of S-sets.

In terms of products of S-sets, we can write the conditions for our category in the form

$$S_{\alpha\beta}S_{\beta\alpha} \subset D, \qquad S_{\alpha\beta}S_{\beta\gamma} \subset S_{\alpha\gamma}, \tag{12}$$

where D is the set of all positions along the principal diagonal of the appropriate square matrix.

A.54. We now construct a family of concrete categories of a certain type. To specify a category $\mathscr{K}$ means to specify all the families $\mathscr{B}_{\alpha\beta}$, or equivalently in the present case, to specify all the sets $S_{\alpha\beta}$. Choosing S_{21} arbitrarily, we then choose S_{12} in such a way that $S_{21}S_{12} \subset D$, $S_{12}S_{21} \subset D$ (we have already described how this is done in Sec. A.52). Suppose S_{jk} has been constructed for all j and k less than n, in such a way that the conditions (12) for a category are satisfied. Then S_{jn} and S_{nj} ($j < n$) are constructed as follows: S_{n1} is chosen arbitrarily, and S_{1n} is chosen to satisfy the conditions $S_{1n}S_{n1} \subset D$, $S_{n1}S_{1n} \subset D$. Suppose S_{jn} and S_{nj} are chosen for all $j < k$ in such a way that (12) holds. The required sets S_{nk} and S_{kn} must satisfy the following conditions implied by (12):

a) $S_{nk}S_{kn} \subset D$, $S_{kn}S_{nk} \subset D$;

b) $S_{jk}S_{kn} \subset S_{jn}$, $S_{kn}S_{nj} \subset S_{kj}$, $S_{in}S_{nk} \subset S_{ik}$, $S_{nk}S_{ki} \subset S_{ni}$;

c) $S_{kn} \supset S_{ki}S_{in}$, $S_{nk} \supset S_{nj}S_{jk}$.

The conditions a) and b) represent "upper bounds" and condition c) "lower bounds" for the sets S_{nk} and S_{kn}. We now show that these conditions are compatible. Suppose, for example, that

$$S_{kn} = \bigcup_{i=1}^{n-1} S_{ki}S_{in}, \qquad S_{nk} = \bigcup_{j=1}^{n-1} S_{nj}S_{jk}. \tag{13}$$

Then, by formula (11) and the induction hypothesis,

$$S_{nk}S_{kn} = \bigcup_{j=1}^{n-1} S_{nj}S_{jk} \bigcup_{i=1}^{n-1} S_{ki}S_{in} = \bigcup_{j=1}^{n-1} \bigcup_{i=1}^{n-1} S_{nj}S_{jk}S_{ki}S_{in}$$

$$\subset \bigcup_{j=1}^{n-1} S_{nj}S_{ji}S_{in} \subset \bigcup_{i=1}^{n-1} S_{ni}S_{in} \subset D,$$

$$S_{jk}S_{kn} = S_{jk} \bigcup_{i=1}^{n-1} S_{ki}S_{in} = \bigcup_{i=1}^{n-1} S_{jk}S_{ki}S_{in} \subset \bigcup_{i=1}^{n-1} S_{ji}S_{in} \subset S_{jn}.$$

This proves the first of the relations a) and the first of the relations b), and it is clear that the remaining relations can be proved by similar arguments. Thus the induction is justified and our construction is correct.

It is possible, of course, to construct a category by using arbitrary S_{nk} and S_{kn} satisfying the conditions a)–c), and not just sets of the special type (13) used to prove the compatibility of these conditions. In this way we obtain a large family of concrete categories, in each of which only the sets S_{n1} are arbitrary, while the remaining sets $S_{\alpha\beta}$ satisfy the extra conditions a)–c).

A.55. We now see that *every category* $\mathscr{K}$ *such that* $\mathscr{B}_{\alpha\beta}\mathscr{B}_{\beta\alpha} \subset \mathscr{B}(D)$ *belongs to the family just constructed.* In fact, the sets S_{n1} and S_{1n} are defined in $\mathscr{K}$ for every n, while the remaining sets S_{nk} and S_{kn} must satisfy the conditions a)–c). But then $\mathscr{K}$ is a category of the family described in Sec. A.54.

It would be interesting to describe the form of the maximal categories of this family.

A.6. Categories and Direct Sums

A.61. Given a category $\mathscr{K}$ with basic spaces $\mathbf{X}_i^p$, algebras $\mathscr{B}_i^p$ and families of operators $\mathscr{B}_{ji}^{qp}$, where $p = 1, \ldots, k_i$ and $q = 1, \ldots, k_q$, we now show how to construct a new category whose basic spaces are direct sums of the spaces $\mathbf{X}_i^p$ and whose basic algebras are the corresponding direct sums of the algebras $\mathscr{B}_i^p$.

Thus let $\mathbf{X}_i$ be the direct sum of the spaces $\mathbf{X}_i^1, \ldots, \mathbf{X}_i^{k_i}$, and let $\mathscr{B}_i$ be the direct sum of the corresponding algebras $\mathscr{B}_i^1, \ldots, \mathscr{B}_i^{k_i}$ (i.e., in the space $\mathbf{X}_i^p$ an operator $\mathbf{A} \in \mathscr{B}_i$ acts like any operator in the algebra $\mathscr{B}_i^p$). To specify an operator $\mathbf{A}_{ji} \in \mathscr{B}_{ji}$, we use the block matrix

$$A_{ji} = \left\| \begin{array}{cccc} A_{ji}^{11} & A_{ji}^{12} & \cdots & A_{ji}^{1k_i} \\ A_{ji}^{21} & A_{ji}^{22} & \cdots & A_{ji}^{2k_i} \\ \cdot & \cdot & \cdots & \cdot \\ A_{ji}^{k_j1} & A_{ji}^{k_j2} & \cdots & A_{ji}^{k_jk_i} \end{array} \right\|, \tag{14}$$

where the block A_{ji}^{qp} corresponds to an arbitrary operator of $\mathscr{K}$ mapping the space $\mathbf{X}_i^p$ into the space $\mathbf{X}_j^q$ ($p = 1, \ldots, k_i; j = 1, \ldots, k_j$). To show that this gives a category, we note that if

$$B_{il} = \left\| \begin{array}{cccc} B_{il}^{11} & B_{il}^{12} & \cdots & B_{il}^{1k_l} \\ B_{il}^{21} & B_{il}^{22} & \cdots & B_{il}^{2k_l} \\ \cdot & \cdot & \cdots & \cdot \\ B_{il}^{k_i1} & B_{il}^{k_i2} & \cdots & B_{il}^{k_ik_l} \end{array} \right\|,$$

then

$$A_{ji}B_{il} = \left\| \begin{array}{cc} A_{ji}^{11}B_{il}^{11} + A_{ji}^{12}B_{il}^{21} + \cdots + A_{ji}^{1k_i}B_{il}^{k_i1} & \cdots \\ \cdots & \cdots \\ \cdots & \cdots \end{array} \right\|,$$

where each sum of products again belongs to the appropriate family of operators, by the definition of the category $\mathscr{K}$. Thus our rule leads to a new category $\overline{\mathscr{K}}$, which we call an *extension* of the category $\mathscr{K}$.

A.62. It turns out that the converse is also true, i.e., if the basic spaces $\mathbf{X}_i$ figuring in a category are direct sums of certain spaces $\mathbf{X}_i^p$ ($p = 1, \ldots, k_i$) and if the corresponding algebras $\mathscr{B}_i$ are direct sums of algebras $\mathscr{B}_i^p$ ($p = 1, \ldots, k_i$) of operators acting in $\mathbf{X}_i^p$, then the whole category is an extension $\overline{\mathscr{K}}$ in the above sense of a category $\mathscr{K}$, constructed from the spaces $\mathbf{X}_i^p$ and algebras $\mathscr{B}_i^p$. In fact, let $\mathscr{K}'$ be a category of the indicated type. Then in an appropriate basis chosen in the subspaces $\mathbf{X}_i^p$, every matrix A_i of an operator of the algebra $\mathscr{B}_i$ has the quasi-diagonal form

$$A_i = \begin{Vmatrix} \boxed{A_i^1} & & & \\ & \boxed{A_i^2} & & \\ & & \ddots & \\ & & & \boxed{A_i^{k_i}} \end{Vmatrix},$$

where A_i^p is a square matrix of order r_i^p $(p = 1, \ldots, k_i)$. Every matrix A_{ji} of an operator of the algebra $\mathscr{B}_{ji}$ is a block matrix of the form (14), where A_{ji}^{qp} is a rectangular matrix with r_j^q rows and r_i^p columns. With each block A_{ji}^{qp} we can associate in a natural way an operator $\mathbf{A}_{ji}^{qp}$ mapping the space $\mathbf{X}_i^p$ into the space $\mathbf{X}_j^q$. Using all such operators, we construct a new category $\mathscr{K}_{ji}^{qp}$ with basic spaces $\mathbf{X}_i^p$, algebras $\mathscr{B}_i^p$ and families $\mathscr{B}_{ji}^{qp}$ of operators $\mathbf{A}_{ji}^{qp}$ specified by the matrices A_{ji}^{qp}. We now show that this collection of objects does in fact define a category.

Let $\mathbf{A}_{ji}^{qp}$ be an operator mapping $\mathbf{X}_i^p$ into $\mathbf{X}_j^q$, and let $\mathbf{A}_{lj}^{rq}$ be an operator mapping $\mathbf{X}_j^q$ into $\mathbf{X}_l^r$. Then the product $\mathbf{A}_{li}^{rp} = \mathbf{A}_{lj}^{rq}\mathbf{A}_{ji}^{qp}$ belongs to the family $\mathscr{B}_{li}^{rp}$. In fact, the category $\mathscr{K}'$ contains the matrix with its qpth block equal to A_{ji}^{qp} and all other blocks equal to zero, as well as the matrix with its rqth block equal to A_{lj}^{rq} and all other blocks equal to zero. The product of these two matrices, which belongs to the category $\mathscr{K}'$, is a matrix with its lith block equal to A_{li}^{rp} and all other blocks equal to zero. Therefore $\mathbf{A}_{li}^{rp} \in \mathscr{B}_{li}^{rp}$, as asserted.

Thus all the conditions for a category are satisfied. It is true that the operators mapping the space $\mathbf{X}_i^p$ into the spaces $\mathbf{X}_i^s$ with the same subscript have not yet been defined. However, all such operators can be set equal to zero without destroying the requirements for a category.

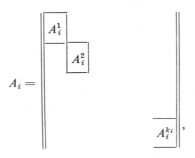

A.63. Since every semisimple algebra of operators acting in a space $\mathbf{X}_i$ allows us to decompose the space $\mathbf{X}_i$ into a direct sum of spaces $\mathbf{X}_i^j$ in which the algebra now acts as a simple algebra, we see that the structure of a general category with semisimple algebras reduces to that of a category with simple algebras (this problem was considered in Sec. A.4). The matrix of every operator $\mathbf{A}_{ji} \in \mathscr{B}_{ji}$ of the category is of the form (14) in an appropriate basis, where each block A_{ji}^{qp} is the matrix of an operator in the family $\mathscr{B}_{ji}^{qp}$ of some category $\mathscr{K}_{ji}^{qp}$ with basic spaces $\mathbf{X}_i^q$, $\mathbf{X}_i^p$ and simple algebras $\mathscr{B}_j^q$, $\mathscr{B}_i^p$. Some blocks of the matrix A_{ji} may be identically zero for all the A_{ji}. If we denote the set of all vanishing blocks by S_{ji}, the question arises of how the sets S_{ji} are related for various indices i and j. A similar problem was considered in Sec. A.5 for the case of one-dimensional blocks. The method used there is also applicable to the present case, and leads to the following result: *If the category determined by the intersection of the jth block row and ith block column of the matrix* $A_{12} \in \mathscr{B}_{12}$ *is of the type* $\mathscr{K}_1$ *or* $\mathscr{K}_3$ *(involving invertible matrices),*† *then all the blocks in the ith block row and jth block column of the matrix* A_{21} *determine zero categories, with the (possible) exception of the block at the intersection of this row and column. If the category in question is of the type* $\mathscr{K}_2$*, then matrices of a category of the type* $\mathscr{K}_2$ *appear in the indicated blocks and have zero products with the given matrix.*

We can now determine the structure of the general category, as in Sec. A.5.

Remark. A. Y. Khelemski (*loc. cit.*) has found the categories corresponding to nilpotent algebras $\mathscr{B}_\alpha$.

† See Sec. A.31 and the footnote on p. 353.

HINTS AND ANSWERS

Chapter 1

1. *Ans.* a) +; b) +.

2. *Ans.* $a_{11}a_{32}a_{23}a_{44}$, $a_{41}a_{12}a_{23}a_{34}$, $a_{31}a_{42}a_{23}a_{44}$.

3. *Ans.* $(-1)^{n(n-1)/2}$.

4. *Hint.* Consider the determinant all of whose elements equal 1.

5. *Ans.* $\Delta = (mq - np)(ad - bc)$.

6. *Hint.* Multiply the first column by 10^4, the second by 10^3, the third by 10^2, the fourth by 10^1, and add them to the last column. Then use Corollary 1.45.

7. *Ans.* $\Delta_1 = -29{,}400{,}000$, $\Delta_2 = 394$.

8. *Hint.* $P(x)$ is obviously a polynomial of degree 4. We first find its leading coefficient, and then determine its roots by making rows of the determinant coincide.
Ans. $P(x) = -3(x^2 - 1)(x^2 - 4)$.

9. *Hint.* Add all the columns to the first.
Ans. $\Delta = [x + (n - 1)a](x - a)^{n-1}$.

10. *Hint.* The determinant on the left is a polynomial of degree n in x_n with roots $x_1, \ldots, x_{n-1}$, and hence can be represented in the form

$$(A + Bx_n) \prod_{k=1}^{n-1} (x_n - x_k).$$

Another representation of the same determinant in the form of a polynomial of degree n in x_n can be obtained by expanding it with respect to its last column. Equating the coefficients of x_n^n and those of x_n^{n-2}, find A and B.

11. *Ans.* $c_1 = 0,\ c_2 = 2,\ c_3 = -2,\ c_4 = 0,\ c_5 = 3.$

12. *Ans.* $\sum M_{j_1,j_2,\ldots,j_k}^{i_1,i_2,\ldots,i_k} \tilde{A}_{j_1,j_2,\ldots,j_k}^{i'_1,i'_2,\ldots,i'_k} = 0$, where $i_1 < i_2 < \cdots < i_k$ and $i'_1 < i'_2 < \cdots < i'_k$ are fixed, and at least one of the i_α differs from the corresponding i'_α (Cauchy).

13. *Hint.* It is sufficient for the corresponding fourth-order determinant to be nonzero.

14. *Hint.* Use the results of Secs. 1.96–1.97.

Chapter 2

1. *Ans.* No, since we cannot multiply by -1 and stay within the set.

2. *Ans.* No, since we cannot add two vectors which are symmetric with respect to the given line and still stay within the set.

3. *Ans.* Yes. In particular, the number $1 \in P$ serves as the "zero vector" of the space P.

4. *Hint.* See Sec. 1.96.

5. *Hint.* Assuming linear dependence of the form

$$\alpha_1 t^{r_1} + \alpha_2 t^{r_2} + \cdots + \alpha_k t^{r_k} \equiv 0,$$

divide by t^{r_1} and differentiate. Then use induction in k.

6. *Hint.* Show that the zero vector also has a unique expansion with respect to the system $e_1, e_2, \ldots, e_n$. From this deduce the linear independence of the vectors of the system.

7. *Ans.* Yes, consisting of a single vector, i.e., any element $x \in P$ different from 1.

8. *Ans.* 1.

9. *Ans.* The intersection is the line of intersection (in the usual sense) of the two planes, while the sum is the whole space.

10. *Hint.* See Sec. 2.34.

11. *Ans.* No. It can be replaced by any other vector of the hyperplane.

12. *Ans.* With the "point" interpretation, the property means that every hyperplane contains the line passing through any two of its points.

13. *Ans.* In general $p + q + 1$, if this number does not exceed the dimension of the whole space.

14. *Ans.* $p + q + r + 2$ if this number does not exceed the dimension of the whole space.

15. *Ans.* With each positive number associate its logarithm.

Chapter 3

1. *Hint.* In a matrix of rank 1 the columns are proportional.

2. *Hint.* We have to write the conditions for a vector y to belong to the subspace $\mathbf{L}$ in such a way that they involve only minors of A of order k. But $y \in \mathbf{L}$ if and only if the matrix B obtained by adding to A the column consisting of the components of the vector y has rank k, or equivalently, if and only if every minor of B of order $k + 1$ vanishes. Expanding every minor of B of order $k + 1$ with respect to elements of the last column, we obtain a system of equations in the components of y, with coefficients which are minors of A of order k.

3. *Hint.* See Secs. 1.51–1.52.

4. *Ans.* $x = (c_1, c_2, c_3, c_4)$, where $c_1 = -16 + c_3 + c_4 + 5c_5$, $c_2 = 23 - 2c_3 - 2c_4 - 6c_5$.

5. *Ans.* If $(\lambda - 1)(\lambda + 2) \neq 0$, then

$$x = -\frac{\lambda + 1}{\lambda + 2}, \qquad y = \frac{1}{\lambda + 2}, \qquad z = \frac{(\lambda + 1)^2}{\lambda + 2}.$$

If $\lambda = 1$, the system has solutions depending on two parameters. If $\lambda = -2$, the system is incompatible.

6. *Ans.* The matrices

$$\left\| \begin{matrix} a_1 & b_1 \\ a_2 & b_2 \\ a_3 & b_3 \end{matrix} \right\| \quad \text{and} \quad \left\| \begin{matrix} a_1 & b_1 & c_1 \\ a_2 & b_2 & c_2 \\ a_3 & b_3 & c_3 \end{matrix} \right\|$$

must have the same rank.

7. *Ans.* The matrices

$$\left\| \begin{matrix} a_1 & b_1 \\ a_2 & b_2 \\ \cdot & \cdot \\ \cdot & \cdot \\ \cdot & \cdot \\ a_n & b_n \end{matrix} \right\| \quad \text{and} \quad \left\| \begin{matrix} a_1 & b_1 & c_1 \\ a_2 & b_2 & c_2 \\ \cdot & \cdot & \cdot \\ \cdot & \cdot & \cdot \\ \cdot & \cdot & \cdot \\ a_n & b_n & c_n \end{matrix} \right\|$$

must have the same rank.

8. *Ans.* $x^{(1)} = (1, -2, 1, 0, 0)$, $x^{(2)} = (1, -2, 0, 1, 0)$, $x^{(3)} = (5, -6, 0, 0, 1)$.

9. *Ans.*

$$x = \begin{Vmatrix} -16 \\ 23 \\ 0 \\ 0 \\ 0 \end{Vmatrix} + \alpha_1 \begin{Vmatrix} 1 \\ -2 \\ 1 \\ 0 \\ 0 \end{Vmatrix} + \alpha_2 \begin{Vmatrix} 1 \\ -2 \\ 0 \\ 1 \\ 0 \end{Vmatrix} + \alpha_3 \begin{Vmatrix} 5 \\ -6 \\ 0 \\ 0 \\ 1 \end{Vmatrix},$$

for example. Here the first column consists of the components of a vector x_0 which is a particular solution of the nonhomogeneous system, while the other columns consist of the components of the vectors $y^{(1)}, y^{(2)}, y^{(3)}$ forming a normal fundamental system of solutions of the corresponding homogeneous system.

10. *Ans.* The rank of A_1 is 3, and there is a basis minor in the upper left-hand corner (for example). The rank of A_2 is 5, and the basis minor is the same as the determinant of the matrix.

11. *Hint.* Move the minor M into the upper left-hand corner and then, by using the procedure of Sec. 3.62, show that all the columns of A starting with the $(r + 1)$st can be made into zero columns.

12. *Hint.* If $P \neq 0$, look for A in the form

$$\begin{Vmatrix} P & 0 & x \\ 0 & 1 & y \end{Vmatrix}.$$

13. *Hint.* The rank of the matrix $\|a_{jk}\|$ is either equal to n or less than n.

14. *Hint.* Use the Kronecker–Capelli theorem.

15. *Hint.* Use the result of Prob. 14.

Chapter 4

1. *Ans.* Also n.

2. *Ans.* c) and g).

3. *Ans.* Yes.

4. *Ans.*

$$\text{a) } A_{(e)} = \begin{Vmatrix} -1 & -1 & 2 \\ 1 & -3 & 3 \\ -1 & -5 & 5 \end{Vmatrix}; \qquad \text{b) } A_{(x)} = \begin{Vmatrix} 2 & 0 & -2 \\ 1 & -1 & 1 \\ 2 & 1 & 0 \end{Vmatrix}.$$

5. *Ans.* $\mathbf{ABAB} \neq \mathbf{A}^2\mathbf{B}^2$.

6. *Ans.* $\mathbf{AB} - \mathbf{BA} = \mathbf{E}$.

7. *Hint.* $(\mathbf{A} + \mathbf{B})^2 = \mathbf{A}^2 + \mathbf{AB} + \mathbf{BA} + \mathbf{B}^2$,

$(\mathbf{A} + \mathbf{B})^3 = \mathbf{A}^3 + \mathbf{A}^2\mathbf{B} + \mathbf{ABA} + \mathbf{AB}^2 + \mathbf{BA}^2 + \mathbf{BAB} + \mathbf{B}^2\mathbf{A} + \mathbf{B}^3$.

8. *Hint.* Use induction.

9. *Ans.* The dimension of the space is nm. For basis operators we can take those corresponding to the matrices A_{ij} ($i = 1, \ldots, n; j = 1, \ldots, m$), where A_{ij} is any matrix whose elements are all zero except for the element in the ith row and jth column.

10. *Ans.*

$$AB = \begin{Vmatrix} 0 & 0 & 0 \\ 0 & 0 & 0 \\ 0 & 0 & 0 \end{Vmatrix}.$$

11. *Ans.*

$$A^n = \begin{Vmatrix} 1 & n \\ 0 & 1 \end{Vmatrix}, \qquad B^n = \begin{Vmatrix} \cos n\varphi & -\sin n\varphi \\ \sin n\varphi & \cos n\varphi \end{Vmatrix}.$$

12. *Ans.* $A = \begin{Vmatrix} a & b \\ c & -a \end{Vmatrix}$, where $bc = -a^2$.

13. *Ans.*

$$\text{a)} \begin{Vmatrix} -9 & -2 & -10 \\ 6 & 14 & 8 \\ -7 & 5 & -5 \end{Vmatrix}; \qquad \text{b)} \begin{Vmatrix} 0 & 0 & 0 \\ 0 & 0 & 0 \\ 0 & 0 & 0 \end{Vmatrix}.$$

15. *Hint.* Use Prob. 14.

16. *Hint.* The three equations for the unknown elements of the matrices A and B lead to equations for three minors of an unknown 2×3 matrix. Now see Chap. 3, Prob. 12.

17. *Hint.* See Sec. 4.54.

18. *Hint.* Express the elements of the minor M in terms of the elements appearing in the first r rows, and then use Theorem 4.75.

19. *Hint.* Use the solution to Prob. 18.

20. *Hint.* See Sec. 4.54.

21. *Ans.*

$$\text{a) } A^{-1} = \begin{Vmatrix} 5 & -2 \\ -1 & 2 \end{Vmatrix}; \qquad \text{b) } B^{-1} = \begin{Vmatrix} 1 & -2 & 7 \\ 0 & 1 & -2 \\ 0 & 0 & 1 \end{Vmatrix}; \qquad \text{c) } C^{-1} = C.$$

23. *Ans.* If A is the zero matrix, then X is arbitrary. If $\det A \neq 0$, then X is the zero matrix. If $\det A = 0$ and A is not the zero matrix, then its rows are proportional. Let α/β be the ratio of the corresponding elements of the first and second rows of the matrix A. Then

$$X = \begin{Vmatrix} -\beta p & \alpha p \\ -\beta q & \alpha q \end{Vmatrix}$$

for any p and q.

24. *Hint*. See Secs. 1.51–1.52.

25. *Ans*. No.

26. *Hint*. Consider the operator $\mathbf{A}_\lambda$ such that

$$\mathbf{A}_\lambda[a_0 + a_1 t + \cdots + a_n t^n] = \lambda a_0 + a_1 + a_2 t + \cdots + a_n t^{n-1}.$$

27. *Hint*. The operator $\mathbf{A}$ carries linearly independent vectors into vectors that are again linearly independent.

28. *Hint*. Apply the equality $\mathbf{AB} = \mathbf{BA}$ to an eigenvector of the operator $\mathbf{A}$.

31. *Hint*. Use the result of Prob. 30.

32. *Hint*. Suitably choosing an operator $\mathbf{B}$ and using Prob. 28, reduce the solution to Prob. 31.

34. *Hint*. Use the factorization of the operator $\mathbf{A}^2 - \mu^2\mathbf{E}$.

35. *Ans*. a) $\lambda_1 = 2, f_1 = (1, 0, 0)$; $\lambda_2 = 1, f_2 = (1, 0, 1)$; $\lambda_3 = -1, f_3 = (0, 1, -1)$; b) $\lambda_1 = -1, f_1 = (1, 0, 0)$; $\lambda_2 = \lambda_3 = 1, f_2 = (1, 0, 1), f_3 = (0, 1, -1)$; c) $\lambda_1 = 2, f_1 = (1, 0, 0)$; d) $\lambda_1 = 1, f_1 = (1, 0, 0, -1)$; $\lambda_2 = 0, f_2 = (0, 1, 0, 0)$.

36. *Hint*. The relation $\mathbf{T}(\mathbf{A}^k) \subset \mathbf{N}(\mathbf{A}^m)$ is necessary and sufficient for the equality $\mathbf{A}^{k+m} = \mathbf{0}$ to hold.

37. *Hint*. Let $f_1, \ldots, f_r$ be a basis for the range of the operator $\mathbf{A}$, so that

$$\mathbf{A}x = \sum_{i=1}^{r} a_i(x) f_i$$

for every $x \in \mathbf{K}_n$. Now let

$$\mathbf{A}_i x = a_i(x) f_i \qquad (i = 1, \ldots, r).$$

Chapter 5

1. *Hint*. The first vector of the new basis is x.

2. *Hint*. Choose a new basis $f_1, f_2, \ldots, f_n$ whose last $n - k$ vectors form a basis for the space $\mathbf{K}'$. Write the condition $x \in \mathbf{K}'$ in the form of a system of equations involving the components of x in the new basis. Use the transformation formulas to construct the corresponding system of equations involving the components of x in the original basis.

3. *Hint.* Use Prob. 2 and the definition of a hyperplane.

4. *Ans.* The matrix of the desired transformation is $C = BA^{-1}$.

5. *Hint.* Let $e_1, e_2, \ldots, e_n$ be an arbitrary basis in $\mathbf{K}_n$, and let

$$L(x) = \sum_{k=1}^{n} l_k \xi_k,$$

where $\xi_1, \xi_2, \ldots, \xi_n$ are the components of the vector x. Begin the formulas for the coordinate transformation with the equation

$$\eta_1 = \sum_{k=1}^{n} l_k \xi_k.$$

6. *Hint.* Use Sec. 4.83 and the invariance of the characteristic polynomial (Sec. 5.53).

7. *Hint.* Choose a basis whose first m vectors lie in the subspace $\mathbf{R}^{(\lambda_0)}$. Show that for this basis the polynomial det $\|A_{(f)} - \lambda E\|$ has the factor $(\lambda - \lambda_0)^m$. Now use the invariance of the characteristic polynomial (Sec. 5.53).

Chapter 6

1. *Ans.* In the basis $e_n, e_{n-1}, \ldots, e_1$.

2. *Hint.* See Sec. 6.44.

3. *Ans.*

$$\begin{Vmatrix} -1 & 1 & 0 & 0 & 0 \\ 0 & -1 & 0 & 0 & 0 \\ 0 & 0 & 2 & 1 & 0 \\ 0 & 0 & 0 & 2 & 0 \\ 0 & 0 & 0 & 0 & 2 \end{Vmatrix}.$$

4. *Ans.* No. $E_2(A) = (\lambda - 2)(\lambda - 1)^2$, $E_2(B) = (\lambda - 1)(\lambda^2 - 5\lambda - 2)$.

5. *Ans.* $E_{n-1}(A_1) = E_{n-1}(A_2) = (1 - \lambda)^n$, $E_{n-2}(A_1) = E_{n-2}(A_2) = 1$;

$$E_{n-1}(A_3) = (n - \lambda)^n,\ E_{n-2}(A_3) = 1;$$

$$E_{n-1}(A_4) = \prod_{k=1}^{n} (\lambda - k),\ E_{n-2}(A_4) = 1.$$

6. *Hint.* $E_{n-1}(A) = (\alpha - \lambda)^n$, $E_{n-2}(A) = 1$.

7. *Ans.* A diagonal matrix with some of the roots of the polynomial $P(\lambda)$ along its principal diagonal.

8. *Ans.* Some of the roots of the polynomial $P(\lambda)$ lie along the principal diagonals of the Jordan blocks, and the sizes of the blocks do not exceed the multiplicities of the corresponding roots.

9. *Hint.* The vectors x, $\mathbf{A}x$ and $\mathbf{A}x^2$ are linearly dependent.

10. *Ans.* Polynomials in $A_m(a)$.

11. *Ans.* Matrices of the form

$$B_{nm} = \begin{Vmatrix} b_1 & b_2 & b_3 & \cdots & b_m & \cdots & b_n \\ 0 & b_1 & b_2 & \cdots & b_{m-1} & \cdots & b_{n-1} \\ \cdot & \cdot & \cdot & \cdots & \cdot & \cdots & \cdot \\ 0 & 0 & 0 & \cdots & b_1 & \cdots & b_{n-m+1} \end{Vmatrix} \qquad (n \geqslant m)$$

or

$$B_{nm} = \begin{Vmatrix} b_1 & b_2 & \cdots & b_n \\ 0 & b_1 & \cdots & b_{n-1} \\ \cdot & \cdot & \cdots & \cdot \\ 0 & 0 & \cdots & b_1 \\ 0 & 0 & \cdots & 0 \\ 0 & 0 & \cdots & 0 \end{Vmatrix} \qquad (n \leqslant m).$$

12. *Ans.* Matrices of the form

$$\begin{Vmatrix} B_{m_1m_1} & B_{m_1m_2} & \cdots & B_{m_1m_k} \\ \cdot & \cdot & \cdots & \cdot \\ B_{m_km_1} & B_{m_km_2} & \cdots & B_{m_km_k} \end{Vmatrix},$$

with the blocks $B_{m_im_j}$ given in the answer to Prob. 11.

13. *Ans.* Matrices of the form

$$\begin{Vmatrix} B_{m_1m_1} & 0 & \cdots & 0 \\ 0 & B_{m_2m_2} & \cdots & 0 \\ \cdot & \cdot & \cdots & \cdot \\ 0 & 0 & \cdots & B_{m_km_k} \end{Vmatrix}.$$

14. *Ans.* To every group of Jordan blocks with the same root of the characteristic polynomial, there corresponds a block of the kind given in the answer to Prob. 12. The remaining elements are all zero.

15. *Ans.* If the multiplicity of each root of the characteristic polynomial equals the size of the corresponding Jordan block, or if the characteristic polynomial coincides with the minimal annihilating polynomial, or if all the elementary divisors except the one with highest index equal 1.

Chapter 7

1. *Ans.* A tensor of order two, with two covariant indices.

2. *Ans.* For example,

$$\eta_1^2 - \eta_2^2 - \eta_3^2,$$

where

$$\eta_1 = \tfrac{1}{2}\xi_1 + \tfrac{1}{2}\xi_2 + \xi_3, \qquad \eta_2 = \tfrac{1}{2}\xi_1 - \tfrac{1}{2}\xi_2, \qquad \eta_3 = \xi_3.$$

3. *Hint.* See Sec. 7.93.

4. *Hint.* See Secs. 4.54 and 7.15.

5. *Ans.* For example,

$$\mathbf{A}(x, y) = \sigma_1\tau_1 + \sigma_2\tau_2 + \sigma_3\tau_3,$$

where σ_i and τ_i $(i = 1, 2, 3)$ are the new components of the vectors x and y. The transformation formulas to the new basis are

$$\sigma_1 = \xi_1 + \xi_2, \qquad \sigma_2 = \xi_2 + 2\xi_3, \qquad \sigma_3 = \xi_3.$$

6. *Hint.* First renumber the variables in such a way that the matrix of the bilinear form $\mathbf{A}(x, y)$ is transformed into a form to which Jacobi's method is applicable.

7. *Hint.* $\|-a_{ik}\|$ must be the matrix of a positive definite form.
Ans.

$$a_{11} < 0, \qquad \begin{vmatrix} a_{11} & a_{12} \\ a_{21} & a_{22} \end{vmatrix} > 0, \ldots, \qquad (-1)^n \det \|a_{ik}\| > 0.$$

8. *Hint.* See the remark to Sec. 7.96.

9. *Hint.* Consider the form on the basis vectors.

10. *Hint.* The last row of the determinant consists of the elements

$$a_k^{(n)} = (-1)^{k-1}\mathbf{A}(e_1, \ldots, e_{k-1}, e_{k+1}, \ldots, e_n) \qquad (k = 1, 2, \ldots, n).$$

11. *Hint.* Use the equation $\mathbf{A}(e_1, e_2) = 1$ to find the first pair of basis vectors. Then construct the subspace $\mathbf{L}$ defined by the equations

$$\mathbf{A}(e_1, x) = 0, \qquad \mathbf{A}(e_2, x) = 0.$$

If the form $\mathbf{A}(x, y)$ does not vanish identically in this subspace, find vectors $e_3, e_4 \in \mathbf{L}$ such that $\mathbf{A}(e_3, e_4) = 1$, and so on.

12. *Hint.* Consider the form

$$\mathbf{A}(x, x) + \varepsilon \sum_{i=1}^{n} \xi_i^2 \qquad (\varepsilon > 0),$$

and apply the criterion of Sec. 7.96.

13. *Hint.* Let

$$x^{(j)} = (\xi_1^{(j)}, \ldots, \xi_n^{(j)}) \qquad (j = 1, \ldots, r)$$

be a basis of the subspace $\mathbf{K}'$. Then $\mathbf{K}''$ consists of the vectors $y = (\eta_1, \ldots, \eta_n)$ satisfying the system

$$\mathbf{A}(x^{(j)}, y) = \sum_{k=1}^{n} \left(\sum_{i=1}^{n} a_{ik} \xi_i^{(j)} \right) \eta_k = 0 \qquad (j = 1, \ldots, r).$$

The matrix of the coefficients of the system is the product of the nonsingular matrix $\|a_{ik}\|$ of the form $\mathbf{A}(x, y)$ and the matrix $\|\xi_i^{(j)}\|$ of rank r. Now use Corollary 4.67.

14. *Ans.* $\mathbf{A}' = \mathbf{A}$.

15. *Hint.* If $y = (\eta_1, \ldots, \eta_n)$ is a solution of the system (44), then

$$(b, y) = (\mathbf{A}x, y) = (x, \mathbf{A}'y) = 0.$$

Conversely, the system (44) is the condition for the vectors y and $a_j = (a_{j1}, \ldots, a_{jn})$ to be conjugate. If $(b, y) = 0$ for all such y, then x lies in the linear manifold spanned by the vectors $a_1, \ldots, a_n$.

16. *Hint.* See Chap. 4, Prob. 37.

17. *Hint.* See Chap. 3, Prob. 1.

18. *Hint.* First consider the case of nonnegative forms of rank 1, using Prob. 17 and then Prob. 16.

Chapter 8

1. *Ans.* No, since axiom b) fails, and so does axiom c) (for $\lambda = -1$).

2. *Ans.* No, since axiom b) fails.

3. *Ans.* Yes. The new definition of the scalar product merely corresponds to a change of units along the coordinate axes.

4. *Hint.* Let e_1, e_2, e_3 denote the vectors directed along three edges of the tetrahedron drawn from a common vertex, and express the other edges of the tetrahedron as vectors.
Ans. 90°.

5. *Ans.* 90°, 60°, 30°.

6. *Ans.*

$$\sqrt{\int_a^b (x(t) + y(t))^2\,dt}\ \begin{cases} \leqslant \sqrt{\int_a^b x^2(t)\,dt} + \sqrt{\int_a^b y^2(t)\,dt}, \\ \geqslant \left| \sqrt{\int_a^b x^2(t)\,dt} - \sqrt{\int_a^b y^2(t)\,dt} \right|. \end{cases}$$

7. *Ans.* $\cos\varphi = \dfrac{1}{\sqrt{n}}$

8. *Ans.* a) $g = (3, 1, -1, -2)$, $h = (2, 1, -1, 4)$; b) $g = (1, 7, 3, 3)$, $h = (-4, -2, 6, 0)$.

9. *Hint.* Use the definition of angle (Sec. 8.33) and the orthogonality of the vector h to all vectors of the subspace $\mathbf{R}'$.

10. *Hint.* Take the scalar product of equation (18), p. 223 with the vector g_0.

11. *Hint.* See Sec. 8.52.

12. *Ans.* $y_1 = \mathbf{i}$, $y_2 = y_3 = \mathbf{0}$, $y_4 = -2\mathbf{j}$, $y_5 = \mathbf{0}$, $y_6 = 5\mathbf{k}$.

13. *Ans.* $(1, 2, 1, 3)$, $(10, -1, 1, -3)$, $(19, -87, -61, 72)$.

14. *Hint.* Assuming that the dimension of $\mathbf{R}''$ is greater than the dimension of $\mathbf{R}'$, consider the vector $e'' \in \mathbf{R}''$ which is orthogonal to the projection of $\mathbf{R}'$ onto $\mathbf{R}''$. Then use Prob. 10.

15. *Ans.* $A_n = \dfrac{(2n)!}{2^n(n!)^2}$.

16. *Ans.* $P_n(-1) = (-1)^n$.

17. *Hint.* Express the coefficients as scalar products.

18. *Hint.* Use the results of Probs. 15 and 16.

19. *Hint.* Expand $Q(t)$ in Legendre polynomials.

Ans. $Q(t) = \dfrac{1}{A_n} P_n(t)$.

20. *Ans.* $\|P_n(t)\|^2 = \dfrac{2}{2n+1}$.

21. *Ans.* $k(\mathbf{A}) = |\det A|$.

22. *Hint.* See Sec. 4.75.

23. *Hint.* This is a question of comparing the altitudes of two hyperparallelepipeds.

24. *Hint.* The inequalities

$$\frac{V[x_1, x_2, \ldots, x_m]}{V[x_1, \ldots, x_{k-1}, x_{k+1}, \ldots, x_m]} \leqslant \frac{V[x_1, x_2, \ldots, x_k]}{V[x_1, x_2, \ldots, x_{k-1}]} \qquad (k = 1, 2, \ldots, m)$$

are easily obtained from the inequality (37). Multiply them all together for $k = 1, 2, \ldots, m$, make appropriate cancellations, and then take the $(m-1)$th root. The geometric meaning of the inequality is the following: The volume of an m-dimensional hyperparallelepiped does not exceed the product of the $(m-1)$th roots of the volumes of its $(m-1)$-dimensional "faces."

25. *Hint.* Write the inequality (38) for $x_{s_1}, x_{s_2}, \ldots, x_{s_r}$, and then multiply these inequalities together for all permissible values of $s_1, s_2, \ldots, s_r$.

26. *Hint.* We must construct a hyperparallelepiped in a 2^m-dimensional space such that the projections of its edges onto each axis have absolute values no greater than M and such that its volume is exactly $M^n n^{n/2}$. For $M = 1$, the matrix A_m of the components of the 2^m-dimensional vectors determining this hyperparallelepiped are given by the following recurrence formula:

$$A_m = \begin{Vmatrix} A_{m-1} & A_{m-1} \\ A_{m-1} & -A_{m-1} \end{Vmatrix}, \qquad A_1 = \begin{Vmatrix} 1 & 1 \\ 1 & -1 \end{Vmatrix}.$$

Comment. For $n \neq 2^m$, the estimate *can be improved.*

27. *Hint.* Given any subspace $\mathbf{G} \subset \mathbf{R}$, let $\mathbf{G}^\perp$ denote the orthogonal complement of $\mathbf{G}$. For every $x \in \mathbf{N}(\mathbf{A})$ and every $z \in \mathbf{R}$,

$$(\mathbf{A}'z, x) = (z, \mathbf{A}x) = 0,$$

and hence $\mathbf{A}'z \in \mathbf{N}^\perp(\mathbf{A})$, i.e., $\mathbf{T}(\mathbf{A}') \subset \mathbf{N}^\perp(\mathbf{A})$, $\mathbf{T}^\perp(\mathbf{A}') \supset \mathbf{N}(\mathbf{A})$. For every $x \in \mathbf{T}^\perp(\mathbf{A})$ and every $y \in \mathbf{R}$,

$$(\mathbf{A}'x, y) = (x, \mathbf{A}y) = 0,$$

and hence $\mathbf{A}'x = 0$, i.e., $x \in \mathbf{N}(\mathbf{A}')$, so that $\mathbf{T}^\perp(\mathbf{A}) \subset \mathbf{N}(\mathbf{A}')$, $\mathbf{T}^\perp(\mathbf{A}') \subset \mathbf{N}(\mathbf{A})$. It follows that $\mathbf{N}(\mathbf{A}) = \mathbf{T}^\perp(\mathbf{A}')$, $\mathbf{N}^\perp(\mathbf{A}) = \mathbf{T}(\mathbf{A}')$. The other assertion is proved similarly.

28. *Hint.* See Sec. 4.77.

29. *Hint.* See Sec. 4.54.

30. *Hint.* The angles of a triangle are uniquely determined by its sides. Alternatively, the symmetric bilinear form $(\mathbf{Q}x, \mathbf{Q}y)$ is uniquely determined by the quadratic form $(\mathbf{Q}x, \mathbf{Q}x)$.

31. *Hint.* A given isogonal operator $\mathbf{A}$ transforms the orthonormal basis $e_1, e_2, \ldots, e_n$ into an orthogonal basis $f_1' = \alpha_1 f_1$, $f_2' = \alpha_2 f_2, \ldots, f_n' = \alpha_n f_n$, where $f_1, f_2, \ldots, f_n$ are unit vectors. Let $\mathbf{Q}$ be the isometric operator carrying the vectors $f_1, f_2, \ldots, f_n$ into $e_1, e_2, \ldots, e_n$. Then the matrix of the isogonal operator $\mathbf{QA}$ is diagonal. Show that the condition $\alpha_i \neq \alpha_j$ allows one to construct a pair of orthogonal vectors which are carried into nonorthogonal vectors by the operator $\mathbf{QA}$.

32. *Hint.* It is sufficient to show that $\mathbf{Q}$ is an isogonal operator (see Prob. 31). Assuming that there is a right angle which is not transformed into a right angle, construct a parallelogram whose area changes as a result of applying the operator $\mathbf{Q}$.

33. *Hint.* Generalize the construction of Prob. 32.

34. *Hint.* Applying the orthogonalization process to the given systems, obtain orthonormal systems $e_1, e_2, \ldots$ and $f_1, f_2, \ldots$. Using Sec. 8.53, show that the formulas expressing the vectors $x_1, x_2, \ldots, x_k$ in terms of $e_1, e_2, \ldots$ are the same as those expressing the vectors $y_1, y_2, \ldots, y_k$ in terms of $f_1, f_2, \ldots$. Then define $\mathbf{Q}$ as the operator which maps the system $e_1, e_2, \ldots$ into the system $f_1, f_2, \ldots$.

35. *Hint.* Consider the finite systems $e_1', e_1'', e_2', e_2'', \ldots, e_k', e_k''$ and $f_1', f_1'', f_2', f_2'', \ldots, f_k', f_k''$ obtained in determining the angles between the subspaces $\mathbf{R}'$, $\mathbf{R}''$ and the subspaces $\mathbf{S}'$, $\mathbf{S}''$. By construction,

$$(e_i', e_i'') = (f_i', f_i'') = \cos \varphi_i \qquad (i = 1, 2, \ldots, k),$$

$$(e_i', e_j') = (f_i', f_j') = 0, \qquad (e_i'', e_j'') = (f_i'', f_j'') = 0 \qquad (i \neq j).$$

Show further that $(e_i', e_j'') = (f_i', f_j'') = 0$ (using Prob. 9). Then use the result of Prob. 34.

36. *Hint.* Use Prob. 11.

37. *Hint.* In the subspaces $\mathbf{L}_1$ and $\mathbf{L}_2$, let $e_1, e_2, \ldots, e_m$ and $f_1, f_2, \ldots, f_m$ be the bases obtained in constructing the angles $\alpha_1, \alpha_2, \ldots, \alpha_m$. In the space $\mathbf{R}$ construct a basis $e_1, e_2, \ldots, e_m, e_{m+1}, \ldots, e_n$ which begins with the vectors obtained by orthogonalizing the vectors $e_1, e_2, \ldots, e_m, f_1, f_2, \ldots, f_m$. Expand the vectors $x_1, x_2, \ldots, x_m, y_1, y_2, \ldots, y_m$ with respect to this basis. Show that the matrices of these expansions each have only one minor of order m, if we disregard minors which are known to vanish. Then use the expression for the volume of a hyperparallelepiped in terms of the minors of the corresponding matrix.

38. *Hint.* See Chap. 3, Prob. 2 and Chap. 4, Prob. 17.

39. *Hint.* Verify the assertion in the special basis whose first k vectors belong to the subspace $\mathbf{L}(x_1, x_2, \ldots, x_k)$. To go over to the general case, use Chap. 4, Prob. 17, showing that $\det \|a_i^{(j)}\| = 1$.

40. *Hint.* First consider the case $k = 2$.

41. *Hint.* Choose a basis in the space $\mathbf{R}$ like that chosen in Prob. 37, and verify that the formula is valid in this basis. Then go over to the general case in the same way as in Prob. 39.

42. *Hint.* See Sec. 4.54.

43. *Hint.* Consider the orthogonal complement $\mathbf{Z}$ of the invariant (with respect to $\mathbf{A}$) subspace $\mathbf{H}$ of all vectors x such that $P(\mathbf{A})x = 0$. The subspace $\mathbf{Z}$ is also invariant with respect to the operator $\mathbf{A}$, and hence with respect to $[P(\mathbf{A})]^{k-1}$. But if $z \in \mathbf{Z}$, then $[P(\mathbf{A})]^{k-1}z \in \mathbf{H}$, so that $[P(\mathbf{A})]^{k-1}z = 0$. From this, deduce that $[P(t)]^{k-1}$ is an annihilating polynomial of the operator $\mathbf{A}$.

Chapter 9

1. *Hint.* Use Sec. 9.45.

2. *Hint.* The operator $\mathbf{B}$ has a basis consisting of eigenvectors $e_1, \ldots, e_n$ with positive eigenvalues $\mu_1, \ldots, \mu_n$. Hence $\mathbf{B}^2 e_i = \mu_i^2 e_i$, and a necessary condition for $\mathbf{B}^2 = \mathbf{A}$ is that the e_i be eigenvectors of the operator $\mathbf{A}$ and that the numbers μ_i^2 coincide with the λ_i. But this is also sufficient for $\mathbf{B}^2 = \mathbf{A}$.

3. *Hint.* First transform the basis in such a way as to diagonalize the matrix of the given operator.

Ans.

$$\sqrt{A} = \begin{Vmatrix} 3 & 2 & 0 \\ 2 & 4 & 2 \\ 0 & 2 & 5 \end{Vmatrix}.$$

4. *Hint.* The operator $\mathbf{A'A}$ is symmetric, and the expression $(\mathbf{A'A}x, x) = (\mathbf{A}x, \mathbf{A}x)$ is nonnegative for arbitrary $x \in \mathbf{R}_n$. If $\mathbf{A}$ is nonsingular, this expression is positive for arbitrary $x \in \mathbf{R}_n$.

5. *Hint.* $\mathbf{Q'} = \mathbf{Q}^{-1}$.

6. *Hint.* The operator $\mathbf{A'A}$ is symmetric and positive (Prob. 4), and hence we can find a symmetric positive operator $\mathbf{S}$ such that $\mathbf{S}^2 = \mathbf{AA'}$. Then construct an operator $\mathbf{Q}$ such that $\mathbf{Q} = \mathbf{S}^{-1}\mathbf{A}$ and show that $\mathbf{Q}$ is isometric.

7. *Hint.* Use Probs. 2 and 5.

8. *Hint.* Let $\mathbf{R'} \subset \mathbf{R}_n$ be the subspace spanned by the eigenvectors of the operator $\mathbf{A'A}$ with nonzero eigenvalues, and let $\mathbf{R''}$ be the orthogonal complement of $\mathbf{R'}$. On $\mathbf{R'}$ let $\mathbf{V}$ equal the isometric component of $\mathbf{A}$ (so that $\sqrt{\mathbf{A'A}}\,\mathbf{V}x = \mathbf{A}x$), and on $\mathbf{R''}$ let $\mathbf{V}x = 0$.

9. *Hint.* Use Chap. 4, Probs. 28–29.

10. *Hint.* Apply the orthogonalization process to the vectors of the Jordan basis of $\mathbf{A}$ (Sec. 6.37).

Chapter 10

1. *Ans.* a) $4\eta_1^2 + \eta_2^2 - 2\eta_3^2$;

$$\eta_1 = \frac{2}{3}\xi_1 - \frac{2}{3}\xi_2 + \frac{1}{3}\xi_3,$$

$$\eta_2 = \frac{2}{3}\xi_1 + \frac{1}{3}\xi_2 - \frac{2}{3}\xi_2,$$

$$\eta_3 = \frac{1}{3}\xi_1 + \frac{2}{3}\xi_2 + \frac{2}{3}\xi_3;$$

b) $10\eta_1^2 + \eta_2^2 + \eta_3^2$;

$$\eta_1 = \frac{1}{3}\xi_1 + \frac{2}{3}\xi_2 - \frac{2}{3}\xi_3,$$

$$\eta_2 = \frac{2}{\sqrt{5}}\xi_1 - \frac{1}{\sqrt{5}}\xi_2,$$

$$\eta_3 = \frac{2}{3\sqrt{5}}\xi_1 + \frac{4}{3\sqrt{5}}\xi_2 + \frac{\sqrt{5}}{3}\xi_3;$$

c) $\eta_1^2 - \eta_2^2 + 3\eta_3^2 + 5\eta_4^2$; $\eta_1 = \frac{1}{2}\xi_1 + \frac{1}{2}\xi_2 + \frac{1}{2}\xi_3 + \frac{1}{2}\xi_4,$

$$\eta_2 = \frac{1}{2}\xi_1 + \frac{1}{2}\xi_2 - \frac{1}{2}\xi_3 - \frac{1}{2}\xi_4,$$

$$\eta_3 = \frac{1}{2}\xi_1 - \frac{1}{2}\xi_2 + \frac{1}{2}\xi_3 - \frac{1}{2}\xi_4,$$

$$\eta_4 = \frac{1}{2}\xi_1 - \frac{1}{2}\xi_2 - \frac{1}{2}\xi_3 + \frac{1}{2}\xi_4;$$

d) $\eta_1^2 + \eta_2^2 + \eta_3^2 - 3\eta_4^2$; $\eta_1 = \frac{\sqrt{2}}{2}\xi_1 + \frac{\sqrt{2}}{2}\xi_2,$

$$\eta_2 = \frac{\sqrt{2}}{2}\xi_3 + \frac{\sqrt{2}}{2}\xi_4,$$

$$\eta_3 = \frac{1}{2}\xi_1 - \frac{1}{2}\xi_2 + \frac{1}{2}\xi_3 - \frac{1}{2}\xi_4,$$

$$\eta_4 = \frac{1}{2}\xi_1 - \frac{1}{2}\xi_2 - \frac{1}{2}\xi_3 + \frac{1}{2}\xi_4.$$

2. *Ans.* A maximum for $x = (\pm 1, 0, 0)$ where $\mathbf{A}(x, x) = 1$. A minimum for $x = (0, 0, \pm 1)$ where $\mathbf{A}(x, x) = \frac{1}{3}$. A minimax for $x = (0, \pm 1, 0)$ where $\mathbf{A}(x, x) = \frac{1}{2}$, i.e., the function $\mathbf{A}(x, x)$ increases if we go along the unit sphere in one direction from the point x and decreases if we go in the other direction.

3. *Hint.* Namely, on the subspace spanned by the corresponding canonical basis vectors.

4. *Hint.* The coefficient λ_k equals the smallest of the maxima of the form $\mathbf{A}(x, x)$ on a system of subspaces, and the coefficient μ_k equals the smallest of the maxima of the form $\mathbf{B}(x, x)$ on the same system of subspaces.

5. *Ans.* $y/x = \pm\frac{1}{2}$.

6. *Ans.* $\mathbf{A}(x, x) = \eta_1^2 + \eta_2^2 + \eta_3^2$, $\mathbf{B}(x, x) = \eta_1^2 + 2\eta_2^2 + 3\eta_3^2$, $\xi_1 = \eta_1 - \eta_2 + 2\eta_3$, $\xi_2 = \eta_2 - \eta_3$, $\xi_3 = \eta_3$.

7. *Hint.* The problem reduces to the uniqueness of the canonical basis of a symmetric operator with distinct eigenvalues.

8. *Hint.* Generalize Sec. 7.44.

9. *Ans.* a) A hyperboloid of one sheet with its axis along the y-axis; b) A hyperboloid of one sheet with its axis along the x-axis; c) A circular paraboloid with its axis along the x-axis; d) A circular paraboloid with its axis along the y-axis, displaced one unit along this axis; e) A hyperbolic paraboloid.

10. *Ans.* a) $x_1^2 + 2y_1^2 + 3z_1^2 = 6$; $3(x - 1) = -x_1 + 2y_1 + 2z_1,$

$$3y = 2x_1 - y_1 + 2z_1,$$

$$3(z + 1) = 2x_1 + 2y_1 - z_1;$$

b) $x_1^2 + 2y_1^2 - 3z_1^2 = 6;$
$$\begin{aligned} 3(x+1) &= -x_1 + 2y_1 + 2z_1,\\ 3(y+1) &= 2x_1 - y_1 + 2z_1,\\ 3z &= 2x_1 + 2y_1 - z_1; \end{aligned}$$

c) $y_1^2 = 2x_1;$
$$\begin{aligned} 3(x-m) &= 2x_1 + 2y_1 + z_1,\\ 3(y+2m) &= 2x_1 - y_1 - 2z_1,\\ 3(z+2m) &= -x_1 + 2y_1 - 2z_1 \end{aligned}$$
(m arbitrary).

11. *Hint.* The semiaxes of the ellipsoid are determined from the canonical coefficients of the corresponding quadratic form. Use the results of Sec. 10.25.

Chapter 11

1. *Hint.* Let $\mathbf{K}'$ be the intersection of the null spaces of all operators belonging to a left ideal $\mathbf{J} \subset \mathbf{B}(\mathbf{K}_n)$, and let r be the dimension of $\mathbf{K}'$. Choose a basis in $\mathbf{K}_n$ whose first r basis vectors lie in $\mathbf{K}'$. Then the first r columns of the matrix of every operator $\mathbf{A} \in \mathbf{J}$ consists entirely of zeros. Let m be the dimension of $\mathbf{J}$, and let $\mathbf{A}_1, \ldots, \mathbf{A}_m$ be linearly independent operators in $\mathbf{J}$. Consider the matrix with $n - r$ columns and mn rows obtained by writing all the matrices $A_1, \ldots, A_m$ on top of each other and omitting the first r (zero) columns. The rank of this matrix is $n - r$, and hence it has $n - r$ basis rows. The linear combinations of these rows give all possible rows consisting of $n - r$ elements. Now use Sec. 4.44.

2. *Hint.* Introducing a nonsingular bilinear form (x, y), consider the set $\mathbf{J}^*$ of all operators $\mathbf{A}^*$ conjugate to the operators $\mathbf{A} \in \mathbf{J}$. This set is a left ideal. Now use Prob. 1.

3. *Ans.* A maximal left ideal of the algebra $\mathbf{B}(\mathbf{K}_n)$ is the set of all operators carrying a fixed vector of the space $\mathbf{K}_n$ into zero. A minimal left ideal is the set of all operators carrying a fixed $(n - 1)$-dimensional subspace of $\mathbf{K}_n$ into zero. A maximal right ideal is the set of all operators carrying the whole space $\mathbf{K}_n$ into a fixed $(n - 1)$-dimensional subspace. A minimal right ideal is the set of all operators carrying the whole space $\mathbf{K}_n$ into a fixed straight line.

4. *Hint.* Let
$$(x, y) = \sum_{j=1}^{n} \xi_j \bar{\eta}_j \qquad \left(x = \sum_{j=1}^{n} \xi_j e_j,\ y = \sum_{j=1}^{n} \eta_j e_j\right)$$
in the basis $e_1, \ldots, e_n$ in which the matrix of the operator $\mathbf{A} \in \mathbf{B}$ takes the form indicated in Sec. 11.85.

5. *Hint.* If a subspace $\mathbf{C}' \subset \mathbf{C}_n$ is invariant (with respect to the algebra $\mathbf{B}$), then so is its orthogonal complement. Expand $\mathbf{C}_n$ as an orthogonal direct sum of irreducible invariant subspaces. Every operator $\mathbf{A} \neq \mathbf{0}$ (of the algebra $\mathbf{B}$) acts as a nonzero operator in at least one of these subspaces.

6. *Hint.* Deduce from the representation of Sec. 11.85 that the commutator of a semisimple but nonsimple matrix algebra **B** intersects **B** in matrices *other than* multiples of the matrix corresponding to the identity operator.

7. *Hint.* Write the desired matrices as block matrices consisting of m^2 blocks. Then write the commutativity condition and use Schur's lemma.

8. *Ans.* For the algebra **B** of all diagonal matrices

$$\left\| \begin{matrix} \lambda_1 & 0 & \cdots & 0 \\ 0 & \lambda_2 & \cdots & 0 \\ \cdot & \cdot & \cdots & \cdot \\ 0 & 0 & \cdots & \lambda_n \end{matrix} \right\|,$$

where $\lambda_1, \lambda_2, \ldots, \lambda_n$ are arbitrary complex numbers. Every matrix algebra $\mathbf{B} = \bar{\mathbf{B}}$ reduces to this form in some basis.

9. *Ans.* Let **B** be the algebra of all operators under which a given system of subspaces, whose direct sum is the whole space $\mathbf{C}_n$, remain characteristic subspaces. Then $\mathbf{B} \subset \bar{\mathbf{B}}$. Every algebra with $\mathbf{B} \subset \bar{\mathbf{B}}$ reduces to this form.

10. *Ans.* The space $\mathbf{C}_n$ is a direct sum of subspaces $\mathbf{C}^{(1)}, \ldots, \mathbf{C}^{(k)}$, and the algebra **B** consists of all operators invariant in each $\mathbf{C}^{(j)}$ $(j = 1, \ldots, k)$. The commutator $\bar{\mathbf{B}}$ consists of all operators which are multiples of the identity operator in each $\mathbf{C}^{(j)}$ $(j = 1, \ldots, k)$.

11. *Hint.* If **B** is a direct sum $\mathbf{B}^{(1)} + \cdots + \bar{\mathbf{B}}^{(k)}$, then $\bar{\bar{\mathbf{B}}} = \bar{\bar{\mathbf{B}}}^{(1)} + \cdots + \bar{\bar{\mathbf{B}}}^{(k)}$.

12. *Ans.* If the multiplicity of each root of the characteristic equation of the operator equals the size of the corresponding Jordan block (see Chap. 6, Prob. 15).

13. *Hint.* If $\mathbf{CB} = \mathbf{B}$, then $\mathbf{CA} = \mathbf{C}$ for some $\mathbf{A} \in \mathbf{B}$. It follows that $\mathbf{C} = \mathbf{CA} = \mathbf{C}(\mathbf{CA}) = \mathbf{C}^2\mathbf{A} = \mathbf{C}^3\mathbf{A} = \cdots$.

15. *Hint.* Let $\mathbf{A}_1, \ldots, \mathbf{A}_m$ be a basis of the algebra **B**. Then, if **B** is not nilpotent, one of the right ideals $\mathbf{A}_1\mathbf{B}, \ldots, \mathbf{A}_m\mathbf{B}$, say $\mathbf{A}_1\mathbf{B}$, is not nilpotent (Prob. 14). Moreover $\mathbf{A}_1\mathbf{B} \neq \mathbf{0}$ (Prob. 13), and the problem reduces to the analogous problem for an algebra of smaller dimension.

16. *Hint.* If $\mathbf{M}_i = \mathbf{M}_{i+1}$, then for every vector $x \in \mathbf{M}_i$ there is an operator $\mathbf{A}_1 \in \mathbf{B}$ such that $\mathbf{A}_1 x \notin \mathbf{M}_i = \mathbf{M}_{i+1}$. Moreover, there is an operator $\mathbf{A}_2 \in \mathbf{B}$ such that $\mathbf{A}_2\mathbf{A}_1 x \in \mathbf{M}_i$, and so on. If $\mathbf{M}_p \neq \mathbf{K}_n$, then for every $x \in \mathbf{M}_{p+1} - \mathbf{M}_p$ there is an operator $\mathbf{A}_p \in \mathbf{B}$ such that $\mathbf{A}_p x \in \mathbf{M}_p - \mathbf{M}_{p-1}$, then an operator $\mathbf{A}_{p-1} \in \mathbf{B}$ such that $\mathbf{A}_{p-1}\mathbf{A}_p \in \mathbf{M}_{p-1} - \mathbf{M}_{p-2}$, and so on, so that $\mathbf{A}_1\mathbf{A}_2 \cdots \mathbf{A}_p x \neq 0$.

17. *Hint.* Use the subspaces $\mathbf{M}_1, \ldots, \mathbf{M}_p$ of Prob. 16.

BIBLIOGRAPHY

Bellman, R., *Introduction to Matrix Algebra*, McGraw–Hill Book Co., Inc., New York (1960).

Gantmakher, F. R., *The Theory of Matrices*, 2 vols., translated by K. A. Hirsch, Chelsea Publishing Co., New York (1959).

Gelfand, I. M., *Lectures on Linear Algebra*, translated by A. Shenitzer, Interscience Publishers, Inc., New York (1961).

Halmos, P. R., *Finite-Dimensional Vector Spaces*, second edition, D. Van Nostrand Co., Inc., Princeton, N.J. (1958).

Hamburger, H. L. and M. E. Grimshaw, *Linear Transformations*, Cambridge University Press, New York (1951).

Hoffman, K. and R. Kunze, *Linear Algebra*, Prentice–Hall, Inc., Englewood Cliffs, N.J. (1961).

Jacobson, N., *Lectures in Abstract Algebra, Vol. 2, Linear Algebra*, D. Van Nostrand Co., Inc., Princeton, N.J. (1953).

Mirsky, L., *An Introduction to Linear Algebra*, Oxford University Press, New York (1955).

Noble, B., *Applied Linear Algebra*, Prentice–Hall, Inc., Englewood Cliffs, N.J. (1969).

Perlis, S., *The Theory of Matrices*, Addison–Wesley Publishing Co., Reading, Mass. (1952).

Shilov, G. E., *An Introduction to the Theory of Linear Spaces*, translated by R. A. Silverman, Prentice–Hall, Inc., Englewood Cliffs, N.J. (1961).

Thrall, R. M. and L. Tornheim, *Vector Spaces and Matrices*, John Wiley and Sons, Inc., New York (1957).

INDEX